CONTROL OF GENE EXPRESSION BY CATECHOLAMINES AND THE RENIN-ANGIOTENSIN SYSTEM

Control of Gene Expression by Catecholamines and the Renin-Angiotensin System

Edited by

HEINZ RUPP

Internal Medicine and Cardiology
Molecular Cardiology Laboratory
Philipps University of Marburg
Karl-von-Frisch-Strasse 1
35044 Marburg, Germany

BERNARD MAISCH

Internal Medical
Molecular Cardiology Laboratory
Philipps University of Marburg
Karl-von-Frisch-Strasse
35033 Marburg, Germany

Reprinted from *Molecular and Cellular Biochemistry*, Volume 212 (2000)

Kluwer Academic Publishers
Dordrecht / Boston / London

Library of Congress Cataloging-in-Publication Data

Control of gene expression by catecholamines and the renin-angiotensin system /
edited by Heinz Rupp, Bernhard Maisch
p. cm. -- (Developments in molecular and cellular
biochemistry ; v. 33)

ISBN 0-7923-7981-0 (acid-free paper)
1. Genetic regulation. 2. Angiotensin II. 3. Catecholamines.
I. Rupp, Heinz. II. Maisch, Bernard. III. Series

QH450 .C66 2000
572.8'65--dc21 00-135010

Distributors for North, Central and South America:
Kluwer Academic Publishers
101 Philip Drive
Assinippi Park
Norwell, Massachusetts 02061 USA
Telephone (781) 871-6600
Fax (781) 681-9045
E-mail <kluwer@wkap.com>

Distributors for all other countries:
Kluwer Academic Publishers Group
Distribution Centre
Post Office Box 322
3300 AH Dordrecht, THE NETHERLANDS
Telephone 31 78 6392 392
Fax 31 78 6546 474
E-mail <services@wkap.nl>

Electronic Services <http://www.wkap.nl>

Printed on acid-free paper

Printed in the Netherlands

Molecular and Cellular Biochemistry:

An International Journal for Chemical Biology in Health and Disease

CONTENTS VOLUME 212, Nos. 1 & 2, September 2000

Control of gene expression by catecholamines and the renin-angiotensin system
Drs. Heinz Rupp and Bernard Maisch

Molecular and Cellular Biochemistry **212:** 1, 2000.

Preface

This special issue of *Molecular and Cellular Biochemistry* focuses on 'Control of Gene Expression by Catecholamines and the Renin-Angiotensin System' in health and disease. In recent years, great progress has been made in the understanding of catecholamine and angiotensin II modulated gene expression. There is also increasing evidence that catecholamine and angiotensin II induced cellular injury not solely arises from classical pathways but also from a perturbed gene expression.

Taking into account that catecholamines and angiotensin II are vital for a balanced gene expression of many cells, the intriguing possibility arises that various diseases are initiated or aggravated by such an imbalance. Catecholamine and angiotensin II influences can be in excess arising from, for example, hypercaloric food intake or psychosocial stress. During early progression of heart failure, sympathetic activity and angiotensin II influences also become increased. Due to beta-adrenergic receptor downregulation, depressed catecholamine influences are expected in the final stage of heart failure. An imbalanced influence of catecholamines and angiotensin II on gene expression leads to disordered molecular structures of the cell and an impaired cell function.

The focused issue is organized into chapters focusing on catecholamines, angiotensin II and the interaction between catecholamines and angiotensin II. Basic biochemical processes are covered in detail and the potential of these pathways for explaining chronic diseases associated with excess catecholamine and angiotensin II influences should become apparent. It is hoped that the focused issue triggers novel research into the development of drugs that are targeted at diseases characterized by an imbalanced gene expression involving catecholamines and angiotensin II.

HEINZ RUPP, Professor of Physiology
BERNHARD MAISCH, Professor of Medicine
Molecular Cardiology Laboratory
Department of Internal Medicine and Cardiology,
Philipps University of Marburg
Marburg, Germany

PART I
CATECHOLAMINE INFLUENCES ON GENE EXPRESSION

Molecular and Cellular Biochemistry **212**: 5–9, 2000.

Mechanisms of transcriptional activation of cAMP-responsive element-binding protein CREB

Philipp Haus-Seuffert and Michael Meisterernst
Institute of Molecular Immunology, Department for Proteinbiochemistry, GSF, München, Germany

Abstract

The CREB-CREM transcription factors are the main gene regulatory effectors of the cAMP signaling pathway. The investigations of this family of transcription factors had a profound impact on the understanding of signaling-induced gene transcription. Here we discuss some key aspects of the underlying biology, review transcriptional activation by CREB proteins through transcription cofactors and present novel insights into the context- and position-specific function of CREB on complex genes. (Mol Cell Biochem **212**: 5–9, 2000)

Key words: transcriptional regulation, gene expression, coactivator, repressor

Biology

The cAMP-pathway is a widely used signaling process that senses and amplifies the response of cells to hormones, growth factors and neurotransmitters. Among the target proteins of protein kinase A are the transcription factors CREB and CREM. Gene regulatory programs induced by CREB control different biological processes (for review see [1]) such as T cell development, spermatogenesis, long term memory but also the regulation of the blood pressure through angiotensin. The latter is the focus of this issue.

Angiotensins

There is increasing evidence, that the protein CREB is involved in the biology of the renin-angiotensin system (RAS), which plays an important role in the regulation of the blood pressure. Expression of the rat angiotensinogen gene is positively influenced by CREB in a cAMP-dependent manner [2]. The stimulating effect of CREB depends on a functional cAMP-responsive element (CRE), located in the 5′-flanking region of the angiotensinogen gene [3]. The expression of testis angiotensin converting enzyme depends on a functional CRE although it does not appear to be responsible for its testis-specific expression [4]. CREB seems to also function as an effector of angiotensin II signaling. The stimulatory effect of angiotensin II depends on a cAMP-responsive element within the fibronectin promoter [5]. Angiotensin II stimulation leads to moderate induction of the CREB protein in human mesangial cells. Angiotensin II increases interleukin-6 expression in a dose-dependent manner, which is mediated through a cAMP-responsive element in the interleukin-6 promoter [6]. As one other example a cAMP-responsive element is found in the tyrosine hydroxylase gene promoter [7]. In this study the CRE proves to be one critical angiotensin II-responsive element in cultured bone adrenal medullary cells.

Regulation of long term memory

A correlation between memory and cAMP pathways became first evident in genetic studies. Flies were trained to discriminate between two different odors, one accompanied with an electric shock, and the other not associated with a shock. Chemically mutagenized flies were used to find mutants which failed to learn the discrimination between the two odors without being affected in other characteristics like lo-

Address for offprints: M. Meisterernst, Institute of Molecular Immunology, Department for Proteinbiochemistry, GSF, Marchioninistrasse 25, D-81377 München, Germany comotion or odor detection [8]. Four mutants were found and the corresponding genes were identified. Remarkably, three out of four mutants affect molecules that are involved in cAMP signaling [9]. Long term – in contrast to short term memory storage depends on transcriptional activity and the synthesis of new proteins [10]. Studies in Drosophila and Aplysia demonstrated that CREB is critically involved in this process [11]. One model implies that CREB-mediated transcription activates a gene expression program that ultimately leads to the production of new synapses between neurons and a prolonged stabilization of the synaptic facilitation (see review [12]).

CREB function in the immune system

ATF/CREB proteins are involved in the development and function of T lymphocytes. The signal transduction pathways in T cells after T-cell receptor engagement, which lead to phosphorylation and activation of CREB, include protein kinases C, RAS, RAF-1, MEK and RSK2 [13]. Functional binding sites for members of the ATF/CREB family were identified in the promoters and enhancers of many T-cell specific genes, including the TCR-α and -β enhancers [14, 15], the CD3δ enhancer [16] and the TCR Vβ promoters [17]. Transgenic models provided strong *in vivo* evidence for the importance of CREB/ATF proteins in the immune system. CREB knock-out mice show defects in the development of specific T-cell lineages [18]. A dominant negative form of CREB under the control of the T-cell specific CD2 promoter/enhancer leads to a profound defect in T cell proliferation after stimulation of the T-cell receptor pathways [19].

CREB structure

The CREB/ATF family consists of a large number of genes that include the factors CREB, CREM, ATF-1, ATF-2, ATF-3 and ATF-4 (also known as CREB2). Various splice variants of each of these proteins have been identified which activate or repress transcription (see review [20]). CREB, first identified [21], is probably one of the most meticulously characterized transcription factors in eukaryotes.

A common feature of all the family members is a basic region leucine zipper (bZIP) domain (Fig. 1). The leucine zipper consists of an α-helical coiled-coil structure, which forms homo- and heterodimers. A particular 'dimerization code' determines which heterodimers are possible [22]. The basic region is responsible for the sequence-specific DNA-binding of the CREB transcription factors to cyclic AMP-response element (CRE). The cognate DNA-recognition motif for the CREB homodimer is a symmetric palindromic motif with sequence 5′-TGACGTCA-3′ [23]. In addition to the core sequence, flanking bases are important for binding of CREB [24].

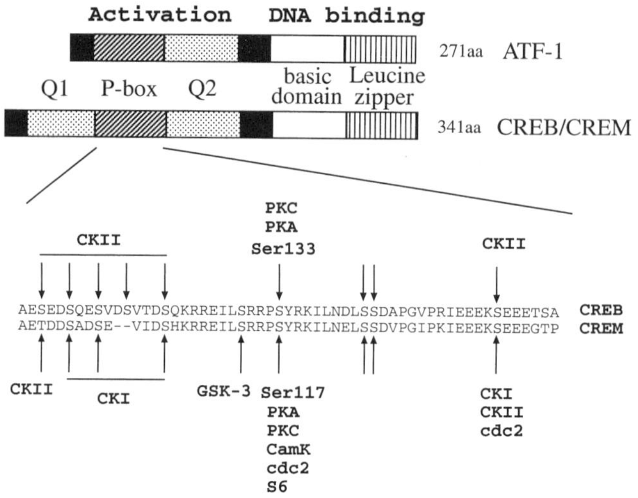

Fig. 1. Structure of ATF-1, CREB and CREM proteins. The P-box, the glutamine-rich domains (Q1 and Q2) and the DNA binding region (leucine zipper and basic domain) are indicated. Sequence alignment of the P-boxes of CREB and CREM. Serine and threonine residues can be phosphorylated by the indicated kinases as marked by arrows.

The members of the CREB family of transcription factors share structural features within their transactivation domains. Transcriptional activation is mediated through two regions (compare Fig. 1). One region contains several recognition motifs for protein kinases. It is therefore called kinase inducible domain (KID) or phosphorylation box (P-box). The transactivation potential of CREB proteins critically depends on their phosphorylation status [25, 26]. The other constitutive activation region, contained in CREB and CREM, consists of two glutamine-rich motifs, called Q1 and Q2, which flank the kinase inducible domain [27]. Mammalian ATF-1 lacks Q1 but contains Q2 [28]. Glutamine-rich regions can be found in many regulatory, coactivator and basal transcription factors and serve as interaction surfaces for other transcription factors. It has been suggested that CREB and CREM require the P-box and at least one glutamine-rich domain [27, 29] to activate transcription.

Several isoforms of each member of the CREB family of transcription factors were identified. Glutamine-rich regions can be removed via alternative splicing, either partially (in Drosophila CREB2b) or completely (in mammalian CREMα, CREMβ and CREMγ), with the consequence that these proteins display repressor function [29]. The insertion of premature stop codons in the CREB gene results in truncated proteins that lack the DNA-binding region and that also function as repressors. In Aplysia, a CREB isoform was identified, which lacks the nuclear localisation signal [30]. This cytoplasmic form (CREB1c) regulates the activity of kinases that phosphorylate nuclear CREB.

Signal-induced activation by CREB

CREB binding sites had been identified in a multitude of inducible promoters. Examples are the somatostatin- [31] and the proenkephaline promoter [32] as well as many others (see reviews [1, 20]) The critical region in CREB for the response to cAMP [25] is the phosphorylation box (P-box) or kinase-inducible domain (KID). As depicted in Fig. 1, the P-box contains several consensus phosphorylation sites for kinases such as PKA, PKC, glycogen synthase kinase-3 and casein kinases (CK) I and II [25] [33]. Upon activation of the adenylate cyclase pathway, the serine at position 133 of CREB (serine 117 in CREM) is phosphorylated by PKA, which enhances the transcriptional activity of the proteins CREB and CREM.

In addition to PKA, other signal transduction pathways target the CREB protein, in order to either increase or decrease its transcriptional activity. For example, the Ca^{2+}-calmodulin-dependent kinase IV (CaMKIV) phosphorylates CREB at Ser133 after membrane depolarization in neuronal cells [34]. Also signal transduction pathways triggered by growth factors and inflammatory cytokines lead to a phosphorylation of CREB (see review [35]). Ca^{2+}-calmodulin-dependent kinase II (CaMKII) phosphorylates CREB at Ser133 and Ser142. Remarkably, phosphorylation of Ser142 by CaMKII neutralizes the activity of CREB [36].

CREB and CREM activate through transcription cofactors

A breakthrough in the understanding of inducible CREB function came from the discovery of the cofactor CBP (CREB binding protein) that interacts specifically with the phosphorylated CREB P-box domain [37]. In the current model, the cofactor CBP and its close relative p300 serve as bridging factors between the activator CREB and the general transcription factors [27, 38]. These cofactors also possess a histone acetyltransferase (HAT) activity, which is thought to play a critical role for gene activation in the chromatin (reviewed in [39]). CBP and p300 bind to other cofactor complexes among them PCAF, SRC-1/NcoA-1, TIF-2/NcoA-2 and pCIP/ACTR which also possess histone acetyltransferase activity (reviewed in [40]). There are indications for the formation of gene- and pathway-specific complexes. For example, binding of CBP to PCAF and pCIP has been reported to be necessary for induced CREB function [41]. The interaction between cofactors and the P-box of CREB and CREM is not always phosphorylation-dependent. A new route for transcriptional activation by CREB and CREM was reported recently, demonstrating the functional interaction between ACT (for activator of CREM in testis) and CREM [42]. ACT appears to be a tissue-specific coactivator for CREM. It possesses an intrinsic activation domain and interacts with CREM in a phosphorylation-independent manner. In addition to the inducible P-box domain, CREB also contains a constitutive activation domain (CAD), which is responsible for the interaction with one or more of the TATA-box associated factors (TAFs), one of which is $TAF_{II}110$ (the drosophila homolog of human $TAF_{II}130$). The constitutive activation domain of CREB can be subdivided into three regions, which are rich in either serine, hydrophobic amino acids, or glutamine. All three regions are necessary for effective interaction with $TAF_{II}110$ in a yeast two-hybrid assay [43].

Transcriptional effects of CREB are context- and position-specific

The biological effects of the cAMP pathway through CREB are entirely based upon a gene expression program initiated by the activator. Hence, unraveling of CREB transcriptional activation is crucial for the understanding of the biological processes. Above we have discussed CREB-structure and -function through cofactors. Additional important parameters for CREB function are the context in which CREs are embedded and the position of CREs relative to the start site of transcription. This has been most clearly demonstrated on the gene encoding the T-cell receptor beta (TCRβ) chain [44]. These studies could have model character for the many other target genes of cAMP-induced CREB proteins and, therefore, will be briefly reviewed here.

The TCRβ gene is an attractive model for the study of promoter and enhancer function. This is mainly based upon the fact that the genome contains many different promoters that can be compared. A functional TCRβ gene is generated through recombination events in which the enhancer is brought into the relative vicinity of one of the many Vβ promoters (although it remains a distal element, several kilobases apart from the promoter). The rearranged TCRβ gene contains CREs in three different positions that seem to fullfill alternative tasks (Fig. 2). Firstly, CREs are contained within the distal TCRβ enhancer [15]. Secondly, in many Vβ promoters one CRE is found in a promoter-proximal position [17], in between position –100 to –40 upstream of the start site of transcription. In our studies of the human Vβ 8.1 promoter we could also detect a third cryptic CRE within the core promoter region [44], located in between position –30 and +11 (compare Fig. 2). This is the region where TFIID and the other general transcription factors bind to the promoter.

In the enhancer CREB appears to be part of a multiprotein enhanceosome [15]. The enhancer is efficiently repressed by overexpression of the 12S form of adenovirus encoded E1A, which is known to compete for the CREB-binding proteins

CBP and p300. One simple scenario would imply that CBP binds to and functions via binding to CREB in the enhancer. However, E1A retained its repression potential even after removal of the CRE. Repression by E1A seems to be rather correlated to the overall enhancer activity. This suggests that CBP is part of, or functions through, the multiprotein enhanceosome rather than through individual activators alone such as CREB [45]. Further evidence for this hypothesis came from experiments with multimerized CREs that proved to act as a poor enhancer element at a distance (unpublished observations). The promoter-upstream (UAS) CRE mainly serves as a platform for the enhancer. This is concluded from the fact that it raises relative enhancer activity but displays little influence on the promoter [44, 45]. Activation requires an intact PKA phosphorylation site in CREB. In contrast, the third functional CRE within the core promoter contributes strongly to Vβ 8.1 promoter activity. This low affinity CRE can be activated through overexpression of CREB, but not through a mutant lacking Ser133. Moreover, replacement of the weak CRE by a consensus CRE efficiently raises promoter activity [44]. Thus, the core CRE is critical for promoter function, whereas the two other CREs help to establish a functional enhancer. Related mechanisms could add a new level of complexity to the control of cAMP-induced gene activation in other biological processes.

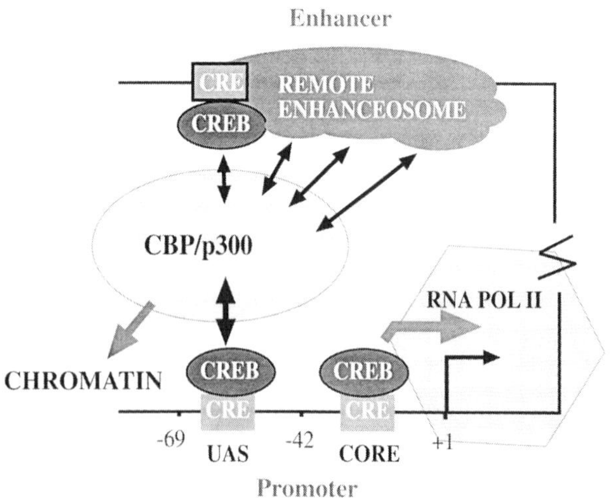

Fig. 2. Position depending function of CREs. The model is based upon the analysis of the TCRβ gene consisting of the distal enhancer and the Vβ 8.1 promoter. CREs (cAMP-responsive elements) are found within the TCRβ enhancer and the Vβ8.1 promoter. The promoter comprises two CREs upstream of the core region (upstream activating sequence (UAS)) and within the core region (positions –30 to +11). Only the core element affects promoter activity (arrow) whereas the two other elements mainly contribute to enhancer-promoter communication. Possible functional interactions between the enhancer and the promoter via the cofactor CBP/p300 are indicated by black arrows. The model implies that upon interaction with the enhanceosome CBP and p300 acetylate histones (grey arrow), which could help to keep the promoter accessible for CREB and other activators that bind to weak interaction sites on the core promoter and subsequently activate transcription.

Acknowledgements

We must apologize to the many researchers whose contributions could not be cited mainly for space limitations. We thank Peter Halle (Switch Biotech Inc., Munich) for providing unpublished observations and Barbara Günzler and Gerhard Mittler (Gene Center, Munich) for their support during preparation of the manuscript.

References

1. Sassone CP: Transcription factors responsive to cAMP. Annu Rev Cell Dev Biol 11: 355–377, 1995
2. Qian JF, Wang TT, Wu XH, Wu J, Ge C, Lachance S, Carriere S, Chan JS: Angiotensinogen gene expression is stimulated by the cAMP-responsive element-binding protein in opossum kidney cells. J Am Soc Nephrol 8: 1072–1079, 1997
3. Wang TT, Chen X, Wu XH, Zhang SL, Chan JS: Molecular mechanism(s) of action of isoproterenol on the expression of the angiotensinogen gene in opossum kidney proximal tubular cells. Kidney Int 55: 1713–1723, 1999
4. Esther CJ, Semeniuk D, Marino EM, Zhou Y, Overbeek PA, Bernstein KE: Expression of testis angiotensin-converting enzyme is mediated by a cyclic AMP responsive element. Lab Invest 77: 483–488, 1997
5. Nahman NJ, Rothe KL, Falkenhain ME, Frazer KM, Dacio LE, Madia JD, Leonhart KL, Kronenberger JC, Stauch DA: Angiotensin II induction of fibronectin biosynthesis in cultured human mesangial cells: Association with CREB transcription factor activation. J Lab Clin Med 127: 599–611, 1996
6. Funakoshi Y, Ichiki T, Ito K, Takeshita A: Induction of interleukin-6 expression by angiotensin II in rat vascular smooth muscle cells. Hypertension 34: 118–125, 1999
7. Kim EL, Esparza FM, Stachowiak MK: The roles of CRE, TRE, and TRE-adjacent S1 nuclease sensitive element in the regulation of tyrosine hydroxylase gene promoter activity by angiotensin II. J Neurochem 67: 26–36, 1996
8. Dudai Y, Jan YN, Byers D, Quinn WG, Benzer S: Dunce, a mutant of Drosophila deficient in learning. Proc Natl Acad Sci USA 73: 1684–1688, 1976
9. Dubnau J, Tully T: Gene discovery in Drosophila: New insights for learning and memory. Annu Rev Neurosci 21: 407–444, 1998
10. Davis HP, Squire LR: Protein synthesis and memory: A review. Psychol Bull 96: 518–559, 1984
11. Dash PK, Hochner B, Kandel ER: Injection of the cAMP-responsive element into the nucleus of Aplysia sensory neurons blocks long-term facilitation. Nature 345: 718–721, 1990
12. Mayford M, Kandel ER: Genetic approaches to memory storage. Trends Genet 15: 463–470, 1999
13. Muthusamy N, Leiden JM: A protein kinase C-, Ras-, and RSK2-dependent signal transduction pathway activates the cAMP-responsive element-binding protein transcription factor following T cell receptor engagement. J Biol Chem 273: 22841–22847, 1998
14. Mayall TP, Sheridan PL, Montminy MR, Jones KA: Distinct roles for P-CREB and LEF-1 in TCR alpha enhancer assembly and activation on chromatin templates *in vitro*. Genes Dev 11: 887–899, 1997

15. Gottschalk LR, Leiden JM: Identification and functional characterization of the human T-cell receptor beta gene transcriptional enhancer: Common nuclear proteins interact with the transcriptional regulatory elements of the T-cell receptor alpha and beta genes. Mol Cell Biol 10: 5486–5495, 1990
16. Gupta A, Terhorst C: CD3-delta enhancer. CREB interferes with the function of a murine CD3-delta A binding factor (M delta AF). J Immunol 152: 3895–3903, 1994
17. Lee MR, Chung CS, Liou ML, Wu M, Li WF, Hsueh YP, Lai MZ: Isolation and characterization of nuclear proteins that bind to T cell receptor V beta decamer motif. J Immunol 148: 1906–1912, 1992
18. Rudolph D, Tafuri A, Gass P, Hammerling GJ, Arnold B, Schutz G: Impaired fetal T cell development and perinatal lethality in mice lacking the cAMP-response element-binding protein. Proc Natl Acad Sci USA 95: 4481–4486, 1998
19. Barton K, Muthusamy N, Chanyangam M, Fischer C, Clendenin C, Leiden JM: Defective thymocyte proliferation and IL-2 production in transgenic mice expressing a dominant-negative form of CREB. Nature 379: 81–85, 1996
20. Lee KA, Masson N: Transcriptional regulation by CREB and its relatives. Biochim Biophys Acta 1174: 221–233, 1993
21. Hoeffler JP, Meyer TE, Yun Y, Jameson JL, Habener JF: Cyclic AMP-responsive DNA-binding protein: Structure based on a cloned placental cDNA. Science 242: 1430–1433, 1988
22. Hai T, Curran T: Cross-family dimerization of transcription factors Fos/Jun and ATF/CREB alters DNA binding specificity. Proc Natl Acad Sci USA 88: 3720–3724, 1991
23. Yamamoto KK, Gonzalez GA, Biggs WD, Montminy MR: Phosphorylation-induced binding and transcriptional efficacy of nuclear factor CREB. Nature 334: 494–498, 1988
24. Deutsch PJ, Hoeffler JP, Jameson JL, Lin JC, Habener JF: Structural determinants for transcriptional activation by cAMP-responsive DNA elements. J Biol Chem 263: 18466–18472, 1988
25. Gonzalez GA, Montminy MR: Cyclic AMP stimulates somatostatin gene transcription by phosphorylation of CREB at serine 133. Cell 59: 675–680, 1989
26. de GR, Ballou LM, Sassone CP: Positive regulation of the cAMP-responsive activator CREM by the p70 S6 kinase: An alternative route to mitogen-induced gene expression. Cell 79: 81–91, 1994
27. Nakajima T, Uchida C, Anderson SF, Parvin JD, Montminy M: Analysis of a cAMP-responsive activator reveals a two-component mechanism for transcriptional induction via signal-dependent factors. Genes Dev 11: 738–747, 1997
28. Rehfuss RP, Walton KM, Loriaux MM, Goodman RH: The cAMP-regulated enhancer-binding protein ATF-1 activates transcription in response to cAMP-dependent protein kinase A. J Biol Chem 266: 18431–18434, 1991
29. Laoide BM, Foulkes NS, Schlotter F, Sassone CP: The functional versatility of CREM is determined by its modular structure. Embo J 12: 1179–1191, 1993
30. Bartsch D, Casadio A, Karl KA, Serodio P, Kandel ER: CREB1 encodes a nuclear activator, a repressor, and a cytoplasmic modulator that form a regulatory unit critical for long-term facilitation. Cell 95: 211–223, 1998
31. Montminy MR, Sevarino KA, Wagner JA, Mandel G, Goodman RH: Identification of a cyclic-AMP-responsive element within the rat somatostatin gene. Proc Natl Acad Sci USA 83: 6682–6686, 1986
32. Hyman SE, Comb M, Lin YS, Pearlberg J, Green MR, Goodman HM: A common trans-acting factor is involved in transcriptional regulation of neurotransmitter genes by cyclic AMP. Mol Cell Biol 8: 4225–4233, 1988
33. Lee CQ, Yun YD, Hoeffler JP, Habener JF: Cyclic-AMP-responsive transcriptional activation of CREB-327 involves interdependent phosphorylated subdomains (retracted by Y Yun, JP Hoeffler, JF Habener. In EMBO J 13: 2736, 1994). Embo J 9: 4455–4465, 1990
34. Bito H, Deisseroth K, Tsien RW: CREB phosphorylation and dephosphorylation: a Ca(2+)- and stimulus duration-dependent switch for hippocampal gene expression. Cell 87: 1203–1214, 1996
35. De CD, Fimia GM, Sassone CP: Signaling routes to CREM and CREB: Plasticity in transcriptional activation. Trends Biochem Sci 24: 281–285, 1999
36. Parker D, Jhala US, Radhakrishnan I, Yaffe MB, Reyes C, Shulman AI, Cantley LC, Wright PE, Montminy M: Analysis of an activator:coactivator complex reveals an essential role for secondary structure in transcriptional activation. Mol Cell 2: 353–359, 1998
37. Chrivia JC, Kwok RP, Lamb N, Hagiwara M, Montminy MR, Goodman RH: Phosphorylated CREB binds specifically to the nuclear protein CBP. Nature 365: 855–859, 1993
38. Nakajima T, Uchida C, Anderson SF, Lee CG, Hurwitz J, Parvin JD, Montminy M: RNA helicase A mediates association of CBP with RNA polymerase II. Cell 90: 1107–1112, 1997
39. Kouzarides T: Histone acetylases and deacetylases in cell proliferation. Curr Opin Genet Dev 9: 40–48, 1999
40. Xu L, Glass CK, Rosenfeld MG: Coactivator and corepressor complexes in nuclear receptor function. Curr Opin Genet Dev 9: 140–147, 1999
41. Korzus E, Torchia J, Rose DW, Xu L, Kurokawa R, McInerney EM, Mullen TM, Glass CK, Rosenfeld MG: Transcription factor-specific requirements for coactivators and their acetyltransferase functions. Science 279: 703–707, 1998
42. Fimia GM, De CD, Sassone CP: CBP-independent activation of CREM and CREB by the LIM-only protein ACT. Nature 398: 165–169, 1999
43. Felinski EA, Quinn PG: The CREB constitutive activation domain interacts with TATA-binding protein-associated factor 110 (TAF110) through specific hydrophobic residues in one of the three subdomains required for both activation and TAF110 binding. J Biol Chem 274: 11672–11678, 1999
44. Halle JP, Haus SP, Woltering C, Stelzer G, Meisterernst M: A conserved tissue-specific structure at a human T-cell receptor beta- chain core promoter. Mol Cell Biol 17: 4220–4229, 1997
45. Haus-Seuffert P, Halle JP, Sanner S, Meisterernst M: Conserved cAMP-responsive element and core promoter complex are critical for specificity of the distal T-cell receptor beta chain enhancer for its native promoter. Gene 236: 209–219, 1999

Molecular and Cellular Biochemistry **212:** 11–17, 2000.

Transcriptional regulation by cAMP in the heart

F.U. Müller, J. Neumann and W. Schmitz

Institut für Pharmakologie und Toxikologie, Westfälische Wilhelms-Universität, Münster, Germany

Abstract

Stimulation of the cAMP-dependent signalling pathway by β-adrenergic catecholamines is an important physiological mechanism to increase contractile force in the heart. In addition to this, long-term β-adrenergic stimulation by elevated catecholamines also influences the expressional control of functionally relevant cardiac regulatory proteins in human heart failure. The regulation of transcription by the cAMP-response element (CRE) is an important mechanism for a cAMP-mediated control of gene expression involved e.g. in spermiogenesis and memory/learning processes. This article discusses recent data leading to the hypothesis that this mechanism also contributes to altered gene regulation in heart failure. (Mol Cell Biochem **212**: 11–17, 2000)

Key words: heart failure, adrenergic receptor, β-adrenoceptor antagonist, cAMP-response element (CRE), cAMP-response element binding protein (CREB), cAMP-response element modulator (CREM)

Introduction

Heart failure develops as a consequence of a variety of cardiac diseases, e.g. ischemic or dilated cardiomyopathy [1]. It is defined as the inability of the heart to adequately cover the organs' demands of blood supply and is associated with a high mortality comparable to malign tumors. Contraction and relaxation of the failing heart are impaired, and the occurrence of arrhythmias is increased. These functional alterations can be explained by diverse alterations in the expression of cardiac regulatory proteins extending from proteins implicated in signal transduction pathways and calcium homeostasis to contractile proteins [2]. As a consequence of neurohumoral activation plasma catecholamines are elevated in patients with severe heart failure [3], and some expressional alterations can be 'restored' and clinical parameters improved by long-term therapy with β-adrenoceptor antagonists [4–7]. Therefore, excessive β-adrenergic stimulation of the cAMP-dependent signalling pathway may contribute a significant part to cardiac gene regulation and the pathophysiology of heart failure. In this article the possible role of a cAMP-dependent transcriptional control mediated by the cAMP-response element in the heart and in the pathogenesis of heart failure is discussed.

Functional and molecular alterations in heart failure

The end-stage failing human heart is often hypertrophied and shows characteristic functional properties *in vitro* in comparison to the non-failing heart. The most prominent change is a decreased responsiveness to cAMP elevating inotropic drugs (e.g. β-adrenoceptor agonists or phosphodiesterase inhibitors) whereas the response to cAMP-independent stimuli (e.g. extracellular calcium, digitalis, calcium sensitizers) is unchanged [8, 9]. Another characteristic phenomenon is the so-called 'negative force frequency relationship' or 'negative Treppe' meaning that an increase of the stimulation frequency fails to increase or even decreases force of contraction in contrast to non-failing myocardium where contractile force is increased [10, 11].

A multitude of cardiac regulatory proteins are differently expressed in the human failing heart. Among these, proteins involved in the cAMP-dependent signalling pathway and the homeostasis of the intracellular calcium have gained much attention since changes of those proteins can possibly explain the impaired function of the failing heart (Fig. 1). A down-regulation and/or uncoupling of β-adrenoceptors was observed

Present address: F.U. Müller, Harvard School of Public Health, Cardiovascular Biology Laboratory, 677 Huntingdon Avenue, 2-129 Boston, MA 02115, USA (until 15 November 2000)
Address for offprints: F.U. Müller, Institut für Pharmakologie und Toxikologie, Westfälische Wilhelms-Universität, Domagkstrasse 12, D-48149 Münster, Germany, e-mail: mullerf@uni-muenster.de

affecting selectively the β_1-subtype in dilated cardiomyopathy and β_1- and possibly β_2-subtype in ischemic cardiomyopathy (for review see ref. [12]). This and an increased expression of inhibitory G-proteins ($G_{i\alpha}$) [13–15] lead to a predominance of inhibitory influences on the activity of the adenylyl cyclase and therefore could explain the smaller β-adrenoceptor-mediated increase of cAMP in failing than in non-failing hearts [16, 17]. Distal from cAMP, the mechanisms explaining the decreased force of contraction after β-adrenergic stimulation in the failing myocardium are less clear. The diastolic reuptake of Ca^{2+} into the sarcoplasmic reticulum is regulated by the sarcoplasmic reticulum Ca^{2+}-ATPase (SERCA) and its regulator phospholamban (PLB) and was found to be reduced in failing myocardium in most studies [18–22] with one exception [23]. Since many groups reported unchanged expression of SERCA and PLB proteins in the failing heart [22–24] reduced Ca^{2+} reuptake may be due to reduced phosphorylation of PLB rather than to altered expression of PLB and SERCA. SERCA activity is inhibited by non-phosphorylated not by phosphorylated PLB [25]. Thus, a reduction in the protein kinase A (PKA)-mediated phosphorylation of PLB at serine 16 could explain a decreased activity of SERCA [26]. However, PKA activity is not changed in heart failure [27, 28] and therefore cannot account for the diminished PLB phosphorylation in the failing human myocardium. The protein phosphatase subtype 1 (PP1) dephosphorylates PLB at serine 16 [29]. Both, activity and mRNA expression of protein phosphatase subtype 1 (PP1) are increased in the failing myocardium [30, 31]. Therefore, increased PP1 activity could lead to decreased PLB phosphorylation, SERCA inhibition and finally decreased contractile response. This concept is supported by recent data showing the development of cardiac hypertrophy in transgenic mice with cardiac-specific overexpression of PP1 [32]. The functional role of a Ca^{2+}/calmodulin-dependent protein kinase II (CamKII) mediated phosphorylation of PLB at threonine 17 is not yet defined, recently the CamKII was reported to be upregulated and its activity was increased in failing myocardium [28, 33]. In summary, there are expressional alterations of different cardiac regulatory proteins involving the cAMP-dependent signalling pathway which can explain impaired cardiac function in heart failure.

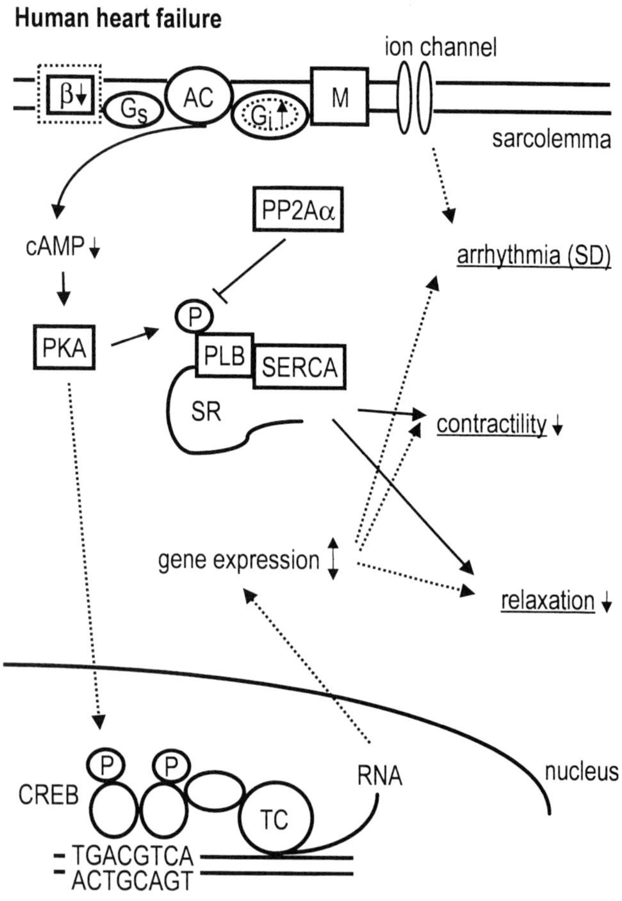

Fig. 1. Simplified scheme of cardiac regulatory proteins and a possible role of a CRE-mediated transcriptional activation in a cardiomyocyte from a human failing heart. The figure is explained in the text. Abbreviations: β – β-adrenoceptor; G_s – stimulatory G-protein; G_i – inhibitory G-protein; AC – adenylyl cyclase; M – muscarinic cholinoceptor; PKA – protein kinase A; PP2Aα – serine/threonine protein phosphatase type 2Aα; PLB – phospholamban; SERCA – sarcoplasmic reticulum Ca^{2+} ATPase; SR – sarcoplasmic reticulum; CREB – CREB-like transcription factors; TC – transcriptional complex. Dotted lines for β and G_i indicate 'normal' expression in the non-failing heart, dotted arrows symbolize possible regulation in the failing human heart.

Regulation of cardiac gene expression by cAMP

At present there is no single specific signaling molecule or transcription factor that can completely account for the expressional and functional changes leading to hypertrophy, cardiomyopathy or heart failure. However, various signalling pathways have been suggested to be involved. While different protein kinases (e.g. stress-activated protein kinases/p38 kinases) were implicated in the response to hypertrophic stress (for review see ref. [34]), the Ca^{2+}-dependent activation of protein phosphatase 2B (calcineurin) induced cardiac hypertrophy via dephosphorylation of the transcription factor NFAT-3 [35]. The clinical value of the latter finding, namely of the use of the calcineurin-inhibiting immunosuppressive agents FK506 or cyclosporin A for the treatment of heart failure, is not yet clear and needs further evaluation [36].

Plasma catecholamines are elevated in patients suffering from heart failure [3] leading to the hypothesis that chronic stimulation of the cAMP-dependent signalling pathway contributes to the regulation of gene expression in the failing heart [37]. Consequently, the treatment of these patients with β-adrenoceptor antagonists was investigated and was shown to considerably improve cardiac performance and mortality [6, 7, 38]. Moreover, both the decrease in β-adrenoceptor density and the increase in the expression of the inhibitory G protein $G_{i\alpha}$ could be restored in these patients by prolonged therapy with the β-blocker metoprolol [4, 5]. The concept was further supported by findings in a rat model of chronic β-adrenergic stimulation [39, 40]. In this model, cardiac hypertrophy and changes in cardiac protein and function similar to human heart failure were induced by prolonged β-adrenergic stimulation using osmotic minipumps for several days. Heart to body weight ratio was increased by 30–40% within 2 days of infusion, and isoproterenol-stimulated force of contraction was decreased in electrically-stimulated cardiac preparations from isoproterenol-treated animals [40]. Furthermore, β-adrenoceptor density and expression of SERCA and PLB on the protein and mRNA levels were diminished [40]. Similar to human heart failure, the expression of the inhibitory G protein $G_{i\alpha}$ was increased on the protein and mRNA levels whereas the stimulatory G protein G_s was unchanged [39, 40]. Using run-on assays the cardiac transcriptional activity of G protein α-subunits was measured in the same model. The transcriptional activity of $G_{i\alpha\text{-}2}$ was increased after isoproterenol infusion and this effect was antagonized by the β-adrenoceptor antagonist propranolol [41, 42]. Therefore, a transcriptional regulation mediated by the β-adrenoceptor may be relevant for the pathophysiology of human heart failure.

Transcriptional regulation by the cAMP-response element

The transcriptional control by the cAMP-response element (CRE) is a major mechanism regulating gene expression by the cAMP-dependent signalling pathway (Fig. 1; for review see refs [43–46]. The CRE is a palindromic 8-bp consensus recognition site (TGACGTCA) typically located within 100 nucleotides of the TATA box. CREs are found in the promoters of numerous genes and multimerization of the CRE enhances the inducibility by cAMP. Transcriptional activation by the CRE is also influenced by the distance between the CRE and the TATA box (greater activity in a more 'proximal' position) and by the nucleotide sequence around the CRE. The structurally related transcription factors cAMP-response element binding protein (CREB), cAMP-response element modulator (CREM) and the activating transcription factor 1 (ATF-1) bind as homo- or heterodimers to the CRE via a carboxy-terminal basic region/leucine zipper domain (bZIP). These proteins, also termed as 'CREB-like' transcription factors, contain a consensus protein kinase A phosphorylation site surrounded by glutamine-rich transactivation domains and are regulated by cAMP *in vivo*. However, there are variants of CREB and CREM generated by alternative splicing [47–53], by internal transcription under control of an intronic promoter [54, 55] or by internal translation [56, 57] lacking the phosphorylation and transactivation domains. These proteins bind to the CRE as homodimers or as heterodimers with activating isoforms and thus prevent transcriptional activation by cAMP.

In most tissues, two alternatively spliced CREB isoforms (CREB327 and CREB341) are expressed which differ by the presence of a 42-bp exon or 14 amino acids, respectively [47]. Both isoforms may act as transcriptional activators, the functional difference between both proteins is not yet clear. The transcriptional activation by CREB is incrementally regulated by the protein kinase A-dependent phosphorylation of serine 119/133 (in CREB327/CREB341), and this phosphorylation seems to be sufficient for full activation [58]. The same site is phosphorylated in response to calcium entry [59] either directly involving calcium calmodulin kinases or indirectly via the calcium-sensitive type I adenylyl cyclase [59, 60]. Phosphorylation *in vivo* at other sites by casein kinase II or glutathion synthase kinase 3 (GSK-3) has been described [61, 62]. However, the relevance of these phosphorylations for the cAMP-inducibility of CREB activity is not yet clear. Mutagenesis of the casein kinase II consensus sites did not affect the inducibility of CREB activity by the protein kinase A [63] and ATF-1 which is structurally similar to CREB is strongly induced by the protein kinase A but contains an alanine instead of serine at the equivalent phosphorylation site [64].

The transcriptional activation by CREB can be regarded as stepwise. In the early 'burst phase', CRE activity is induced within 15–20 min after cAMP elevation which coincides with CREB phosphorylation at serine 133 and with the translocation of the C-subunit of the cAMP-dependent protein kinase A as a rate-limiting step [65]. The next phases are characterized by attenuation and refractoriness of transactivation by CREB, even in the presence of a cAMP-elevating agonist [66]. Protein phosphatases type 1 or 2A may be involved in the attenuation since inhibition of these phosphatases by specific inhibitors, e.g. by okadaic acid, suppressed this phase [66]. Several mechanisms may contribute to long-term alterations of the CRE-mediated transcriptional regulation in which cells are unresponsive to further stimulation by the cAMP-dependent signalling pathway. A downregulation of the phosphorylating C-subunit of the protein kinase A was suggested to be responsible for decreased CREB phosphorylation in this phase [67]. Moreover, the expression of the inducible cyclic AMP early repressor (ICER) [54, 55] or of CREB [68] was observed. This could alter the balance between activators and repressors at the CRE. In summary, in the early phase of the CRE-mediated and cAMP-induced transcriptional regulation, the protein kinase A-dependent phosphorylation of activating isoforms of CREB-like factors is the major mechanism whereas after prolonged stimulation the inducibility is influenced by a downregulation of the C-subunit and/or an altered ratio between activating and inhibiting CRE-binding factors.

Whereas a transcriptional regulation by the CRE has been implicated in spermiogenesis and memory/learning [43–46], the role in the cardiovascular system, namely in the heart and in the pathophysiology of heart failure, is less clear. The expression and phosphorylation of CREB327 and CREB341 was described in primary cultivated chicken embryonic cardiomyocytes [69] and in failing and non-failing human hearts [70]. In electrophoretic mobility shift assays human cardiac ventricular nuclear extracts showed a CRE-specific DNA-binding activity that was inhibited and supershifted by antibodies specific for whole CREB and for CREB phosphorylated at serine 119/133 [70]. CREs are present in various genes which are or may be of relevance for the cardiovascular system (see Table 1). A CRE-mediated transcriptional regulation has been shown for the β-adrenoceptors, Kv1.5, TGF- β3, c-fos and ANP whereas the function of CREs in the other genes remains to be elucidated. Some of these genes, e.g. β-adrenoceptors, PLB, protein kinase A and protein phosphatase 2Aα are directly involved in the cAMP-dependent signalling pathway and regulation of cardiac force of contraction. Other genes, e.g. the transcription factor c-fos or the transforming growth factor β3 (TGF-β3) regulate gene expression and growth. So, it is conceivable that alterations in the CRE-mediated gene transcription contribute to the cardiac gene regulation and to the pathophysiology of human heart failure.

A link between the pathophysiology of heart failure and attenuation of the CRE-mediated gene transcription was suggested by data on transgenic mice with heart-specific overexpression of a dominant-negative mutant of CREB (dnCREB) where serine 133 was replaced by alanine [71]. Due to this mutation, dnCREB cannot be phosphorylated by the protein kinase A and cannot transactivate [58]. The dnCREB-overexpressing mice developed signs of heart failure and died within several weeks after birth. Thus, it may be speculated that attenuation of the CRE-mediated transcriptional activation contributes to the pathophysiology of heart failure. Our group identified and characterized a short CREM-isoform, CREM-IbΔC-X, in the human heart exhibiting a similar function as dnCREB. Thus, this protein may be involved in the pathophysiology of heart failure in contrast to the mutant dnCREB that is not naturally expressed [57]. CREM-IbΔC-X lacks the domains for phosphorylation and transactivation, and a frame-shift at the first exon-exon-border leads to translation of a truncated CREM-protein lacking the DNA-binding domain. Two further proteins translated from internal translational start sites were described to act as repressors at the CRE similar to dnCREB and other CREM-repressors [57].

At present it is not clear whether there are alterations in the function and/or expression of cardiac CREB-like factors in human heart failure or after stimulation of the cAMP-dependent signalling pathway in the heart. We investigated the time course of CRE-mediated transcriptional activation in cardiomyocytes transiently transfected with a luciferase reporter construct under control of tetramerized CREs (Fig. 2). There was a peak activity after 8 h stimulation with 10 μM forskolin, an activator of the adenylyl cyclase. Then, luciferase

Table 1. Examples of genes putatively regulated by cAMP-response elements

Promoter	Species	Reference
β_1-adrenergic receptor	human	76
β_2-adrenergic receptor	human	77
Phospholamban	chicken, rat	78, 79
Proteinkinase Aregulatory subunit	pig	80
Protein phosphatase 2Aα	human	81
eNOS	human	82
Kv1.5	rat	83
ANP	human	84
TGF-β3	human	85
c-fos	human	86

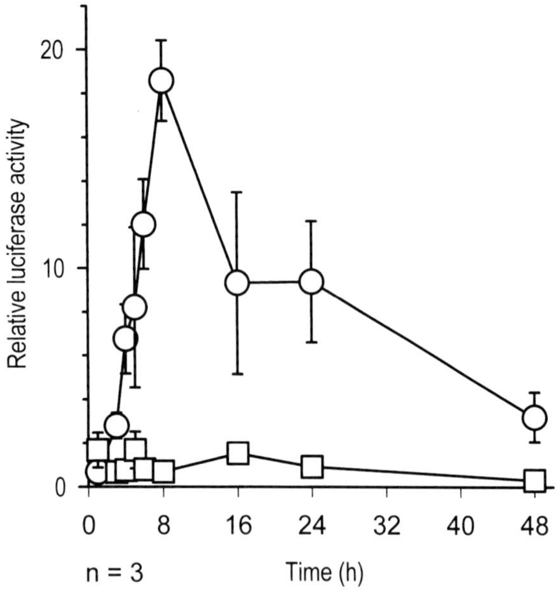

Fig. 2. Time course of CRE-mediated transactivation in cardiomyocytes. Primary chicken embryonic cardiomyocytes were prepared according to ref. [85]. Cells were cultivated at a density of 200,000 cells per cm^2 and were transiently transfected with 0.2 μg per 200,000 cells of a CRE-controlled reporter construct using the Superfect reagent (Qiagen, Hilden, Germany) as described [57]. The reporter construct expresses firefly luciferase under control of four tetramerized CREs from the somatostatin promoter cloned upstream to the minimal promoter of the Herpes simplex virus thymidin kinase gene [88]. Cells were stimulated with 10 μM forskolin (circles) or treated with solvent (DMSO, squares) for the indicated intervals and luciferase activity was determined as described [57]. The medium (with our without forskolin) was renewed daily. Values were referred to the control (1 h DMSO) and are presented as mean ± S.E.M.

activity declined to basal values in the presence of forskolin. This initial experiment shows that (i) there is a CRE-mediated transcriptional activation in the heart and that (ii) this response is attenuated by prolonged stimulation of the cAMP-dependent signalling pathway similar to other tissues. In homology to the dnCREB-mice, an attenuation of the CRE-response might contribute to expressional changes in the failing heart.

The underlying mechanism of this attenuation in the heart is not clear. Activation of the cAMP-dependent signalling system seems to regulate CREB mRNA which might be explained by the presence of CREs in the CREB promoter [72]. In chicken cardiomyocytes, an accumulation of CREB-mRNA was found after 30 min exposure to forskolin, an activator of the adenylyl cyclase [69] while prolonged β-adrenergic stimulation in rats *in vivo* by a 4-day infusion of isoproterenol decreased both mRNAs encoding CREB327 and CREB341 [73]. Nevertheless, the expression of CREB proteins was quantified in ventricular nuclear extracts from human failing and non-failing hearts and was not different in both groups [74]. Thus, expressional alterations in human heart failure cannot be simply explained by decreased expression of CREB. Representing another possible mechanism of attenuation, ICER mRNA was increased in rat neonatal cardiomyocytes after stimulation with dibutyryl-cAMP [75]. In contrast to this, we precipitated different CREM proteins from failing human hearts using a polyclonal antibody recognizing all known CREM-isoforms. From the molecular weight, these proteins were assigned to CREMα/β and to one protein internally translated from CREM-IbΔC-X, but there was no ICER protein detectable [57]. It remains to be elucidated whether the expression of other CREM isoforms is changed in the failing myocardium.

In summary, recent data suggest a contribution of a CRE-mediated transcriptional control to the pathophysiology of heart failure: (i) functionally relevant expressional changes of cardiac regulatory proteins are associated with increased β-adrenergic stimulation; (ii) CREB-like transcription factors were identified in human, chicken and rat hearts; (iii) inhibition of CRE-mediated transcriptional activation in dnCREB-mice led to dilated cardiomyopathy and heart failure; and (iv) prolonged stimulation of the cAMP-dependent signalling pathway by forskolin attenuated CRE-mediated transcriptional activation in cardiomyocytes. The characterization of CREB-like factors and the CRE-mediated transactivation in human failing and non-failing cardiac tissues and in transgenic models is a challenge for further studies possibly resulting in novel pharmacological or genetic approaches for the treatment of heart failure.

References

1. Colucci WS, Braunwald E: Pathophysiology of heart failure. In: E. Braunwald (ed). Heart Disease, 5. Auflage. W.B. Saunders Company, Philadelphia, 1997, pp 394–420
2. Mittmann C, Eschenhagen T, Scholz H: Cellular and molecular aspects of contractile dysfunction in heart failure. Cardiovasc Res 39: 267–275, 1998
3. Daly PA, Sole MJ: Myocardial catecholamines and the pathophysiology of heart failure. Circulation 82(suppl I): I35–I43, 1990
4. Motomura S, Deighton NM, Zerkowski HR, Doetsch N, Michel MC, Brodde OE: Chronic beta 1-adrenoceptor antagonist treatment sensitizes beta 2-adrenoceptors, but desensitizes M2-muscarinic receptors in the human right atrium. Br J Pharmacol 101: 363–369, 1990
5. Sigmund M, Jakob H, Becker H, Hanrath P, Schumacher C, Eschenhagen T, Schmitz W, Scholz H, Steinfath M: Effects of metoprolol on myocardial β-adrenoceptors and $G_{i\alpha}$-proteins in patients with congestive heart failure. Eur J Clin Pharmacol 51: 127–132, 1996
6. Packer M, Bristow MR, Cohn JN, Colucci WS, Fowler MB, Gilbert EM, Shusterman NH (U.S. Carvedilol Heart Failure Study Group): The effect of carvedilol on morbidity and mortality in patients with chronic heart failure. N Engl J Med 334: 1349–1355, 1996
7. Anonymous: Effect of metoprolol CR/XL in chronic heart failure: Metoprolol CR/XL Randomised Intervention Trial in Congestive Heart Failure (MERIT-HF) (see comments). Lancet 353: 2001–2007, 1999
8. Schmitz W, Scholz H, Erdmann E: Effects of alpha- and beta-adrenergic agonists, phosphodiesterase inhibitors and adenosine on isolated human heart muscle preparations. Trends Pharmacol Sci 8: 447–450, 1987
9. Bristow MR, Ginsburg R, Minobe W, Cubicciotti RS, Sageman WS, Lurie K, Billingham ME, Harrison DC, Stinson EB: Decreased catecholamine sensitivity and beta-adrenergic receptor density in failing human hearts. N Engl J Med 307: 205–211, 1982
10. Mulieri LA, Hasenfuss G, Leavitt B, Allen P, Alpert NR: Altered myocardial force-frequency relation in human heart failure. Circulation 85: 1743–1750, 1992
11. Davies CH, Davia BS, Bennet JG, Pepper JR, Poole-Wilson PA, Harding SE: Reduced contraction and altered frequency response of isolated ventricular myocytes from patients with heart failure. Circulation 92: 2540–2549, 1995
12. Brodde OE: β_1- and β_2-adrenoceptors in the human heart: Properties, function and alterations in chronic heart failure. Pharmacol Rev 43: 203–242, 1991
13. Feldman AM, Cates AE, Veazey WB, Hershberger RE, Bristow MR, Baughman KL, Baumgartner WA, Van Dop C: Increase of the 40,000-mol wt pertussis toxin substrate (G protein) in the failing human heart. J Clin Invest 82: 189–197, 1988
14. Neumann J, Schmitz W, Scholz H, von Meyerinck L, Döring V, Kalmár P: Increase in myocardial G_i-proteins in heart failure. Lancet 2: 936–937, 1988
15. Böhm M, Gierschik P, Jakobs KH, Pieske B, Schnabel P, Ungerer M, Erdmann E: Increase of $G_{i\alpha}$ in human hearts with dilated but not ischemic cardiomyopathy. Circulation 82: 1249–1265, 1990
16. Bristow MR, Hershberger RE, Port JD, Minobe W, Rasmussen R: β_1- and β_2-adrenergic receptor-mediated adenylate cyclase stimulation in non-failing and failing human ventricular myocardium. Mol Pharm 35: 295–303, 1989
17. Danielsen W, von der Leyen H, Meyer W, Neumann J, Schmitz W, Scholz H, Starbatty J, Stein B, Döring V, Kalmár P: Basal and isoprenaline-stimulated cAMP content in failing versus non-failing human cardiac preparations. J Cardiovasc Pharmacol 14: 171–173, 1989
18. Pieske B, Kretschmann B, Meyer M, Holubarsch C, Weirich J, Posival H, Minami K, Just H, Hasenfuss G: Alterations in intracellular calcium handling associated with the inverse force-frequency relation in human dilated cardiomyopathy. Circulation 92: 1169–1178, 1995
19. Limas CJ, Olivari MT, Goldenberg IF, Levine TB, Benditt DG, Simon A: Calcium uptake by cardiac sarcoplasmic reticulum in human dilated cardiomyopathy. Cardiovasc Res 21: 601–605, 1987

20. Hasenfuss G, Reinecke H, Studer R, Meyer M, Pieske B, Holtz J, Holubarsch C, Posival H, Just H, Drexler H: Relation between myocardial function and expression of sarcoplasmic reticulum Ca^{2+}-ATPase in failing and non-failing human myocardium. Circ Res 75: 434–442, 1994
21. O'Brien PJ, Gwathmey JK: Myocardial Ca^{2+}- and ATPase-cycling imbalances in end-stage dilated and ischemic cardiomyopathies. Cardiovasc Res 30: 394–404, 1995
22. Flesch M, Schwinger RH, Schnabel P, Schiffer F, van Gelder I, Bavendiek U, Südkamp M, Kuhn-Regnier F, Böhm M: Sarcoplasmic reticulum Ca^{2+}-ATPase and phospholamban mRNA and protein levels in end-stage heart failure due to ischemic or dilated cardiomyopathy. J Mol Med 74: 321–332, 1996
23. Movsesian AM, Bristow MR, Krall J: Ca^{2+} uptake by cardiac sarcoplasmic reticulum from patients with idiopathic dilated cardiomyopathy. Circ Res 65: 1141–1144, 1989
24. Linck B, Boknik P, Eschenhagen T, Müller FU, Neumann J, Nose M, Jones LR, Schmitz W, Scholz H: Messenger RNA expression and immunological quantification of phospholamban and SR-Ca^{2+}-ATPase in failing and non-failing human hearts. Cardiovasc Res 31: 625–632, 1996
25. Sham SJK, Jones LR, Morad M: Phospholamban mediated β-adrenergic enhanced Ca^{2+} uptake in mammalian ventricular myocytes. Am J Physiol 261: H1344–H1349, 1991
26. Schwinger RHG, Münch G, Bölck B, Karczewski P, Krause EG, Erdmann E: Reduced Ca^{2+}-sensitivity of SERCA 2a in failing human myocardium due to reduced serin-16 phospholamban phosphorylation. J Mol Cell Cardiol 31: 479–491, 1999
27. Böhm M, Reiger B, Schwinger R, Erdmann E: cAMP concentrations, cAMP dependent protein kinase activity, and phospholamban in non-failing and failing myocardium. Cardiovasc Res 28: 1713–1719, 1994
28. Kirchhefer U, Schmitz W, Scholz H, Neumann J: Activity of cAMP-dependent protein kinase and Ca^{2+}/calmodulin-dependent protein kinase in failing and non-failing human hearts. Cardiovasc Res 42: 254–261, 1999
29. McDougall LK, Jones LR, Cohen PK: Identification of the major protein phosphatases in mammalian cardiac muscle which dephosphorylate phospholamban. Eur J Biochem 196: 725–734, 1991
30. Neumann J, Eschenhagen T, Jones LR, Linck B, Schmitz W, Scholz H, Zimmermann N: Increased expression of cardiac phosphatases in patients with end-stage heart failure. J Mol Cell Cardiol 29: 265–272, 1997
31. Neumann J, Maas R, Bokník P, Jones LR, Zimmermann N, Scholz H: Pharmacological characterization of protein phosphatase activities in preparations from failing human hearts. J Pharmacol Exp Ther 289: 188–193, 1999
32. Huke S, Bokník P, Knapp J, Linck B, Lüss H, Müller FU, Schmitz W, Suzuki Y, Vahlensieck U, Neumann J, DePaoli-Roach AA: Cardiac overexpression of catalytic subunit of phosphatase 1. Z Kardiol 88(suppl 1): (abstr) 47, 1999
33. Hoch B, Meyer R, Hetzer R, Krause EG, Karczewski P: Identification and expression of δ-isoforms of the multifunctional Ca^{2+}/calmodulin-dependent protein kinase in failing and non-failing human myocardium. Circ Res 84: 713–721, 1999
34. Force T, Hajjar R, Del Monte F, Rosenzweig A, Choukroun G: Signaling pathways mediating the response to hypertrophic stress in the heart. Gene Expr 7: 337–348, 1999
35. Molkentin JD, Lu JR, Antos CL, Markham B, Richardson J, Robbins J, Grant SR, Olson EN: A calcineurin-dependent transcriptional pathway for cardiac hypertrophy. Cell 93: 215–228, 1998
36. Olson EN, Molkentin JD: Prevention of cardiac hypertrophy by calcineurin inhibition: Hope or hype? Circ Res 84: 623–632, 1999
37. Katz AM: The cardiomyopathy of overload: an unnatural growth response in the hypertrophied heart. Ann Int Med 121: 363–371, 1994
38. Anonymous: The cardiac insufficiency bisoprolol study II (CIBIS-II): A randomised trial. Lancet 353 (9146): 9–13, 1999
39. Mende U, Eschenhagen T, Geertz B, Schmitz W, Scholz H, Schulte am Esch J, Sempell R, Steinfath M: Isoprenaline-induced increase in the 40/41 kDa pertussis toxin substrates and functional consequences on the contractile response in rat heart. Naunyn-Schmiedeberg's Arch Pharmacol 345: 44–50, 1992
40. Eschenhagen T, Mende U, Diederich M, Nose M, Schmitz W, Scholz H, Schulte am Esch J, Warnholtz A, Schäfer H: Long-term β-adrenoceptor mediated upregulation of $G_{i\alpha}$ and $G_{o\alpha}$ mRNA levels and pertussis toxin-sensitive guanine nucleotide-binding proteins in rat heart. Mol Pharm 42: 773–783, 1992
41. Müller FU, Boheler KR, Eschenhagen T, Schmitz W, Scholz H: Isoprenaline stimulates gene transcription of the inhibitory G protein α-subunit $G_{i\alpha\text{-}2}$ in rat heart. Circ Res 72: 696–700, 1993
42. Müller FU, Eschenhagen T, Reidemeister A, Schmitz W, Scholz H: *In vivo* β-adrenergic stimulation leads to biphasic regulation of $G_{i\alpha\text{-}2}$ gene transcriptional activity in rat heart. J Mol Cell Cardiol 26: 869–875, 1994
43. Meyer TE, Habener JF: Cyclic adenosine 3',5'-monophosphate response element binding protein (CREB) and related transcription-activating deoxy-ribonucleic acid-binding proteins. Endocrine Rev 14: 269–290, 1993
44. Delmas V, Molina CA, Lalli E, de Groot R, Foulkes NS, Masquilier D, Sassone-Corsi P: Complexity and Versatility of the transcriptional response to cAMP. Rev Physiol Biochem Pharmacol 124: 1–28, 1994
45. Janknecht R, Hunter T: Transcription: A growing coactivator network. Nature 383: 22–23, 1996
46. Montminy M: Transcriptional regulation by cyclic AMP. Annu Rev Biochem 66: 807–822, 1997
47. Hoeffler JP, Meyer TE, Waeber G, Habener JF: Multiple adenosine 3', 5'- monophosphate response element DNA-binding proteins generated by gene diversification and alternative exon splicing. Mol Endocrinol 4: 920–930, 1990
48. Foulkes NS, Borrelli E, Sassone-Corsi P: CREM gene: Use alternative DNA-binding domains generates multiple antagonists of cAMP-induced transcription. Cell 64: 739–749, 1991
49. Foulkes NS, Mellström B, Benusiglio E, Sassone-Corsi P: Developmental switch of CREM function during spermatogenesis: From antagonist to activator. Nature 355: 80–84, 1992
50. Waeber G, Habener JF: Novel testis germ cell-specific transcript of the CREB gene contains an alternatively spliced exon with multiple inframe stop codons. Endocrinology 131: 2010–2015, 1992
51. Walker WH, Sanborn BM, Habener JF: An isoform of transcription factor CREM expressed during spermatogenesis lacks the phosphorylation domain and represses cAMP-induced transcription. Proc Natl Acad Sci USA 91: 12423–12427, 1994
52. Walker WH, Girardet C, Habener JF: Alternative exon splicing controls a transcriptional switch from activator to repressor isoforms of transcription factor CREB during spermatogenesis. J Biol Chem 271: 20145–20150, 1996
53. Gellersen B, Kempf R, Telgmann R: Human endometrial stromal cells express novel isoforms of the transcriptional modulator CREM and regulate ICER in the course of decidualization. Mol Endocrinol 11: 97–113, 1997
54. Molina CA, Foulkes NS, Lalli E, Sassone-Corsi P: Inducibility and negative autoregulation of CREM: An alternative promoter directs the expression of ICER, an early response repressor. Cell 75: 875–886, 1993
55. Stehle JH, Foulkes NS, Molina CA, Simmoneaux V, Pévet P, Sassone-Corsi P: Adrenergic signals direct rhythmic expression of transcriptional repressor CREM in the pineal gland. Nature 365: 314–320, 1993
56. Girardet C, Walker WH, Habener JF: An alternatively spliced poly-

cistronic mRNA encoding cyclic adenosine 3′,5′-monophosphate (cAMP)-responsive transcription factor CREB (cAMP response element binding protein) in human testis extinguishes expression of an internally translated inhibitor CREB isoform. Mol Endocrinol 10: 879–891, 1996
57. Müller FU, Bokník P, Knapp J, Neumann J, Vahlensieck U, Oetjen E, Scheld HH, Schmitz W: Identification and expression of a novel isoform of the cAMP-response element modulator in the human heart. FASEB J 12: 1191–1199, 1998
58. Gonzalez GA, Montminy M: Cyclic AMP stimulates somatostatin gene transcription by phosphorylation at serine 133. Cell 59: 675–680, 1989
59. Sheng M, Thompson MA, Greenberg ME: CREB: A Ca^{2+}-regulated transcription factor phosphorylated by calmodulin-dependent kinases. Science 252: 1427–1430, 1991
60. Impey S, Wayman G, Wu Z, Storm DR: Type I adenylyl cyclase functions as a coincidence detector for control of cyclic AMP response element-mediated transcription. Mol Cell Biol 14: 8272–8281, 1994
61. deGroot RP, den Hertog J, Vandenheede JR, Goris J, Sassone-Corsi: Multiple and cooperative phosphorylation events regulate the CREM activator function. EMBO J 12: 3903–3911, 1993
62. Fiol CJ, Williams JS, Chou CH, Wang QM, Roach PJ, Andrisani OM: A secondary phosphorylation of CREB341 at Ser 129 is required for the cAMP-mediated control of gene expression. J Biol Chem 269: 32187–32193, 1994
63. Gonzalez GA, Menzel P, Leonard J, Fischer WH, Montminy MR: Characterization of motifs which are critical for activity of the cyclic AMP-responsive transcription factor CREB. Mol Cell Biol 11: 1306–1312, 1991
64. Liu F, Thompson MA, Wagner S, Greenberg ME, Green MR: Activating transcription factor-1 can mediate Ca(2+)- and cAMP-inducible transcriptional activation. J Biol Chem 268: 6714–6720, 1993
65. Hagiwara M, Brindle P, Harootunian A, Armstrong R, Rivier J, Vale W, Tsien R, Montminy MR: Coupling of hormonal stimulation and transcription via the cyclic AMP-responsive factor CREB is rate limited by nuclear entry of protein kinase A. Mol Cell Biol 13: 4852–4859, 1993
66. Hagiwara M, Alberts A, Brindle P, Meinkoth J, Feramisco J, Deng T, Karin M, Shenolikar S, Montminy M: Transcriptional attenuation following cAMP induction requires PP-1-mediated dephosphorylation of CREB. Cell 70: 105–113, 1992
67. Armstrong R, Wen W, Meinkoth J, Taylor S, Montminy M: A refractory phase in cyclic AMP-responsive transcription requires down regulation of protein kinase A. Mol Cell Biol 15: 1826–1832, 1995
68. Walker WH, Fucci L, Habener JF: Expression of the gene encoding transcription factor cyclic adenosine 3′,5′-monophosphate (cAMP) response element-binding protein (CREB): Regulation by follicle stimulating hormone-induced cAMP signaling in primary rat Sertoli cells. Endocrinology 136: 3534–3545, 1995
69. Goldspink PH, Russell B: The cAMP response element binding protein is expressed and phosphorylated in cardiac myocytes. Circ Res 74: 1042–1049, 1994
70. Müller FU, Bokník P, Horst A, Knapp J, Linck B, Schmitz W, Vahlensieck U, Böhm M, Deng MC, Scheld HH: The cAMP response element binding protein is expressed and phosphorylated in the human heart. Circulation 92: 2041–2043, 1995
71. Fentzke RC, Korcarz CE, Lang RM, Lin H, Leiden JM: Dilated cardiomyopathy in transgenic mice expressing a dominant-negative CREB transcription factor in the heart. J Clin Invest 101: 2415–2426, 1998
72. Meyer TE, Waeber G, Lin J, Beckmann W, Habener JF: The promoter of the gene encoding 3′,5′-cyclic adenosine monophosphate (cAMP) response element binding protein contains cAMP response elements: Evidence for positive autoregulation of gene transcription. Endocrinol 132: 770–780, 1993
73. Müller FU, Bokník P, Horst A, Knapp J, Linck B, Schmitz W, Vahlensieck U, Walter A: *In vivo* isoproterenol stimulation leads to downregulation of the mRNA encoding the cAMP response element binding protein in the rat heart. Biochem Biophys Res Comm 215: 1043–1049, 1995
74. Müller FU, Bokník P, Knapp J, Lüss H, Neumann J, Vahlensieck U, Böhm M, Deng MC, Scheld HH, Schmitz W: Quantification of the cAMP response element binding protein in ventricular nuclear protein from failing and non-failing human hearts. Biochem Biophys Res Comm 236: 351–354, 1997
75. Folco EJ, Koren G: Degradation of the inducible cAMP early repressor (ICER) by the ubiquitin-proteasome pathway. Biochem J 328: 37–43, 1997
76. Collins S, Ostrowski J, Lefkowitz RJ: Cloning and sequence analysis of the human β_1-adrenergic receptor 5′-flanking promoter region. Biochim Biophys Acta 1172: 171–174, 1993
77. Collins S, Altschmied J, Herbsman O, Caron MG, Mellon PL, Lefkowitz RJ: A cAMP response element in the β_2-adrenergic receptor gene confers transcriptional autoregulation by cAMP. J Biol Chem 265: 19330–19335, 1990
78. Toyofuku T, Zak R: Characterization of cDNA and genomic sequences encoding chicken phospholamban. J Biol Chem 266: 5375–5383, 1991
79. Johns DC, Feldman AM: Identification of a highly conserved region at the 5′ flank of the phospholamban gene. Biochem Biophys Res Commun 188: 927–933, 1992
80. Nowak I, Seipel K, Schwarz M, Jans DA, Hemmings BA: Isolation of a cDNA and characterization of the 5′ flanking region of the gene encoding the type I regulatory subunit of the cAMP-dependent protein kinase. Eur J Biochem 167: 27–33, 1987
81. Khew-Goodall Y, Mayer RE, Maurer F, Stone SR, Hemmings BA: Structure and transcriptional regulation of protein phosphatase 2A catalytic subunit genes. Biochemistry 30: 89–97, 1990
82. Zhang R, Min W, Sessa WC: Functional analysis of the human endothelial nitric oxide synthase promoter. Sp1 and GATA factors are necessary for basal transcription in endothelial cells. J Biol Chem 270: 15320–15326, 1995
83. Mori Y, Matsubara H, Folco E, Siegel A, Koren G: The transcription of a mammalian voltage-gated potassium channel is regulated by cAMP in a cell-specific manner. J Biol Chem 268: 26482–26493, 1993
84. Cornelius T, Holmer SR, Müller FU, Riegger GAJ, Schunkert H: Regulation of the gene for the rat atrial natriuretic peptide after acute imposition of left ventricular overload. Hypertension 30: 1348–1355, 1997
85. Lafyatis R, Lechleider R, Kim SJ, Jakowlew S, Roberts AB, Sporn MB: Structural and functional characterization of the transforming growth factor β3 promoter: A cAMP-responsive element regulates basal and induced transcription. J Biol Chem 265: 19128–19136, 1990
86. Sassone-Corsi P, Visvader J, Ferland L, Mellon PL, Verma IM: Induction of proto-oncogene fos transcription through the adenylate cyclase pathway: Characterization of a cAMP-responsive element. Genes Dev 2: 1529–1538, 1988
87. Eschenhagen T, Friedrichsen M, Gsell S, Hollmann A, Mittmann C, Schmitz W, Scholz H, Weil J, Weinstein LS: Regulation of the human G_i alpha-2 gene promotor activity in embryonic chicken cardiomyocytes. Bas Res Cardiol 91(suppl 2): 41–46, 1996
88. Oetjen E, Diedrich T, Eggers A, Eckert B, Knepel W: Distinct properties of the cAMP-responsive element of the rat insulin I gene. J Biol Chem 269: 27036–27044, 1994

Molecular and Cellular Biochemistry 212: 19–28, 2000.

Regulation of tumor growth and metastasis of human melanoma by the CREB transcription factor family

Didier Jean and Menashe Bar-Eli

Department of Cancer Biology, The University of Texas M.D. Anderson Cancer Center, Houston, TX, USA

Abstract

The purpose of this study was to determine the role of CREB and its associated proteins in melanoma progression. We used MeWo human melanoma cells transfected with a dominant negative construct of CREB, KCREB. KCREB has a mutation in its DNA-binding domain and can not bind the CRE element. Expression of KCREB yields proper heterodimerization with CREB and its associated proteins, but the proteins associated with KCREB do not confer the same degree of transcriptional activity as they would in the case of wild-type CREB. Here, we demonstrate that expression of KCREB in MeWo melanoma cells leads to a decrease in their tumorigenicity and metastatic potential in nude mice. We identified two mechanisms that explain at least partially this effect of KCREB. The first, is one in which CREB and its associated proteins play an essential role in invasion. We showed that the invasive properties of KCREB-transfected MeWo cells were reduced due to the downregulation of the CRE-dependent expression of the type IV collagenase MMP-2 and the adhesion molecule MCAM/MUC18. In the second mechanism, CREB and its associated proteins act as survival factors for human melanoma cells. Here we demonstrated that expression of KCREB in MeWo cells rendered them susceptible to apoptosis induced by thapsigargin, which in turn increased the intracellular level of Ca^{2+}. Thapsigargin induced CREB and ATF-1 phosphorylation and activated CRE-dependent transcription in MeWo cells. Collectively, our data demonstrate that CREB and its associated proteins play an important role in tumor growth and metastasis of human melanoma. (Mol Cell Biochem **212**: 19–28, 2000)

Key words: melanoma, transcription factors, CREB, invasion, apoptosis

Introduction

As it progresses through a multistage process, malignant melanoma in humans switches from melanocyte to nevi, to radial growth and subsequently to a vertical growth (metastatic phenotype), and these switches are associated with molecular and genetic changes [1–3]. The mechanisms that mediate this differential expression of genes during melanoma progression remain largely unknown.

By their ability to regulate the expression of several genes, transcription factors could play an important role during melanoma progression. Recently, it has been demonstrated that the progression of human melanoma is associated with loss of the AP-2 transcription factor [4–6]. This loss caused changes in the expression of the tyrosine kinase receptor c-KIT and the adhesion molecule MCAM/MUC18, both of which contribute to the acquisition of the metastatic phenotype [4, 5]. The transcription factor CREB (cAMP response element-binding protein) could also be involved in melanoma progression. Indeed, previous studies have demonstrated that melanocyte proliferation and differentiation are positively regulated by agents that increase cAMP [7–9]. In addition, CREB expression correlates directly with the metastatic potential of murine melanoma cells [10]. The other CREB family transcription factor member, ATF-1 (activating transcription factor 1), is not detected in normal melanocytes but

Address for offprints: M. Bar-Eli, Department of Cancer Biology, The University of Texas M.D. Anderson Cancer Cancer, 1515 Holcombe Boulevard, Box 173, Houston, TX 77030, USA is easily found in metastatic melanoma cells [11]. Furthermore, CREB is involved in melanoma resistance to radiation [12].

The CREB family belongs to the basic leucine zipper class of transcription factors (bZIP) [13]. Its other members are CREM and ATF-1, the latter of which shares sequence and functional homology with CREB [14]. CREB is a nuclear protein of 43 kDa that binds to CRE (cAMP response element): 5′-TGACGTCA-3′ as a homodimer or heterodimer (with members of ATF and AP1 transcription factor families) [15]. However, to stimulate CRE-dependent transcription, CREB must be activated through phosphorylation. One of the known mechanism of CREB transcriptional activation is phosphorylation of Ser-133 by the catalytic subunit of c-AMP-activated protein kinase A [16]. Many other Ser/Thr kinases can phosphorylate CREB, including protein kinase C [17], Ca^{2+}/calmodulin-dependent protein kinases such as CaMKI, CaMKII [18], and CaMKIV [19], and Ras-dependent p105 kinase [20], $p90^{rsk}$ [11], and Rsk 2 [21]. Furthermore, for its transcriptional activity, CREB requires the coactivator CBP (CREB-binding protein) [22].

Based on these studies, we investigated here the involvement of CREB in melanoma progression. Our approach was to transfect tumor cells with a dominant-negative form of CREB, KCREB, and assay various properties *in vitro* and *in vivo*. KCREB has a single base pair substitution in its DNA-binding domain that causes a change at position 287 from Arg to Leu. This mutation prevents the binding of KCREB to the CRE-element [23]. KCREB can associate with CREB and other proteins that interact with CREB, including members of the ATF and AP-1 transcription factor families and the coactivator CBP. But, because of the mutation in the DNA-binding domain, heterodimers formed with KCREB do not possess the same degree of transcriptional activity as they would in the case of the wild-type CREB. We reasoned that overexpression of KCREB in a cell should quench CREB and its associated proteins. Indeed, the ability of KCREB to inhibit CRE-dependent transcription was demonstrated previously in F9 teratocarcinoma and CA-77 thyroid carcinoma cells [23, 24].

By using KCREB-transfected cells, we demonstrate that CREB plays an important role in human melanoma progression. KCREB expression reduced the tumorigenicity and the metastatic potential of MeWo melanoma cells in nude mice. We show that CREB regulates two important properties of melanoma cells. First, the invasive properties of KCREB-transfected MeWo cells were reduced, due to the down-regulation of the CRE-dependent expression of the type IV collagenase MMP-2 and the adhesion molecule MCAM/MUC18. Second, our results suggest that CREB and its associated proteins act as survival factors for human melanoma cells. Expression of KCREB in MeWo cells rendered them susceptible to apoptosis induced by thapsigargin, which increases the intracellular level of Ca^{2+}. We verified that thapsigargin induced CREB and ATF-1 phosphorylation and activated CRE-dependent transcription in MeWo cells. These data indicate that, through their transcriptional activities, CREB and its associated proteins play an important role in the progression of malignant melanoma.

Materials and methods

Cells

The human melanoma MeWo cell line was established in culture from a lymph node metastasis of a melanoma patient [25] and was kindly provided by Dr. S. Ferrone (New York Medical College, NY, USA). MeWo cells are tumorigenic and metastatic in nude mice [26]. Transfection of MeWo melanoma cells with the pRSV-KCREB construct and the isolation of the two clones K-2 and K-10 that express KCREB were described previously [12]. Cells were maintained in culture as adherent monolayers in RPMI, supplemented with 10% fetal bovine serum (Summit, Ft. Collins, CO, USA), sodium pyruvate, nonessential amino acids, L-glutamine, and penicillin-streptomycin (Life Technologies Inc.). The transfected cells MeWo-Neo, K-2, and K-10 were maintained in the same medium containing G418 (Life Technologies Inc.) at 200 μg/ml. All cells were grown at 37°C with 5% CO_2.

Animals

Male athymic nude mice (BALB/c background) were obtained from the Animal Production Area of the NCI-Frederick Cancer Research and Development Center (Frederick, MD, USA). The mice were housed in laminar flow cabinets under specific pathogen-free conditions and used at 8 weeks of age. Animals were maintained in facilities approved by the American Association for Accreditation of Laboratory Animal Care and in accordance with current regulations and standards of the U.S. Department of Agriculture, Department of Health and Human Services, and the National Institutes of Health. Their use in these experiments was approved by the institutional Animal Care and Use Committee.

Northern blot analysis

mRNA was extracted from cultured cells by FastTrack 2.0 (Invitrogen Co., San Diego, CA, USA). Three micrograms of purified mRNA was separated on 1% denaturing formaldehyde/agarose gels, transferred to GeneScreen nylon membrane (NEN Life Science Products), and UV cross-linked with 120,000 mJ/cm^2 using a UV Stratalinker 1800 (Stratagene, La Jolla, CA, USA). Hybridizations were performed as described previously [27], and filters were washed three times with 30 mM NaCl, 3 mM sodium citrate, pH 7.2, containing 0.1% sodium dodecyl sulfate. The DNA probes used in these analyses were a 1.3-kilobase pair *Pst*I cDNA fragment corresponding to rat glyceraldehyde 3-phosphate dehydrogenase (GAPDH), a 1-kb *EcoR*I cDNA fragment from human MMP-2, and a 1.2-kb *Sst*I-*Pst*I cDNA fragment from human MCAM/MUC18. Each cDNA fragment was purified by agarose gel electrophoresis, recovered using Qiaquick (Qiagen, Chatsworth, CA, USA), and radiolabeled by the random primer technique using [α-^{32}P]deoxyribonucleotide triphosphates. mRNA expression was quantified in the linear range of the film using the ImageQuantÙ software program (Molecular Dynamics, Sunnyvale, CA, USA). Each sample measurement was calculated as the ratio of the area of the specific mRNA transcript to the GAPDH transcript.

Western blot analysis

Melanoma cells were seeded onto 100-mm Petri dishes in culture medium without G418. Thapsigargin treatment (1 μM) was begun when cultures reached 70–80% cell confluence. Total cell extracts were prepared in Triton lysis buffer (25 mM Tris-HCl, pH 7.5 containing 150 mM NaCl, 1% Triton X-100, 5 mM EDTA, 1 mM phenylmethylsulfonyl fluoride, 20 μM leupeptin, 0.15 unit/ml aprotinin, 1 mM Na_3VO_4, and 10 mM NaF). For nuclear extract, nuclei were first isolated and lysed in Triton lysis buffer. For both extracts, after removal of cell debris by centrifugation, the protein content was quantified using the BCA protein assay (Pierce). Proteins of total cell extract (40 μg) and nuclear extract (20 μg) were separated on 10% SDS-polyacrylamide gel electrophoresis in reducing conditions and electrophoretically transferred to Immobilon P transfer membrane (Millipore, Bedford, MA, USA). The blots were probed with relevant antibodies at a dilution of 1/5000 except for the antibody to ATF-1 (1/500). Anti-CREB and anti-phosphorylated-CREB (antibody 5322) were kindly provided by Dr. Marc R. Montminy (Harvard Medical School, Boston, MA, USA) [28]. The antibody to ATF-1, antibody A1, was kindly provided by Dr. Michael R. Green (University of Massachusetts Medical Center, Worcester, MA, USA) [14], and an anti-β-actin antibody was purchased from Sigma. Immunoreactive materials were detected with a secondary antibody conjugated to horseradish peroxidase and an enhanced chemiluminescence system (Amersham Pharmacia Biotech).

Zymography

Gelatinolytic activity was determined on gelatin-impregnated gels. Conditioned media were prepared from 5×10^3 cells and separated on gelatin-impregnated SDS-PAGE under non-reducing conditions. After washes, gels were incubated for 16 h at 37ºC in 50 mM Tris, pH 7.6, containing 0.2 M NaCl, 5 mM $CaCl_2$ and 0.02% Brij 35. At the end of the incubation, gels were stained using Coomassie G250.

DNA transfection and reporter assay

Melanoma cells (1.5×10^5) were transfected with 3 μg of the Somat-*Bgl*II CAT construct and 50 ng of pRL-CMV vector (Promega, Madison, WI, USA) using the Lipofectin reagent (Life Technologies Inc.). The Somat-*Bgl*II CAT construct (kindly provided by Dr. Marc R. Montminy) contained the somatostatin gene promoter from –71 to 153 linked to the CAT reporter gene [29]. The pRL-CMV vector, which contained the *Renilla* luciferase gene driven by the CMV promoter, was introduced to correct for variations of transfection efficiency. For cotransfection experiments, pRc/RSV (Invitrogen, San Diego, CA, USA) and pRSV-KCREB plasmids were added in the indicated amounts to a total concentration of 6 μg. pRSV-KCREB (kindly provided by Dr. Richard H. Goodman (Oregon Health Sciences, Portland, OR, USA) contained a full-length CREB cDNA with a single base pair substitution in the DNA-binding domain that causes a change at position 287 from Arg to Leu [23]. Thapsigargin was applied 24 h after transfection, and the cells were incubated for an additional 24 h. Cell extracts were prepared in a reporter lysis buffer (Promega) and were first assayed for *Renilla* luciferase activity using the dual-luciferase reporter assay system (Promega). Samples were normalized to the same relative luciferase activity and assayed for CAT activity [30]. The conversion of chloramphenicol to the acetylated form was quantified by scanning densitometry of an autoradiograph with a personal densitometer (Molecular Dynamics, Sunnyvale, CA, USA). Each assay was repeated at least three times; there was less than 10% variation among individual transfections.

Detection of apoptosis

Cells were incubated in the presence of 1 μM thapsigargin (Tg) or 1 μM staurosporine (St) for 48 h. Detached cells and adherent cells were collected. For flow cytometry analysis after propidium iodide staining, cells were resuspended in PBS containing 3 mM sodium citrate, 0.1% Triton X-100, and 50 mg/ml propidium iodide and incubated for 2–4 h. The propidium iodide-stained cells were subjected to flow cytometric analysis on an EPICS Profile flow cytometer (Coulter Corp., Miami, FL, USA). For transmission electron microscopy analysis, cells were fixed in a cacodylate buffer (0.1 M) containing 3% glutaraldehyde and 2% paraformaldehyde for 1 h, washed, and treated with 0.1% tannic acid for 20 min. The samples were then incubated in 1% buffered osmium tetroxide for 1 h and stained with 1% aqueous uranyl acetate for 1 h [31]. The samples were dehydrated in increasing concentrations of ethanol and then infiltrated and embedded in Spurr's low viscosity medium. The blocks were polymerized in a 60°C oven overnight. Thin sections were cut, stained with uranyl acetate and lead citrate, and examined using a Jeol 1200-EX transmission electron microscope (Jeol, Peabody, MA, USA).

Invasion assay

Filters (8 μm pore size, nucleopore, Pleasanton, CA, USA) were coated with Matrigel at 25 μg/filter and placed in modified Boyden chambers. Fibroblast-conditioned medium was placed in the lower compartment of the Boyden chamber as a source of chemoattractants. Cells (2×10^5) were placed in the upper compartment and after incubation for 6 h at 37°C, cells were stained on the lower surface of the filter with Dif-Quick (American Scientific Products, McFaw, IL, USA). The number of cells was quantified with an image analyzer (Optomax V) attached to an olympus CK2 microscope.

Tumor cell injection

To prepare tumor cells for inoculation, cells in exponential growth phase were harvested by brief exposure to 0.25% trypsin, 0.2% EDTA solution (w/v). The flask was sharply tapped to dislodge the cells, and supplemented medium was added. The cell suspension was pipetted to produce a single-cell suspension. The cells were washed and resuspended in Ca^{2+}- and Mg^{2+}-free HBSS to the desired cell concentration. Cell viability was determined by trypan blue exclusion, and only single-cell suspensions of more than 90% viability were used. Subcutaneous tumors were produced by injecting 1×10^6 tumor cells in 0.2 ml of HBSS over the right scapular region. Growth of subcutaneous tumors was monitored for 60 days.

For experimental lung metastasis, 1×10^6 tumor cells in 0.2 ml of HBSS were injected into the lateral tail vein of nude mice. The mice were killed after 80 days, and the lungs were removed, washed in water, and fixed with Bouin's solution for 24 h to facilitate counting of tumor nodules as we described previously [32]. The number of surface tumor nodules was counted under a dissecting microscope. Sections of the lungs were stained with hematoxylin and eosin to confirm that the nodules were melanoma and to identify micrometastasis.

Results

Involvement of CREB in tumorigenicity and metastasis of human melanoma cells

To analyze the role of CREB in tumor growth and metastasis of melanoma cells, we have used MeWo human melanoma cells that ectopically express a dominant-negative of CREB, KCREB, that had been mutated in the DNA-binding domain and had lost its ability to bind the CRE element [23, 24]. The isolation and the characterization of the neo-transfected cells and the two KCREB-transfected clones, K-2 and K-10, were described previously [12]. The K-2 and K-10 clones displayed a 10-fold decrease in radiation resistance and 5-fold reduction in their ability to form colonies in soft agar compared with control MeWo parental and neo-transfected cells [12].

To determine the tumorigenicity of the KCREB transfected cells, 1×10^6 cells were injected s.c. over the right scapular region of BALB/c nude mice and the tumor diameter was determined 60 days after injection. As the results summarized in Table 1 show, the two KCREB-transfected clones K-2 and K-10 formed smaller tumors than the control cells, MeWo parental (MeWo-P) and neo-transfected (MeWo-Neo) cells.

Next, we analyzed the metastatic potential of KCREB-transfected cells in an experimental lung metastasis assay. To that end, 1×10^6 cells were injected into the lateral tail vein of BALB/c nude mice. The mice were killed 80 days after injection and the lung tumor colonies were counted. As shown in Table 1, transfection of MeWo cells with KCREB significantly reduced their potential to produce lung metastasis. The median number of metastases was lower in mice injected with KCREB-transfected cells (12 and 4 for K2 and K10 respectively), than in mice injected with control cells (27 and 22 for MeWo-P and MeWo-neo respectively). Furthermore, the lower metastatic potential of K-10 as compared with K-2 cells could be explained by the higher KCREB activity in these cells [12].

Table 1. Tumor growth and experimental lung metastasis of human melanoma cells transfected with KCREB in nude mice

Cell lines	Mean tumor diameter (cm)[a]	Lung metastasis median (range)[b]
MeWo-P	1.3	27 (0–70)
MeWo-Neo	NT	22 (11–74)
K-2	0.6	12 (0–59)
K-10	0.6	4 (1–22)

[a] 1×10^6 cells were injected s.c. and the tumor size was monitored after 60 days; [b] 1×10^6 cells were injected into the lateral tail vein and the number of lung tumor colonies was counted after 80 days. NT – not tested.

Effect of CREB on the invasive properties of human melanoma cells

Several molecules involved in melanoma invasion have CRE-elements in their promoter such as the type IV collagenase MMP-2 (gelatinase A) [33] and the cell adhesion molecule MCAM/MUC18 [34]. MMP-2 activity directly modulates melanoma cell-adhesion and spreading to extracellular matrix [35, 36], while MCAM/MUC18 expression correlates directly with the metastatic ability of melanoma cells and its ectopic expression in primary cutaneous melanoma cells rendered them tumorigenic and metastatic in nude mice [37]. By the presence of CRE elements in their promoters, these genes could be a target for regulation by the CREB transcription factor.

We investigated the expression of MMP-2 and MCAM/MUC18 at the mRNA level in MeWo cells following transfection with KCREB. Northern blot analysis showed that KCREB-transfected cells displayed lower levels of the 3.1-kb MMP-2 transcript and 3.3-kb MCAM/MUC18 transcript than control cells (Fig. 1A). MMP-2 and MCAM/MUC18 expression were inhibited by 2.5-fold and by 3.5-fold in K-2 and K-10 cells, respectively, as compared with MeWo-P and MeWo-neo cells.

To verifiy that the downregulation of MMP-2 expression in KCREB-transfected cells was reflected in MMP-2 activity, we analyzed supernatants from control and KCREB-transfected cells for collagenase activity by zymography. Collagenase activity in the supernatants was normalized to cell number. As shown in Fig. 1B, the activity of the 72-kDa collagenase (MMP-2) was significantly lower in K-2 and K-10 cells than in control MeWo-P and MeWo-Neo cells. It is possible that both the latent and the activated forms of MMP-2 were reduced in the KCREB-transfected cells. The activity of the 92-kDa collagenase (MMP-9) that does not contain CRE element in its promoter remained unchanged and served as an internal control for equal loading.

Chloramphenicol acetyltransferase (CAT) activity driven by the MMP-2 promoter was inhibited by 14- and 45-fold in KCREB-transfected K-10 and K-2 cells, respectively, as compared with control cells, suggesting that KCREB's regulation of MMP-2 expression and activity occurred at the transcriptional level ([38], data not shown).

Because we observed a downregulation of MMP-2 and MCAM/MUC18 expression in KCREB-transfected cells, we next analyzed the effect of KCREB expression on invasive properties of melanoma cells. To that end, KCREB-transfected cells were assayed for their potential to penetrate through the basement membrane, an important component in the process of tumor invasion and metastasis. Migration of cells through Matrigel-coated filters was monitored by using fibroblast-conditioned medium as a source of chemoattractant. The results depicted in Fig. 1C show a 10–20-fold decrease in the number of cells that invaded Matrigel-coated filters in K-2 and K-10 cells as compared with MeWo-P and MeWo-Neo cells.

These results show that expression of KREB inhibited the ability of melanoma cells to migrate and invade the basement membrane, possibly by decreasing MMP-2 and MCAM/MUC18 expression, and suggest that CREB plays an impor-

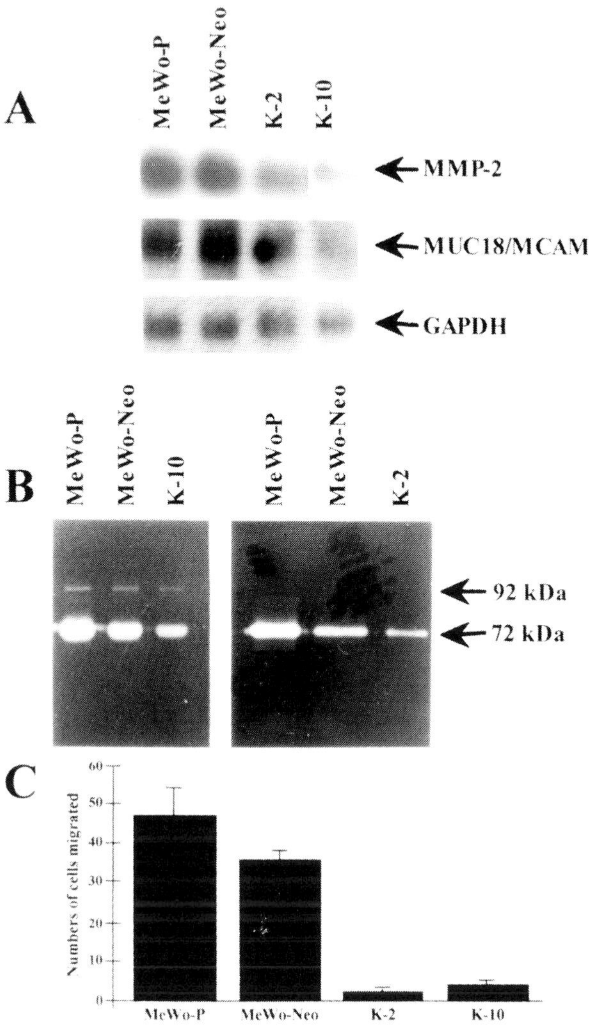

Fig. 1. (A) Northern blot analysis of KCREB-transfected cells for the expression of MMP-2 or MCAM/MUC18. MMP-2 and MCAM/MUC18 mRNA levels were decreased in K-2 and K-10 when compared with MeWo-P and MeWo-neo cells. The same blot was hybridized with G3PDH (GAPDH) probe to verify that intact RNA in similar amounts was loaded in each lane. (B) Analysis by zymography of collagenase activity of KCREB-transfected cells. The 72-kDa collagenase activity corresponding to MMP-2 was inhibited in cells that expressed KCREB, while the 92-kDa collagenase activity corresponding to MMP-9 was unchanged. (C) Effect of KCREB expression on migration through Matrigel-coated filters. KCREB-transfected cells exhibited decreased ability to migrate through Matrigel-coated filters. Results are expressed as the average number of migrated cells from 3 independent experiments ± S.E.M.

tant role in the regulation of the invasive properties of melanoma cells [38].

Role of CREB as a survival factor for human melanoma cells

Resistance to apoptosis induced by exogenous signals is an important property of tumor cells. Previous studies demonstrated that the induction of apoptosis by diverse exogenous signals depends on an elevation of cytosolic Ca^{2+} [39, 40]. Because CREB mediates both cAMP and Ca^{2+} transcriptional responses [16, 18], we investigated the role of CREB in the resistance of melanoma cells to apoptosis induction. To induce apoptosis, we used thapsigargin (Tg), which inhibits endoplasmic reticulum-dependent Ca^{2+}-ATPase and thereby increases cytosolic Ca^{2+} [41]. Furthermore, Tg has been shown to trigger apoptosis in melanoma cells [42].

The effect of Tg treatment on MeWo-P, MeWo-Neo and the two KCREB-transfected clones (K-2 and K-10) was analyzed by flow cytometry analysis after propidium iodide staining. Figure 2 A is a summary of at least three independent flow cytometry analyses. After treatment with 1 µM Tg for 48 h, KCREB-transfected cells showed a higher percentage of hypodiploid cells than the control cells. Cells with hypodiploid DNA content were increased by 2.2- and 3.3-fold in K-2 and K-10 cells, respectively, as compared with MeWo-P and MeWo-Neo cells.

Hypodiploid DNA content is indicative of apoptotic cells. To further confirm that Tg-treatment induced apoptosis in KCREB-transfected cells, we analyzed the cellular morphology of K-10 cells. Transmission electron microscopy analysis of K-10 cells (Figs 2B and 2C) after treatment with 1 µM Tg for 48 h, showed typical apoptotic morphology with plasma and nuclear membrane blebbing, chromatin condensation, and cytoplasmic vacuolarization (Fig. 2B). In addition, nuclear changes consistent with apoptosis, such as condensation and segregation of chromatin into compact masses aligning with the inner side of the nuclear membrane were also apparent (Fig. 2C). Control cells did not show these changes.

To verify that the higher sensitivity of KCREB-transfected cells to Tg-induced apoptosis was specific for the increase of intracellular Ca^{2+}, we treated the cells for 48 h with 1 µM staurosporine (St), which inhibits various kinases and induces apoptosis in most nucleated mammalian cells [43]. As shown in Fig. 2A, MeWo-P, MeWo-Neo, K-2 and K-10 cells had the same sensitivity to St-induced apoptosis. Furthermore, agents that stimulate adenylate cyclase or cAMP analogs (such as forskolin and dibutyryl-cAMP) did not cause apoptosis in these cells (data not shown), indicating that Tg-induced apoptosis was mediated via the Ca^{2+} and not the cAMP pathway.

These results demonstrated that KCREB expression in melanoma cells decreased their resistance to Tg-induced

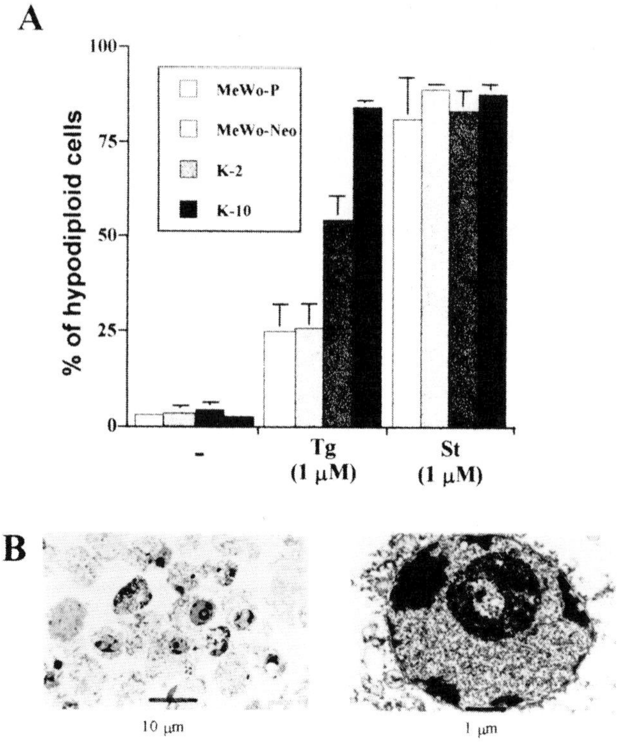

Fig. 2. Apoptosis analysis. (A) Analysis by flow cytometry of cells treated with 1 µM Tg or 1 µM St for 48 h and stained with propidium iodide. The percentage of cells with a hypodiploid DNA content indicative of apoptosis was higher in KCREB-transfected cells than in control cells. (B) and (C) Cellular morphology analysis by transmission electron microscopy of K-10 cells treated for 48 h with 1 µM Tg. K-10 cells showed morphology typical of apoptosis.

apoptosis. The higher sensitivity to Tg-induced apoptosis of K-10 cells than K-2 cells was consistent with the higher KCREB activity in these cells [12].

To understand the mechanism by which KCREB expression increased the sensitivity to Tg-induced apoptosis in melanoma cells, we next investigated the activation of CREB and the induction of CRE-dependent transcription by Tg. Previous studies showed that CREB was activated by phosphorylation on its ser[133] after an increase in intracellular Ca^{2+} [18]. We therefore verified first that Tg can induce CREB phosphorylation. Western blot analysis with α-p-CREB, an antibody raised against the ser[133]-phosphorylated peptide of CREB, that recognizes only the activated form of CREB [28], was performed on total extract of MeWo-P cells treated for 30 min with 1 µM Tg. After treatment with Tg, α-p-CREB detected two major proteins characterized by molecular masses of 38 and 43 kDa (Fig. 3A). On the basis of their molecular weight, the 43-kDa protein could correspond to CREB and the protein of 38 kDa could correspond to ATF-1. Indeed, the transcription factor ATF-1 shares high homology with the CREB peptide sequence used to raise the α-p-CREB antibody and was previously described to be activated by an increase of intracellular Ca^{2+} [14]. To verify this hypothesis, we analyzed nuclear extracts of MeWo-P cells treated for 30 min with 1 μM Tg, by Western blot with antibodies specific to CREB (α-CREB) or ATF-1 (α-ATF-1). The α-CREB antibody detected a protein that comigrated with the 43-kDa protein, and the α-ATF-1 antibody recognized a protein that comigrated with the 38-kDa protein (Fig. 3B).

Because Tg treatment induced phosphorylation of CREB and ATF-1, we next analyzed its ability to activate CRE-dependent transcription. To that end, we used the CRE-dependent vector (Somat-*Bgl*II-CAT) in which the CAT reporter gene is driven by the somatostatin gene promoter spanning nucleotides –71 to +53 (29). MeWo-P cells were transfected with the Somat-*Bgl*II-CAT vector, and the CAT activity was measured before and after 24 h treatment with 1 μM Tg. Figure 3C shows that Tg-treatment increased the activity of the CRE-dependent promoter in MeWo-P cells by 14.7-fold. On the other hand, in the K-2 KCREB-transfected cells, only a 9.5-fold induction was observed (Fig. 3C). This result suggests that KCREB inhibited the activation of the CRE-dependent transcription by Tg. To further confirm these results, we performed cotransfection experiments with the KCREB expression vector and the Somat-*Bgl*II-CAT reporter vector in MeWo-P cells treated with 1 μM Tg for 24 h. KCREB expression inhibited the activity of the CRE-dependent promoter in a dose-dependent manner (Fig. 3D).

To search for a possible mechanism by which CREB and its associated proteins might protect MeWo cells from Tg-induced apoptosis, we next analyzed the effect of KCREB on the expression of Bcl-2 and Bcl-2-related proteins such as Bcl-x, Bax, and Bad. We did not observe any changes in Bcl-2 and Bcl-2-related proteins in control or KCREB-transfected cells before or after treatment with Tg ([44], data not shown).

Collectively, these data show that CREB and ATF-1 are involved in the resistance of MeWo cells to Tg-induced apoptosis and suggest that they act as survival factors for human melanoma cells [44].

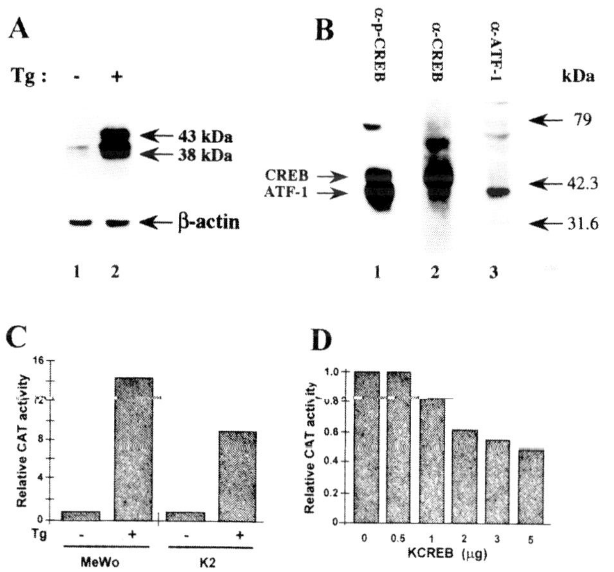

Fig. 3. (A) Western blot analysis of the phosphorylation of CREB after Tg treatment. Two major proteins of 43 kDa and 38 kDa were detected by the α-p-CREB antibody after Tg treatment of MeWo-P cells for 30 min. (B) Identification of the 43 and 38-kDa proteins by Western blot. The α-CREB antibody detected a protein that comigrated with the 43-kDa protein, and the α-ATF-1 antibody recognized a protein that comigrated with the 38-kDa protein. (C) Regulation of a CRE-dependent promoter (Somat-*Bgl*II) by Tg. Treatment with Tg for 24 h activated CRE-dependent transcription in MeWo-P cells. Tg-activation of the CRE-dependent transcription was reduced in K-2 cells as compared with MeWo-P cells. (D) Effect of KCREB expression on Tg-activation of CRE dependent transcription. Cotransfection of KCREB inhibited the activation by Tg of the CRE-dependent promoter in a dose-dependent manner.

Discussion

Several previous studies have suggested that the transcription factor CREB could be involved in human melanoma progression [7–12]. In these studies, we provided several possible molecular mechanisms by which CREB and its associated proteins might affect tumor growth and metastasis of human melanoma. By using the dominant-negative form of CREB, KCREB, we demonstrated that CREB is involved in the regulation of two important properties of melanoma cells. First, CREB regulates their invasive properties by controlling CRE-dependent expression of the type IV collagenase MMP-2 and the adhesion molecule MCAM/MUC18. Second, CREB and its associated proteins act as survival factors for human melanoma cells, as inhibition of CREB transcriptional activity in melanoma cells rendered them susceptible to apoptosis induced by thapsigargin, which increases the intracellular level of Ca^{2+}.

To inhibit the transcriptional activity of CREB in MeWo melanoma cells, we used a dominant negative of CREB, KCREB, which is mutated in its DNA-binding domain and so has lost its ability to bind DNA [23]. KCREB can, however, associate with CREB and other proteins that interact with CREB. Consequently, the transcriptional inhibitory effect of KCREB could be due to inhibition of CREB or CREB-associated proteins. Through its leucine zipper dimerization motif, CREB and KCREB could form heterodimers with others members of the ATF family of transcription factors. Heterodimerization with CREB was demonstrated for the two members involved in the cAMP transduction pathway, CREM (45) and ATF-1 [14]. The expression of CREM multiple splice variants are not detected in melanoma cells, but ATF-1 is easily found in metastatic melanoma cells [11]. We demonstrated that ATF-1 is expressed in MeWo cells by Western blot with a specific antibody. CREB and ATF-1 share functional homology, both are phosphorylated, and both induce CRE-dependent transcription in response to an increase in cAMP or Ca^{2+} intracellular levels [14, 16, 18]. In HeLa cells, CREB and ATF-1 form mainly heterodimers [46], and so KCREB should inhibit ATF-1 transcriptional activity in MeWo cells. Formation of heterodimer with CREB was also shown with another member of the ATF family, ATF-2 [47], and with members of the AP-1 family [48, 49]. The functional significance of these pairings are unknown. Another CREB-associated protein is the coactivator CBP (CREB-binding protein) [22]. CBP regulates the function of multiple transcription factors, and a possible role of CBP in cellular transformation and differentiation was suggested [50]. We can't exclude the possibility that some of the effects of KCREB were due to the inhibition of CBP. Indeed, it was previously shown that nuclear receptor inhibition of AP-1 activity was the apparent result of competition for limiting amounts of CBP in cells, since the transcriptional activities of both nuclear receptors and AP-1 require CBP [51]. We are currently investigating the effect of a more specific inhibitor of CREB transcriptional activity, the single chain Fv (scFv) antibody fragment (containing the antigen-binding variable domains of the light and heavy chains connected by a peptide spacer) directed against CREB. We are also using the decoy approach with oligonucleotides containing a cAMP-response element (CRE) binding site.

Among the mechanisms by which KCREB might inhibit melanoma metastasis is through its ability to regulate invasion. During tumor growth and metastasis, cells need to modify and to migrate through the surrounding connective tissue. The motility of melanoma cells through the basement membrane depends on their ability to express proteases that degrade extracellular matrix components and adhesion molecules mediating cell-to-cell and cell-to-matrix interactions. The type IV collagenase MMP-2 and the adhesion molecule MCAM/MUC18 had been demonstrated to play an important role in melanoma invasion [35–37]. The downregulation of the expression of these two molecules by KCREB could explain the low invasive properties of KCREB-transfected cells. Furthermore, the downregulation of MMP-2 expression and activity by KCREB suggests a role for CREB in angiogenesis. It has been shown that MMP-2 is involved in the initiation and promotion of angiogenesis [52]. We showed that CREB transcriptional activity is necessary to maintain MMP-2 activity. Therefore, CREB transcriptional activity could be important to allow a proper vascularization of the tumor and its ability to degrade type IV collagen, which is important for tumor growth and metastasis. Interestingly, the expression of vascular endothelial growth factor (VEGF), one of the most important angiogenic factors, is upregulated by cAMP [53] and downregulated by an inhibitor of protein kinase A [54]. Furthermore, ATF-1 and CREB bind to the CRE element present within the VEGF promoter [55]. A role of CREB in VEGF transcriptional regulation supports the notion of CREB involvement in angiogenesis and, hence, in melanoma progression.

Resistance to apoptosis is another important property of tumor cells. In the present study, we showed that KCREB expression decreases resistance of melanoma cells to apoptosis induced by an agent that increases the intracellular level of Ca^{2+}. We verified that Tg induced CREB and ATF-1 phosphorylation and activated CRE-dependent transcription in MeWo cells. We showed that KCREB expression inhibits the induction of CRE-dependent transcription by Tg. These data suggest that CREB and its associated proteins act as a survival factor for melanoma cells. Several other studies support the role of CREB as a negative regulator of apoptosis. The protective effect of cAMP-mediated signal transduction has been described for human cancer cells including melanoma [56, 57]. Using transgenic mice that express the same dominant-negative form of CREB (KCREB) used in our studies, Barton *et al.* [58] have demonstrated that thymocytes and T cells display a profound proliferative defect characterized by G1 cell-cycle arrest and subsequent apoptotic death in response to a number of different activation signals, including Tg. Downregulation of CREB activity during apoptosis was observed in different cellular models [59, 60]. These studies show that CREB could act as a survival factor in diverse types of cells. One mechanism by which CREB proteins can rescue cells from apoptosis is by up-regulation of Bcl-2 expression. Indeed, CREB has been shown to function as a positive regulator of the Bcl-2 gene via a direct binding to a CRE element within the Bcl-2 promoter [61]. We did not observe any changes in the expression of Bcl-2 or Bcl-2-related proteins (Bcl-x, Bax, and Bad) in control or KCREB-transfected cells following Tg treatment, suggesting that phosphorylation of CREB or ATF-1 by Tg did not change the expression of these proteins in melanoma cells [44]. Von Knethen *et al.* [62] have suggested that CREB can act as a survival factor by regulating indirectly p53 accumulation. This hypothesis is currently under investigation in our laboratory.

Based on our studies that CREB transcriptional activity plays an important role in melanoma progression, we conclude that it is possible to use inhibitors of CREB transcriptional activity or CREB activation pathway to inhibit tumor growth and metastasis of human melanoma.

Acknowledgements

We thank Dr. M.R. Montminy for the gift of the CRE-dependent promoter-CAT vector and for antibodies against CREB and p-CREB, Dr. M.R. Green for the antibody against ATF-

1, Dr. R.H. Goodman for the KCREB expression vector, and Dr. Z. Ronai for the KCREB-transfected cells. This work was supported by National Institutes of Health grants CA 64137 and CA 76098.

References

1. Clark WH, Elder DE, Guerry D, Epstein MN, Greene MH, Van Horn M: A study of tumor progression: The precursor lesions of superficial spreading and nodular melanoma. Hum Pathol 15: 1147–1165, 1984
2. Bar-Eli M: Molecular mechanisms of melanoma metastasis. J Cell Physiol 173: 275–278, 1997
3. Luca MR, Bar-Eli M: Molecular changes in human melanoma metastasis. Histol Histopath 13: 1225–1231, 1998
4. Huang S, Jean D, Luca M, Tainsky M, Bar-Eli M: Loss of AP-2 results in downregulation of c-KIT and enhancement of melanoma tumorigenicity and metastasis. EMBO J 17: 4358–4369, 1998
5. Jean D, Gershenwald JE, Huang S, Luca M, Hudson M, Tainsky M, Bar-Eli M: Loss of AP-2 results in up-regulation of MCAM/MUC18 and an increase in tumor growth and metastasis of human melanoma cells. J Biol Chem 273: 16501–16508, 1998
6. Karjalainen JM, Kellokoski JK, Eskelinen MJ, Alhava EM, Kosma VM: Downregulation of transcription factor AP-2 predicts poor survival in stage I cutaneous malignant melanoma. J Clin Oncol 16: 3584–3591, 1998
7. Halaban R, Pomerantz SH, Marshall S, Lambert DT, Lerner AB: Regulation of tyrosinase in human melanocytes grown in culture. J Cell Biol 97: 480–488, 1983
8. Halaban R, Pomerantz SH, Marshall S, Lerner AB: Tyrosinase activity and abundance in Cloudman melanoma cells. Arch Biochem Biophys 230: 383–387, 1984
9. Halaban R, Ghosh S, Baird A: bFGF is the putative natural growth factor for human melanocytes. *In Vitro* Cell Dev Biol 23: 47–52, 1987
10. Rutberg SE, Goldstein IM, Yang YM, Stackpole CW, Ronai Z: Expression and transcriptional activity of AP-1, CRE, and URE binding proteins in B16 mouse melanoma subclones. Mol Carcinog 10: 82–87, 1994
11. Bohm M, Moellmann G, Cheng E, Alvarez-Franco M, Wagner S, Sassone-Corsi P, Halaban R: Identification of p90RSK as the probable CREB-Ser133 kinase in human melanocytes. Cell Growth Differ 6: 291–302, 1995
12. Yang YM, Dolan LR, Ronai Z: Expression of dominant negative CREB reduces resistance to radiation of human melanoma cells. Oncogene 12: 2223–2233, 1996
13. Landschulz WH, Johnson PF, McKnight SL: The leucine zipper: A hypothetical structure common to a new class of DNA binding proteins. Science 240: 1759–1764, 1988
14. Liu F, Thompson MA, Wagner S, Greenberg ME, Green MR: Activating transcription factor-1 can mediate Ca(2+)-and cAMP-inducible transcriptional activation. J Biol Chem 268: 6714–6720, 1993
15. Borrelli E, Montmayeur JP, Foulkes NS, Sassone-Corsi P: Signal transduction and gene control: The cAMP pathway. Crit Rev Oncol 3: 321–338, 1992
16. Gonzalez GA, Montminy MR: Cyclic AMP stimulates somatostatin gene transcription by phosphorylation of CREB at serine 133. Cell 59: 675–680, 1989
17. Xie H, Rothstein TL: Protein kinase C mediates activation of nuclear cAMP response element-binding protein (CREB) in B lymphocytes stimulated through surface Ig. J Immunol 154: 1717–1723, 1995
18. Sheng M, Thompson MA, Greenberg ME: CREB: A Ca(2+)-regulated transcription factor phosphorylated by calmodulin-dependent kinases. Science 252: 1427–1430, 1991
19. Matthews RP, Guthrie CR, Wailes LM, Zhao X, Means AR, McKnight GS. Calcium/calmodulin-dependent protein kinase types II and IV differentially regulate CREB-dependent gene expression. Mol Cell Biol 14: 6107–6116, 1994
20. Ginty DD, Bonni A, Greenberg ME: Nerve growth factor activates a Ras-dependent protein kinase that stimulates c-fos transcription via phosphorylation of CREB. Cell 77: 713–725, 1994
21. Xing J, Ginty DD, Greenberg ME: Coupling of the RAS-MAPK pathway to gene activation by RSK2, a growth factor-regulated CREB kinase. Science 273: 959–963, 1996
22. Kwok RP, Lundblad JR, Chrivia JC, Richards JP, Bachinger HP, Brennan RG, Roberts SG, Green MR, Goodman RH: Nuclear protein CBP is a coactivator for the transcription factor CREB. Nature 370: 223–226, 1994
23. Walton KM, Rehfuss RP, Chrivia JC, Lochner JE, Goodman RH: A dominant repressor of cyclic adenosine 3′,5′-monophosphate (cAMP)-regulated enhancer-binding protein activity inhibits the cAMP-mediated induction of the somatostatin promoter *in vivo*. Mol Endocrinol 6: 647–655, 1992
24. Woloshin PI, Walton KM, Rehfuss RP, Goodman RH, Cone RD: 3′,5′-Cyclic adenosine monophosphate-regulated enhancer binding (CREB) activity is required for normal growth and differentiated phenotype in the FRTL5 thyroid follicular cell line. Mol Endocrinol 6: 1725–1733, 1992
25. Fogh J, Wright WC, Loveless JD: Absence of HeLa cell contamination in 169 cell lines derived from human tumors. J Natl Cancer Inst 58: 209–214, 1977
26. Ishikawa M, Dennis JW, Man S, Kerbel RS: Isolation and characterization of spontaneous wheat germ agglutinin-resistant human melanoma mutants displaying remarkably different metastatic profiles in nude mice. Cancer Res 48: 665–670, 1988
27. Luca M, Xie S, Gutman M, Huang S, Bar-Eli M: Abnormalities in the CDKN2 (p16INK4/MTS-1) gene in human melanoma cells: Relevance to tumor growth and metastasis. Oncogene 11: 1399–1402, 1995
28. Hagiwara M, Brindle P, Harootunian A, Armstrong R, Rivier J, Vale W, Tsien R and Montminy MR: Coupling of hormonal stimulation and transcription via the cyclic AMP-responsive factor CREB is rate limited by nuclear entry of protein kinase A. Mol Cell Biol, 13: 4852–4859, 1993
29. Montminy MR, Sevarino KA, Wagner JA, Mandel G, Goodman RH: Identification of a cyclic-AMP-responsive element within the rat somatostatin gene. Proc Natl Acad Sci U S A 83: 6682–6686, 1986
30. Gorman CM, Moffat LF, Howard BH: Recombinant genomes which express chloramphenicol acetyltransferase in mammalian cells. Mol Cell Biol 2: 1044–1051, 1982
31. Muller HK, Bucana CD, Kripke ML, Cox PA, Saijo S, Strickland FM: Ultraviolet irradiation of murine skin alters cluster formation between lymph node dendritic cells and specific T lymphocytes. Cell Immunol 157: 263–276, 1994
32. Radinsky R, Fidler IJ, Price JE, Esumi N, Tsan R, Petty CM, Bucana CD, Bar-Eli M: Terminal differentiation and apoptosis in experimental lung metastases of human osteogenic sarcoma cells by wild type p53. Oncogene 9: 1977–1983, 1994
33. Huhtala P, Chow LT, Tryggvason K: Structure of the human type IV collagenase gene. J Biol Chem 265: 11077–11082, 1990
34. Lehmann JM, Riethmuller G, Johnson JP: MUC18, a marker of tumor progression in human melanoma, shows sequence similarity to the neural cell adhesion molecules of the immunoglobulin superfamily. Proc Natl Acad Sci USA 86: 9891–9895, 1989
35. Ray JM, Stetler-Stevenson WG: Gelatinase A activity directly modulates melanoma cell adhesion and spreading. EMBO J 14: 908–917, 1995
36. Luca M, Huang S, Gershenwald JE, Singh RK, Reich R, Bar-Eli M: Expression of interleukin-8 by human melanoma cells up-regulates MMP-2 activity and increases tumor growth and metastasis. Am J Pathol 151: 1105–1113, 1997

37. Xie S, Luca M, Huang S, Gutman M, Reich R, Johnson JP, Bar-Eli M: Expression of MCAM/MUC18 by human melanoma cells leads to increased tumor growth and metastasis. Cancer Res 57: 2295–2303, 1997
38. Xie S, Price JE, Luca M, Jean D, Ronai Z, Bar-Eli M: Dominant-negative CREB inhibits tumor growth and metastasis of human melanoma cells. Oncogene 15: 2069–2075, 1997
39. McConkey DJ, Nicotera P, Hartzell P, Bellomo G, Wyllie AH, Orrenius S: Glucocorticoids activate a suicide process in thymocytes through an elevation of cytosolic Ca^{2+} concentration. Arch Biochem Biophys 269: 365–370, 1989
40. McConkey DJ, Orrenius S: The role of calcium in the regulation of apoptosis. J Leukoc Biol 59: 775–783, 1996
41. Thastrup O, Cullen PJ, Drobak BK, Hanley MR, Dawson AP: Thapsigargin, a tumor promoter, discharges intracellular Ca(2+) stores by specific inhibition of the endoplasmic reticulum Ca(2+)-ATPase. Proc Natl Acad Sci U S A 87: 2466–2470, 1990
42. Muthukkumar S, Nair P, Sells SF, Maddiwar NG, Jacob RJ, Rangnekar VM: Role of EGR-1 in thapsigargin-inducible apoptosis in the melanoma cell line A375-C6. Mol Cell Biol 15: 6262–6272, 1995
43. Ishizaki Y, Cheng L, Mudge AW, Raff MC: Programmed cell death by default in embryonic cells, fibroblasts, and cancer cells. Mol Biol Cell 6: 1443–1458, 1995
44. Jean D, Harbison M, McConkey DJ, Ronai Z, Bar-Eli M: CREB and its associated proteins act as survival factors for human melanoma cells. J Biol Chem 273: 24884–24890, 1998
45. Loriaux MM, Brennan RG, Goodman RH: Modulatory function of CREB.CREM alpha heterodimers depends upon CREM alpha phosphorylation. J Biol Chem 269: 28839–28843, 1994
46. Hurst HC, Totty NF, Jones NC: Identification and functional characterisation of the cellular activating transcription factor 43 (ATF-43) protein. Nucleic Acids Res 19: 4601–4609, 1991
47. Abdel-Hafiz HA, Chen CY, Marcell T, Kroll DJ, Hoeffler JP: Structural determinants outside of the leucine zipper influence the interactions of CREB and ATF-2: Interaction of CREB with ATF-2 blocks E1a-ATF-2 complex formation. Oncogene 8: 1161–1174, 1993
48. Benbrook DM, Jones NC: Heterodimer formation between CREB and JUN proteins. Oncogene 5: 295–302, 1990
49. Hai T, Curran T: Cross-family dimerization of transcription factors Fos/Jun and ATF/CREB alters DNA binding specificity. Proc Natl Acad Sci USA 88: 3270–3274, 1991
50. Goldman PS, Tran VK, Goodman RH: The multifunctional role of the co-activator CBP in transcriptional regulation. Recent Prog Hormone Res 52: 103–120, 1997
51. Kamei Y, Xu L, Heinzel T, Torchia J, Kurokawa R, Gloss B, Lin SC, Heyman RA, Rose DW, Glass CK, Rosenfeld MG: A CBP integrator complex mediates transcriptional activation and AP-1 inhibition by nuclear receptors. Cell 85: 403–414, 1996
52. Sang QX: Complex role of matrix metalloproteinases in angiogenesis. Cell Res 8: 171–177, 1998
53. Acarregui MJ, Penisten ST, Goss KL, Ramirez K, Snyder JM: Vascular endothelial growth factor gene expression in human fetal lung *in vitro*. Am J Respir Cell Mol Biol 1: 14–23, 1999
54. Bianco C, Tortora G, Baldassarre G, Caputo R, Fontanini G, Chine S, Bianco AR, Ciardiello F: 8-Chloro-cyclic AMP inhibits autocrine and angiogenic growth factor production in human colorectal and breast cancer. Clin Cancer Res 3: 439–448, 1997
55. D'Arcangelo G, Habas R, Wang S, Halegoua S, Salton SR: Activation of codependent transcription factors is required for transcriptional induction of the vgf gene by nerve growth factor and Ras. Mol Cell Biol 16: 4621–4631, 1996
56. Tortora G, Di Isernia G, Sandomenico C, Bianco R, Pomatico G, Pepe S, Bianco AR, Ciardiello F: Synergistic inhibition of growth and induction of apoptosis by 8-chloro-cAMP and paclitaxel or cisplatin in human cancer cells. Cancer Res 57: 5107–5111, 1997
57. Tortora G, Caputo R, Damiano V, Bianco R, Pepe S, Bianco AR, Jiang Z, Agrawal S, Ciardiello F: Synergistic inhibition of human cancer cell growth by cytotoxic drugs and mixed backbone antisense oligonucleotide targeting protein kinase A. Proc Natl Acad Sci USA 94: 12586–12591, 1997
58. Barton K, Muthusamy N, Chanyangam M, Fischer C, Clendenin C, Leiden JM: Defective thymocyte proliferation and IL-2 production in transgenic mice expressing a dominant-negative form of CREB. Nature 379: 81–85, 1996
59. Bussfeld D, Bacher M, Moritz A, Gemsa D, Sprenger H: Expression of transcription factor genes after influenza A virus infection. Immunobiology 198: 291–298, 1997
60. Sikora E, Grassilli E, Radziszewska E, Bellesia E, Barbieri D, Franceschi C: Transcription factors DNA-binding activity in rat thymocytes undergoing apoptosis after heat-shock or dexamethasone treatment. Biochem Biophys Res Commun 197: 709–715, 1993
61. Wilson BE, Mochon E, Boxer LM: Induction of bcl-2 expression by phosphorylated CREB proteins during B- cell activation and rescue from apoptosis. Mol Cell Biol 16: 5546–5556, 1996
62. Von Knethen A, Lotero A, Brune B: Etoposide and cisplatin induced apoptosis in activated RAW 264.7 macrophages is attenuated by cAMP-induced gene expression. Oncogene 17: 387–394, 1998

Molecular and Cellular Biochemistry **212**: 29–34, 2000.

CRE-decoy oligonucleotide-inhibition of gene expression and tumor growth*

Yoon S. Cho-Chung, Yun Gyu Park, Maria Nesterova, Youl Nam Lee and Yee Sook Cho

Cellular Biochemistry Section, Laboratory of Tumor Immunology and Biology, National Cancer Institute, NIH, Bethesda, MD, USA

Abstract

Nucleic acid molecules with high affinities for a target transcription factor can be introduced into cells as decoy *cis*-elements to bind these factors and alter gene expression. This review discusses a synthetic single-stranded palindromic oligonucleotide, which self-hybridizes to form a duplex/hairpin and competes with cAMP response element (CRE) enhancers for binding transcription factors. This oligonucleotide inhibits CRE- and Ap-1-directed gene transcription and promotes growth inhibition *in vitro* and *in vivo* in a broad spectrum of cancer cells, without adversely affecting normal cell growth. Evidence presented here suggests that the CRE-decoy oligonucleotide can provide a powerful new means of combating cancers, viral diseases, and other pathological conditions by regulating the expression of cAMP-responsive genes. (Mol Cell Biochem **212**: 29–34, 2000)

Key words: cAMP, transcription factor-decoy oligonucleotides, CRE, Ap-1, p53, tumor growth, gene expression

Introduction

A major focus of cellular and molecular research has been to develop a means to regulate gene expression in an effort to treat and cure a variety of diseases and abnormal physiological conditions. The importance of such research dramatically increases as the Human Genome Project continues to identify genes at an accelerated pace.

Prokaryotic repressors can function as negative regulators of eukaryotic promoters [1, 2]. This observation suggests that displacement of activating proteins might provide a general strategy for gene-specific repression in eukaryotes. Several approaches have been undertaken to control eukaryotic gene expression through such displacement.

In one approach, *trans*-dominant mutants, which retain the ability to bind to *cis*-regulatory DNA sequences but lack functional transcriptional activation domains, are generated. These mutant transcription factors compete with their functional, wild-type counterparts for binding to the enhancer sequences and prevent the activation or repression of the target gene. Although this strategy has been successful *in vitro* [3] the generation of such mutants is not always possible and the requirement of gene therapy procedures limits its application *in vivo*.

In another approach, promoter competition is used whereby plasmids containing *cis*-acting elements, in common with the targeted gene, are introduced in high copy number into cells [4]. At high copy number, a majority of the transcription factors can be competitively bound away from the natural enhancer sequences, with gene expression accordingly regulated. Because these plasmids must be maintained uniformly in a large number of cells, this approach also has been limited.

Another alternative is to employ oligonucleotides (ODNs) to form triple helices with enhancer elements. Pyrimidine ODNs were found to bind in a sequence-specific manner to homopurine sites in duplex DNA by triple helix formation while maintaining sufficient specificity and affinity to compete with site-specific DNA binding proteins for overlapping target sites [5]. However, such enhancer-inhibitory effects of

Address for offprints: Y.S. Cho-Chung, National Cancer Institute, Building 10, Room 5B05, Bethesda, MD 20892-1750, USA

ODN-directed triple helix has not been shown in intact cells *in vitro* or *in vivo*.

A more successful oligonucleotide-based approach has been the use of a synthetic double-stranded oligonucleotide (ODN) containing an enhancer element. ODNs with modified phosphodiester bonds, such as phosphorothioate, methyl phosphate, phosphoramidite, or methyl phosphonate derivatives, and second generation ODNs can be routinely synthesized in large amounts and are relatively resistant to nucleases [6, 7]. Because of their increased stability in the cell, these compounds have been used as antisense agents [8, 9]. It has been shown that double-stranded phosphorothioate oligonucleotides (PS-ODN) can penetrate cells, bind sequence-specific DNA binding proteins, and interfere with eukaryotic transcription *in vivo* [10, 11].

The CRE (cyclic AMP response element)-transcription factor complex is a pleiotropic activator that participates in the induction of a wide variety of cellular and viral genes [12]. Because the CRE *cis*-element, TGACGTCA [12] is palindromic, a synthetic single-stranded oligonucleotide composed of the CRE sequence self-hybridizes to form a duplex/hairpin. This review revisits the reported ability of the transcription factor decoy-ODN CRE-decoy to penetrate cells, bind transcription factors, and interfere with gene transcription *in vivo*. Significantly, the CRE-decoy is harmless to normal cells, but it is a potent inhibitor to cancer cell growth *in vitro* and *in vivo* [13].

Interference with CRE DNA-protein complex formation

A 24-mer CRE-decoy comprising palindrome triplet copies of TGACGTCA can compete for binding the CRE-sequence-specific DNA binding proteins [13]. In an electrophoretic mobility shift assay, using a ^{32}P-labeled CRE probe, nuclear protein binding was inhibited by unlabeled CRE-palindrome ODN, but not by a two-base mismatched control ODN, nonsense-sequence palindrome ODN containing no CRE. An ODN containing Oct-1, Ap-1, or Sp-1 sequences was also unable to compete.

The CRE-decoy can also penetrate cells and compete with cellular CRE-elements for binding transcription factors *in vivo* [13]. Nuclear extracts from cells treated with the 24-mer CRE ODN demonstrated a marked decrease in formation of the CRE-protein complex, as evidenced in a mobility shift assay. An antibody against CREB (CRE-binding protein) [14] caused a supershift, indicating the presence of CREB protein within the labeled protein-DNA complexes, while the control ODN treatment did not affect CRE-protein complex formation. A Western blot analysis for CREB protein in untreated, CRE oligo-treated, and control oligo-treated cells demonstrated no change in the level of CREB protein, indicating that the CRE-decoy treatment did not affect CREB protein levels in the cell. The CRE-decoy ODN successfully competed with the cellular CRE-enhancer for binding of the CRE-sequence-specific DNA-binding proteins in a wide variety of cell types, including breast, prostate, and lung carcinoma cells, as well as non-cancerous cells [13].

Inhibition of basal and induced CRE-directed transcription

The CRE-decoy can modulate the transcriptional activity of sequence-specific DNA binding proteins *in vivo* [13]. Transfection of MCF-7 breast cancer cells with somatostatin Δ-71-CAT [15] and CRE-decoy resulted in a greater than 90% inhibition of CRE-directed transcription compared to cells transfected with Δ-71-CAT alone. Addition of the nonsense-sequence palindromic ODN or the two-base mismatched control ODN, had no inhibitory effect on CAT activity. Moreover, forskolin-stimulated CAT activity was also almost completely abolished by CRE ODN treatment. Significantly, the CRE-decoy is capable of inhibiting both basal and cAMP-stimulated CAT activities.

A group of cAMP-responsive genes, such as somatostatin and phosphoenol pyruvate carboxykinase (PEPCK), contain a CRE that lies within the first 150 bp of the 5′-flanking region of the gene [12]. Therefore, this element could be regarded as a basal enhancer, in addition to functioning as an inducible enhancer [16]. A role for the CRE as a basal transcription element was suggested in deletion analysis of the PEPCK promoter-regulatory region [17]. When the CRE was deleted from the promoter, the basal level of gene transcription was reduced and the responsiveness of the promoter to cAMP in hepatoma cells was abolished. CRE binding protein binds to the CRE in a cAMP-independent manner [12, 14]. This binding may stimulate basal transcription by interacting with proximal promoter element, such as the TATA box binding factor and/or RNA polymerase II [12]. The cAMP-induced phosphorylation of CREB could lead to a higher-order complex formation with the basic transcription factor. Importantly, as shown above, the CRE-oligonucleotide interfered with both the basal and cAMP-induced transcription of an exogenously supplied CRE-containing gene.

The CRE-decoy inhibited expression of the endogenous cAMP-responsive genes. CRE-decoy treatment brought about a marked reduction in the mRNA levels of PEPCK and the catalytic (C_α) and regulatory (RI_α) subunits of cAMP-dependent protein kinase (PKA) in both MCF7 breast cancer cells and MCF-10A noncancerous cells [13]. In contrast, the control oligonucleotide had no effect.

The observation that the CRE-decoy inhibits basal expression of RI_{α} and C_{α} genes strongly indicates that the CRE-decoy can indeed compete with the *cis*-CRE element in binding CREB. Because the CRE-decoy can interfere with CREB binding to the *cis*-element, it is expected that the oligonucleotide could produce a more profound effect on the mRNA reduction under cAMP-induced conditions. Thus, the CRE-decoy ODN can interfere with both basal and cAMP-induced expression of the endogenous CRE-containing genes.

Interference with Ap-1 DNA-protein binding and c-*fos* induction

CREB is known to associate with (e.g. heterodimerize) a variety of other transcription factors (e.g. member of the Jun/Fos family) [18]. The products of the proto-oncogenes *jun* and *fos* bind as a heterodimeric complex to a DNA sequence element TRE (Ap-1) binding site [19], whereas CREB-1 homodimer and CREB-2/ATF heterodimer bind the CRE sequence [18]. However, Jun/Jun homodimer bind to both CRE and Ap-1 [18] and CREB-2 (ATF-2)/Jun heterodimer binds CRE [18, 20], and c-*fos* is cAMP-inducible [12]. These data clearly demonstrate Ap-1 and CRE cross-talk.

Whether CRE-decoy treatment affects Ap-1 binding was examined. Nuclear extracts from cells treated with the CRE-decoy demonstrated a marked reduction in formation of the Ap-1 DNA-protein complex in the mobility shift assay compared to control (saline-treated) cells. Two-base mismatched control ODN treatment had no effect on the Ap-1 DNA-protein complex formation. By comparison, CRE-decoy ODN had no effect on Sp-1 or Oct-1 DNA-protein binding. The CRE-decoy brought about a marked decrease in the TPA-inducible mRNA level of c-*fos* in response to cAMP [12], while the control ODNs had no effect. A cAMP unresponsive gene, such as PKC-α, was not affected by CRE-decoy treatment. These results show that CRE-decoy treatment resulted in inhibition of transcription factor binding at two different *cis*-elements, the CRE and the Ap-1.

CRE-decoy oligonucleotide inhibition of cancer cell growth *in vitro* and *in vivo*

The ability of CRE-decoy treatment to interfere with cell growth was examined as there are many cAMP-regulated genes and they are ubiquitously expressed. Surprisingly, CRE-decoy treatment resulted in selective growth inhibition of cancer cells without adversely affecting normal cell growth [13]. The 24-mer CRE ODN produced potent growth inhibition in a variety of cancer cells. In contrast, the CRE-decoy had little or no effect on the growth of noncancerous cells, MCF-10A (human mammary epithelian cell), L132 (human lung epithelial cell), Hs68 (human newborn foreskin fibroblasts) and NIH/3T3 fibroblasts. The growth inhibition of cancer cells was achieved at nanomolar concentrations of CRE ODN (IC_{50}, 100–200 nM) without obvious cytotoxicity and accompanied by changes in cell morphology and the appearance of apoptotic nuclei (programmed cell death). Growth inhibition was CRE-sequence specific, as the two-base mismatched control ODN or the nonsense-sequence palindromic ODN had little (< 30%) or no growth inhibitory effect.

CRE-decoy treatment also inhibited *in vivo* tumor growth. Treatment of nude mice bearing HT-15 human MDR colon carcinoma with 24-mer CRE ODN (0.1 mg/kg, i.p., daily, 5×/week for 4 weeks) resulted in greater than 85% inhibition of tumor growth without causing systemic toxicity [13]. In comparison, the two-base mismatched control ODN had no growth-inhibitory effect. Growth inhibition may have been due to actions other than blockade of CRE-gene transcription, as nonspecific binding of oligonucleotide or its degradation products to biological targets has also been shown [21]. However, the binding of decoy ODN at the transcription factor DNA-binding domain is clearly related to the inhibition of cell growth. First, in undifferentiated F9 teratocarcinoma cells, a cell line that is unresponsive to cAMP, the CRE-decoy treatment did not affect growth [13]. This suggests that the decoy may act as a growth inhibitor, at least in part, through binding to CREB, since the CRE is nonfunctional in F9 cells although CREB is present [22]. Second, KCREB, a CREB mutant that changes a single amino acid in the DNA-binding domain, is known not to bind to native CRE sequences [23], and cancer cells overexpressing KCREB exhibited decreased cell growth, as compared to parental cells, and showed little or no response to the decoy oligonucleotide treatment [13].

Cross-talk between CRE and p53

The possible involvement of the p53 signaling pathway in the growth-inhibitory effect of the CRE-decoy oligonucleotide was examined. The CRE-decoy oligonucleotide treatment of MCF-7 cells, which contain wild-type p53, resulted in an increase in p53 protein levels without changing mRNA levels [24]. Pulse-chase experiments demonstrated that the accumulation of p53 is related to increased stability of the protein [24]. Transient transcription assays showed that decoy oligonucleotide treatment induces p53-luciferase activity [24]. P21Cip1/WAF1 protein was also induced by CRE-decoy treatment, and the induction of p21 accompanied a reduction in cdk2- and cyclin E-associated kinase activity, and pRB hypophosphorylation [24]. These results indicate that the stabilization and activation of p53 may contribute to growth inhibition promoted by CRE-transcription factor decoy oligonucleotide in MCF-7 breast cancer cells [24].

The cross-talk between cAMP- and p53-generated signals has been shown in the induction of steroidogenesis (differentiation) and apoptosis in granulosa cells [25, 26]. An inverse relationship between Bcl-2 and p53 has been shown in several breast cancer cell lines. For example, MCF-7 breast cancer cells contain very high levels of Bcl-2 but very low levels of p53 [27]. Furthermore, a recent study using a CRE-decoy oligonucleotide showed that sensitivity to DNA-damaging drugs is regulated by both cAMP-induced gene activation and the p53-dependent pathway [28]. The cross-talk between cAMP and p53 may also exist at the binding of CREB and p53 to a transcription coactivator, phospho-CREB binding protein (CBP) [29]. CREB interacts with its coactivator CBP [30, 31], which is involved in the transcriptional activation of other genes, such as p53 [35], Ap-1, and retinoic acid receptors [33]. As it is likely that competition for a common coactivator like CBP may in part explain the cross-talk (which is commonly observed between disparate signaling pathways), CRE–decoy oligonucleotide, through reducing the CREB binding to CBP, may render CBP more accessible to p53, thereby resulting in increased binding of p53 to CBP. If this is the case, the increased stability of p53 could be explained, at least in part, by its increased binding to CBP.

Conclusions and perspectives

Evidence presented here shows that the synthetic single stranded CRE ODN of palindrome structure functioned as effective and stable transcription factor-decoy to alter gene expression *in vivo*. Importantly, the CRE-decoy achieved gene-specific regulation *in vivo* leading to relative inhibition of cancer cell growth without adversely affecting the growth of normal cells.

The specificity of the growth-inhibitory effect of the CRE-decoy on cancer cells is supported by several lines of evidence. The CRE-decoy promoted growth inhibition, *in vitro* and *in vivo*, of cancer cells but not normal cells, whereas mismatched control ODN or nonsense-sequence palindromic ODN did not inhibit growth. Administration of CRE-decoy ODN, but not mismatched ODNs, markedly inhibited CRE DNA-protein complex formation, CRE-directed transcription activity, and endogenous cAMP-responsive gene expression in both cancer cells and normal cells. Cellular uptake of decoy ODN and control ODN was similar for cancer cells and normal cells, and the specific growth-inhibitory effect toward cancer cells correlated with induction of phenotypic change and apoptosis [13].

Recent evidence has indicated that DNA hairpin formation may represent an additional level of transcriptional control. A 23-bp synthetic ODN of the human enkephalin gene enhancer has been shown to undergo a reversible conformational change from a duplex to a cruciform structure of two hairpins [34, 35]. Within the enkephalin enhancer, mutations that stabilize or destabilize a potential cruciform structure increased or decreased levels of transcription, respectively, without affecting protein binding [36].

Structural changes in DNA linked to transcriptional regulation have been further described in other genes, both prokaryotic and eukaryotic. Sequence-specific regulatory proteins, such as TFIIIA [37] and Ap-1 [38] can bend synthetic ODN enhancers *in vitro*. Bending has been shown to actively influence gene transcription [39]. Cruciforms inserted at transcriptional start sites decreased transcription through negative supercoiling-induced cruciform formation [40]. A ubiquitous eukaryotic non-histone nuclear protein, HMG1 (high mobility group-1 protein), specifically binds to cruciforms in a sequence-independent manner [44], and the viral RNA polymerase of bacteriophage N4 requires a hairpin structure in the template strand of the promotor for recognition [42].

These studies raise questions as to whether a secondary structure of DNA other than the linear duplex structure plays a role in transcriptional controls and whether such a structure can serve as a more effective transcription factor decoy. The experimental data obtained with the CRE-palindrome decoy ODN (duplex/hairpin structure) strongly support this view. The exogenously supplied decoy ODN effectively competed with the native CRE-enhancer (linear DNA) resulting in inhibition of the basal expression of the CRE-containing genes *in vivo* [13]. Such inhibition resulted in a block of not only the CRE-PKA pathway, but also of the Ap-1-PKC pathway by inhibiting c-*fos* gene expression that is CRE-responsive, and selective inhibition of tumor cell proliferation without affecting the growth of noncancerous cells [13]. Because the probability of hairpin formation from a stable duplex of cellular *cis*-transcription element is small at neutral pH, and because the hairpin ODN may have a higher affinity for the transcription factor than the linear DNA, the exogenously supplied hairpin decoy ODN would be a more efficient competitor for the binding of cellular transcription factors than a linear duplex decoy, and thus can interfere with transcription more efficiently *in vivo*.

CREB is a member of the CREB/ATF family of transcription factors, and its transactivation activity increases after phosphorylation by cAMP-dependent protein kinase [14]. More recently, it has become apparent that CREB is an *in vivo* substrate for a variety of other kinases including calmodulin kinases II and IV [43] or RSK2 [44], implying that the CREB/ATF family of transcription factors can activate CRE-transcription in response to cAMP, Ca^{2+}, and growth factor stimulation [22, 45–47]. That the growth factor-stimulated pathways are quiescent in non-cancerous cells could explain, at least in part, the tumor cell-specific growth inhibition demonstrated by the CRE-decoy.

In addition to providing an effective means of controlling cancer cell proliferation, the CRE-decoy ODN will be useful for many other applications. A large number of genes are transcriptionally regulated through cAMP response elements. Many of these genes encode for regulatory molecules that control the expression of other genes. For example, CREB interacts with its co-activator CBP [30, 31], which is involved in the transcriptional activation of other genes, such as p53 [32], Ap-1 and RARs [33]. In addition, the CREB/FTF family is known to associate with (e.g. heterodimerize) a variety of other transcription factors (e.g. members of the Jun/Fos family) mediating transcription in response to an epidermal growth factor/Ras/Raf signaling pathway [48].

One of the many possible examples for further use of CRE-decoys is in the regulation of hepatitis B virus infection. The X gene product (i.e., pX) of the hepatitis B virus is an important transactivator of a variety of viral and cellular genes [49]. pX has been shown to interact with transcription factors that bind to CREs such as CREB and ATF. The use of CRE-decoys may prevent pX action and thus interfere with the hepatitis B virus life cycle.

The HTLV-I promoter also contains three imperfect repeats of asymmetric CRE's that govern the response of the promoter to the transactivator protein Tax and to varying cAMP levels [50–52]. The critical role of CRE's and CREB has been shown in Tax-mediated transactivation of the HTLV-I promoter [50, 53, 54]. The decoy oligonucleotide technology may not only offer great promise as a tool for defining cellular regulatory processes, but also for treating cancer, viral diseases, and other pathological conditions.

Acknowledgements

We thank Dr. Sudhir Agrawal (Hybridon Inc., Milford, MA, USA) for providing the oligonucleotides.

References

1. Hu MC-T, Davidson N: The inducible *lac* operator-repressor system is functional in mammalian cells. Cell 48: 555–566, 1987
2. Brown M, Figge J, Hansen U, Wright C, Jeang K-T, Khoury G, Livingston DM, Roberts TM: *Lac* repressor can regulate expression from a hybrid SV40 early promoter containing a *lac* operator in animal cells. Cell 49: 603–612, 1987
3. Friedman AD, Triezenberg SJ, McKnight SL: Expression of a truncated viral trans-activator selectively impedes lytic infection by its cognate virus. Nature 335: 452–454, 1988
4. Wang XF, Calame K: SV40 enhancer-binding factors are required at the establishment but not the maintenance step of enhancer-dependent transcriptional activation. Cell 47: 241–247, 1986
5. Maher LJ, Wold B, Dervan PB: Inhibition of DNA binding proteins by oligonucleotide-directed triple helix formation. Science 245: 725–730, 1989
6. Zon G: Oligonucleotide analogues as potential chemotherapeutic agents. Pharm Res 5: 539–549, 1988
7. Agrawal S, Jiang Z, Zhao Q, Shaw D, Cai Q, Roskey A, Channavajjala L, Saxinger C, Zhang R: Mixed-backbone oligonucleotides as second generation antisense oligonucleotides: *In vitro* and *in vivo* studies. Proc Natl Acad Sci USA 94: 2620–2625, 1997
8. Crooke ST: Therapeutic applications of oligonucleotides. Annu Rev Pharmacol Toxicol 32: 329–376, 1992
9. Agrawal S, Zhao Q: Antisense therapeutics. Curr Opin Chem Biol 2: 519–528, 1998
10. Bielinska A, Shivdasani RA, Zhang L, Nabel GJ: Regulation of gene expression with double-stranded phosphorothioate oligonucleotides. Science 250: 997–1000, 1990
11. Morishita R, Gibbons GH, Horiuchi M, Ellison KE, Nakajima M, Zhang L, Kaneda Y, Ogihara T, Dzau VJ: A novel molecular strategy using *cis* element 'decoy' of E2F binding site inhibits smooth muscle proliferation *in vivo*. Proc Natl Acad Sci USA 92: 5855–5859, 1995
12. Roesler WJ, Vandenbark GR, Hanson RW: Cyclic AMP and the induction of eukaryotic gene transcription. J Biol Chem 263: 9063–9066, 1988
13. Park YG, Nesterova M, Agrawal S, Cho-Chung YS: Dual blockade of cyclic AMP responsive element-(CRE) and AP-1-directed transcription by CRE-transcription factor decoy oligonucleotide. J Biol Chem 274: 1573–1580, 1999
14. Montminy MR, Bilezikjian LM: Binding of a nuclear protein to the cyclic-AMP response element of the somatostatin gene. Nature 328: 175–178, 1987
15. Montminy MR, Sevarino KA, Wagner JA, Mandel G, Goodman RH: Identification of a cyclic-AMP-responsive element within the rat somatostatin gene. Proc Natl Acad Sci USA 83: 6682–6686, 1986
16. Maniatis T, Goodbourn S, Fischer JA: Regulation of inducible and tissue-specific gene expression. Science 236: 1237–1245, 1987
17. Short JM, Wynshaw-Boris A, Short HP, Hanson RW: Characterization of the phosphoenolpyruvate carboxykinase (GTP) promoter-regulatory region. II. Identification of cAMP and glucocortcoid regulatory domains. J Biol Chem 261: 9721–9726, 1986
18. Habener JF: Cyclic AMP response element binding proteins: A cornucopia of transcription factors. Mol Endocrinol 4: 1087–1094, 1990
19. Rauscher FJ, Sambucetti LC, Curran T, Distel RJ, Spieglman BM: Common DNA binding site for Fos protein complexes and transcription factor AP-1. Cell 52: 471–480, 1988
20. Ellis MJC, Hurst HC, Goodbourn SA: A novel cyclic AMP response element-binding protein-1 (CREB-1) splice product may down-regulate CREB-1 activity. J Mol Endocrinol 14: 191–198, 1995
21. Stein CA: Does antisense exist? Nature Med 1: 1119–1121, 1995
22. Gonzalez GA, Montminy MR: Cyclic AMP stimulates somatostatin gene transcription by phosphorylation of CREB at serine 133. Cell 59: 675–680, 1989
23. Walton KM, Rehfuss RP, Chrivia JC, Lochner JE, Goodman RH: A dominant repressor of cyclic adenosine 3′,5′-monophosphate (cAMP)-regulated enhancer-binding protein activity inhibits the cAMP-mediated induction of the somatostatin promoter *in vivo*. Mol Endocrinol 6: 647–655, 1992
24. Lee YN, Park YG, Choi YH, Cho YS, Cho-Chung YS: CRE-transcription factor decoy oligonucleotide inhibition of MCF-7 breast cancer cells: Cross talk with p53 signaling pathway. Biochemistry 39: 4863–4868, 2000
25. Amsterdam A, Keren-Tal I, Aharoni D: Cross-talk between cAMP and p53-generated signals in induction of differentiation and apoptosis in steroidogenic granulosa cells. Steroids 61: 252–256, 1996

26. Keren-Tal I, Suh B-S, Dantes A, Lindner S, Oren M, Amsterdam A: Involvement of p53 expression in cAMP-mediated apoptosis in immortalized granulosa cells. Exp Cell Res 218: 283–295, 1995
27. Haldar S, Negrini M, Monne M, Sabbioni S, Croce CM: Down-regulation of *bcl*-2 by p53 in breast cancer cells. Cancer Res 54: 2095–2097, 1994
28. von Knethen A, Lotero A, Brune B: Etoposide and cisplatin induced apoptosis in activated RAW 264.7 macrophages is attenuated by cAMP-induced gene expression. Oncogene 387–394, 1998
29. Chrivia JC, Kwok RPS, Lamb N, Hagiwara M, Montminy MR, Goodman RH: Phosphorylated CREB binds specifically to the nuclear protein CBP. Nature 365: 819–823, 1993
30. Kwok RPS, Lundblad JR, Chrivia JC, Richards JP, Bachinger HP, Brennan RG, Roberts SGE, Green MR, Goodman RH: Nuclear protein CBP is a coactivator for the transcription factor CREB. Nature 370: 223–226, 1994
31. Arias J, Alberts AS, Brindle P, Claret FX, Smeal T, Karin M, Feramisco J, Montminy M: Activation of cAMP and mitogen responsive genes relies on a common nuclear factor. Nature 370: 226–229, 1994
32. Gu W, Shi XL, Roeder RG: Synergistic activation of transcription by CBP and p53. Nature 387: 819–523, 1997
33. Agadir A, Shealy YF, Hill DL, Zhang X: Retinyl methyl ether down-regulates activator protein 1 transcriptional activation in breast cancer cells. Cancer Res 57: 3444–3450, 1997
34. McMurray CT, Wilson WD, Douglass JO: Hairpin formation within the enhancer region of the human enkephalin gene. Proc Natl Acad Sci USA 88: 666–670, 1991
35. Gacy AM, McMurray CT: Hairpin formation within the human enkephalin enhancer region. Kinetic Anal Biochem 33: 11951–11959, 1994
36. Comb M, Mermod N, Hyman SE, Pearlberg J, Ross ME, Goodman HM: Proteins bound at adjacent DNA elements act synergistically to regulate human proenkephalin cAMP inducible transcription. EMBO J 7: 3793–3805, 1988
37. Bazett-Jones DP, Brown ML: Electron microscopy reveals that transcription factor TFIIIA bends 5S DNA. Mol Cell Biol 9: 336–341, 1989
38. Kerppola TK, Curran T: DNA bending by Fos and Jun: The flexible hinge model. Science 354: 1210–1214, 1991
39. Perez-Martin J, Espinoza M: Protein-induced bending as a transcriptional switch. Science 260: 805–807, 1993
40. Horwitz MSZ, Loeb LA: An *E. coli* promoter that regulates transcription by DNA superhelix-induced cruciform extrusion. Science 241: 703–705, 1988
41. Bianchi ME, Beltrame M, Paonessa G: Specific recognition of cruciform DNA by nuclear protein HMG1. Science 243: 1056–1059, 1989
42. Glucksmann MA, Markiewicz P, Malone C, Rothman-Denes LB: Specific sequences and a hairpin structure in the template strand are required for N4 viron RNA polymerase promoter recognition. Cell 70: 491–500, 1992
43. Sun P, Enslen H, Myung PS, Maurer RA: Differential activation of CREB by Ca^{2+}/calmodulin-dependent protein kinases type II and type IV involves phosphorylation of a site that negatively regulates activity. Genes Dev 8: 2527–2539, 1994
44. Xing J, Ginty DD, Greenberg ME: Coupling of the RAS-MAPK pathway to gene activation by RSK2, a growth factor-regulated CREB kinase. Science 273: 959–963, 1996
45. Sheng M, Thomson MA, Greenberg ME: CREB: A Ca(2+)-regulated transcription factor phosphorylated by calmodulin-dependent kinases. Science 252: 1427–1430, 1991
46. Ginty DD, Bonni A, Greenberg ME: Nerve growth factor activates a ras-dependent protein kinase that stimulates c-*fos* transcription via phosphorylation of CREB. Cell 77: 713–725, 1994
47. Tan Y, Rouse J, Zhang A, Cariati S, Cohen P, Comb MJ: FGF and stress regulate CREB and ATF-1 via a pathway involving p38 MAP kinase and MAPKAP kinase-2. EMBO J 15: 4629–4642, 1996
48. Nilsson M, Ford J, Bohm S, Toftgard R: Characterization of a nuclear factor that binds juxtaposed with ATF3/Jun on a composite response element specifically mediating induced transcription in response to an epidermal growth factor/Ras/Raf signaling pathway. Cell Growth Diff 8: 913–920, 1997.
49. Williams JS, Andrisani OM: The hepatitis B virus X protein targets the basic region-leucine zipper domain of CREB. Proc Natl Acad Sci USA 92: 3819–3823, 1995
50. Jeang K-T, Boros I, Brady J, Radonovich M, Khoury G: Characterization of cellular factors that interact with the human T-cell leukemia virus type I p40-responsive 21-base-pair sequence. J Virol 62: 4499–4509, 1988
51. Nakamura M, Niki M, Obtani K, Sugamura K: Differential activation of the 21-base-pair enhancer element of human T-cell leukemia virus type I by its own *trans*-activator and cyclic AMP. Nucleic Acids Res 17: 5207–5221, 1989
52. Poteat HT, Kadison P, McGuire K, Park L, Park RE, Sodroski JG, Haseltine WA: Response of the human T-cell leukemia virus type 1 long terminal repeat to cyclic AMP. J Virol 64: 1604–1611, 1989
53. Niki M, Ohtani K, Nakamura M, Sugamura K: Multistep regulation of enhancer activity of the 21-base-pair element of human T-cell leukemia virus type I. J Virol 66: 4348–4357, 1992
54. Bodor J, Walker W, Flemington E, Spetz A-L, Habener J: Modulation of Tax and PKA-mediated expression of HTLV-I promoter via cAMP response element binding and modulator proteins CREB and CREM, FEBS Lett 377: 413–418, 1995

Molecular and Cellular Biochemistry 212: 35–43, 2000.

Attenuation of macrophage apoptosis by the cAMP-signaling system

Andreas von Knethen and Bernhard Brüne
Department of Medicine IV-Experimental Division, Faculty of Medicine, University of Erlangen-Nürnberg, Germany

Abstract

Previous studies revealed that expression and activation of cyclooxygenase-2 (Cox-2) conveyed a protective principle in murine macrophages, thus attenuating pro-apoptotic actions of chemotherapeutic agents or programmed cell death as a result of massive nitric oxide (NO) generation. Expression of Cox-2 was achieved by treatment of cells with lipopolysaccharide/interferon-γ or nontoxic doses of NO releasing agents. We reasoned E-type prostanoid formation, and in turn an intracellular cAMP increase as the underlying protective mechanism. To prove our hypothesis, we analyzed the effects of lipophilic cAMP-analogs on NO, cisplatin, or etoposide induced apoptosis in RAW 264.7 macrophages. Selected apoptotic parameters comprised DNA fragmentation (diphenylamine assay), annexin V staining of phosphatidylserine, caspase activity (quantitated by the cleavage of a fluorogenic caspase-3-like substrate Ac-DEVD-AMC), and mitochondrial membrane depolarisation ($\Delta\Psi$). Western blots detected accumulation of the tumor suppressor protein p53, relocation of cytochrome c to the cytosol, and expression of the anti-apoptotic protein Bcl-x_L. Prestimulation with lipophilic cAMP-analogs attenuated apoptosis with the notion that cell death parameters were basically absent. To verify gene induction by cAMP in association with protection we established activation of cAMP response element binding protein (CREB) by gel-shift analysis and moreover, treated macrophages with oligonucleotides containing a cAMP-responsive element (CRE) in order to scavenge CREB. Decoy oligonucleotides, but not control oligonucleotides, attenuated cAMP-evoked protection and reestablished pro-apoptotic parameters. We conclude that gene induction by cAMP protects macrophages towards apoptosis that occurs as a result of excessive NO formation or addition of chemotherapeutica. Attenuating programmed cell death by the cAMP-signaling system may be found in association with Cox-2 expression and tumor formation. (Mol Cell Biochem **212**: 35–43, 2000)

Key words: cAMP, CRE, Cox-2, NO, apoptosis

Abbreviations: GSNO – S-nitrosoglutathione; Cox-2 – cyclooxygenase-2; db-cAMP – dibutyryl-cAMP; LPS – lipopolysaccharide; IFN-γ – interferon-γ; CRE – cAMP responsive element; CREB – cAMP repsonsive element binding protein; NO – nitric oxide; Ac-DEVD-AMC – N-acetyl-Asp-Glu-Val-Asp-7-amino-4-methylcoumarin; $DiOC_6(3)$ – 3,3′-dihexyloxacarbocyanide iodide; PARP – poly (ADP-ribose) polymerase

Introduction

Apoptosis, a synonym for genetically determined and thus naturally occurring cell death, is essential for the maintenance of tissue homeostasis [1]. It removes genetically altered, toxicologically damaged, or immunologically marked cells [2]. Hence, dysregulation of apoptosis has been implicated in a plethora of human diseases with either too little or an exaggerated rate of cell death [3–6]. Apoptosis is characterized by defined morphological and biochemical features. Examples comprise chromatin condensation, DNA degradation known as DNA laddering, as well as alterations in the expression of Bcl-2 family members, activation of cysteine proteases known as caspases, and mitochondrial alterations such as cyctochrome c release or membrane depolarization ($\Delta\Psi$ breakdown) [7–12]. In several cases accumulation of the

Address for offprints: B. Brüne, Faculty of Medicine, University of Erlangen-Nürnberg, Loschgestrasse 8, 91054 Erlangen, Germany tumor suppressor p53 participates in eliciting apoptosis [13]. Accumulated p53 appears as a common denominator of various stress conditions, i.e. oncogene activation, hypoxia, nitric oxide, DNA damage, and others, thereby ensuring that affected cells are efficiently taken care of. The pivotal role of p53 in maintaining genomic integrity has earned it the nickname 'guardian of the genome' [14, 15].

Bacterial LPS is a strong immunostimulator and can elicit a shock syndrome [16]. In macrophages and other cell types LPS primarily induces the expression of an impressive number of genes and triggers the release of inflammatory cytokines and radicals such as O_2^- as well as NO. Macrophages are a major component of the innate immune defense system directed against pathogens or tumor cells and represent one of the main cellular sources of cytokines and oxygen- as well as nitrogen-based radicals [17, 18]. Radicals are used during host defense with the implication to elicit apoptotic or necrotic target cell death. Control of macrophage responsiveness may be of primary importance for the host and may limit proinflammatory responses to protect the host from excessive destructive processes during inflammation and infection. Macrophage function is therefore orchestrated by compensatory mechanisms as a result of cytokine or radical formation that can be both produced and utilized by macrophages in processes known as autocrine regulatory pathways.

Activation of macrophages often is associated with cyclooxygenase-2 (Cox-2) expression [19–22]. To achieve macrophage activation stimulation with lipopolysaccharide/interferon-γ (LPS/IFN-γ) is a standard protocol that in turn provokes Cox-2 expression. Cox-2, also known as prostaglandin (PG) H synthase catalyzes a rate-limiting step in the formation of prostaglandin endoperoxides [23, 24]. Prostanoids function as extracellular or intracellular messengers and classical examples such as $PGF_2\alpha$ or PGE_2, activate adenylyl cyclase via the interaction with plasma membrane G-protein coupled receptors, thereby causing a cAMP increase [25, 26].

Using murine (RAW 264.7 cells) or human macrophages as a test system we studied the apoptotic response towards toxic doses of NO releasing compounds (NO donors) or chemotherapeutic agents. Macrophage cell death elicited by cisplatin or etoposide as well as endogenously produced or exogenously supplied NO was associated with typical apoptotic features. Selected pro-apoptotic markers have been chromatin condensation, DNA laddering, caspase-activation, cytochrome c relocation, Bcl-x_L-downregulation, annexin V staining, and p53 accumulation. Attenuating p53 accumulation by p53 antisense encoding plasmids or enforced Bcl-2 overexpression abrogated NO-evoked apoptosis [27].

By analyzing macrophage programmed cell death in further detail, we realized that desensitization towards NO- or chemotherapeutic-elicited apoptosis occurs upon preactivation with a combination of LPS and IFN-γ or from prestimulation with a nondestructive dose of NO donors. During our studies it became apparent that induction of Cox-2 represents a critical regulator of macrophage apoptosis. Resting macrophages do not express Cox-2, whereas LPS/IFN-γ or a nontoxic dose of GSNO caused protein expression within 6–12 h [28]. A protective role of Cox-2 was assured in Cox-2 overexpressing cells with the notion that protection was antagonized by the Cox-2 selective inhibitor NS 398 and was lost in antisense Cox-2 transfected macrophages [28]. Obligatory for Cox-2 expression appears the activation of the nuclear transcription factor NF-κB (p50/p65-heterodimer formation). Degradation of I-κBα and activation of a luciferase reporter construct, containing four copies of the NF-κB-site derived from the murine Cox-2 promoter, confirmed NF-κB activation by NO. Furthermore, a NF-κB decoy approach attenuated not only cell protection but also restored DNA fragmentation and p53 accumulation in response to a high dose GSNO [29, 30]. These examinations provided evidence for an anti-apoptotic role of NO, transmitted by NF-κB activation. Furthermore, NO also activated the transcription factor AP-1 that additionally was required for Cox-2 expression and protection. Taking into account that protection demanded an active Cox-2 it was our intention to study the Cox-2 downstream signaling cascade, thus promoting cell protection.

Here and in other more recent studies we sought to elucidate how an active Cox-2 with the concomitant formation of cAMP attenuates apoptosis in macrophages [31, 32]. Our experiments suggest that cAMP via activation of the transcription factor CREB provokes protection and imply that CREB-associated signaling pathways convey the predominant apoptosis suppressive principle.

Materials and methods

Materials

Diphenylamine and LPS (*E. coli* serotype 0127:B8) were purchased from Sigma, Deisenhofen, Germany. Recombinant murine interferon-γ, and dibutyryl-cAMP were provided by Roche, Mannheim, Germany. RPMI 1640, cell culture supplements, and fetal calf serum were ordered from Biochrom, Berlin, Germany. Ac-DEVD-AMC was bought from Biomol, Hamburg, Germany. All other chemicals were of the highest grade of purity and commercially available.

Cell culture

The mouse monocyte/macrophage cell line RAW 264.7 was maintained in RPMI 1640 supplemented with 100 U/ml penicillin, 100 μg/ml streptomycin, and 10% heat-inactivated fetal calf serum (complete RPMI). All experiments were performed using complete RPMI. NO donors and dibutyryl-cAMP were dissolved in bidistilled water. Appropriate vehicle controls were performed.

Quantitation of DNA fragmentation

DNA fragmentation was measured with the diphenylamine assay as previously reported [33]. Briefly, following incubations cells were scraped off the culture plates, resuspended in 250 µl 10 mM Tris, 1 mM EDTA, pH 8.0 (TE-buffer), and incubated with an additional volume lysis-buffer (5 mM Tris, 20 mM EDTA, pH 8.0, 0.5% Triton X-100) for 30 min at 4°C. After lysis, the intact chromatin (pellet) was separated from DNA fragments (supernatant) by centrifugation for 15 min at 13000 × g. Pellets were resuspended in 500 µl TE-buffer and samples were precipitated by adding 500 µl 10% trichloroacetic acid at 4°C. Samples were pelleted at 1400 × g for 10 min and the supernatant was removed. After addition of 300 µl 5% trichloroacetic acid, samples were boiled for 15 min. DNA contents were quantitated using the diphenylamine reagent [34]. The percentage of DNA fragmentation was calculated as the ratio of the DNA content in the supernatant to the amount in the pellet.

Nuclear protein extraction

Preparation of crude nuclear extract was basically as described [35]. Briefly, following cell activation for the times indicated, 4 × 10^6 RAW 264.7 macrophages were washed in 1 ml of ice-cold PBS, centrifuged at 1000 × g for 5 min, resuspended in 400 µl ice-cold hypotonic buffer (10 mM HEPES/KOH, 2 mM $MgCl_2$, 0.1 mM EDTA, 10 mM KCl, 1 mM DTT, 0.5 mM PMSF, pH 7.9), left on ice for 10 min, vortexed, and centrifuged at 15 000 × g for 30 sec. Pelleted nuclei were gently resuspended in 50 µl ice-cold saline buffer (50 mM HEPES/KOH, 50 mM KCl, 300 mM NaCl, 0.1 mM EDTA, 10% glycerol, 1 mM DTT, 0.5 mM PMSF, pH 7.9), left on ice for 20 min, vortexed, and centrifuged at 15 000 × g for 5 min at 4°C. Aliquots of the supernatant, that contain nuclear proteins where frozen in liquid nitrogen and stored at –70°C. Protein was determined using a Bio-Rad II Kit.

Electrophoretic mobility shift assays

An established EMSA-method, with slight modifications, was used [36]. Nuclear protein (5 µg) was incubated for 20 min at room temperature with 20 µg bovine serum albumin, 2 µg poly(dI-dC) (Pharmacia, Freiburg, Germany), 2 µl buffer D (20 mM HEPES/KOH, 20% glycerol, 100 mM KCl, 0.5 mM EDTA, 0.25% NP-40, 2 mM DTT, 0.5 mM PMSF, pH 7.9), 4 µl buffer F (20% Ficoll-400, 100 mM HEPES/KOH, 300 mM KCl, 10 mM DTT, 0.5 mM PMSF, pH 7.9), and 20,000 cpm of a (^{32}P)-labeled oligonucleotide in a final volume of 20 µl. DNA-protein complexes were resolved at 180 V for 4 h in a taurine-buffered, native 6% polyacrylamide-gel, dried, and visualized (autoradiography using a Fuji X-ray film). Oligonucleotide probes were labeled by a filling reaction using the Klenow fragment (Roche, Mannheim, Germany). One pmol oligonucleotide was labeled with 50 µCi of (α-^{32}P)-dCTP (3000 Ci/mmol, Amersham, Braunschweig, Germany), cold nucleotides (dATP, dTTP, dGTP from GIBCO, Eggenstein, Germany), purified on a CHROMA SPIN-10 column (Clontech, Heidelberg, Germany), and stored at –20°C until use. The oligonucleotide sequence corresponding to the CRE-site (consensus sequence underlined) of the c-fos promoter was used [37]:

5′-GAT ATT GCC <u>TGA CGT CA</u>G AG-3′
3′-TAA CGG <u>ACT GCA GT</u>C TCT CG-5′

GSNO synthesis

S-nitrosoglutathione was synthesized and characterized as previously described [38].

Transfection

RAW 264.7 cells were exposed to CRE oligonucleotides, containing the CRE-consensus site derived from the c-fos promoter (as described for gel shift analysis) and an unrelated oligonucleotide as a control. 5′-terminal fluorescein labeled oligonucleotides were stabilized by phosphothiorate linkages (Eurogentec, Seraing, Belgium). One day before transfection cells were seeded at a density of 1 × 10^6 cells/well into 6-well plates. Oligonucleotides (3 µM) were added 24 h prior to cell stimulation. After changing the medium, cell stimulation was performed as indicated. Transfection efficiency was determined by counting labeled cells by fluorescence-microscopy.

Statistical analyses

Each experiment was performed at least 3 times and statistical analysis was performed using the two tailed Student's *t*-test. Normal distribution of all data is assured. Results are expressed as mean values ± S.D. Otherwise representative data are shown.

Results

Apoptosis and its inhibition by cAMP

In corroboration with earlier reports, we noticed NO-mediated apoptosis in RAW 264.7 macrophages. NO was generated by various, chemically distinct NO donors such as S-nitrosoglutathione (GSNO, 1 mM) that promoted 30–35% DNA fragmentation during an 8-h incubation period. Variations in the amount of individual NO donors needed to promote apoptosis is related to the kinetics of NO release and may further refer to the redox-species being formed. GSNO is considered the most physiological NO donor that releases either NO-radical or nitrosonium ion. In our experiments DNA cleavage under control conditions was around 5% and was quantitated using the diphenylamine assay. DNA fragmentation is regarded a late apoptotic parameter and requires incubation periods of roughly 8 h. Other markers that defined apoptosis, especially p53 accumulation and caspase activation, occurred much earlier [27]. Moreover, the absence of lactate dehydrogenase in the cell supernatant underscored plasma membrane integrity during these incubations, supporting the notion that cell demise resembles apoptosis rather than necrosis.

Based on our previous studies we intended to ask whether lipophilic cAMP-analogs such as 8-Br-cAMP or db-cAMP protect macrophages from entering an apoptotic pathway. In a first set of experiments we time-dependently exposed RAW 264.7 macrophages to 1 mM db-cAMP prior to initiation of apoptosis by 1 mM GSNO. Cyclic-AMP analogs are supposed to enter the cell and to be effective at those concentrations. Evidently, GSNO evoked DNA fragmentation of roughly 35% remained unaffected when cAMP-analogs were preincubated for 30 min up to 4 h. With a 6-h lasting preincubation period DNA fragmentation started to decline and attenuation of NO-elicited apoptosis became prominent with 12–15 h of preincubation (Fig. 1A). At late time points, suppression of apoptosis was complete, since fragmentation values comparable to controls were noted for the combination of db-cAMP/GSNO.

In another set of experiments we determined the effective dose of lipophilic cAMP-analogs that was required to achieve protection (Fig. 1B). For these studies we preexposed macrophages for 15 h to increasing concentrations of db-cAMP, ranging from 50 μM to 1 mM. Thereafter, we used 1 mM GSNO to initiate apoptosis. Up to a concentration of 250 μM db-cAMP apoptosis proceeded unaltered, slowed down with 0.5 mM of the cAMP-analog, and was completely blocked in the presence of 1 mM db-cAMP. Our studies indicate that attenuation of macrophage apoptosis by lipophilic cAMP-analogs demanded preincubation periods of at least 12 h and concentrations of 1 mM. Similar results were obtained when NO was replaced by the chemotherapeutic agents cisplatin or etoposide [32].

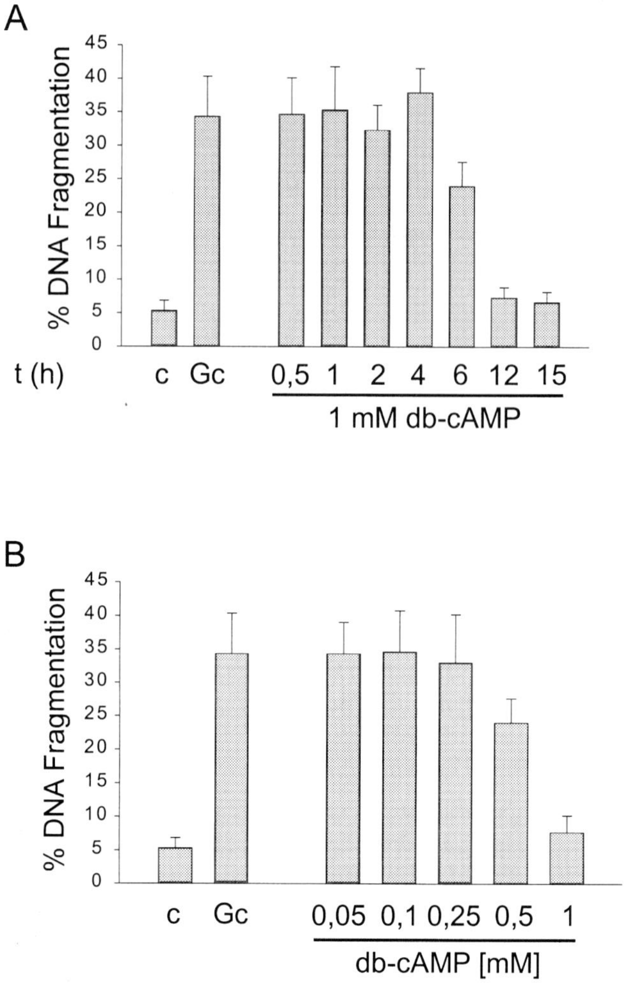

Fig. 1. DNA fragmentation and its inhibition by db-cAMP pretreatment. RAW 264.7 macrophages were prestimulated with dibutyryl-cAMP prior to the addition of 1 mM GSNO. DNA fragmentation was measured after 8 h and quantitated with the diphenylamine assay; c: control without stimulation; Gc: GSNO-stimulation. (A) Cells were preexposed to 1 mM db-cAMP for 30 min up to 15 h prior to the addition of GSNO; (B) Cells were preincubated with increasing concentrations of db-cAMP for 15 h. Data are mean values ± S.D. of 3 individual experiments. For details and statistical analysis see Materials and methods.

In related studies we examined the impact of cAMP-analogs on the occurrence of accompanying apoptotic signals such as p53 accumulation, caspase activation, cytochrome c relocation, mitochondrial $\Delta\Psi$ breakdown, or alterations in the expression of Bcl-2 family members [31]. During NO- or cisplatin-evoked apoptosis Western blot analysis revealed no or only minor expression of p53 in controls but showed massive protein accumulation within a 2-h treatment in response to individual agonists. In some analogy, caspase activation was noticed in response to NO or cisplatin after 4 h and was determined by the cleavage of DEVD-AMC or poly (ADP-ribose) polymerase (PARP). Moreover, apoptotic cells appeared annexin V positive after 4 h. Six h after agonist addition we observed cytochrome c relocation from mitochondria to the cytosol and decreased expression of the anti-apoptotic protein Bcl-x_L. Mitchondrial $\Delta\Psi$ breakdown as measured by the $DiOC_6(3)$-fluorescence or DNA-laddering were noticed after 8 h. Of note, under the influence of 1 mM db-cAMP, preincubated for 15 h, the pro-apoptotic accompanying signals remained basically absent [31]. Table 1 summarizes apoptotic parameters that emerged in RAW macrophages in response to 1 mM GSNO, indicates their relative occurrence after agonist addition, and depicts the inhibitory impact of 1 mM db-cAMP, supplied 15 h prior to agonist addition.

We conclude that preactivation of macrophages for an extended time with cAMP-analogs inhibits NO- or cisplatin-induced apoptosis. Interestingly, the pro-apoptotic signaling cascade was attenuated at an early point, thus not allowing p53 accumulation.

Molecular actions in response to cAMP

The observation that protection from apoptosis by db-cAMP demanded a long preexposure period implied cAMP-mediated gene activation as the underlying mechanism. Assuming that the transcription factor CREB is involved, we performed gel-shift analysis to prove activation of CREB in response to lipophilic-AMP analogs. In a first set of experiments we looked for the time-dependency of CREB activation. We observed a very minor response in controls, but noticed CREB activation 30 min after 1 mM db-cAMP addition (Fig. 2A). CREB activation became more prominent at 1–2 h and slowly declined after a 4–6 h lasting exposure period.

In extending experiments we determined the dose-response of CREB activation towards db-cAMP. Therefore, macrophages were exposed to 250 µM up to 1 mM of db-cAMP for 1 or 2 h and CREB activation was determined by gel-shift analysis (Fig. 2B). Activation was low with the addition of 250 µM db-cAMP, became significantly stronger with 500 µM of the cAMP-analog, while the strongest response was elicited with 1 mM db-cAMP being added. In

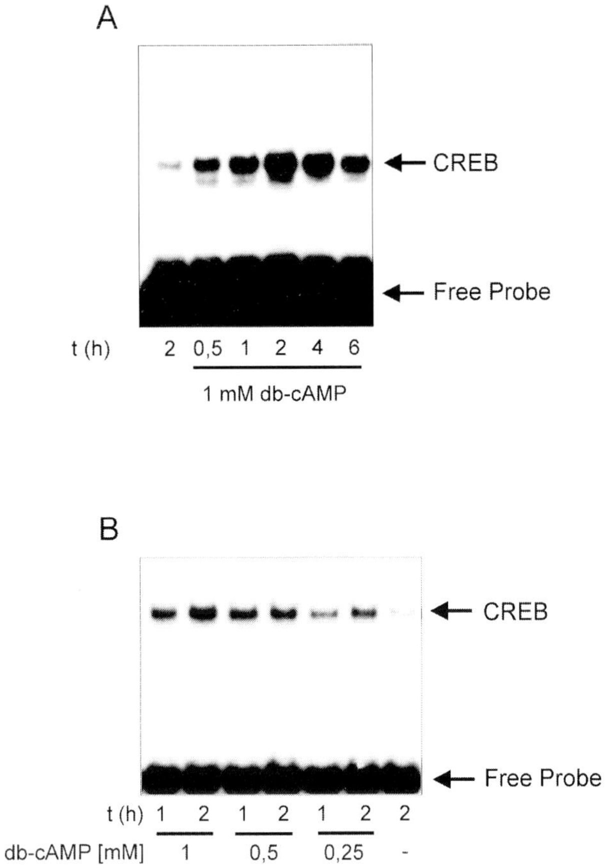

Fig. 2. Activation of CREB in response to db-cAMP. RAW 264.7 macrophages were prestimulated with dibutyryl-cAMP prior to gel shift analysis in order to determine activation of the transcription factor CREB. (A) Cells were preexposed to 1 mM db-cAMP for 30 min up to 6 h; (B) Increasing concentration of db-cAMP (250 µM to 1 mM) were preincubated for 1 or 2 h. Gels are representative of three similar experiments. For details see Materials and methods.

Table 1. Occurrence of apoptotic parameters and their inhibition by db-cAMP

Selected parameter	Method used for detection	Occurrence after adding 1 mM GSNO	Impact of 1 mM db-cAMP
p53 accumulation	WB	2 h	absent
Caspase activation	DEVD-AMC-cleavage, PARP-cleavage	4 h	absent
Annexin V staining	flow-cytometry	4 h	absent
Cytochrome c	WB	6 h	absent
Expression of Bcl-x_L	WB	6 h	as controls
$\Delta\Psi$ breakdown	$DiOC_6(3)$	8 h	absent
DNA fragmentation	diphenylamine-assay	8 h	absent

RAW 264.7 macrophages were exposed to 1 mM GSNO to initiate apoptosis and to study inhibition of cell death by 1 mM db-cAMP, preincubated for 15 h. Accumulation of p53, expression of Bcl-x_L, caspase-evoked cleavage of PARP, and relocation of mitochondrial cytochrome c were determined by Western blot analysis (WB). Breakdown of mitochondrial membrane potential ($\Delta\Psi$) and the occurrence of phosphatidylserine at the outer leaflet of the plasma membrane was measured by flow cytometry using $DiOC_6(3)$ and annexin V-FITC. Caspase-activation was followed by the cleavage of DEVD-AMC and the release of the fluorocrome aminomethylcoumarin (AMC). DNA fragmentation was quantitated by the diphenylamine-assay.

corroboration with above mentioned results CREB activation at 2 h was consistently stronger compared to a 1-h lasting exposure period. In some agreement with suppression of apoptosis we noticed that 1 mM db-cAMP was needed to provoke strongest CREB activation and moreover, was required for efficient attenuation of cell death. Having established that cAMP promotes CREB activation we went on to demonstrate the transcriptional relevance in lipophilic cAMP-mediated protection by using a decoy approach. Oligonucleotides containing a CRE site were added to macrophages in order to scavenge the activated transcription factor CREB (decoy-approach). Alternatively, oligonucleotides with an altered base sequence were used (control-approach). In macrophages that were exposed to 3 μM of fluorescence-labeled oligonucleotides for 24 h we determined roughly 50% of all cells to be positively affected. After oligonucleotide incorporation, cells were exposed to 1 mM db-cAMP in order to promote protection towards NO donors.

Following these experimental modulations, apoptotic DNA fragmentation was initiated with 1 mM GSNO (Fig. 3). Addition of CRE-decoy oligonucleotides reversed protection, thus restoring DNA fragmentation, comparable to controls, in response to NO generation. In contrast, control oligonucleotides did not counteract the inhibitory principle evoked by lipophilic cAMP treatment. Decoy-oligonucleotides turned out to be similar active when the NO donor was replaced by the chemotherapeutic agent cisplatin (data not shown). In good agreement to DNA fragmentation we noticed an effect of decoy-oligonucleotides on the p53 response. CRE-decoy oligonucleotides restored p53 accumulation in response to NO donors or cisplatin, while control oligonucleotides left lipophilic cAMP-initiated suppression of p53 unaltered.

Conclusively, inducible protection as a result of cAMP-action along with an attenuated p53 response is antagonized by oligonucleotides that contain a CRE-binding site. Obviously, cAMP-evoked gene activation resembles a protective principle in macrophages, thereby regulating susceptibility towards cell demise.

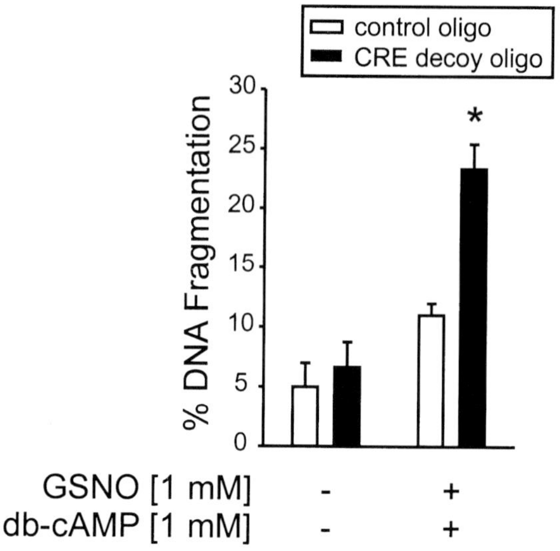

Fig. 3. CRE decoy oligonucleotides attenuate cAMP-mediated protection. RAW 264.7 macrophages (1×10^6 cells) were incubated with decoy (CRE) or control olignucleotides for 24 h as described under Materials and methods. After changing the medium cells were prestimulated with 1 mM dibutyryl-cAMP for 15 h. Thereafter, DNA fragmentation was initiated by the addition of 1 mM GSNO. DNA fragmentation was quantitated after 8 h by the diphenylamine assay as described. Data are mean values ± S.D. of 3 individual experiments (*$p \leq 0.05$ vs. control oligonucleotides).

Discussion

Preactivation of macrophages confers resistance towards apoptotic death that is elicited by NO donors or chemotherapeutic agents. Protection is achieved by LPS/IFN-γ or low level NO prestimulation that ultimately accounts for Cox-2 expression and activation [28–30]. Furthermore, prevention of cell death is provided by stimulating cells with lipophilic cAMP-analogs for 15 h. Under conditions of cell protection we noticed that typical pro-apoptotic markers such as caspase-3 activation, cytochrome c relocation, ΔΨ breakdown, p53 accumulation, DNA fragmentation, annexin V staining, or alterations in Bcl-x_L expression were absent. Importantly, decoy oligonucleotides containing a CRE-binding site reversed inducible resistance, thus supporting the notion that cAMP-mediated gene activation is demanded to convey protection [31, 32]. We conclude that upregulation of Cox-2 with the concomitant formation of cAMP-elevating prostanoids accounts for macrophage deactivation and prevention of apoptosis. Figure 4 summarizes major signaling components as a result of LPS/IFN-γ or low level NO stimulation that provoke Cox-2 expression, thus leading to cAMP formation, and ultimately attenuating macrophage apoptosis.

Expression of Cox-2 in the sequence of events leading to colorectal cancer seem established and genetic evidence links Cox-2 to tumorigenesis [39–42]. Mating tumor suppressor APC gene knock-out mice which spontaneously develop intestinal polyps, with Cox-2 deficient mice resulted in suppression of polyp formation [43]. In addition, polyp formation was reduced by Cox-2 specific inhibitors, thus implying that Cox-2 expression is required for the development of some types of cancer [44]. Because apoptotic resistant cells may be prone to mutagenic events, inhibition of naturally occurring cell death is one potential explanation for Cox-2-evoked tumorigenesis [42].

We have shown by the use of the specific Cox-2 inhibitor NS 398, that Cox-2, which is also induced by prestimulation of macrophages with lipophilic cAMP-analogs [45] is not responsible for db-cAMP-mediated protection against apoptosis [31]. Therefore, expression of Cox-2 in macrophages

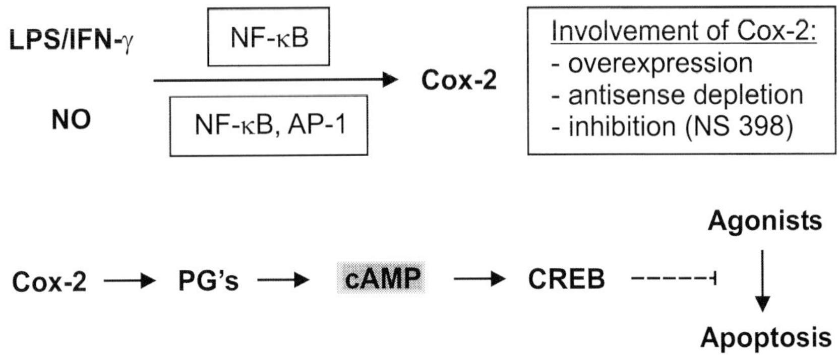

Fig. 4. Signaling pathways that promote protection in RAW macrophages. A summary of pathways that are initiated in response to NO- or LPS/IFN-γ-treatment, thereby attenuating agonist-promoted apoptosis in macrophages. Components that convey protection comprise upregulation and/or activation of the transcription factors NF-κB and AP-1, expression of cyclooxygenase-2 (Cox-2), prostanoid formation (PG's), cAMP generation, and activation of CREB. The involvement of Cox-2 in establishing protection is advised by overexpression, antisense depletion, or pharmacological inhibition by NS 398. For details, see the text.

is associated with prostaglandin E_2 formation which in turn may increase intracellular cAMP [46]. Deactivation of several macrophage responses by cAMP-elevating manipulations such as adenyly cyclase stimulation, phosphodiesterase inhibition, and/or Cox-2 activation are known and considered a standard regime to suppress an early phase of the innate immune response that is closely associated with macrophage stimulation [47, 48]. In addition, the action of cAMP-elevating agents in suppressing apoptosis is known for several other systems, i.e. for hepatocytes, neutrophils, thymocytes, natural-killer cell-derived cytotoxicity, or neurons [49–54].

Mechanistic considerations on cAMP-mediated signal transduction in conferring apoptotic resistance are largely unknown. Under inflammatory conditions increased levels of cAMP-elevating prostaglandins, acting via adenylyl cyclase may efficiently balance apoptotic initiating and counteracting pathways in cells of the monocyte/macrophage lineage. Our assumption that cAMP is involved in establishing cellular resistance was endorsed by applying lipophilic cAMP-analogs that blocked all apoptotic features. Cellular responses towards cAMP are mediated by activation of cAMP-dependent protein kinases. The inactive kinase holoenzyme dissociates after the binding of cAMP and liberates the catalytic subunit which in turn phosphorylates multiple substrates. Phosphorylation may affect structural components or enzyme activities, or alternatively, may promote alterations in gene transcription by phosphorylation and activation of the transcription factor CREB [55]. Phosphorylated CREB binds to the promoter region of various genes and initiates transcription [56]. Protection towards apoptosis by cAMP-elicited signal transduction has been shown, among others, for radiation resistance of melanoma cells, FAS/CD-95-mediated apoptosis of primary hepatocytes, delayed apoptosis in human neutrophils at a site upstream of caspase-activation, glucocorticoid-induced apoptosis and activation of caspases in thymocytes, ceramide-induced apoptosis of U937 cells, and neutrotrophin i.e. nerve growth factor supported survival of neurons [49–51, 53, 54, 57, 58].

Our study verifies a macrophage protective mechanism to convey resistance towards NO donors or chemotherapeutic agents by lipophilic cAMP-analog pretreatment. The successful use of decoy oligonucleotides with their ability to scavenge activated CREB, thereby abrogating protection and reestablishing p53 accumulation confirmed the assumption of cAMP-derived gene induction. Although cAMP-responsive gene products remain elusive, they seem to participate in the regulation of p53 thereby affecting a rather upstream signaling component during NO-initiated apoptosis.

Our results on the involvement of CREB during macrophage protection are in line with most recent observations that CREB-mediated gene expression was both necessary for nerve growth factor-dependent survival and sufficient on its own to promote survival of sympathetic neurons [51]. As Bcl-2 was activated by a CREB-dependent transcription mechanism the authors concluded that survival is promoted by a pathway involving CREB family transcription factor-dependent gene expression of Bcl-2 . This is at variance with the situation seen in macrophages. Protection evoked by cAMP suppresses accumulation of p53, one of the most upstream pro-apoptotic signals in macrophages. As seen in previous studies, endorsed expression of Bcl-2 also facilitates protection, however, at a point downstream of p53 accumulation [12, 59, 60]. In line, the action of CREB as a survival factor for human melanoma cells is independent of any changes in Bcl-2 or related proteins [58]. Therefore, cAMP-evoked gene expression that halts the apoptotic program in neurons and other cells such as macrophages or human melanoma cells obviously shows some variation. Interestingly, it has been documented, that the anti-apoptotic Akt/PKB-survival promoting pathway stimulates gene expression via CREB

[61]. Obviously, independent survival promoting pathways merge at the level of CREB. Unfortunately, the identity of CREB-sensitive genes and corresponding proteins that contribute to protection in general remain so far unknown.

For macrophages the regulation of apoptosis may contribute to inflammatory conditions such as sepsis or atherosclerosis. Therefore, it is pertinent to gain insights into underlying signaling mechanisms. Understanding transducing mechanisms that confer inducible resistance towards NO or chemotherapeutic agents in immune competent cells such as macrophages by cAMP-mediated gene activation may open avenues for pharmacologic intervention. Along that line, CREB expression directly correlates with the metastatic potential of some melanoma cells which may refer to tumorigenicity and metastasis [62]. Knowing potential downstream targets of CREB-dependent gene expression will allow to direct therapeutic modalities to control tumor growth, with the intention to reestablish pro-apoptotic signaling pathways that are being offset by the cAMP-signaling system.

Acknowledgements

We thank Sabine Häckel for expert technical assistance. This work was supported by grants from Deutsche Forschungsgemeinschaft and Deutsche Krebshilfe.

References

1. Eastman A, Rigas JR: Modulation of apoptosis signaling pathways and cell cycle regulation. Semin Oncol 26: 7–16, 1999
2. Hinds PW, Weinberg RA: Tumor suppressor genes. Curr Opin Genet Dev 4: 135–141, 1994
3. Hale AJ, Smith CA, Sutherland LC, Stoneman VEA, Longthorne VL, Culhane AC *et al.*: Apoptosis: Molecular regulation of cell death. Eur J Biochem 236: 1–26, 1996
4. Lee JM, Bernstein A: Apoptosis, cancer and the p53 tumour suppressor gene. Cancer Metastasis Rev 14: 149–161, 1995
5. Ridgway WM, Weiner HL, Fathman CG: Regulation of autoimmune response. Curr Opin Immunol 6: 946–955, 1994
6. Fadeel B, Orrenius S, Zhivotovsky B: Apoptosis in human disease: A new skin for the old ceremony? Biochem Biophys Res Commun 266: 699–717, 1999
7. Koopman G, Reutelingsperger CP, Kuijten GA, Keehnen RM, Pals ST, van Oers MH: Annexin V for flow cytometric detection of phosphatidylserine expression on B cells undergoing apoptosis. Blood 84: 1415–1420, 1994
8. Martin SJ, Reutelingsperger CP, McGahon AJ, Rader JA, van Schie RC, LaFace DM *et al.*: Early redistribution of plasma membrane phosphatidylserine is a general feature of apoptosis regardless of the initiating stimulus: Inhibition by overexpression of Bcl-2 and Abl. J Exp Med 182: 1545–1556, 1995
9. Nicholson DW: Caspase structure, proteolytic substrates, and function during apoptotic cell death. Cell Death Diff 6: 1028–1042, 1999
10. Petit PX, O'Connor JE, Grunwald D, Brown SC: Analysis of the membrane potential of rat- and mouse-liver mitochondria by flow cytometry and possible applications. Eur J Biochem 194: 389–397, 1990
11. Okada S, Zhang H, Hatano M, Tokuhisa T: A physiologic role of Bcl-xL induced in activated macrophages. J Immunol 160: 2590–2596, 1998
12. Brockhaus F, Brüne B: U937 apoptotic cell death by nitric oxide: Bcl-2 downregulation and caspase activation. Exp Cell Res 238: 33–41, 1998
13. May P, May E: Twenty years of p53 research: Structural and functional aspects of the p53 protein. Oncogene 18: 7621–7636, 1999
14. Agarwal ML, Taylor WR, Chernov MV, Chernova OB, Stark GR: The p53 network. J Biol Chem 273: 1–4, 1998
15. Levine AJ: p53, the cellular gatekeeper for growth and division. Cell 88: 323–331, 1997
16. Watson RW, Redmond HP, Bouchier-Hayes D: Role of endotoxin in mononuclear phagocyte-mediated inflammatory responses. J Leukoc Biol 56: 95–103, 1994
17. Lander HM: An essential role for free radicals and derived species in signal transduction. FASEB J 11: 118–124, 1997
18. James PE, Grinberg OY, Swartz HM: Superoxide production by phagocytosing macrophages in relation to the intracellular distribution of oxygen. J Leukoc Biol 64: 78–84, 1998
19. Herschman HR, Reddy ST, Xie W: Function and regulation of prostaglandin synthase-2. Adv Exp Med Biol 407: 61–66, 1997
20. Hempel SL, Monick MM, He B, Yano T, Hunninghake GW: Synthesis of prostaglandin H synthase-2 by human alveolar macrophages in response to lipopolysaccharide is inhibited by decreased cell oxidant tone. J Biol Chem 269: 32979–32984, 1994
21. Lee SH, Soyoola E, Chanmugam P, Hart S, Sun W, Zhong H *et al.*: Selective expression of mitogen-inducible cyclooxygenase in macrophages stimulated with lipopolysaccharide. J Biol Chem 267: 25934–25938, 1992
22. O'Sullivan MG, Huggins EMJ, Meade EA, DeWitt DL, McCall CE: Lipopolysaccharide induces prostaglandin H synthase-2 in alveolar macrophages. Biochem Biophys Res Commun 187: 1123–1127, 1992
23. Williams CS, Mann M, DuBois RN: The role of cyclooxygenases in inflammation, cancer, and development. Oncogene 18: 7908–7916, 1999
24. Marnett LJ, Rowlinson SW, Goodwin DC, Kalgutkar AS, Lanzo CA: Arachidonic acid oxygenation by COX-1 and COX-2. Mechanisms of catalysis and inhibition. J Biol Chem 274: 22903–22906, 1999
25. Smith WL, Garavito RM, DeWitt DL: Prostaglandin endoperoxide H synthases (cyclooxygenases)-1 and -2. J Biol Chem 271: 33157–33160, 1996
26. Coleman RA, Smith WL, Narumiya S: International Union of Pharmacology classification of prostanoid receptors: properties, distribution, and structure of the receptors and their subtypes. Pharmacol Rev 46: 205–229, 1994
27. Brüne B, von Knethen A, Sandau KB: Nitric oxide (NO): An effector of apoptosis. Cell Death Diff 6: 969–975, 1999
28. von Knethen A, Brüne B: Cyclooxygenase-2: an essential regulator of NO-mediated apoptosis. FASEB J 11: 887–895, 1997
29. von Knethen A, Callsen D, Brüne B: NF-kappaB and AP-1 activation by nitric oxide attenuated apoptotic cell death in RAW 264.7 macrophages. Mol Biol Cell 10: 361–372, 1999
30. von Knethen A, Callsen D, Brüne B: Superoxide attenuates macrophage apoptosis by NF-kappaB and AP-1 activation that promotes cyclooxygenase-2 expression. J Immunol 163: 2858–2866, 1999
31. von Knethen A, Brockhaus F, Kleiter I, Brüne B: NO-Evoked macrophage apoptosis is attenuated by cAMP-induced gene expression. Mol Med 5: 672–684, 1999

32. von Knethen A, Lotero A, Brüne B: Etoposide and cisplatin induced apoptosis in activated RAW 264.7 macrophages is attenuated by cAMP-induced gene expression. Oncogene 17: 387–394, 1998
33. McConkey DJ, Nicotera P, Hartzell P, Bellomo G, Wyllie AH, Orrenius S: Glucorticoids activate a suicide process in thymocytes through an elevation of cytosolic Ca^{2+} concentration. Arch Biochem Biophys 269: 365–370, 1989
34. Burton K: A study of the conditions and mechanisms of the diphenylamine reaction for the estimation of deoxyribonucleic acid. Biochem J 62: 315–323, 1956
35. Schoonbroodt S, Legrand Poels S, Best Belpomme M, Piette J: Activation of the NF-kappaB transcription factor in a T-lymphocytic cell line by hypochlorous acid. Biochem J 321: 777–785, 1997
36. Camandola S, Leonarduzzi G, Musso T, Varesio L, Carini R, Scavazza A *et al.*: Nuclear factor kappa B is activated by arachidonic acid but not by eicosapentaenoic acid. Biochem Biophys Res Commun 229: 643–647, 1996
37. Sassone-Corsi P, Visvader J, Ferland L, Mellon PL, Verma IM: Induction of proto-oncogene fos transcription through the adenylate cyclase pathway: characterization of a cAMP-responsive element. Genes Dev 2: 1529–1538, 1988
38. Hart TW: Some observations concerning the S-nitroso and S-phenylsulphonyl derivates of L-cysteine and glutathione. Tetrahedron Lett 26: 2013–2016, 1985
39. Eberhart CE, Coffey RJ, Radhika A, Giardiello FM, Ferrenbach S, DuBois RN: Up-regulation of cyclooxygenase 2 gene expression in human colorectal adenomas and adenocarcinomas. Gastroenterology 107: 1183–1188, 1994
40. Kargman SL, O'Neill GP, Vickers PJ, Evans JF, Mancini JA, Jothy S: Expression of prostaglandin G/H synthase-1 and -2 protein in human colon cancer. Cancer Res 55: 2556–2559, 1995
41. Sano H, Kawahito Y, Wilder RL, Hashiramoto A, Mukai S, Asai K *et al.*: Expression of cyclooxygenase-1 and -2 in human colorectal cancer. Cancer Res 55: 3785–3789, 1995
42. Levy GN: Prostaglandin H synthases, nonsteroidal antiinflammatory drugs, and colon cancer. FASEB J 11: 234–247, 1997
43. Oshima M, Dinchuk JE, Kargman SL, Oshima H, Hancock B, Kwong E *et al.*: Suppression of intestinal polyposis in Apc delta716 knockout mice by inhibition of cyclooxygenase 2 (COX-2). Cell 87: 803–809, 1996
44. Sheng H, Shao J, Kirkland SC, Isakson P, Coffey RJ, Morrow J *et al.*: Inhibition of human colon cancer cell growth by selective inhibition of cyclooxygenase-2. J Clin Invest 99: 2254–2259, 1997
45. Lo CJ, Fu M, Lo FR, Cryer HG: Cyclooxygenase 2 (COX-2) gene activation is regulated by cyclic adenosine monophosphate. Shock 13: 41–45, 2000
46. Zhang F, Warskulat U, Wettstein M, Schreiber R, Henninger HP, Decker K *et al.*: Hyperosmolarity stimulates prostaglandin synthesis and cyclooxygenase-2 expression in activated rat liver macrophages. Biochem J 312: 135–143, 1995
47. Vial D, Arbibe L, Havet N, Dumarey C, Vargaftig B, Touqui L: Down-regulation by prostaglandins of type-II phospholipase A2 expression in guinea-pig alveolar macrophages: A possible involvement of cAMP. Biochem J 330: 89–94, 1998
48. Kawase T, Orikasa M, Suzuki A: Phorbol ester-like action of staurosporine on the cAMP response to prostaglandin E2 in two macrophage-like cell lines at distinct differentiation stages. Cell Signal 4: 479–485, 1992
49. Fladmark KE, Gjertsen BT, Doskeland SO, Vintermyr OK: Fas/APO-1(CD95)-induced apoptosis of primary hepatocytes is inhibited by cAMP. Biochem Biophys Res Commun 232: 20–25, 1997
50. Parvathenani LK, Buescher ES, Chacon-Cruz E, Beebe SJ: Type I cAMP-dependent protein kinase delays apoptosis in human neutrophils at a site upstream of caspase-3. J Biol Chem 273: 6736–6743, 1998
51. Riccio A, Ahn S, Davenport CM, Blendy JA, Ginty DD: Mediation by a CREB family transcription factor of NGF-dependent survival of sympathetic neurons. Science 286: 2358–2361, 1999
52. Torgersen KM, Vaage JT, Levy FO, Hansson V, Rolstad B, Tasken K: Selective activation of cAMP-dependent protein kinase type I inhibits rat natural killer cell cytotoxicity. J Biol Chem 272: 5495–5500, 1997
53. Walker BA, Rocchini C, Boone RH, Ip S, Jacobson MA: Adenosine A2a receptor activation delays apoptosis in human neutrophils. J Immunol 158: 2926–2931, 1997
54. Stefanelli C, Stanic I, Bonavita F, Flamigni F, Pignatti C, Guarnieri C *et al.*: Inhibition of glucocorticoid-induced apoptosis with 5-aminoimidazole-4-carboxamide ribonucleoside, a cell-permeable activator of AMP-activated protein kinase. Biochem Biophys Res Commun 243: 821–826, 1998
55. Nakajima T, Uchida C, Anderson SF, Parvin JD, Montminy M: Analysis of a cAMP-responsive activator reveals a two-component mechansim for transcriptional induction via signal-dependent factors. Genes Dev 11: 738–747, 1997
56. Arias J, Alberts AS, Brindle P, Claret FX, Smeal T, Karin M *et al.*: Activation of cAMP and mitogen responsive genes relies on a common nuclear factor. Nature 370: 226–229, 1994
57. Scheid MP, Foltz IN, Young PR, Schrader JW, Duronio V: Ceramide and cyclic adenosine monophosphate (cAMP) induce cAMP response element binding protein phosphorylation via distinct signaling pathways while having opposite effects on myeloid cell survival. Blood 93: 217–225, 1999
58. Yang YM, Dolan LR, Ronai Z: Expression of dominant negative CREB reduces resistance to radiation of human melanoma cells. Oncogene 12: 2223–2233, 1996
59. Messmer UK, Reed UK, Brüne B: Bcl-2 protects macrophages from nitric oxide-induced apoptosis. J Biol Chem 271: 20192–20197, 1996
60. Brockhaus F, Brüne B: p53 accumulation in apoptotic macrophages is an energy demanding process that precedes cytochrome c release in response to nitric oxide. Oncogene 18: 6403–6410, 1999
61. Du K, Montminy M: CREB is a regulatory target for the protein kinase Akt/PKB. J Biol Chem 273: 32377–32379, 1998
62. Jean D, Harbison M, McConkey DJ, Ronai Z, Bar-Eli M: CREB and its associated proteins act as survival factors for human melanoma cells. J Biol Chem 273: 24884–24890, 1998

Molecular and Cellular Biochemistry **212**: 45–50, 2000.

Catecholamines induce IL-10 release in patients suffering from acute myocardial infarction by transactivating its promoter in monocytic but not in T-cells

Ulrike Riese,[1] Susanne Brenner,[1] Wolf-Dietrich Döcke,[2] Susanna Prösch,[3] Petra Reinke,[4] Michael Oppert,[4] Hans-Dieter Volk[2] and Cornelia Platzer[1]

[1]*Institute of Anatomy II, Medical School, Friedrich Schiller University, Jena;* [2]*Institute of Medical Immunology;* [3]*Institute of Virology;* [4]*Department of Nephrology and Internal Intensive Care Medicine, Medical School Charité, Humboldt University, Berlin, Germany*

Abstract

The anti-inflammatory cytokine IL-10 is up-regulated in response to TNF-α suggesting a control mechanism of inflammation. In addition, we recently found systemic IL-10 release in response to acute stress reactions in the absence of any systemic inflammation. *In vitro* and *in vivo* studies in experimental models suggest that catecholamines induce IL-10 release via a cyclic adenosine monophosphate/protein kinase A (cAMP/PKA) dependent pathway. Here we studied patients for plasma IL-10 after acute myocardial infarction, a very stressful event without significant signs of systemic inflammation. In fact, the activation of the sympathetic system initiated by cardiac infarction was accompanied by a temporary systemic release of IL-10. Catecholamine induced IL-10 may be released by different cells. Recently, we demonstrated that catecholamines directly stimulate the IL-10 promoter/enhancer via a cAMP/PKA pathway in monocytic cells. A cAMP responsive element (CRE) was identified as major target. Here we show that there is no influence of catecholamines on the IL-10 promoter activity in T-cells. In contrast to monocytic cells, in T-cells cAMP-induced PKA-dependent phosphorylation of the CRE-binding protein 1 (CREB-1) seems to play a marginal role in IL-10 induction, which was reflected by a low cAMP-dependent IL-10-promoter/enhancer stimulation in reporter gene assays. Thus, catecholamines are directly involved in the regulation of IL-10 expression in monocytic but not in T-cells after acute stressful conditions. (Mol Cell Biochem **212**: 45–50, 2000)

Key words: IL-10, IL-10 promoter, cAMP, catecholamines, myocardial infarction, T-cells

Introduction

Pro-inflammatory cytokines (IL-1β, IL-6, IL-8, TNF-α) play an important role in the antimicrobial defense, in wound healing and other diseases regulating not only inflammation but also cellular adhesion, production of growth factors and vasoactive substances. In acute myocardial infarction (AMI) an elevated production of pro-inflammatory cytokines by myocytes, inflammatory mononuclear cells infiltrating the border zone and endothelial cells was found, which represents a part of an intrinsic myocardial stress response system to tissue injury and may contribute not only to wound healing but also to secondary injury [1–6]. A systemic increase of the anti-inflammatory cytokines IL-1 receptor antagonist and

Address for offprints: C. Platzer, Institute of Anatomy II, Friedrich Schiller University Jena, Teichgraben 7, 07743 Jena, Germany

IL-10 has been described as well, whereas their peak levels were closely correlated and were higher in severe AMI and in non-survivors [7]. Studies in cardiopulmonary bypass surgery, however, demonstrate that the myocardium is not the source of systemic IL-10 increase due to cardiac ischemia/reperfusion [8].

In addition to direct feedback loops within the immune system, the neuroendocrine system including the activation of the stress axis, plays an important control function. Neurohumoral changes during myocardial infarction, among others, consist of a significant increase in plasma catecholamines, which play a central role in the pathogenesis and/or progression of the disease. Moreover, plasma norepinephrine levels have been found to predict poor prognosis in heart failure patients [9, 10].

Whereas the hypothalamo-pituitary-adrenal (HPA) axis and the immunosuppressive action of corticosteroids are well investigated, the mechanism of catecholamine-related immune alterations have not yet been fully elucidated. Recently, we have demonstrated that catecholamines induce *de novo* IL-10 synthesis by monocytes *in vitro* and are responsible for the rapid systemic IL-10 release after sympathetic activation *in vivo*, thus contributing to immunodepression after brain injury and neurosurgery in the absence of systemic inflammation [11]. Moreover, we showed that catecholamines directly stimulate the IL-10 promoter in monocytic cells via the β2-adrenoreceptor cAMP-dependent signaling pathway and CRE4, playing the major role in mediating the stimulatory effect of cAMP [12, 13].

Here, in a prospective clinical study we detected simultaneously increased plasma levels of epinephrine, norepinephrine and IL-10 in patients with AMI during the first two days after admission to the emergency unit. By comparing IL-10 promoter activity in THP-1 and Jurkat cells after epinephrine- and norepinephrine-stimulation, we demonstrate that in contrast to monocytic cells, T-cells do not contribute to the catecholamine induced systemic IL-10 release. In addition, cAMP-dependent CREB-1-phosphorylation seems to play a marginal role in IL-10 induction in T-cells.

Materials and methods

Patients, catecholamine and IL-10 determination

Ten patients (7 males, 3 females, mean age 60 years, range 46–75 years) without known infectious or inflammatory disorders, with the clinical diagnosis of acute myocardial infarction and typical electrocardiogram changes for an anterior (n = 3) or posterior (n = 7) infarct were included in the study. The diagnosis was confirmed with serum analysis of cardiac enzymes and coronary angiograms. Four patients underwent systemic thrombolysis, three patients received a coronary angioplasty, one patient needed a percutaneous transluminal coronary angioplasty after non-successful thrombolysis. Three patients received catecholamine infusions because of cardiogenic shock. Plasma catecholamine concentrations were determined using high-performance liquid chromatography as described by Ganhao *et al.* [14]. IL-10 was estimated by the UltraSensitiv-ELISA from Biosource (Nivelles, Belgium). The study was approved by the institutional Review Board and informed consent was obtained from all patients.

Cells, plasmids, transient transfection and reporter gene assay

THP-1 and Jurkat cells were maintained in RPMI 1640 with 10% FCS certified for low endotoxin and free of mycoplasma (Biochrom, Berlin, Germany). Transfections with plasmid pGL2-1308, containing the human IL-10 promoter/enhancer fragment (GeneBank accession No. Z30175) upstream of the luciferase gene and luciferase were carried out basically as described [12]. Using the following modified electrophoration conditions 1200 μF, 100 ohm, 300 V for THP-1 cells and 350 V for Jurkat cells we achieved comparable transfection efficiencies of 21% (SD 7%) for THP-1 and 17% (SD 4%) for Jurkat cells, as estimated by co-transfection with GFP-expression plasmid. Transfected cells were stimulated with N6,2-O-dibutyryl-cAMP (dbcAMP), epinephrine (Epinephrine-Hydrochloride) or norepinephrine (trade name 'Arterenol', all from Sigma-Aldrich Chemie, Deisenhofen, Germany), harvested and lysed after 24 h [12]. The stimulation was expressed as the ratio of stimulated vs. basal luciferase activity obtained in the same experimental setting.

Oligonucleotide-based affinity chromatography and Western blotting

Whole-cell extracts of 1×10^7 untreated, epinephrine- or dbcAMP-treated THP-1 and Jurkat cells were prepared as described [15]. The cell extracts were incubated with 100 μg of M280-Streptavidin Dynabeads (Dynal, Oslo, Norway) carrying multimers of oligonucleotides derived from the IL-10 promoter and containing CRE4 (–476 to –320). The following primers flanked by *BamHI*-sites were used to obtain ds-oligonucleotide fragments by PCR: 5′-CGGGATCCTTCATCCCAGGTTGG-3′ and 5′-CGGGATCCAGTTGATTTCCTGGG-3′. The PCR-fragments were ligated via *BamHI*-sites, biotinylated and coupled to the beads according to the manufacturer's recommendations. After incubation of cell extracts with the beads for 1 h at room temperature and three washes with low salt buffer (50 mM sodium acetate pH 7.5; 50 mM NaCl), bound proteins were eluted with high salt buffer (50 mM sodium acetate pH 7.5; 1 M NaCl) for

30 min at room temperature. Sodium dodecyl sulphate polyacrylamide gel electrophoresis (SDS-PAGE) and Western Blot analyses were performed with eluates prepared from whole-cell extracts containing 1 mg protein or directly with whole-cell extracts (100 μg protein per lane). For the detection of phospho-CREB-1 the Enhanced chemiluminescence kit (Pierce, Rockford, IL, USA), the anti-phospho-CREB-1 antibody from Upstate Biotechnology (Lake Placid, NY, USA) and a peroxidase-conjugated anti-rabbit antibody (Pierce) were used. Pre-stained molecular weight standard to visually monitor electrophoresis and transfer was derived from Sigma-Aldrich Chemie (Deisenhofen, Germany), the precise molecular weight of phospho-CREB-1 was verified with a biotinylated standard from Amersham Pharmacia (Freiburg, Germany).

Results

In a prospective clinical study, 10 patients with acute myocardial infarction were monitored for plasma epinephrine and norepinephrine levels as well as for IL-10 from intensive care unit-admission up to day 14 after the acute event. Catecholamine plasma concentrations were enhanced at admission and normalized until day 4. Simultaneously, IL-10 plasma levels were increased up to 6-fold and dropped to normal levels until day 7 to 14 (Fig. 1).

Based on the fact that monocytes/macrophages may be an important cellular source of plasma IL-10 *in vivo*, we recently demonstrated the direct activation of the IL-10 promoter by epinephrine and norepinephrine in a concentration-dependent manner in THP-1 cells. We showed that catecholamines mediated their effects on IL-10 induction in monocytic cells via β2-adrenoreceptors and the cAMP pathway [13]. We examined now whether T-cells contribute to the catecholamine-induced IL-10 release. Therefore, we carried out transient transfection and subsequent luciferase reporter gene assays using the native IL-10 promoter/enhancer plasmid (pGL2-1308) with the T-cell line Jurkat in comparison to monocytic THP-1 cells. In contrast to THP-1 cells, where we achieved a 10 fold and 5.5 fold IL-10 promoter stimulation by epinephrine and norepinephrine, respectively, in Jurkat cells we did not observe any significant activation after 24 h (Figs 2A and 2B), nor at any other time point between 2 and 48 h (not shown).

To evaluate the general relevance of the PKA pathway for IL-10 promoter activation in T-cells we compared the stimulatory capacity of dbcAMP in THP-1 and in Jurkat cells (Fig. 2C). Whereas in THP-1 cells we obtained a 29 fold average promoter stimulation with 1 mM dbcAMP, we found only marginal stimulation rates in Jurkat cells (up to 3.5 fold). Similar results with epinephrine-, norepinephrine- and dbcAMP-stimulation were obtained with a second T-cell line (CCRF-CEM, not shown).

The inactivation of the CRE4 site, located in the enhancer upstream of the IL-10 promoter, causes a significantly decreased promoter responsiveness to catecholamines and dbcAMP. In electrophoretic mobility shift and supershift

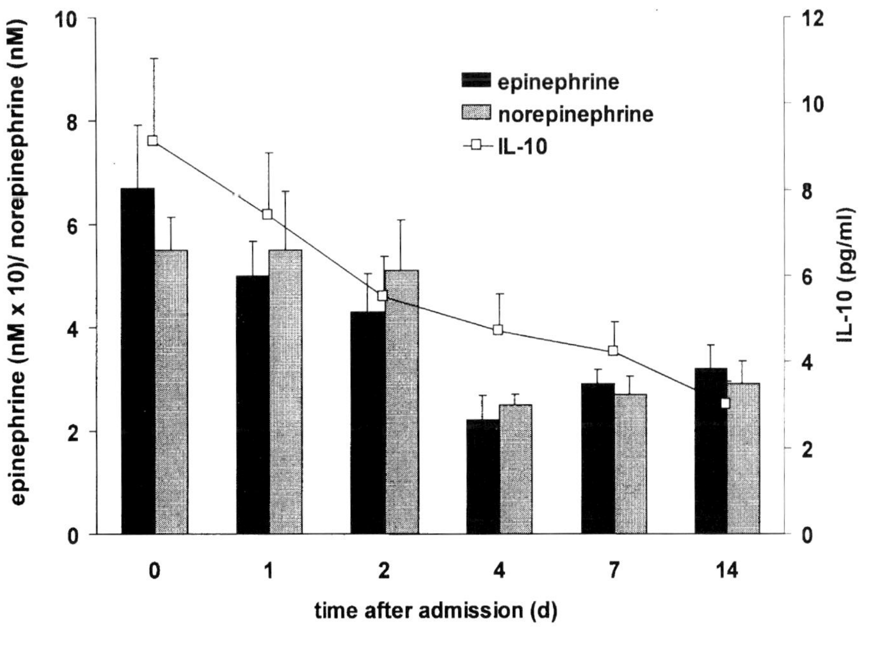

Fig. 1. Mean values and S.E.M. of plasma epinephrine, norepinephrine and IL-10 levels from 10 patients suffering from acute myocardial infarction at different time points after admission. Normal levels for epinephrine, norepinephrine and IL-10 were ~ 0.5 nM, ~ 1.5 nM and < 3 pg/ml, respectively.

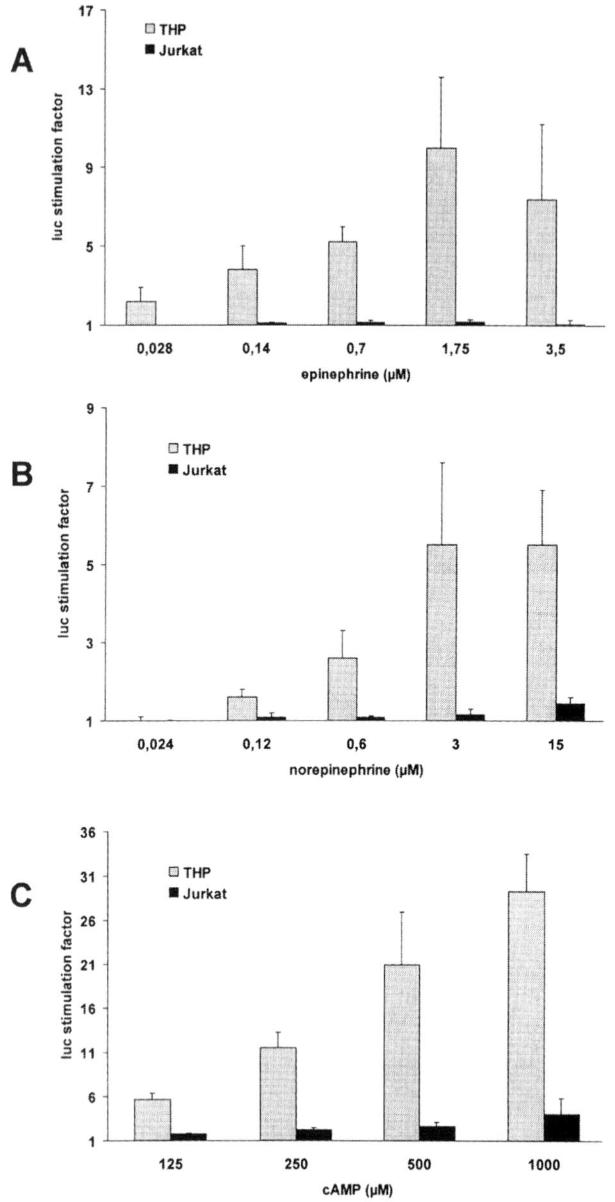

Fig. 2. Dose-dependent effect of (A) epinephrine, (B) norepinephrine and (C) dbcAMP on IL-10 promoter activity in pGL2-1308 (WT) transfected THP-1 and Jurkat cells. Mean values and S.D. from 4–30 experiments are shown.

assays we showed an enhanced binding of CREB-1 and activating transcription factor-1 (ATF-1) to CRE4 in nuclear extracts from catecholamine and dbcAMP treated THP-1 cells compared to control extracts [12, 13]. These data emphasize the functional importance of CRE4 for epinephrine/norepinephrine and cAMP-mediated effects on IL-10 promoter activation. Therefore, we were interested in the binding activity of CRE4 in protein extracts from Jurkat cells. By an oligonucleotide-based affinity chromatography method we showed an enhanced quantity of phosphorylated CREB-1 bound to CRE4 from THP-1 extracts after epinephrine- and dbcAMP-stimulation compared to unstimulated extracts (Fig. 3B, lanes 3, 4 and 2, respectively). In contrast, with extracts from Jurkat cells we could not detect any binding of phosphorylated CREB-1, even after stimulation with epinephrine and dbcAMP (Fig. 3A, lanes 2 and 3). If the extracts from both cell lines were compared by western blotting, we found a significant lower content of phosphorylated CREB-1 in stimulated Jurkat than in THP-1 cells (Fig. 3A, lanes 4–6 and Fig. 3B, lane 1, unstimulated THP-1 extract not shown), which may account for the lack of binding of this transcription factor to CRE4 and consequently, the failed detection by oligonucleotide-based affinity chromatography. Beside detection of p43-CREB-1 Western blots with whole-cell extracts and anti-Phospho-CREB-1 (anti-P-CREB-1) antibody revealed cross-reaction with some other proteins migrating at 35–38 kDa. They apparently belong to the CREB/ATF family of transcription factors (e.g. p38 ATF-1) and seem to be more abundant in Jurkat than in THP-1 cells.

Discussion

Based on the observation that highly stressful events, that are known to be associated with an increased catecholamine release, may also lead to an elevation of systemic IL-10 levels without preceding inflammation [7, 11, 16], we investigated plasma catecholamine and IL-10 levels in a prospective clinical study on 10 patients with AMI. We found markedly enhanced norepinephrine and IL-10, as well as slightly enhanced epinephrine plasma levels during the first two days, which normalized between day 4 and 14. The range of estimated concentrations (and the respective standard deviations) were very high because of the short half-life of plasma catecholamines and IL-10; and of the different time intervals between the acute event and the delivery to the emergency unit, as well.

It is well recognized that the immune response is influenced by a variety of neural and neuroendocrine mechanisms. The relationship between sympathetic nervous system activation and the regulatory effect of catecholamines on cytokine release from monocytes/macrophages is a well studied phenomenon [17–20]. There are various effects of catecholamines on T-cells. Catecholamines act on thymocyte differentiation and proliferation by influencing the thymic microenvironment and stimulating thymic epithelial cells to produce IL-6 [21, 22]. They enhance the mobilization, modulate the migration and homing of T-cells by altering adhesion molecule expression [23, 24]. Catecholamines trigger apoptosis of lymphocytes [25].

In the present study we asked, whether epinephrine and norepinephrine are capable of stimulating IL-10 promoter/enhancer in T-cells directly, as we could verify for monocytic cells recently. In contrast to monocytic cells, we did not ob-

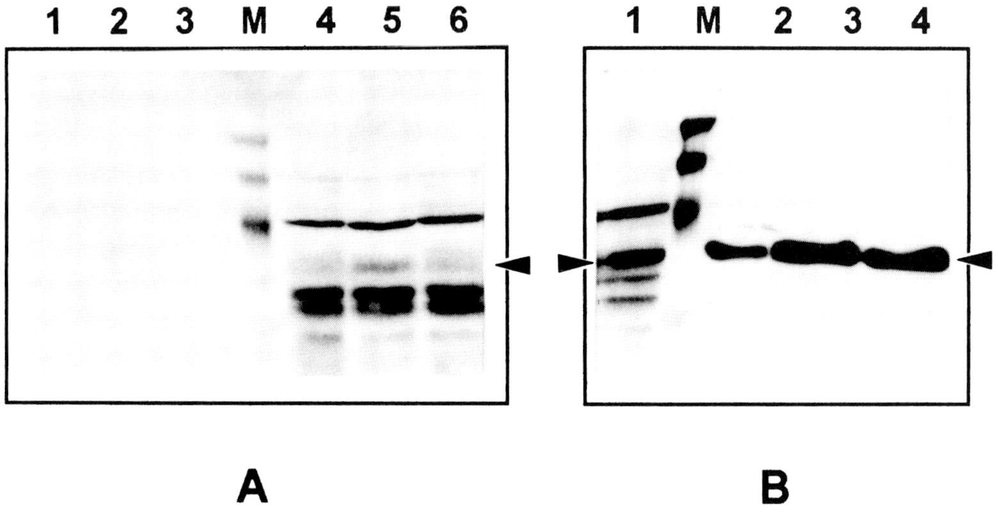

Fig. 3. (A) Western blot detection of phospho-CREB-1 in eluats from oligo-affinity chromatography with CRE4 containing fragments (lanes 1–3) and in whole cell extracts (lanes 4–6) from unstimulated (lanes 1 and 4), epinephrine- (lanes 2 and 5) and cAMP- (lanes 3 and 6) stimulated Jurkat cells. (B) Phospho-CREB-1 detection in whole-cell extracts from cAMP stimulated THP-1 cells (lane 1) and in CRE4-eluats from unstimulated (lane 2), epinephrine- (lane 3) and cAMP- (lane 4) stimulated THP-1 cells. Phospho-CREB-1 bands (43 kDa) are indicated by arrows. M-pre-stained molecular weight standard with the approximate MW 116, 88,5 and 64 kDa, as given by the manufacturer. The precise molecular weight of phospho-CREB-1 was estimated in an independent experiment with a biotinylated standard.

serve any promoter stimulation by catecholamines and only marginal stimulation by dbcAMP in T-cells using luciferase reporter gene assays. Moreover, we could not detect any binding of phosphorylated CREB-1 to CRE4 with extracts from epinephrine- and dbcAMP-stimulated Jurkat cells by an oligonucleotide-based affinity chromatography method. The significant lower content of phosphorylated CREB-1 in stimulated Jurkat comparing to THP-1 cells appears in the context with the low and failed stimulation by dbcAMP and catecholamines, respectively. Other proteins cross-reacting with the anti-P-CREB-1 antibody probably belong to the CREB/ATF family of transcription factors (ATF-1 p38, ATF-2 p65) and seem to be more abundant in Jurkat than in THP-1 cells.

In monocytic cells epinephrine and norepinephrine transactivate the IL-10 promoter via β2-adrenoreceptors and the cAMP pathway involving PKA and the enhanced binding of CREB-1/ATF-1 to the CRE4 enhancer element [13]. Lymphocytes also bear β-adrenoreceptors [25], but apparently use different mechanisms for CREB activation compared with monocytic cells. CREB phosphorylation, which is a key event in the expression of genes during T-cell activation, is realized in lymphoid cells by at least two different signal transduction pathways involving mitogen-activated protein kinase (MAPK)/p38/stress-activated protein kinase 2 (SAPK2) and CaMKIV, a member of the Ca^{2+}-calmodulin-dependent protein kinase cascade [26–28]. The Ca^{2+}-calmodulin-dependent protein kinase cascade responds to elevation of intracellular Ca^{2+} levels and is particularly abundant in T-cells. There is an intensive cross-talk between the members of this and other signaling cascades, including those that involve PKA and MAP kinases. Many data are published regarding the role of PKA, negatively regulating T-cell activation [29–31]. On the other hand, after preactivation of peripheral blood mononuclear cells by Ca^{2+} ionophore and PMA (4β-phorbol 12-myristate 13-acetate), IL-10 expression in CD8+ cells was strongly enhanced in the presence of cAMP-elevating substances, thus suggesting a prominent role for PKA in the switch of these cells to a Th2 like profile [32]. In lymphoid Jurkat cells we obtained a strong IL-10 promoter stimulation with PMA/ionomycin, which could be moderately enhanced in the presence of dbcAMP (data not shown). Whether PMA/ionomycin stimulation results in an increased activation and binding of CREB-1 to the CRE4 on the IL-10 promoter, remains to be investigated. Preliminary data suggest the relevance of C/EBP (CAAT-enhancer binding protein) for cAMP-induced IL-10 expression. C/EBP is an important and abundant transcriptional regulator of genes induced in activated monocytes/macrophages, but not in T-cells [33].

Taken together, our results support the idea of a close cross talk between the neuroendocrine and the immune systems and emphasize the immunosuppressive potential of catecholamines endogenously produced or administered as a component of therapeutic treatment, mainly on monocytes. The effects of the second component of stress response via the HPA axis and glucocorticoids on IL-10 promoter activation in monocytes and lymphocytes will be addressed in future experiments.

Acknowledgements

This work was supported by a grant from the Deutsche Forschungsgemeinschaft to C.P. Pl 163/4-2. U.R. and S.B. contributed equally to this work.

References

1. Pudil R, Pidrman V, Krejsek J, Gregor J, Tichy M, Andrys C, Drahosova M: Cytokines and adhesion molecules in the course of acute myocardial infarction. Clin Chim Acta 280: 127–134, 1999
2. Kato K, Matsubara T, Iida K, Suzuki O, Sato Y: Elevated levels of pro-inflammatory cytokines in coronary artery thrombi. Int J Cardiol 70: 267–273, 1999
3. Irwin MW, Mak S, Mann DL, Qu R, Penninger JM, Yan A, Dawood F, Wen WH, Shou Z, Liu P: Tissue expression and immunolocalization of tumor necrosis factor-alpha in postinfarction dysfunctional myocardium. Circulation 99: 1492–1498, 1999
4. Gwechenberger M, Mendoza LH, Youker KA, Frangogiannis NG, Smith CW, Michael LH, Entman ML: Cardiac myocytes produce interleukin-6 in culture and in viable border zone of reperfused infarctions. Circulation 99: 546–551, 1999
5. Marx N, Neumann FJ, Ott I, Gawaz M, Koch W, Pinkau T, Schomig A: Induction of cytokine expression in leukocytes in acute myocardial infarction. J Am Coll Cardiol 30: 165–170, 1997
6. Wan S, DeSmet JM, Barvais L, Goldstein M, Vincent JL, LeClerc JL: Myocardium is a major source of pro-inflammatory cytokines in patients undergoing cardiopulmonary bypass. J Thorac Cardiovasc Surg 112: 806–811, 1996
7. Shibata M, Endo S, Inada K, Kuriki S, Harada M, Takino T, Sato N, Arakawa N, Suzuki T, Aoki H, Suzuki T, Hiramori K: Elevated plasma levels of interleukin-1 receptor antagonist and interleukin-10 in patients with acute myocardial infarction. J Interferon Cytokine Res 17: 145–150, 1997
8. Wan S, LeClerc JL, Schmartz D, Barvais L, Huynh CH, Deviere J, DeSmet JM, Vincent JL: Hepatic release of interleukin-10 during cardiopulmonary bypass in steroid-pretreated patient. Am Heart J 133: 335–339, 1997
9. Anker SD: Catecholamine levels and treatment in chronic heart failure. Eur Heart J 19: F56–F61, 1998
10. Packer M: Beta-blockade in heart failure. Basic concepts and clinical results. Am J Hypertens 11: 23S–37S, 1998
11. Woiciechowsky C, Asadullah K, Nestler D, Eberhardt B, Platzer C, Schöning B, Glöckner F, Lanksch WR, Volk HD, Döcke WD: Sympathetic activation triggers systemic IL-10 release in immunodepression induced by brain injury. Nat Med 4: 808–813, 1998
12. Platzer C, Fritsch E, Elsner T, Lehmann MH, Volk HD, Prösch S: Cyclic adenosine monophosphate responsive elements are involved in the transcriptional activation of the human interleukin-10 gene in monocytic cells. Eur J Immunol 29: 3098–3104, 1999
13. Platzer C, Döcke WD, Volk HD, Prösch S: Catecholamines trigger IL-10 release in acute systemic stress reaction by direct stimulation of its promotor/enhancer activity in monocytic cells. J Neuroimmunol: 2000 (in press)
14. Ganhao MF, Hattingh J, Hurwitz ML, Pitts NI: Evaluation of a simple plasma catecholamine extraction procedure prior to high-performance liquid chromatography and electrochemical detection. J Chromatogr 564: 55–66, 1991
15. Lischke A, Moriggl R, Brändlein S, Berchthold S. Kammer W, Sebald W, Groners B, Liu X, Henninghausen L, Friedrich K: The interleukin-4 receptor activates STAT5 by a mechanism that relies upon common g-chain. J Biol Chem 273: 31222–31229, 1998
16. Döcke WD, Felix S, Baumann G, Volk HD: Rapid IL-10 release and signs for immunodepression after acute myocardial infarction. Crit Care Med 25: (abstr) A50, 1997
17. Suberville S, Bellocq A, Fouqueray B, Philippe C, Lantz O, Perez J, Baud: Regulation of interleukin-10 production by beta-adrenergic agonists. Eur J Immunol 26: 2601–2605, 1996
18. Van der Poll T, Coyle SM, Barbosa K, Braxton CC, Lowry SF: Epinephrine inhibits tumor necrosis factor-a and potentiates interleukin 10 production during human endotoxemia. J Clin Invest 97: 713–719, 1996
19. Kavelaars A, van de Pol M, Zijlstra J, Heijnen CJ: Beta 2-adrenergic activation enhances interleukin-8 production by human monocytes. J Neuroimmunol 77: 211–216, 1997
20. Siegmund B, Eigle A, Hartmann G, Hacker U, Endres S: Adrenaline enhances LPS-induced IL-10 synthesis: Evidence for protein kinase A-mediated pathway. Int J Immunopharmacol 20: 57–69, 1998
21. von Patay B, Kurz B, Mentlein R: Effect of transmitters and co-transmitters of the sympathetic nervous system on interleukin-6 synthesis in thymic epithelial cells. Neuroimmunomodulation 6: 45–50, 1999
22. Kurz B, Feindt J, von Gaudecker B, Kranz A, Loppnow H, Mentlein R: β-adrenoceptor-mediated effects in rat cultured thymic epithelial cells. Br J Pharmacol 120: 1401–8, 1997
23. Mills PJ, Karnik RS, Dillon E: L-selectin expression affects T-cell circulation following isoproterenol infusion in humans. Brain Behav Immun 11: 333–342, 1997
24. Carlson SL, Fox S, Abell KM: Catecholamine modulation of lymphocyte homing to lymphoid tissues. Brain Behav Immun 11: 307–320, 1997
25. Bergquist J, Tarkowski A, Ewing A, Ekman R: Catecholaminergic suppression of immunocompetent cells. Immunol Today 19: 562–567, 1998
26. Rolli M, Kotlyarov A, Sakamoto KM, Gaestel M, Neininger A: Stress-induced stimulation of early growth response gene-1 by p38/stress-activated protein kinase 2 is mediated by a cAMP-responsive promoter element in a MAPKAP kinase 2-independent manner. J Biol Chem 274: 19559–19564, 1999
27. Hsueh YP, Liang HE, Ng SY, Lai MZ: CD28-costimulation activates cyclic AMP-responsive element-binding protein in T lymphocytes. J Immunol 158: 85–93, 1997
28. Soderling TR: The Ca-calmodulin-dependent protein kinase cascade. Trends Biochem Sci 24: 232–236, 1999
29. Li L, Yee C, Beavo JA: CD3- and CD28-dependent induction of PDE7 required for T-cell activation. Science 283: 848–851, 1999
30. Wayman GA, Tokumitsu H, Soderling TR: Inhibitory cross-talk by cAMP kinase on the calmodulin-dependent protein kinase cascade. J Biol Chem 272: 16073–16076, 1997
31. Aandahl EM, Aukrust P, Skalhegg BS, Muller F, Froland SS, Hansson V, Tasken K: Protein kinase A type I antagonist restores immune responses of T-cells from HIV-infected patients. FASEB J 12: 855–862, 1998
32. Shin HC, Benbernou N, Fekkar H, Esnault S, Guenounou M: Regulation of IL-17, IFN-gamma and IL-10 in human CD8(+) T-cells by cyclic AMP-dependent signal transduction pathway. Cytokine 10: 841–850, 1998
33. Berrier A, Siu G, Calame K: Transcription of a minimal promoter from the NF-IL6 gene is regulated by CREB/ATF and SP1 proteins in U937 promonocytic cells. J Immunol 161: 2267–2275, 1998

Molecular and Cellular Biochemistry **212**: 51–60, 2000.

Regulation of tyrosine hydroxylase gene transcription by the cAMP-signaling pathway: Involvement of multiple transcription factors

Jinkyu Lim, Chunying Yang, Seok Jong Hong and Kwang-Soo Kim
Molecular Neurobiology Laboratory, McLean Hospital, Harvard Medical School, Belmont, MA, USA

Abstract

The conversion of L-tyrosine to 3,4-dihydroxy-L-phenylalanine by tyrosine hydroxylase (TH) is the first and rate-limiting step in biosynthesis of catecholamine neurotransmitters. TH gene expression is regulated in a cell type-specific and cAMP-dependent manner. Evidence from this laboratory and others indicates that the cAMP response element (CRE), residing at –45 to –38 bp upstream of the transcription initiation site, is essential for both basal and cAMP-inducible transcription of the TH gene. To understand the control mechanisms of TH gene transcription in greater detail, we sought to identify and characterize the transcription factors involved in recognition and activation of the CRE of the TH gene. Remarkably, electrophoretic mobility shift assay and antibody supershift experiments indicated that all three major CRE-binding protein factors, i.e. CREB, ATF1, and CREM, may participate in forming specific DNA/protein complexes with the CRE of the TH gene. To address the transcriptional activation function of individual factors, we replaced the TH CRE with a GAL4-binding site and cotransfected this modified TH promoter-reporter gene with an effector plasmid that encodes GAL4-fused transcription factor. Our results indicate that CREB but not ATF1 can support basal promoter activity while both can robustly induce the promoter activity in response to co-expression of the catalytic subunit of cAMP-dependent protein kinase (PKA). We further show that the coactivator CBP up-regulates PKA-mediated activation of the TH promoter and, if tethered to the TH promoter by a GAL4-fusion, can robustly transactivate the TH promoter even in the absence of PKA. Collectively, our results suggest that multiple CRE-binding factors interact with the CRE and regulate, in conjunction with the coactivator CBP, the transcriptional activity of the TH gene. (Mol Cell Biochem **212**: 51–60, 2000)

Key words: tyrosine hydroxylase, gene regulation, cAMP response element (CRE), cAMP response element binding protein (CREB), CREB binding protein (CBP), activator transcription factor (ATF)

Introduction

Tyrosine hydroxylase (TH; EC 1.14.16.2.) is the first and rate-limiting enzyme in the biosynthesis of catecholamine (CA) neurotransmitters and is responsible for the conversion of tyrosine to 3,4-dihydroxyphenylalanine (L-DOPA) [1]. Accordingly, TH gene expression may play a crucial role in regulating neurotransmission of CA neurotransmitters. TH is selectively expressed in CA-synthesizing and secreting cells, including dopaminergic, noradrenergic and adrenergic neurons in the central nervous system and sympathetic ganglia and adrenal chromaffin cells in the periphery. Expression of the TH gene is subject to many physiological stimuli including those linked to the cAMP-signaling pathway (reviewed in [2, 3]). In numerous cAMP-inducible eukaryotic genes, an octamer DNA motif with the nucleotide sequence 5′-

Present address: J. Lim, Department of Animal Science and Biotechnology, College of Agriculture, Kyungbuk National University, Taeku, South Korea; C. Yang, Department of Biochemistry, St. Jude Children's Research Hospital, 332 North Lauderdale, Memphis, TN 38105, USA
Address for offprints: K.-S. Kim, Molecular Neurobiology Laboratory, McLean Hospital, Harvard Medical School, 115 Mill Street, Belmont, MA 02478, USA

TGACGTCA-3′, termed cAMP response element (CRE), mediates transcriptional induction by the cAMP-regulated signaling pathway [4, 5].

The 5′ flanking sequence of the rat TH gene contains a consensus CRE motif located at –38 to –45 bp upstream of the transcription initiation site [6]. This laboratory and others have previously demonstrated that the cAMP-regulated signaling pathway, via the CRE, regulates TH gene expression. In TH-expressing cell lines such as the human neuroblastoma SK-N-BE(2)C and rat PC12 [7, 8] or mouse CATH.a and PATH.a [9], site-directed mutation of the CRE significantly diminished both basal and cAMP-inducible transcription of the rat TH gene. Co-expression of the catalytic subunit of PKA (PKA_c) dramatically increased the transcriptional activity of the rat TH gene in a dose-dependent manner, while co-expression of the specific inhibitor (PKI) blocks cAMP-stimulated induction and reduces basal transcriptional activity [10]. Furthermore, TH gene expression was found to be significantly attenuated at the transcriptional level in several PKA-deficient PC12 cell lines [11]. These studies strongly suggest that PKA, via the CRE, mediates cAMP-inducible transcription of the TH gene. The cAMP signaling pathway critically regulates transcription of the dopamine β-hydroxylase gene as well, suggesting that it may co-regulate catecholamine-synthesizing enzyme genes [10, 12]. In support of this, treatment of the specific PKA inhibitor N-[2-(p-bromocinnamylamine) ethyl]-5-isoquinoline sulfonamide (H-89) significantly diminished both basal and cAMP-induced transcription of all catecholamine-specific genes (i.e. TH, DBH, and PNMT) and resulted in a significant decreases of the intracellular levels of norepinephrine and epinephrine in the primary cultured bovine chromaffin cells [13].

Transcriptional regulation by the cAMP-signaling pathway is mediated by a family of transcription factors which bind to the CRE. These transcription factors have a similar DNA-binding structural motif, so-called leucine zipper, and include the CRE-binding protein (CREB), the CRE-modulatory protein (CREM), and ATF-1 (reviewed in [14]). Different protein kinases are known to phosphorylate these transcription factors, which appear to be critical in modulation of their transcriptional activities [14]. For instance, phosphorylation at serine 133 of CREB or at serine 63 of ATF1 induces recruitment of the coactivators CREB-binding protein (CBP) and p300, resulting in a prominent activation of the target genes [15–17]. Recent evidence from different laboratories has shown that these coactivators regulate transcription of a wide range of target genes by bridging the interacting transcription factors to the basal transcriptional machinery including THIIB [15], TBP [18], RNA polymerase II [19–21]. In addition, p300/CBP have recently been shown to modulate chromatin/nucleosome structure by acetylating histones, resulting in facilitation of access of transcription factors to the promoter sequences [22, 23].

To understand TH gene regulation by the cAMP-regulated pathway in greater detail, it is of great interest to determine which transcription factors bind to the CRE and how they interact with additional transcriptional factors. To address these questions, we sought to identify protein factors that interact with the CRE of the rat TH gene using the electrophoretic mobility shift assay and antibody supershift experiments. Surprisingly, using nuclear proteins prepared from TH-expressing SK-N-BE(2)C and Cath.a as well as TH-negative HeLa cell lines, our analyses show that all three major CRE-binding proteins, i.e. CREB, ATF1, and CREM, may interact with the TH CRE. We next tested whether individual transcription factors can stimulate the transcriptional activity of the TH gene in response to PKA in the context of the TH promoter using Gal4-fusion protein assay. We also tested whether the coactivator CBP may regulate the TH promoter activity. Together with previous study showing that an isoform of CREM regulates TH gene transcription [24], this study strongly suggests that multiple transcription factors including CREB and ATF1, in conjunction with the coactivator CBP/p300, may control transcriptional regulation of the TH gene in basal and/or cAMP-induced conditions.

Materials and methods

Reporter plasmids

The TH2.4CAT reporter gene contains 2.4 kb of the rat TH upstream sequences, fused to the bacterial chloramphenicol acetyltransferase (CAT) gene [11]. The reporter plasmid TH2.4(GAL4)CAT was constructed in two steps. First, the consensus CRE sequence (TGACGTCA) of the TH promoter was changed to a nonfunctional one (TGAAGTCG) by a site-directed mutagenesis procedure as described using an oligonucleotide of the sequence 5′-GCCAGGCTG-AAGTCGAAGCCCCT-3′ [8]. Then, a single copy of GAL 4 binding site with the nucleotide sequence (5′-CGGAGG-ACTGTCCTCCG-3′) was subcloned at the Bgl II site residing at –168 bp from the start site. The sequences of the upstream regions of these reporter genes were verified by DNA sequencing analyses.

Effector plasmids

The plasmid pGAL4-CREB encodes the chimeric protein composed of the full length CREB protein fused to the GAL4 DNA binding domain, while the plasmid pGAL4-CREB-(SA^{119}) encodes the same chimeric protein with a mutation at the $serine^{119}$ residue (described as pNEXδ-WT and pNEXδ-SA^{119}, respectively, in [25]). pGAL4-ATF1 and pGAL4-ATF2 likewise encode full-length ATF1 and ATF2 fused to the GAL4 DNA binding domain under the control of the SV40 promoter and enhancer [26]. pGAL4-CBP 1,678–2,441 encodes a chimeric protein of the GAL4 DNA binding domain and the C-terminal portion of the CBP, which encompasses a potential transactivation domain [16]. pRc/RSV-CBP is an expression vector encoding full-length mouse CBP under the control of Rous Sarcoma Virus promoter/enhancer [16]. The expression plasmid for the catalytic subunit of PKA (PKA_c) was described before [11].

Cell culture and transient transfection assay

Human neuroblastoma SK-N-BE(2)C [27, 28] and HeLa cell lines were maintained in Dulbecco's modified Eagle's medium (Life Technologies Inc.) containing 10% heat-inactivated fetal calf serum (Hyclone Laboratory), 100 units/ml of penicillin (Life Technologies Inc.), and 100 mg/ml of streptomycin (Life Technologies Inc.). Cath.a cells [29] were grown in RPMI 1640 medium (Life Technologies Inc.), supplemented with 10% heat-inactivated horse serum (Hyclone Laboratory), and 5% heat-inactivated fetal bovine serum. SK-N-BE(2)C cells were transfected using the calcium phosphate co-precipitation method as described [30]. Briefly, cells were grown to 50–60% confluent in a 60 mm dish and were transfected with 1 μg of each reporter construct, 1 μg of pRSV-βgal, varying amounts of pGAL4-fusion protein plasmid, and the inert plasmid, pUC19, to a total of 5.5 μg of DNA. When indicated, 0.5 μg of PKA_c-expressing plasmid was included based on our previous results [11]. For experiments described in Figs 4 and 5, a total of 13.5 μg DNA was transfected to accommodate increasing amounts of pRc/RSV-CBP. Transfected cells were collected 36 h after transfection, and activities of chloramphenicol acetyl transferase (CAT) and β-galactosidase were determined as described elsewhere [30]. The CAT activity was normalized by the β-galactosidase activity to correct for differences in transfection efficiency among different DNA precipitates.

Electrophoretic mobility shift assay (EMSA) and antibody supershift experiment

Nuclear extracts were made from SK-N-BE(2)C and HeLa cell lines based on a described procedure [31]. Sense and antisense strands of oligonucleotides, which encompass the cAMP response element of the TH gene [7], were annealed and labeled by kination using [γ-^{32}P]ATP. Specific antibodies were preincubated with nuclear extracts on ice for 2 h prior to adding the binding mixture. EMSA and antibody supershift experiments were performed using 30,000–50,000 cpm of labeled probe (approximately 0.05–0.1 ng) and nuclear extracts (20–30 μg) in a final volume of 20 μl of 12.5% glycerol, 12.5 mM HEPES (pH 7.9), 4 mM Tris-HCl (pH 7.9), 60 mM KCl, 1 mM EDTA, and 1 mM DTT with 1 μg of poly(dI-dC) as described [32]. For experiments described in Figs 1 and 2, nuclear extracts and labeled probe were added at the last step, respectively. It was noted that adding the labeled probe at the last step reduced the formation of a non-specific band (C3). Antibodies used were against CREB (a gift from Dr. M. Montminy), ATF-1, ATF-2, ATF-3 (Santa-Cruz), c-fos, c-jun (Protooncogene), and CREM (Upstate

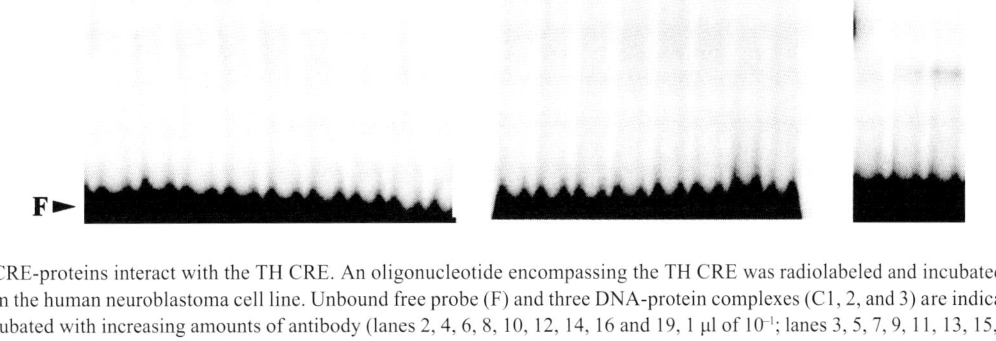

Fig. 1. Multiple CRE-proteins interact with the TH CRE. An oligonucleotide encompassing the TH CRE was radiolabeled and incubated with nuclear extracts isolated from the human neuroblastoma cell line. Unbound free probe (F) and three DNA-protein complexes (C1, 2, and 3) are indicated. Each nuclear extract was preincubated with increasing amounts of antibody (lanes 2, 4, 6, 8, 10, 12, 14, 16 and 19, 1 μl of 10^{-1}; lanes 3, 5, 7, 9, 11, 13, 15, 17 and 20, 1 μl of 10^{0}; lanes 1 and 18, no antiserum) prior to binding reaction. Antibody against CREB (lanes 2, 3, 19 and 20), ATF1 (lanes 4 and 5), ATF2 (lanes 6 and 7), ATF3 (lanes 8 and 9), ATF4 (lanes 10 and 11), c-fos (lanes 12 and 13), c-jun (lanes 14 and 15), and CREM (lanes 16 and 17) are used.

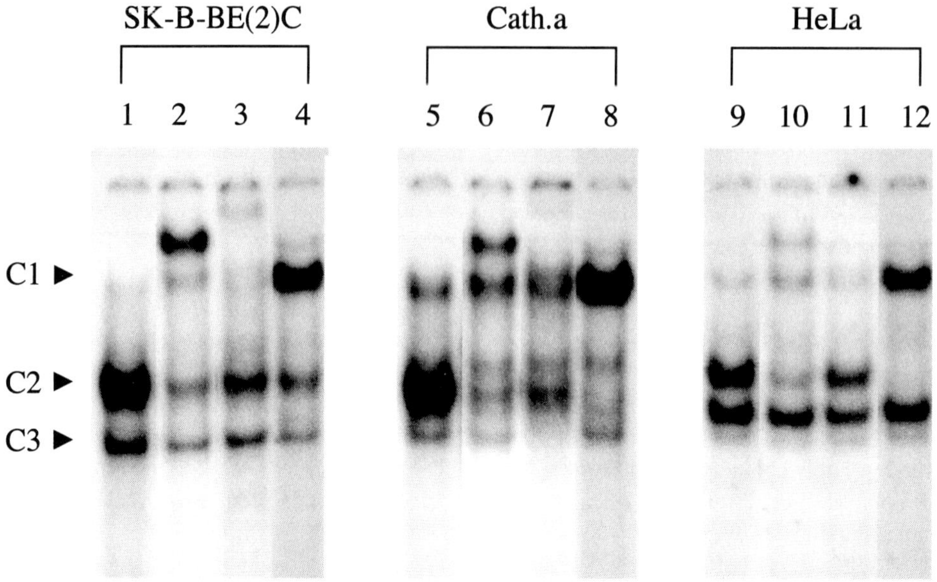

Fig. 2. Antibody supershift experiments of the TH CRE oligonucleotide with nuclear extracts isolated from SK-N-BE(2)C, Cath.a, and HeLa cell lines. Antibody supershift assays are performed using 20–25 μg of different nuclear extracts. Antibodies against CREB (lanes 2, 6 and 10), CREM (lanes 3, 7 and 11), and ATF1 (lanes 4, 8 and 12) are preincubated prior to incubation with the radiolabeled probe. Co-incubation of different nuclear extracts with each antibody resulted in the generation of a supershifted band at a similar location with different intensity.

Biotech). Additional antibody against CREB (SC240; Santa-Cruz) was also used in antibody supershift experiments. DNA-protein complexes and DNA-protein-antibody complexes were resolved on nondenaturing 6% polyacrylamide gels as described [30, 32].

Results

CREB, ATF1 and CREM may participate in forming protein/DNA complexes with the CRE of the TH gene

Previous studies from this and other laboratories have demonstrated that cAMP signaling pathway critically controls transcriptional regulation of the TH gene via the CRE site. To identify potential transcription factor(s) that bind to the CRE and control TH gene transcription, the electrophoretic mobility shift assay (EMSA) and antibody supershift experiments were performed. Three DNA-protein complexes (C1, C2, and C3) were formed between the oligonucleotide encompassing the TH CRE and nuclear extracts of SK-N-BE(2)C cells in EMSA, suggesting the possibility that more than one protein factor may control the transcriptional activity of the CRE of the TH gene (Fig. 1, lane 1). Antibodies against three CRE-binding proteins (ATF1, CREB, and CREM) and structurally related proteins such as ATF2, ATF3, ATF4, c-fos, and c-jun were used to examine components of CRE/protein complexes. Coincubation of antibody against CREB with nuclear extracts of SK-N-BE(2)C cells mostly diminished formation of C2 and resulted in formation of a supershifted band (Fig. 1, lanes 19 and 20). Interestingly, incubation of specific antibody against ATF1 also diminished most of the C2 complex and resulted in formation of a supershifted band at the position of C1 (Fig. 1, lanes 4 and 5). In addition, coincubation with antibody against CREM diminished C1 and C2 and resulted in formation of a supershifted band at higher location (Fig. 1, lane 16). Antibodies against CREM and CREB worked in a concentration-dependent manner; while antibody against CREM affected the formation of DNA-protein complexes at lower concentration, antibody against CREB worked better at higher concentration. Antibody against ATF1 worked equally well at both concentrations used in this assay. Notably, all three antibodies commonly blocked formation of C2. One possible interpretation is that C2 complex represents heterodimer between CREB, CREM, and/or ATF1, and thus can be recognized by these antibodies. An alternate, but mutually nonexclusive explanation is that these antibodies show some degree of cross-reactivity with CREB, CREM, and/or ATF1. Formation of C3 was not affected by any of these antibodies, suggesting that this complex may be formed by nonspecific binding. Consistent with this possibility, formation of C3 was not efficiently blocked by molar excess of cold oligonucleotides compared to those of C1 and C2 in the competition assay (data not shown). Incubation with antibodies against ATF2, ATF3, ATF4, c-fos, or c-jun apparently did not change the pattern of formation of DNA-protein complexes and no supershifted band was generated. Notably, co-incubation of another antibody against CREB (SC-240) neither significantly changed the pattern of DNA/protein complexes nor produced a supershifted band (Fig. 1, lanes 2 and

3), indicating that this antibody supershift experiment may result in different patterns depending on the properties of individual antibodies. Collectively, these results indicate that all three major CRE-binding transcription factors (i.e. CREB, ATF1, and CREM) are candidate transcription factors that bind to the CRE and regulate TH gene transcription *in vivo*.

To test whether these CRE-binding factors are similarly expressed in different cell lines, we next performed antibody supershift experiments using nuclear extracts of another TH-expressing Cath.a and non-expressing HeLa cell lines (Fig. 2). In these experiments, we incubated the radiolabeled oligonucleotide probe after all other binding reagents including nuclear extracts and antibody were mixed (See Materials and methods). While the same pattern of three complexes (C1, C2, and C3) were formed, nonspecific complex C3 was reduced and C2 was increased in this procedure (compare lane 1 of Figs 1 and 2). When nuclear extracts of SK-N-BE(2)C were co-incubated with specific antibodies against CREB, CREM, and ATF1, supershifted bands were generated at the same positions as in Fig. 1 and formation of C2 was diminished to a lesser degree, compared to Fig. 1. When nuclear extracts of Cath.a cell line were used, formation of C1 and C2 was more robust than those of SK-N-BE(2)C. Co-incubation of Cath.a nuclear extracts with specific antibodies similarly diminished formation of C2 and generated supershifted bands with an almost identical pattern to SK-N-BE(2)C. When HeLa nuclear extracts were co-incubated with these antibodies, supershifted bands were generated in a similar fashion. However, supershifted band generated by CREM-specific antibody was much weaker in HeLa, indicating that CREM protein may exist in a lower concentration in HeLa compared to SK-N-BE(2)C and Cath.a.

While CREB appears to support both basal and cAMP-inducible promoter activity, ATF1 primarily supports cAMP-inducible promoter activity of the TH gene

We next sought to investigate whether the transcription factors found to interact with the TH CRE can transactivate the basal and/or cAMP-inducible promoter activity of the TH gene. Because CREM is primarily involved in negative gene regulation [33] and one of its isoforms has been shown to repress the TH promoter activity [24], we have focused on ATF1 and CREB in this study. Based on the previous finding that the CRE is critical for both basal and cAMP-inducible promoter activity of the TH gene [8, 11], we have modified the reporter plasmid (TH2.4CAT) which contains the 2.4 kb upstream promoter fused to the reporter gene by two steps: (1) the CRE was rendered nonfunctional by introducing a double mutation to the CRE, and (2) a GAL4 binding site was inserted at –168 bp using the unique Bgl II site (Fig. 3A). As expected, this modified reporter construct (TH2.4GAL4CAT) drove CAT activity at a level no higher than the promoter-less plasmid, pBLCAT3 [34]. Thus, this new reporter construct can be used for testing the transactivating function of a certain transcription factor which is fused to the GAL4 DNA-binding domain [35]. Expression plasmids encoding a hybrid protein, consisting of CREB, a mutant form of CREB (CREB^m), ATF1, and ATF2 proteins fused to the GAL4 DNA binding domain, were used as the transactivator (Fig. 3A). In pGAL4-CREB^m, the serine residue was changed to alanine at amino acid position 119 which is essential for phosphorylation of CREB by PKA. Co-transfection of pGAL4-CREB increased the CAT activity driven by TH2.4GAL4CAT reporter construct by approximately 4-fold (Fig. 3B). Co-expression of the coding sequence of GAL4 DNA-binding domain (1–147) only did not affect the CAT activity driven by TH2.4GAL4CAT [35], ascribing the observed increase of CAT activities by pGAL4-CREB to CREB. Interestingly, co-transfection of pGAL4-CREB^m increased the reporter gene activity to a similar degree (approximately 3-fold), suggesting that the serine residue at 119 amino acid position may not be critically required for supporting basal transcription of the TH gene. pGAL4-ATF1 increased the promoter activity of the TH2.4GAL4CAT construct only marginally, if any. Thus, CREB and ATF1 seem to exhibit differential ability to support basal transcriptional activity of the TH promoter. In a control experiment, co-transfection of pGAL4-ATF2 containing the full length coding region of ATF2 modestly transactivated the reporter gene activity by 2.5-fold (Fig. 3B).

To address whether these transcription factors can activate TH gene transcription in response to cAMP-signaling pathway, we next co-transfected an expression plasmid encoding catalytic subunit of PKA (PKA$_c$) together with the reporter construct and each effector plasmid. When PKA$_c$ was co-expressed, GAL4-CREB and GAL4-ATF1 dramatically increased the reporter gene activity by 112 and 158-fold, respectively (Fig. 3C). These results support the idea that CREB and ATF1 participate in PKA-inducible transcription of the TH gene. PKA$_c$-induced transactivation by GAL4-CREB was mostly diminished (from 112 to 6-fold) by introducing a mutation (Ser to Ala) at the 119 position of the CREB protein, demonstrating that phosphorylation at this residue by PKA is crucial for cAMP-induced activation of the TH gene. In addition, PKA$_c$ barely affected transactivation function of pGAL4-ATF2, showing its specific activation of CREB and ATF1.

Transcriptional cofactor CBP may be an important component of cAMP-inducible transcription of the TH gene and is able to transactivate the promoter activity when tethered to the proximal promoter area

To test the role of CBP in cAMP-regulated activation of TH gene transcription, we co-transfected a CBP-expression plas-

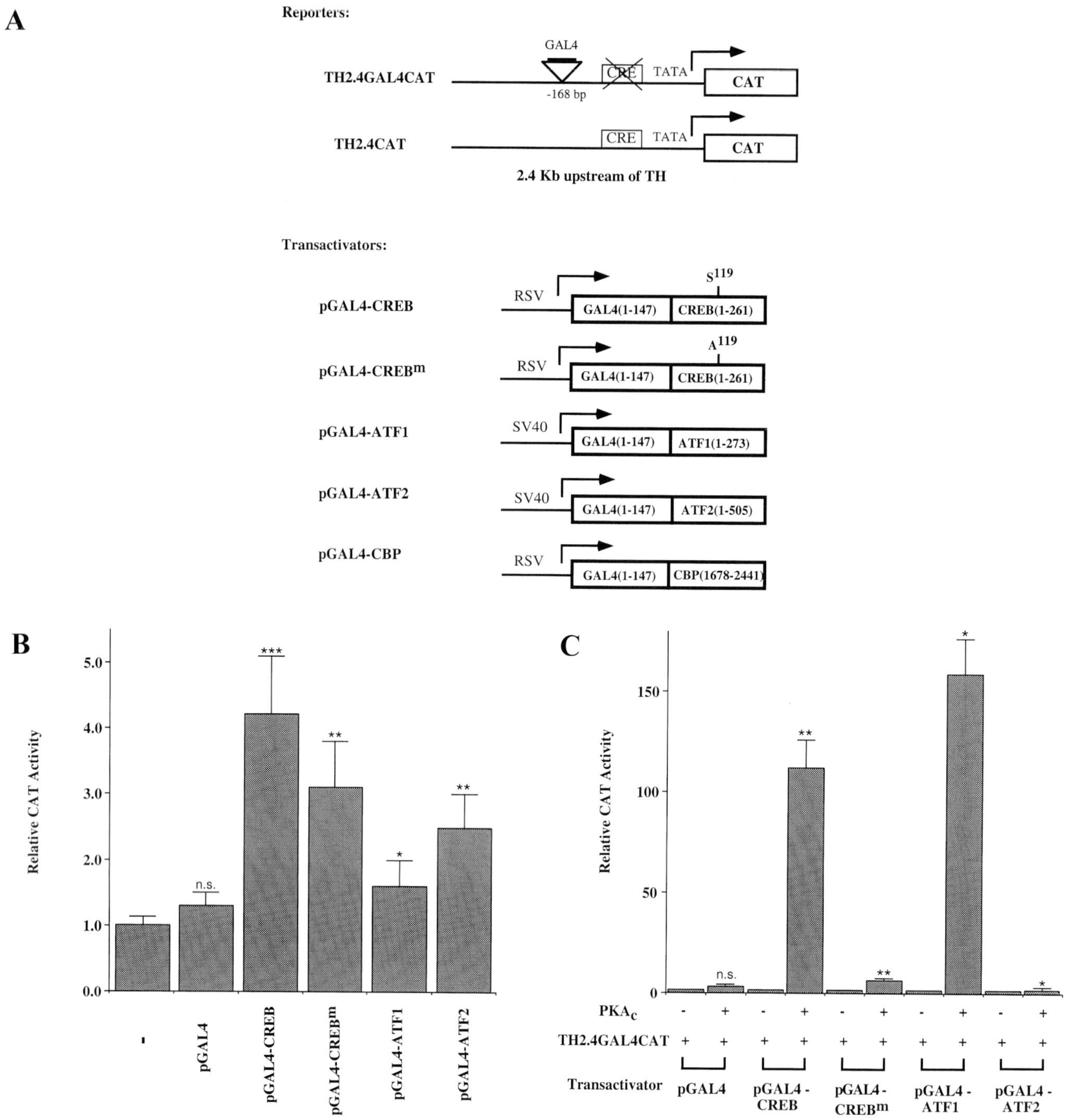

Fig. 3. ATF1 and CREB transactivate the basal and PKA-mediated promoter activities of the TH gene. (A) Diagram of reporter and transactivator plasmids. TH2.4GAL4CAT plasmid containing a GAL4 binding site instead of a nonfunctional CRE is used as the reporter plasmid. Transcriptional start site is de noted by a bent arrow. Various plasmids encoding GAL4-fused transactivators are schematically shown with the promoters used to express these fusion molecules. (B) SK-N-BE(2)C cells were transiently co-transfected with TH2.4GAL4CAT reporter construct and the transactivator plasmids indicated below. The molar ratio of the reporter plasmid to the transactivator was 1 to 3. Samples were harvested 48–72 h after transfection, and relative CAT activities were determined as described in Materials and methods. To compare the fold transactivation directly, the basal CAT activity driven by reporter construct without the transactivator plasmid was set to 1. The significance of the effects of the transactivator plasmids is indicated as follows: n.s. – not significant; *p < 0.02; **p < 0.01; ***p < 0.001. (C) Co-transfection assay was performed in SK-N-BE(2)C cells with addition of 0.5 μg of expression plasmid of the catalytic subunit of PKA (PKA_c), when indicate by +. Bars represent the normalized CAT activities driven by TH2.4GAL4CAT and each transactivator-expression plasmid, relative to those without PKA_c. The relative values are presented as mean value ± S.E.M. from 6 independent samples. The significance of the effects of the PKA_c-expression plasmid on CAT gene expression in the presence of each transactivator plasmid is indicated as follows: n.s. – not significant; *p < 0.02; **p < 0.01.

mid along with the reporter plasmid TH2.4CAT (Fig. 4). In agreement with our previous observation [10], co-expression of PKA_c robustly transactivated the intact promoter activity of the TH gene (64-fold) in the absence of pCBP. Co-transfection of pCBP further increased the CAT activity driven by TH2.4CAT by an additional 2-fold (from 62 to 128-fold) in a dose-dependent manner showing that increased availability of CBP can further up-regulate PKA-induced transcription of the TH gene. We next tested whether CBP similarly can increase PKA-induced transactivation function of CREB. As shown in Fig. 5, co-transfection of pCBP further increased the CAT activity driven by pTH2.4GAL4CAT and pGAL4-CREB from 98 to 187-fold in a dose-dependent manner. Similarly, pCBP also increased the reporter activity driven by pTH2.4GAL4CAT and pGAL4-ATF1 in a dose-dependent manner (data not shown).

Finally, we co-transfected pTH2.4GAL4CAT and pGAL4-CBP to the SK-N-(BE)2C cell line (Fig. 6). Strikingly, in the absence of ectopic expression of PKA, the full-length CBP protein fused to GAL4 was able to transactivate the modified TH promoter by 192-fold. This result show that tethering CBP to the TH promoter by GAL4 fusion alleviates the requirement of PKA activity for transactivation of the promoter activity and thus strongly suggest that CBP is an essential component in PKA-induced activation of TH transcription. In addition, co-expression of PKA_c further increased the reporter activity up to 562-fold indicating that the transactivation function of CBP can be further facilitated by its phosphorylation. Alternatively, it is also possible that the promoter activity of the reporter construct is stimulated through other elements such as the AP-1 site.

Discussion

In this study, we sought to identify transcription factors involved in the transcriptional regulation of the CRE of the TH gene. Previous studies have shown that the CRE, which resides at –38 to –45 bp upstream of the transcription initiation site, is crucial for not only cAMP-mediated induction but also basal transcription of the TH gene [7, 10, 36]. We demonstrate that all three major CRE-binding proteins, i.e. CREB,

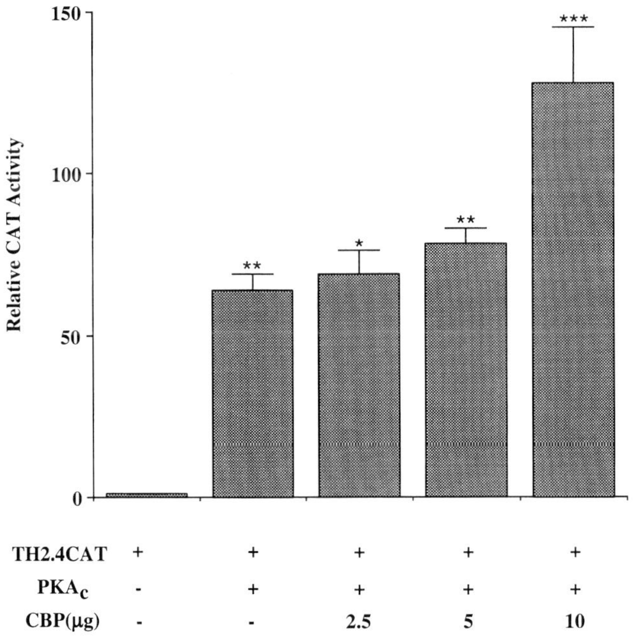

Fig. 4. CBP enhances the PKA-induced promoter activity of the TH gene. SK-N-BE(2)C cells were transfected with TH2.4CAT, PKA_c, and increasing amounts of pRc/RSV-CBP (CBP) as shown. The total amount of DNA was kept constant at 13.5 µg using the inert plasmid pUC19. The bars represent the normalized CAT activities relative to that driven by TH2.4CAT only, and are shown as mean value ± S.E.M. from 6 independent samples. The significance of the effects of the CBP-expressing plasmids, relative to the control, on PKA_c-induced expression of the TH-CAT fusion gene is indicated as follows: *p < 0.02; **p < 0.01; ***p < 0.001. The transient transfection experiments were performed at least twice more in triplicate with similar patterns.

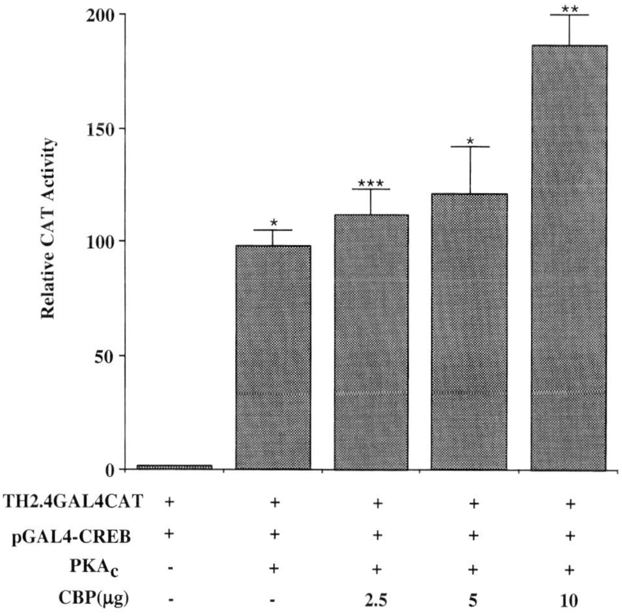

Fig. 5. Expression of CBP is able to enhance the transactivation function of pGAL-CREB in the presence of PKA_c. SK-N-BE(2)C cells were transfected with TH2.4GAL4CAT, PKA_c, and increasing amount of pRc/RSV-CBP (CBP) as shown. Transfection assays were performed as in Fig. 4 with a total 13.5 µg of DNA and the relative CAT activities are presented as mean value ± S.E.M. from 6 independent samples. The significance of the effects of the CBP-expressiong plasmids, relative to the control, on PKA_c-induced CAT expression in the presence of pGAL-CREB and TH2.4GAL4CAT is indicated as follows: *p < 0.02; **p < 0.01; ***p < 0.001.

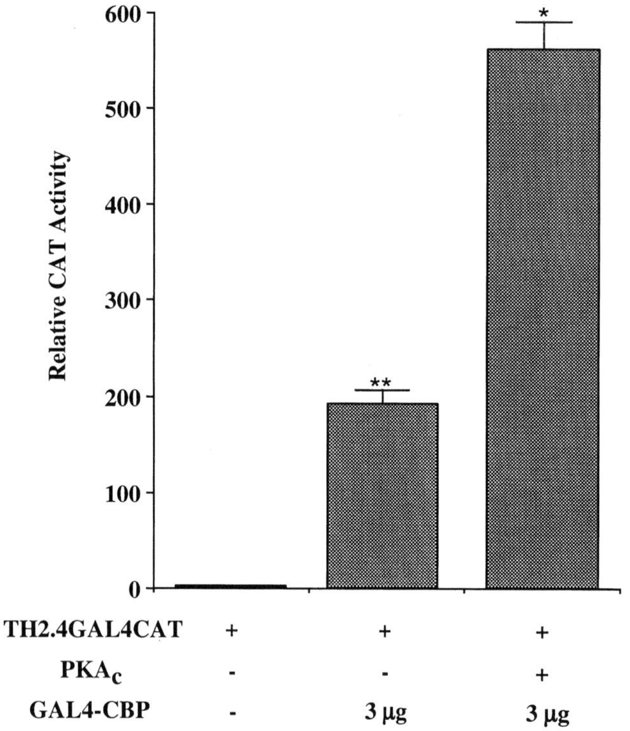

Fig. 6. GAL4-CBP can dramatically increase the CAT activity driven by TH2.4GAL4CAT. SK-N-BE(2)C cells were transfected with TH2.4GAL4-CAT, PKA_c, and GAL4-CBP. The total amount of DNA was kept constant at 5 μg. The bars represent the normalized CAT activities relative to that driven by TH2.4GAL4CAT only, and are shown as mean value ± S.E.M. from 6 independent samples. The significance of the effects of GAL4-CBP and PKA_c-expressing plasmids on CAT expression driven by TH2.4GAL4-CAT is indicated as follows: $*p < 0.02$; $**p < 0.01$. The transient transfection experiments were performed at least twice more in triplicate, resulting in similar patterns.

ATF1, and CREM, may participate in forming complexes with the CRE of the TH gene in TH-expressing human neuroblastoma SK-N-BE(2)C and mouse central catecholaminergic Cath.a cell lines. In contrast, EMSA of the present study did not detect AP-1 protein complex (e.g. c-fos and c-jun) or other related proteins (e.g. ATF2, 3, and 4) in the CRE/protein complexes. Previously, an isoform of the CREM family proteins, the inducible cAMP early repressor (ICER), has been shown to be induced by forskolin stimulation in PC12 cells and by reserpine treatment of rats in adrenal glands [24]. In addition, Liu *et al.* [37] recently showed that CREB and CREM contributed to formation of complexes with the TH CRE in the olfactory bulb dopaminergic neurons. Thus, our present data agree with these studies and further suggest that all three major CRE-binding transcription factors may contribute to formation of DNA/protein complexes with the CRE and regulation of the TH gene.

We also tested whether ATF1 and CREB can support basal and cAMP-inducible transcriptional activity of the TH gene promoter. To this end, based on our previous analysis of the structure/function relationship of the TH CRE [8], we made the CRE of the TH-reporter construct be nonfunctional by introducing a double mutation and then inserted a GAL4-binding site (Fig. 3A). Because this mutation rendered the whole 2.4 kb upstream promoter to be as inactive as the promoter-less construct, we were able to assess the transactivation function of each transcription factor which is fused to the GAL4 DNA-binding domain. These co-transfection assays in SK-N-BE(2)C cells showed that CREB modestly activated the basal TH promoter activity (approximately 4-fold) while ATF1 minimally activated it (approximately 1.6-fold). Interestingly, a mutant form of CREB containing alanine at 119 amino acid position instead of serine similarly activated the basal TH promoter activity (approximately 3-fold), indicating that the basal transactivation function of CREB may not depend on phosphorylation by PKA at this position. It is to be noted that exogenous expression of GAL4-CREB did not restore the promoter activity of the modified reporter construct (TH2.4GAL4CAT) to the level of the wild type promoter activity of TH2.4CAT (Fig. 3; [8]). This apparent discrepancy can be explained by several possibilities. First, in our modified reporter construct, the GAL4-binding site was located at a more distal position (–168 bp) compared to the CRE (–45 to –38 bp) of the wild type promoter. The transactivator may not work efficiently at this new site compared to the original site. However, our previous work does not support this possibility because the CRE worked as efficient as the wild type when located within a window of approximately 200 bp upstream of the transcription start site [8]. Secondly, GAL4-fusion of CREB may have rendered it a less efficient transactivator compared to CREB, possibly due to its structural changes. Thirdly, CREB may be able to transactivate the TH CRE efficiently only in concert with additional CRE-binding proteins. For instance, since CREB and ATF1 can form heterodimers, one possibility is that these heterodimeric forms have more potent transactivation function to the TH promoter.

In contrast to their limited ability to support the basal activity of the modified TH promoter, both GAL4-CREB and GAL4-ATF1 were shown to dramatically up-regulate the reporter activity in the presence of PKA (Fig. 3C). Mutation of the serine residue to alanine at 119 amino acid position diminished the PKA-mediated transactivation by CREB from 112 to 6-fold, indicating that this serine residue of CREB is critical for PKA-mediated activation of TH transcription. In addition, our data showed that coexpression of the full length CBP further up-regulated the transactivation function of both GAL4-CREB and GAL4-ATF1 in a dose-dependent manner (Fig. 4; data not shown). Furthermore, co-expression of GAL4-CBP dramatically activated the CAT activity driven by TH2.4GAL4CAT by 192-fold in the absence of exogenous PKA expression (Fig. 6). Recently, transcriptional coactivator CBP/p300 have been shown to critically regulate transcription by interacting with the basal transcriptional machinery including THIIB [16], TBP [18], RNA polymerase II [19–21] as well as by modulating chromatin/nucleosome structure [22, 23]. Additionally, CBP/p300 have been shown to interact with ever increasing numbers of transcription factors including YY1 [38], MyoD [39], nuclear factor 1 [40], and Phox2a/Arix [41]. This converging evidence suggests that CBP/p300 may coordinate development stage-, second messenger-, and cell type-specific control of a wide range of genes.

In summary, we showed that multiple members of CRE-binding proteins may interact with the TH CRE in TH-positive cells and are capable of differentially supporting basal and cAMP-inducible promoter activity of the TH gene. Together with a previous study showing that an isoform of CREM, ICER, is induced in certain physiological conditions and down-regulates TH gene transcription [24], this study strongly suggests that multiple transcription factors including CREB and ATF1, in conjunction with the coactivator CBP/p300, may intricately control transcriptional regulation of the TH gene in basal and/or cAMP-induced conditions.

Acknowledgements

The authors thank Dr. David Stone at McLean Hospital, Harvard Medical School, for critical reading of the manuscript. We also thank Drs. M. Montminy, R.H. Goodman, M. Green and B. Kaang for CREB-specific antibody, CBP-expressing plasmids, ATF1- and ATF2-expressing plasmids and CREB-expressing plasmids, respectively. This work was supported by NIH grants MH48866.

References

1. Nagatsu T, Levitt M, Udenfriend S: Tyrosine hydroxylase; the initial step in norepinephrine biosynthesis. J Biol Chem 239: 2910–2917, 1964
2. Zigmond RE, Schwarzschild MA, Rittenhouse AR: Acute regulation of tyrosine hydroxylase by nerve activity and by neurotransmitters via phosphorylation. Annu Rev Neurosci 12: 415–461, 1989
3. Kumer SC, Vrana KE: Intricate regulation of tyrosine hydroxylase activity and gene expression. J Neurochem 67: 443–462, 1996
4. Roesler WJ, Vandenbark GR, Hanson RW: Cyclic AMP and the induction of eukaryotic gene transcription. J Biol Chem 263: 9063–9066, 1988
5. Goodman R: Regulation of neuropeptide gene expression. Annu Rev Neurosci 13: 111–127, 1990
6. Lewis EJ, Harrington CA, Chikaraishi DM: Transcriptional regulation of the tyrosine hydroxylase gene by glucocorticoid and cAMP. Proc Natl Acad Sci USA 84: 3550–3554, 1987
7. Kim KS, Lee MK, Carroll J, Joh TH: Both the basal and inducible transcription of the tyrosine hydroxylase gene are dependent upon a cAMP response element. J Biol Chem 268: 15689–15695, 1993
8. Tinti C, Yang C, Seo H, Conti B, Kim C, Joh TH, Kim KS: Structure/function relationship of the cAMP response element in tyrosine hydroxylase gene transcription. J Biol Chem 272: 19158–19164, 1997
9. Lazaroff M, Patankar S, Yoon SO, Chikaraishi DM: The cyclic AMP response element directs tyrosine hydroxylase expression in catecholaminergic central and peripheral nervous system cell lines from transgenic mice. J Biol Chem 270: 21579–21589, 1995
10. Kim KS, Tinti C, Song B, Cubells JF, Joh TH: Cyclic AMP-dependent protein kinase regulates basal and cAMP-stimulated but not phorbol ester-stimulated transcription of the tyrosine hydroxylase gene. J Neurochem 63: 834–842, 1994
11. Kim KS, Park DH, Wessel TC, Song B, Wagner JA, Joh TH: A dual role of the cAMP-dependent protein kinase on tyrosine hydroxylase gene expression. Proc Natl Acad Sci USA 90: 3471–3475, 1993
12. Ishiguro H, Kim KT, Joh TH, Kim KS: Neuron-specific expression of the human dopamine β-hydroxylase gene requires both the cAMP-response element and a silencer region. J Biol Chem 268: 17987–17994, 1993
13. Hwang O, Park SY, Kim KS: Inhibition of PKA with H-89 causes reduction in basal gene expression of the catecholamine enzymes. J Neurochem 68: 2241–2247, 1997
14. Montminy M: Transcriptional regulation by cyclic AMP. Annu Rev Biochem 66: 807–822, 1997
15. Chrivia JC, Kwok RP, Lamb N, Hagiwara M, Montminy MR, Goodman RH: Phosphorylated CREB binds specifically to the nuclear protein CBP. Nature 365: 855–859, 1993
16. Kwok RP, Lundblad JR, Chrivia JC, Richards JP, Bachinger HP, Brennan RG, Roberts SG, Green MR, Goodman RH: Nuclear protein CBP is a coactivator for the transcription factor CREB. Nature 370: 223–226, 1994
17. Lundblad JR, Kwok RP, Laurance ME, Harter ML, Goodman RH: Adenoviral E1A-associated protein p300 as a functional homologue of the transcriptional co-activator CBP. Nature 374: 85–88, 1995
18. Dallas PB, Yaciuk P, Moran E: Characterization of monoclonal antibodies raised against p300: both p300 and CBP are present in intracellular TBP complexes. J Virol 71: 1726–1731, 1997
19. Cho H, Orphanides G, Sun X, Yang XJ, Ogryzko V, Lees E, Nakatani Y, Reinberg D: A human RNA polymerase II complex containing factors that modify chromatin structure. Mol Cell Biol 18: 5355–5363, 1998
20. Kee BL, Arias J, Montminy MR: Adaptor-mediated recruitment of RNA polymerase II to a signal-dependent activator. J Biol Chem 271: 2373–2375, 1996
21. Nakajima T, Uchida C, Anderson SF, Lee CG, Hurwitz J, Parvin JD, Montminy M: RNA helicase A mediates association of CBP with RNA polymerase II. Cell 90 :1107–1112, 1997
22. Ogryzko VV, Schiltz RL, Russanova V, Howard BH, Nakatani Y The transcriptional coactivators p300 and CBP are histone acetyltransferases. Cell 87: 953–959, 1996
23. Wolffe AP, Pruss D: Targeting chromatin disruption: Transcription regulators that acetylate histones. Cell 84: 817–819, 1996
24. Tinti C, Conti B, Cubells J, Kim KS, Baker H, Joh, TH: Expression of ICER in PC12 and rat adrenal gland: A role for tyrosine hydroxylase gene transcription repression. J Biol Chem 271: 25375–25381, 1996
25. Kaang BK, Kandel ER, Grant SG: Activation of cAMP-responsive genes by stimuli that produce long-term facilitation in Aplysia sensory neurons. Neuron 10: 427–435, 1993
26. Liu F, Green MR: A specific member of the ATF transcription factor family can mediate transcription activation by the adenovirus E1a protein. Cell 61:1217–1224, 1990
27. Ross RA, Biedler JL, Spengler BA, Reis DJ: Neurotransmitter-synthesizing enzymes in 14 human neuroblastoma cell lines. Cell Mol Neurobiol 1: 301–311, 1981
28. Ciccarone V, Spengler BA, Meyers MB, Biedler JL, Ross RA: Phenotypic diversification in human neuroblastoma cells: Expression of distinct neural crest lineages. Cancer Res 49: 219–225, 1989

29. Suri A, Fung BP, Tischler AS, Chikaraishi DM: Catecholaminergic cell lines from the brain and adrenal glands of tyrosine hydroxylase-SV40 T antigen transgenic mice. J Neurosci 13: 1280–1291, 1993
30. Seo H, Yang CY, Kim H-S, Kim KS: Multiple protein factors interact with the cis-regulatory elements of the proximal promoter in a cell-specific manner and regulate transcription of the dopamine β-hydroxylase gene. J Neurosci 16: 4102–4112, 1996
31. Dignam JD, Lebovitz RM, Roeder RG: Accurate transcription initiation by RNA polymerase II in a soluble extract from isolated mammalian nuclei. Nucleic Acids Res 11: 1475–1489, 1983
32. Kim HS, Seo H, Brunet JF, Kim KS: Noradrenergic-specific transcription of the dopamine β-hydroxylase gene requires synergy of multiple cis-regulatory elements including at least two Phox2a-binding sites. J Neurosci 18: 8247–8260, 1998
33. Foulkes NS, Borrelli E, Sassone-Corsi P: CREM gene: Use of alternative DNA-binding domains generates multiple antagonists of cAMP-induced transcription. Cell 64: 739–749, 1991
34. Luckow B, Schutz G: CAT constructions with multiple unique restriction sites for the functional analysis of eukaryotic promoters and regulatory elements. Nucleic Acids Res 15: 5490, 1987
35. Sadowski I, Ptashne M: A vector for expressing GAL4(1-147) fusion in mammalian cells Nucleic Acids Res 17: 7539, 1989
36. Lazaroff M, Patankar S, Yoon SO, Chikaraishi DM: The cyclic AMP response element directs tyrosine hydroxylase expression in catecholaminergic central and peripheral nervous system cell lines from transgenic mice. J Biol Chem 270: 21579–21589, 1995
37. Liu N, Cigola E, Tinti C, Jin BK, Conti B, Volpe BT, Baker H: Unique regulation of immediate early gene and tyrosine hydroxylase expression in the odor-deprived mouseolfactory bulb. J Biol Chem 274: 3042–3047, 1999
38. Galvin KM, Shi Y: Multiple mechanisms of transcriptional repression by YY1. Mol Cell Biol 17: 3723–3732, 1997
39. Sartorelli V, Huang J, Hamamori Y, Kedes L: Molecular mechanisms of myogenic coactivation by p300: Direct interaction with the activation domain of MyoD and with the MADS box of MEF2C. Mol Cell Biol 17: 1010–1026, 1997
40. Leahy P, Crawford DR, Grossman G, Gronostajski RM, Hanson RW: CREB binding protein coordinates the function of multiple transcription factors including nuclear factor I to regulate phosphoenolpyruvate carboxykinase (GTP) gene transcription. J Biol Chem 274: 8813–8822, 1999
41. Swanson DJ, Adachi M, Lewis EJ: The homeodomain protein arix promotes protein kinase A-dependent activation of the dopamine beta-hydroxylase promoter through multiple elements and interaction with the coactivator cAMP-response element-binding protein-binding protein. J Biol Chem 275: 2911–2923, 2000

Molecular and Cellular Biochemistry **212**: 61–70, 2000

Norepinephrine transporter expression and function in noradrenergic cell differentiation

Maya Sieber-Blum and Zeguang Ren

Department of Cell Biology, Neurobiology and Anatomy, Medical College of Wisconsin, Milwaukee, WI, USA

Abstract

The classical view of norepinephrine transporter (NET) function is the re-uptake of released norepinephrine (NE) by mature sympathetic neurons and noradrenergic neurons of the *locus ceruleus* (LC; [1–3]). In this report we review previous data and present new results that show that NET is expressed in the young embryo in a wide range of neuronal and non-neuronal tissues and that NET has additional functions during embryonic development. Sympathetic neurons are derived from neural crest stem cells. Fibroblast growth factor-2 (FGF-2), neurotrophin-3 (NT-3) and transforming growth factor-β1 (TGF-β1) regulate NET expression in cultured quail neural crest cells by causing an increase in NET mRNA levels. They also promote NET function in both neural crest cells and presumptive noradrenergic cells of the LC. The growth factors are synthesized by the neural crest cells and therefore are likely to have autocrine function. In a subsequent stage of development, NE transport regulates differentiation of noradrenergic neurons in the peripheral nervous system and the LC by promoting expression of tyrosine hydroxylase (TH) and dopamine-β-hydroxylase (DBH). Conversely, uptake inhibitors, such as the tricyclic antidepressant, desipramine, and the drug of abuse, cocaine, inhibit noradrenergic differentiation in both tissues. Taken together, our data indicate that NET is expressed early in embryonic development, NE transport is involved in regulating expression of the noradrenergic phenotype in the peripheral and central nervous systems, and norepinephrine uptake inhibitors can disturb noradrenergic cell differentiation in the sympathetic ganglion (SG) and LC. (Mol Cell Biochem **212**: 61–70, 2000)

Key words: neural crest, sympathetic ganglion, *locus ceruleus*, norepinephrine, norepinephrine transporter, tyrosine hydroxylase, dopamine-beta-hydroxylase, desipramine, cocaine

Introduction

Our interest in NET was sparked by the fortuitous observation that the tricyclic antidepressant, desipramine, inhibits the differentiation of cultured neural crest cells into sympathetic neuroblasts [4]. This observation suggested that reverse must be true as well, that NE promotes noradrenergic differentiation. Subsequent experiments confirmed that NE promotes expression of TH and DBH [5]. Physiological relevance is established by the observations of Strudel *et al.* [6] and Rothman *et al.* [7] that the notochord synthesizes NE and other neurotransmitters at the time when migrating neural crest cells have acquired NE transport activity and are localized near the notochord. Moreover, serum and embryo extract, which at that time were components of our culture medium, contain nanomolar quantities of NE [5].

In the peripheral nervous system, NET is expressed early in development, most likely by neural crest stem cells [8]. Whereas our initial work focused on the neural crest, we report here that similar regulatory mechanisms act in the differentiation of noradrenergic neurons of the LC. FGF-2 and NT-3 cause an increase in NET mRNA levels and NET function. In both tissues, NE promotes expression of the noradrenergic biosynthetic enzymes, TH and DBH. By contrast, uptake inhibitors such as desipramine and cocaine block TH and DBH expression. These observations strongly indicate that NE transport plays an important role in the development of noradrenergic neurons.

Present address: Z. Ren, Department of Surgery, Emory University, Atlanta, GA 30322, USA
Address for offprints: M. Sieber-Blum, Department of Cell Biology, Neurobiology and Anatomy, Medical College of Wisconsin, 8701 Watertown Plank Road, Milwaukee, WI 53226, USA

Materials and methods

Neural crest primary cultures

Neural crest cell cultures were prepared as described previously [9, 10]. Thirty-five mm culture plates were coated with collagen, laminin (28 µg per plate) and fibronectin (100 µg plasma fibronectin from horse serum). The last 6 segments of Hamburger-Hamilton [11] stage 14 quail embryos (49 h of incubation) were excised and isolated. At 24 h post-explantation, the neural tubes were removed, leaving the emigrated neural crest cells on the substratum. The defined culture medium was exchanged every day. The cultures were incubated at 37°C in a humidified atmosphere of 5%CO_2 and 10% O_2. TGF-β1 (Upstate Biotechnology, Lake Placid, NY, USA) was used at 1 ng/ml [12], FGF-2 (Upstate Biotechnology, Lake Placid, NY, USA) at 2.5 ng/ml and hrNT-3 (gift of Genentech Inc., South San Francisco CA and from Promega, Madison, WI, USA) at 10 ng/ml. Where indicated anti-NT-3 blocking antibody (mouse monoclonal IgG, 2 mg/ml; gift of Y.-A. Barde and I. Bartke [13]) were added to the cultures at 1:500 and pan-TGF-β blocking antibodies (rabbit IgG, 1 mg/ml, Research Biochemicals International, Natick, MA, USA) at 1:250.

Neural crest colony assay

Neural crest cell colony assays were performed as described [9, 10, 12]. Briefly, cells from 18–24 h primary explants were resuspended by trypsinization and the cell density adjusted to 500 cells per ml. One ml-aliquots were then placed into each of 10 collagen- and fibronectin-coated plates per experimental group. The cells attached to the substratum within half an hour. One hour after plating a second ml of culture medium, that in experimental groups contained a 2 × concentration of growth factor(s), was added to each plate. The culture medium and growth factors were replaced every other day. At day 10, the cultures were processed for norepinephrine uptake as detailed below. NE (*d,l*-arterenol; Sigma) and cocaine (ecgonine methyl ester benzoate) hydrochloride (Sigma) were added daily as indicated.

LC colony assay

See Results section.

Norepinephrine uptake and autoradiography

Neural crest cells with a functional norepinephrine transporter were identified *in situ* in colony assays exactly as described previously [5]. Briefly, the cultures were rinsed with Hanks' balanced salt solution (HBSS) containing 1% bovine serum albumin (BSA). They were then incubated for 2 h at 37°C with 0.5 ml of 0.5 µM [^{3}H]norepinephrine (specific activity, 40.8 Ci/mmol) in HBSS that also contained 1 mM ascorbic acid and 0.1 mM monoamine oxidase inhibitor, pargyline (Sigma). Subsequently, uptake of radioactive norepinephrine was terminated by rinsing the cultures 3 times with HBSS that contained 24 mM non-radioactive norepinephrine (*d,l*-arterenol; Sigma), fixed with 4% paraformaldehyde in calcium-magnesium-free PBS for 20 min at room temperature, and rinsed again. The cultures were then dried in a stream of cold air, coated in the dark with NTB2 emulsion (Kodak) for 2.5 min, and air-dried in the dark. After 10 days of exposure at 4°C, autoradiograms were developed with D-19 (Kodak) for 2.5 min, rinsed with 1% acetic acid and fixed with Rapid Fix (Kodak). Plates were mounted with mineral oil and a coverslip. The total number of colonies per plate and the number of colonies that contained grain-positive cells were scored and the data statistically analyzed by 1-factor analysis of variance.

RT-PCR

Primers were synthesized by Operon Technologies Inc. (Alameda, CA, USA). The following primer pairs were used. qNET; GGATTGATGCGGCTACTCAGA and GCTTCCATGCCACCCATAGA (product length, 358 bp). Glyceraldehyde-3-phosphate dehydrogenase (GAPDH); ACG CCA TCA CTA TCT TCC AG and TCAGCTCAGGGATGACTTTC (product length, 458 bp; ref. [14]).

Total RNA was isolated using the RNA STAT-60 kit (TEL-TEST Inc., Friendswood, TX, USA). For RT-PCR, the 'Access RT-PCR System' (Promega, Madison, WI, USA) was used. Template RNA was reverse transcribed with AMV RT for 45 min at 48°C, followed by 2 min of inactivation of RT and denaturation at 94°C. The conditions for PCR were as follows. Denaturation, 94°C, 30 sec; annealing, 60°C, 1 min; extension, 68°C, 2 min; cycle number, 36.

PCR products were analyzed by agarose gel (2%) electrophoresis, sequenced and found to be correct. For controls, the PCR reaction was carried out without a template and without reverse transcriptase. The PCR conditions were optimized in pilot experiments to assure linear amplification.

Immunocytochemistry

NET rabbit antiserum #43408 (used at 1:2000; ref. [15]) is an anti-peptide antibody that was raised against an external epitope of human NET, which is highly conserved in qNET. The slides with cryosections were rinsed with PBS, incubated with 5% normal goat serum in PBS for 30 min, and subsequently incubated over night at 10°C with primary antibodies in PBS containing 5% goat serum and 0.1% Triton-X100. For multiple stains, the primary antibodies were pooled. The sections/cultures were rinsed three times for 10 min each with PBS. The secondary antibodies were diluted in 5% normal goat serum in PBS, added to the sections/cultures and incubated in the dark for 1 h at room temperature. For multiple stains the secondary antibodies were pooled. After 3 more rinses at 10 min each, the slides were mounted with 50% glycerol (containing 1 mg/ml paraphenylene diamine (PPD) in PBS, pH 8.5) and a cover slip and were observed with the confocal microscope.

Results

NET expression during neural crest cell differentiation

We have established by RT-PCR (Fig. 1) and immunocytochemistry [8] that NET is expressed at early stages of quail neural crest cell differentiation. The neural crest stem cell gives rise to sympathetic neuroblasts among other cell types [9, 16]. Many cultured neural crest cells express NET protein at high levels as early as 6 h after explantation of the neural tube [8]. In 60 h-old embryos, early migratory, and possibly even pre-migratory, neural crest cells express NET protein. Moreover, the dorsal aspect of the neural tube, from which neural crest cells delaminate, is intensely NET immunoreactive [8].

In 84 h-old embryos, several neural crest derivatives are NET immunoreactive. They include the sympathetic ganglion, the dorsal root ganglion, presumptive Schwann cells in the ventral root (Fig. 2), as well as presumptive enteric neurons in the gut and presumptive smooth muscle cells in the cardiac outflow tract [17].

NET expression in noncrest embryonic tissues

NET is also expressed in a variety of tissues that are not of neural crest origin. Notably, many components of the cardiovascular system express NET. In addition to the crest-derived cardiac outflow tract mesenchyme, the lining of arteries (e.g. the dorsal aorta) and veins (e.g. subcardial vein), the epicardium and the cardiac trabeculae are strongly NET-immunoreactive [8, 17]. Additional tissues include the notochord and the neural tube. The notochord expresses NET protein in the 60 h embryo, but immunoreactivity is significantly lower in the 84 h old embryo. NET immunoreactivity in the neural tube is limited to the roof plate in the 60 h old embryo, but covers the entire cross section of the neural tube in the 84 h embryo [8, 17]. Furthermore, the dermomyotome of the somite is strongly immunoreactive in both the 60 and 84 h embryo. Myotome-derived cells that are in the process of migrating into the limb bud continue to express NET protein.

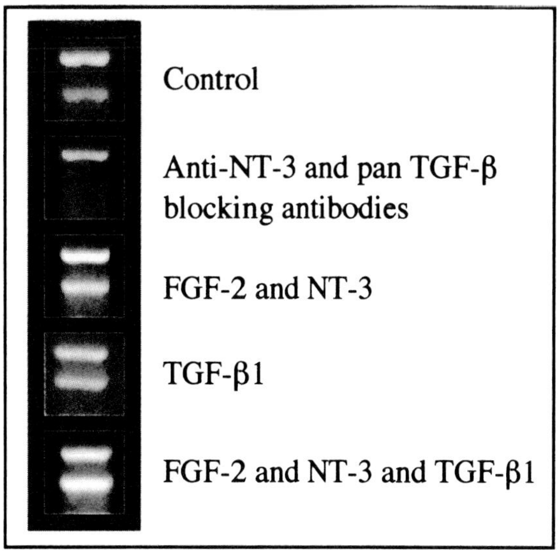

Fig. 1. NET mRNA levels in the presence and absence of growth factors. Neural crest cells in primary culture were grown in the presence and absence of FGF-2, NT-3, TGF-β1 and blocking antibodies. The top band in each panel is GAPDH RT-PCR product, whereas the bottom band is NET RT-PCR product. When the cells were grown in defined culture medium in the absence of noncrest cells, NET mRNA was detected. This suggested an autocrine mechanism (top panel). When the cells were grown in the presence of anti-NT-3 blocking antibodies and pan-TGF-β blocking antibodies, the NET band was significantly reduced (second panel), supporting the notion of autocrine NT-3 and TGF-β. In the presence of FGF-2 plus NT-3 (third panel) or TGF-β1 (fourth panel), NET mRNA levels were significantly higher compared to the antibody control. When all three factors were present (fifth panel), there was a further increase, indicating that the growth factors act in an additive manner. When ethidium bromide fluorescence was quantified, the following NET to GAPDH ratios were obtained. Control, 0.78 ± 0.008; blocking antibodies, 0.29 ± 0.019; FGF-2 and NT-3, 0.92 ± 0.006; TGF-β1, 0.86 ± 0.01; FGF-2 and NT-3 and TGF-β1, 1.34 ± 0.018.

Regulation of NET expression by autocrine growth factors

Our previous data indicate that NT-3, in particular the combination of FGF-2 and NT-3, and TGF-β1 promote NE transport activity in an additive manner in neural crest cell colony assays [8, 12]. To determine whether NET function is regulated at the message level, neural crest cell primary explants were cultured in the presence and absence of the three growth factors. Both the FGF-2/NT-3 combination and TGF-β1 caused an increase in NET mRNA levels (Fig. 1 and ref. [8]). The two effects were additive. Interestingly, there was a NET band present even with RNA from control cultures, which were grown for 24 h in a chemically defined culture medium in the absence of added growth factors and in the absence of

Fig. 2

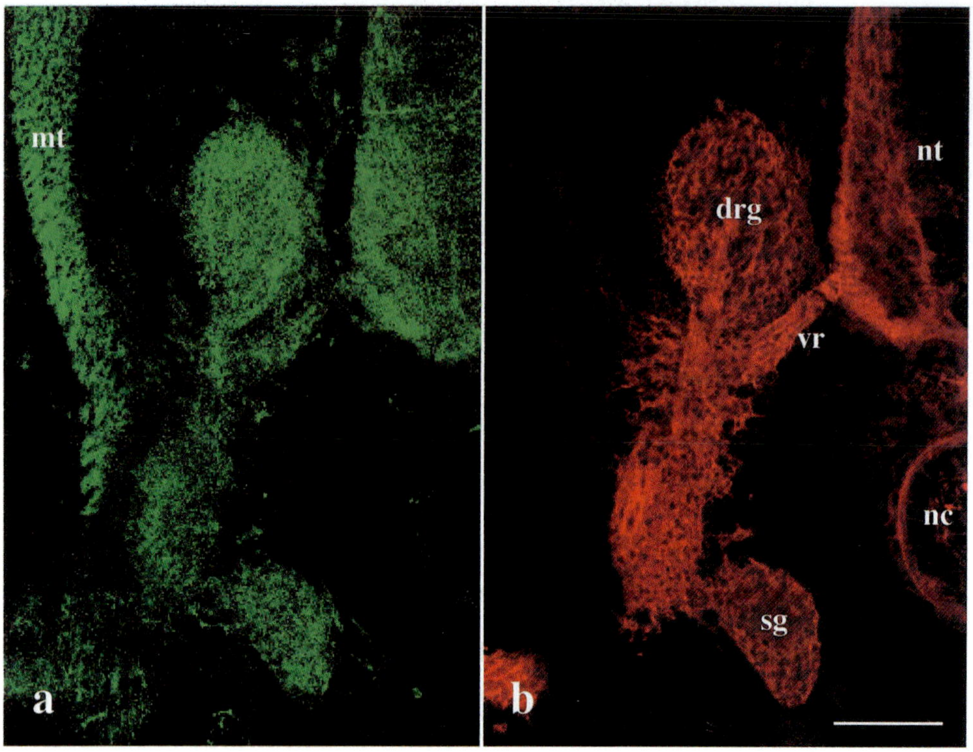

Fig. 3

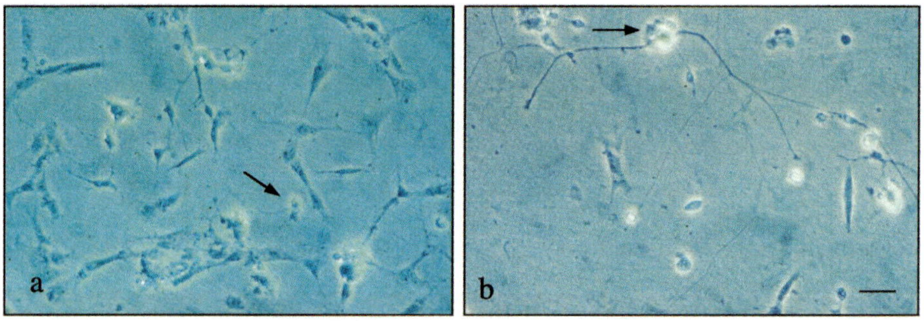

Fig. 10

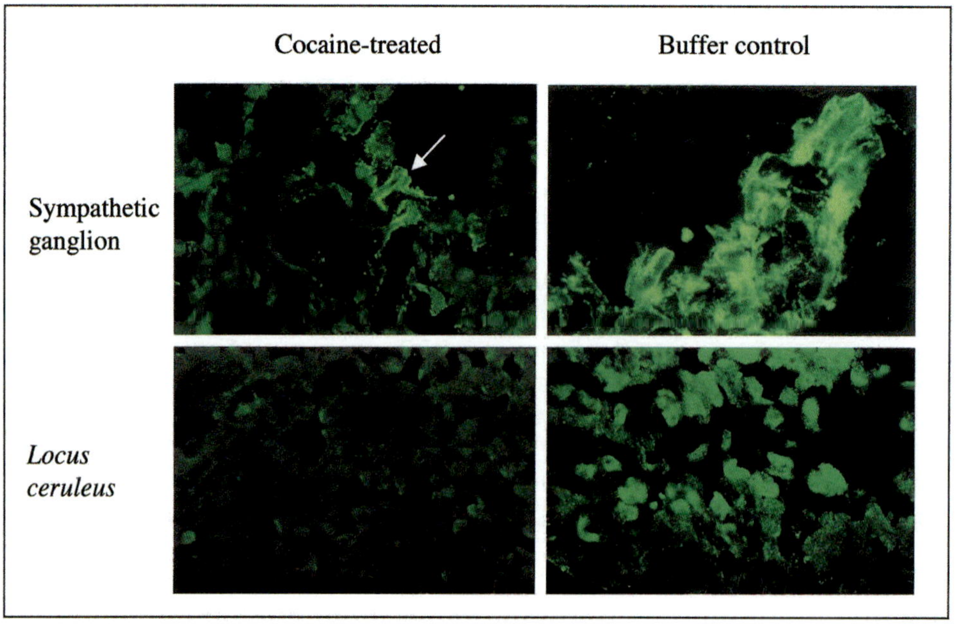

non-crest cells (Fig. 1, top panel). This observation suggested either autocrine expression of the candidate growth factors or NET expression prior to the onset of neural crest cell migration. To distinguish between these two possibilities, we amplified RNA with growth factor-specific primers. PCR products were obtained not only for NT-3, but also for TGF-β1, TGF-β2 and TGF-β3, indicating that all four growth factors are produced by neural crest cells and may therefore have autocrine function [8]. The data were confirmed at the protein level in both neural crest cell cultures and in the embryo [8]. As mentioned above, however, the dorsal neural tube and early migratory neural crest cells already express NET protein . We interpret the data therefore that NET expression in neural crest cells is regulated by autocrine growth factors that can act even before onset of their emigration from the neural tube.

←

Fig. 2. NET protein expression in the 84 h-old quail embryo. (a) NET fluorescein-immunoreactivity in cryosections of the trunk area. (b) HNK-1 stain (rhodamine fluorescence) in the same section. HNK-1 is characteristic of neural crest cells, but also recognizes N-CAM and some other glyscosaminoglycans. Several neural crest-derived and noncrest tissues are strongly NET immunoreactive: the dorsal root ganglion (drg), sympathetic ganglion (sg), presumptive Schwann cells in the ventral root (vr) and spinal nerve, the myotome (mt) and the neural tube (nt). Bar, 200 μm.

Fig. 3. Phase contrast photomicrographs of LC cells in high-density culture. (a) When presumptive LC tissue is taken from day 6 old quail embryos, most cells have a flattened stellate shape. A small minority had a neuronal morphology with a small cell body (arrow). The latter die within a few days in culture. (b) In contrast, when tissue was taken from day 9 old embryos, the portion of flattened cells was reduced. Instead, many large diameter multipolar neuroblasts were present (arrow). The latter take up NE by high affinity transport (Fig. 5) and express NET protein (not shown). Bar, 25 μm.

Fig. 10. Inhibition of DBH expression in the sympathetic ganglion and in the *locus ceruleus in vivo*. Quail embryos were treated with either 5 μM cocaine or phosphate buffered saline once each on embryonic days 3–8. Ten μl of 5 mM cocaine or 10 μl buffer were placed on top of the extraembryonic blood circulation in the vitelline membrane. Since the volume of a quail egg is 10 ml, this lead to a 5 μM cocaine concentration. The rationale for daily drug applications was the entrapment of the drug in the egg yolk. On day 9, embryos that did not have and gross morphological anomalies or developmental delays were sacrified and further processed. The areas of the sympathetic ganglia at trunk levels and of the LC were dissected, fixed with 4% paraformaldehyde, cryosectioned and then processed for DBH immunocytochemistry. Fluorescence in both the SG and LC was significantly reduced compared to buffer controls. Virtually no DBH-immunoreactive cell bodies were found in cocaine treated sympathetic ganglia. Rare fluorescent axons were observed (arrow). By contrast, virtually all cells in the buffer treated SG were intensely fluorescent. Equivalent results were observed with LC cells. Measurements of immunofluorescence of individual cells in all 4 groups indicated a 4–5-fold decrease in DBH immunofluorescence in the presence of the drug. This result indicated that cocaine, which blocks NE uptake, prevents the NE-induced enhancement of DBH expression.

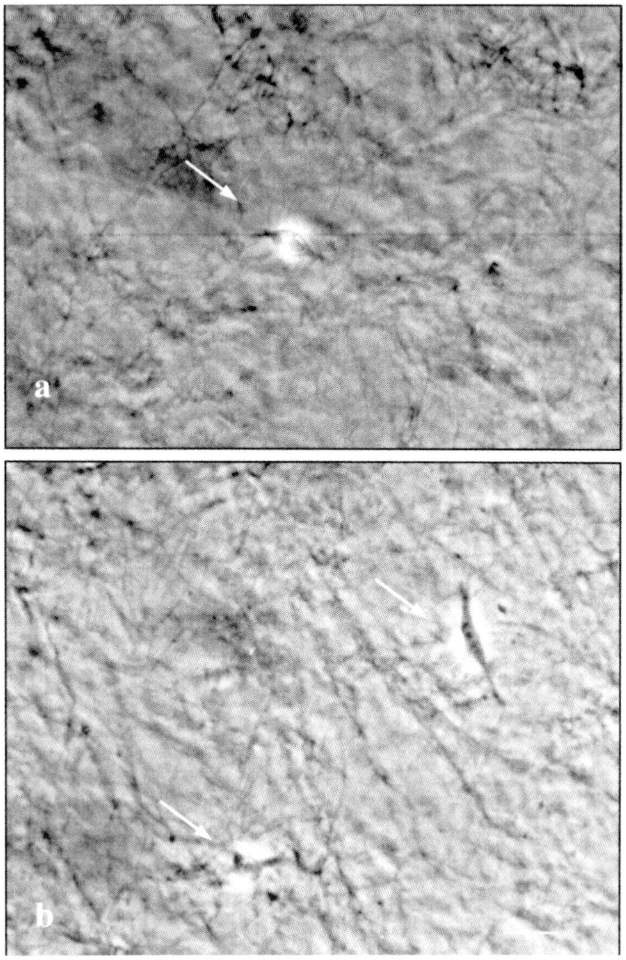

Fig. 4. LC cells at clonal density. For colony assays, LC cells are seeded at 250–500 cells per 35 mm culture plate. (a) LC cell (arrow) 1 h after plating (arrow). Fibers of the collagen substratum are visible in the background,. (b) Two daughter cells (arrows) 24 h after plating. Bar, 25 μm.

Norepinephrine transport promotes, and uptake inhibitors block, noradrenergic differentiation in neural crest cells

Tricyclic antidepressants and the drug of abuse, cocaine, block the differentiation of cultured neural crest cells into sympathetic neuroblasts [4, 5]. Both types of drug block the norepinephrine transporter. The drug-mediated inhibition of noradrenergic cell differentiation was dose-dependent, potency-dependent and transporter-specific [4]. Thus weak norepinephrine transporters elicited a weaker inhibition of noradrenergic differentiation than potent NE uptake inhibitors. Citalopram, an inhibitor of serotonin transport did not affect noradrenergic differentiation. The local anesthetic, lidocaine, did not affect noradrenergic differentiation either, indicating that general membrane disturbances are not responsible for the observed effect. The drug treatment was nontoxic, as it was reversible [4]. Taken together, these data indicate that the mechanism by which NE transport inhibi-

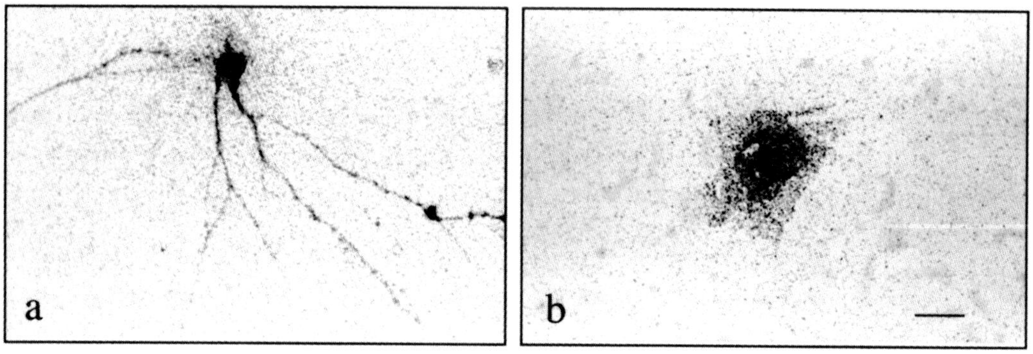

Fig. 5: NE transport-positive LC cells. LC cells from day 6 embryos were plated at clonal density for colony assays. At day 10 of culture, they were incubated with tritiated NE and processed for autoradiography (see Materials and methods section). There were two types of grain-positive cells ; (a) multipolar neuronal cells with a large cell body and (b) flattened cells. The flattened cells are thought to be progenitor cells because they express low levels of neuron-specific beta-III tubulin (data not shown). However, they are excluded form the analysis of subsequent experiments (Figs. 6 and 7). Bar, 25 μm.

tors block the differentiation of neural crest cells into sympathetic neurons is NET specific. Conversely, these observations also led to the hypothesis that NE uptake promotes noradrenergic differentiation. We were able to show this in neural crest colony assays [5]. Moreover, characterization of the quail NE transport activity indicated that the apparent Km is 1.3 μM, which is similar to that of the human NET and indicates high-affinity uptake [5].

Colony assays with quail neural crest cells indicated that norepinephrine promotes DBH and TH expression in a dose-dependent manner, as judged by immunocytochemistry [4, 12]. In the presence of NE, the number of colonies that contained DBH-immunoreactive cells was greater than in the buffer control, whereas the total number of colonies per plate remained unchanged. The fact that there is an increase in DBH and TH immunoreactivity suggested that NE promotes DBH and TH gene expression, not enzyme activity. Both desipramine, an inhibitor of NE transport, and cocaine, which among other actions also blocks NE uptake, caused an inhibition of the NE-induced increase in DBH-positive colonies [5, 12].

Application of NE to cultured neural crest cells caused an increase in calcium transients [12]. This observation suggests that the signaling pathway through which NE transport causes TH and DBH expression involves intracellular free calcium. It is conceivable that the co-transport of sodium during NE transport causes changes in the membrane potential that are sufficient to cause the opening of calcium channels and a

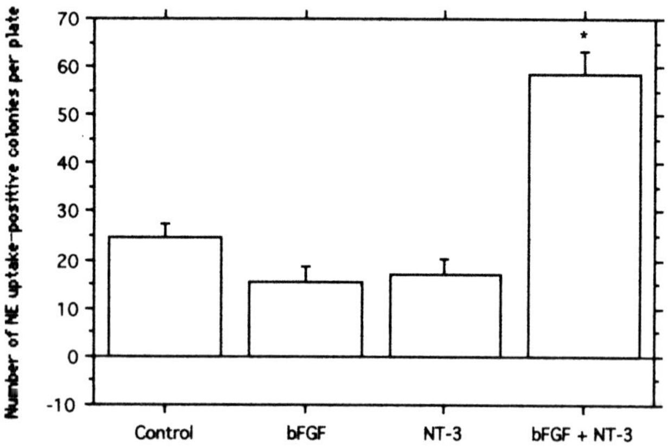

Fig. 6. Synergy between FGF-2 and NT-3 in promoting the ability of LC cells to transport NE. In the presence of FGF-2 or NT-3, the portion of uptake-positive colonies per plate was similar to that in the control (no growth factors). In control plates most 'colonies' consisted of single cells due to the absence of a mitotic stimulus. In the presence of both FGF-2 and NT-3, there was an approximately 3-fold, statistically significant increase in the portion of uptake-positive colonies, indicating that the two factors synergize in promoting high-affinity NE uptake.

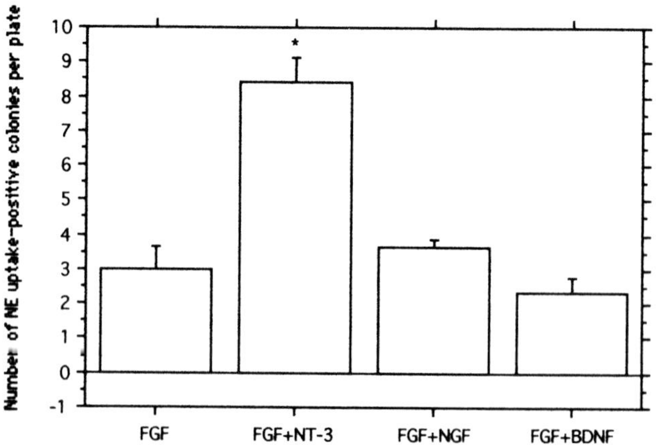

Fig. 7. Neurotrophin selectivity in LC cells. In the presence of FGF-2 plus NT-3, the portion of NE uptake-positive colonies was significantly increased over the control in LC colony assays. When FGF-2 was combined with either NGF or BDNF, no increase in the number of uptake positive colonies per plate was observed, indicating that NT-3 selectively synergizes with FGF-2 in promoting NE transport activity.

subsequent calcium influx [12]. However, this notion remains to be established and the downstream targets of calcium remain to be elucidated.

The locus ceruleus *colony assay*

In order to determine whether the mechanisms that promote NET activity in the LC are similar to the ones in the SG, we have developed a colony assay for noradrenergic differentiation of LC neurons. To isolate the LC anlage, the trunk and facial region of day 6 quail embryos are removed with electrolytically sharpened tungsten needles, retaining the mesencephalon and rostral rhombencephalon. In two parallel cuts that are perpendicular to the first cuts, the rostral pons that contains the LC anlage is dissected. By inserting a tungsten needle into the IVth ventricle, and pushing dorsally, the neural tube is splayed open. The alar plate-derived tissue is subsequently removed, leaving a small piece of tissue of the floor of the IVth ventricle, which contains the LC anlage. Pieces of tissue from several embryos are pooled, rinsed with phosphate-buffered saline and incubated with collagenase/Dispase (1 U/ml; Boehringer-Mannheim) for 10 min at 37°C. The tissue is subsequently triturated with a small bore pipet, the cell suspension filtered through a Nytex screen and then centrifuged. The resuspended cells consist of virtually 100% single cells. When plated on a collagen- and laminin-coated plates, they consist primarily of flattened stellate cells, but also contain small neuronal-like cells (Fig. 3a). The latter die within a few days in culture. Day 6 embryos were chosen as starting material, because older embryos already contain a significant number of mature neurons (Fig. 3b). Moreover, the onset of catecholamine synthesis and storage has been shown to be at day 9 of incubation in the chick [18] and quail embryo (Fig. 10). In clonal culture, the cells divide approximately every 24 h (Fig. 4). After several days in culture, flattened cells as well as mature neuronal cells are capable of high affinity NE transport, as judged by uptake of tritiated norepinephrine and subsequent autoradiography (Fig. 5). Moreover, these cells express NET as detected by immunocytochemistry (data not shown). The flattened NE uptake-positive cells are thought to be neuronal precursors because they express low levels of neuron-specific beta-III tubulin (data not shown). However, flattened uptake-positive cells were excluded from analysis (Figs 6 and 7). For colony assays, cultures are incubated for 5–10 days in a defined culture medium [10], supplemented with 0.5% chick embryo extract [10]. At the end of the culture period, cultures are processed for either radioactive NE uptake and autoradiography or for immunocytochemistry. The total number of colonies per plate and the number of colonies per plate that contain the cell type of interest (uptake-positive, DBH-immunoreactive) are scored in each control and experimental group. Each group consists of 8 10 culture plates. The data are statistically analyzed by 1-factor analysis of variance.

FGF-2 and NT-3 promote NE uptake in locus ceruleus *cells* in vitro

When LC cells were grown in colony assays, the combination of FGF-2 and NT-3, but not either factor alone, caused an increase in the number of NE-uptake positive colonies per plate, whereas the total number of colonies per plate remained unchanged (Fig. 6). NE uptake-positive cells were visualized by high affinity uptake of tritiated norepinephrine and subsequent autoradiography. This observation indicated that, like in the neural crest cells, the combination of FGF-2 and NT-3 promotes the ability to transport NE in the LC.

In the LC, as in the SG, NT-3 is the only neurotrophin of the NGF family of factors that promotes NE uptake. When cultured in the presence of FGF-2 and either nerve growth factor (NGF) or brain derived neurotrophic factor (BDNF), the number of NE uptake-positive colonies per plate was not significantly different from those in the control (FGF-2 alone; Fig. 7). This observation indicated that NT-3, but not NGF or BDNF, in combination with FGF-2 promotes expression or activation of NET in LC cells.

Norepinephrine promotes, and cocaine blocks, DBH expression in cultured locus ceruleus *cells*

NE transport promotes expression of TH and DBH in neural crest-derived sympathetic neuroblasts [5, 12]. To determine whether NE transport promotes DBH in LC cells as well, colony assays were performed with presumptive LC cells from day 6 quail embryos. NE caused a dose-dependent increase in the portion of colonies that contained DBH-positive cells, while the total number of colonies per plate remained unchanged (Fig. 8). This result indicated that NE does promote DBH expression in LC cells, as it does in neural crest-derived sympathetic neuroblasts.

The promoting effect of NE was inhibited by cocaine in a dose-dependent manner. In the presence of 2 μM cocaine, there was an approximately 80% decrease in the portion of colonies that contained DBH-immunoreactive cells (Fig. 9). There was a further decrease with higher cocaine concentrations. The total number of colonies per plate remained unchanged and the cells looked healthy.

Cocaine blocks noradrenergic differentiation in the sympathetic ganglion and in the locus ceruleus

In order to determine whether NE affects noradrenergic neuron differentiation *in vivo*, we treated quail embryos with 5 μM

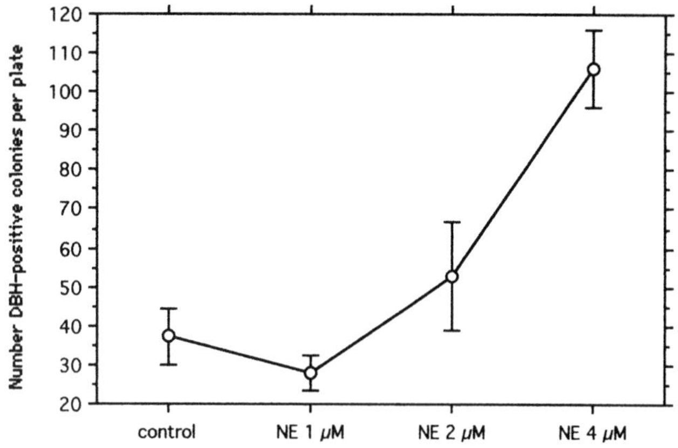

Fig. 8. NE-mediated increase in DBH expression. In the presence of NE, the portion of DBH-positive colonies per plate increase in a dose dependent manner, as measured by immunocytochemistry. The 4 μM value is statistically significantly different from the control ($p < 0.001$). This observation indicated that NE promotes expression of DBH.

cocaine once a day on embryonic days 3–8. On day 9, the embryos were terminated, the areas of the sympathetic ganglia and the LC dissected, cryosectioned and stained for DBH immunoreactivity. In both the sympathetic ganglia and the LC, DBH immunoreactivity was significantly reduced after drug treatment compared to buffer controls (Fig. 10). Measurements of fluorescence intensity in individual cells of the SG indicated an almost 5-fold reduction. The relative intensity of fluorescence in the SG was 125.7 ± 6.2 in buffer-treated control animals, but was reduced to 27.5 ± 1.7 in cocaine treated animals. The embryos did not show any gross

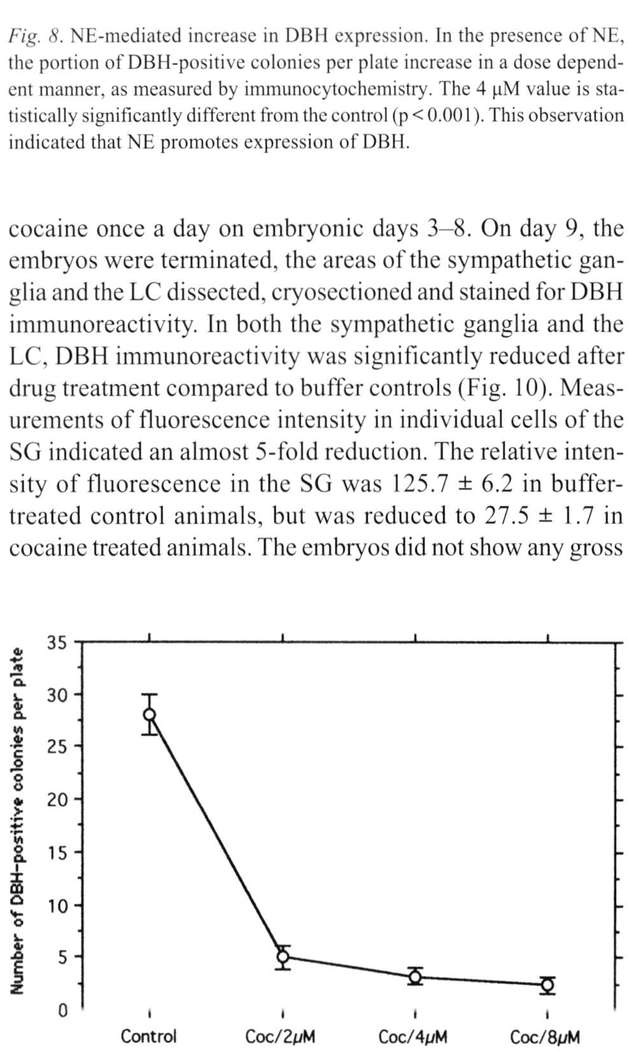

Fig. 9. Cocaine-mediated decrease in DBH expression. LC cells in colony assays were grown in the presence of FGF-2 (2.5 ng/ml), NT-3 (10 ng/ml) and NE (4 μM) and in the presence and absence of cocaine. There was a cocaine dose-dependent, and statistically significant ($p < 0.001$), decrease in the portion of the colonies that contained DBH-immunoreactive cells. This observation supported the notion that NE transport, not receptor activation, promotes DBH expression. Moreover, the data indicate that cocaine blocks noradrenergic LC differentiation *in vitro*.

anatomic abnormalities or delays in development. The data thus indicate that cocaine blocks the differentiation of both types of noradrenergic neurons *in vivo*.

Discussion

Taken together, there are four major conclusions:

1. FGF-2 and NT-3 regulate NET expression during quail embryonic development. Using a functional assay (uptake and autoradiography of radioactive norepinephrine in colony assays) we have shown for neural crest cells (which give rise to sympathetic neurons) and for LC cells, that the combination of FGF-2 and NT-3 selectively promotes NE uptake. Furthermore, RT-PCR has shown in the neural crest that the FGF-2/NT-3 combination causes an increase in NET message levels. It remains to be determined whether this holds true for LC cells as well. The receptors for NT-3, termed TrkC, TrkA and p75NTR, are expressed in neural crest cells [19] and in the LC of the rat [20] chick [21] and quail embryo (our data, not shown). We have shown that NT-3 is produced by the neural crest cells and therefore most likely has autocrine function. In the LC, NT-3 could be either target-derived or autocrine.

2. NET is expressed considerably earlier and by many more tissues than was expected. Presumptive sympathetic neurons show NET activity when they arrive in the vicinity of the dorsal aorta [7], where they settle and aggregate into the primary sympathetic ganglia. Our data show that NET mRNA and NET protein are expressed much earlier, in early migratory neural crest cells and possibly also in premigratory neural crest cells. As the leading edge of the cohort of migrating neural crest cells consists primarily of two classes of pluripotent stem cell [16], this indicates that at least some neural crest stem cells express NET. At trunk levels, NET expression is not limited to neural crest cells and sympathetic ganglia. Several other crest-derived and some noncrest tissues also express the transporter. Among neural crest derivatives, the dorsal root ganglion, presumptive Schwann cells in the ventral root and spinal nerve, mesenchymal cells (presumptive smooth muscle cells) in the cardiac outflow tract and mesenchymal cells in the gut (presumably presumptive enteric neurons) are NET-immunoreactive. Among noncrest tissues at trunk axial levels, both the neural tube and the notochord express NET. In the 64 h old embryo, the notochord is strongly positive, whereas NET immunoreactivity in the neural tube is limited to the dorsal aspect, where neural crest cells delaminate. By contrast, in the 80 h-old embryo, the NET immunoreactivity is down regulated in the notochord, but is strong throughout the cross-section of the neural tube. NET expression is also widespread in the cardiovascular system. The lining of arteries (dorsal aorta; [8]) and veins (subcardial vein; [17]) is consistently and strongly

immunoreactive. It remains to be determined whether it is the endothelial cells that express NET. Moreover, the heart itself expresses NET. In addition to the neural crest-derived cells in the outflow tract, NET localizes also to the cardiac trabeculae and the epicardium. NET expression in the nervous and cardiovascular systems is relevant with regard to the recent observation by Shannon *et al.* [22] that a mutation in NET leads to orthostatic intolerance, which presents with tachycardia, fatigue and fainting.

It is important to note that NET protein expression does not necessarily translate into transporter function. In the neural crest for instance, it has been shown that the cells acquire the capacity to transport NE when they have migrated to the dorsal aorta [7]. By contrast, NET protein is expressed already at initiation of neural crest cell migration. Because of the widespread expression of the NET, it will be of great interest to determine whether NET can play another role in addition to NE transport, such as having a receptor-like function.

3. NE promotes expression of the noradrenergic biosynthetic enzymes, DBH and TH in neural crest via a desipramine-sensitive and cocaine-sensitive signaling pathway. Moreover, our present data indicated that NE promotes, and cocaine inhibits, DBH expression in noradrenergic LC neurons as well. Since both types of drug inhibit NE transport, but not adrenergic receptor activation, it is thought that NE transport is involved. Addition of NE to cultured neural crest cells causes an increase in intracellular free calcium [12]. Since there is concomitant co-transport of sodium ions during NE transport [23], it is conceivable that the change in membrane potential is sufficiently pronounced to cause opening of calcium channels and subsequent calcium influx. It is of interest that both DBH and TH expression are enhanced by NE. This is in contrast to BMP2, which is synthesized in the dorsal aorta and promotes DBH expression via a MASH-1 and phox2a mediated pathway [24]. BMP-2 is synthesized in the dorsal aorta and may thus act slightly later than NE. Since the DBH and TH promoter contain both common and unique responsive elements [25], it is conceivable that parallel signaling pathways need to be activated for optimal noradrenergic gene expression.

4. The observation that NE uptake inhibitors inhibit the NE-mediated increase in DBH and TH expression both in the sympathetic nervous system and in the LC may have clinical relevance. Although it has been established that children who have been exposed to the drugs *in utero* have no gross developmental and behavioral anomalies [26], there are several reports of more subtle developmental deficiencies. Children born to mothers that have used antidepressants during pregnancies or taken cocaine have a higher prevalence for sudden infant death syndrome, which has been attributed to a delay in LC development [27–31]. Children that have been exposed to these drugs prenatally also have transient postnatal autonomic disturbances. Other adverse effects become more evident in later childhood. They manifest in poor organization skills, reading problems, and below average mathematical skills [30]. Ten to 45% of women cared for at urban teaching hospitals take cocaine during pregnancy [31]. In a different study it was found that up to 6% of infants born at private suburban hospitals in the U.S. had cocaine metabolites in their meconium [31]. The observation that tricyclic antidepressants and cocaine inhibit noradrenergic differentiation may therefore be relevant to drug-related public health issues.

Acknowledgements

This work was supported by USPHS grant NS38281 from the National Institute of Neurological Disorders and Stroke and a research grant from the Medical College of Wisconsin (MSB). We thank Peter Pörzgen and Susan Amara for cloning a partial quail NET transporter cDNA, Randy Blakely and Sally Schroeter for a generous gift of the anti-NET antibody, and Yves Barde and Ilse Bartke for anti-NT-3 blocking antibodies.

References

1. Axelrod J: The metabolism, storage, and release of catecholamines. Rec Prog Horm Res 21: 597–619, 1965
2. Iversen LL: The Uptake and Storage of Noradrenaline in Sympathetic Nerves. Cambridge University Press, New York, 1967
3. Snyder SH: Putative neurotransmitters in the brain: Selective neuronal uptake, subcellular localization, and interactions with centrally acting drugs. Biol Psychiatry 2: 367–389, 1970
4. Sieber-Blum M: Inhibition of the adrenergic phenotype in cultured neural crest cells by norepinephrine uptake inhibitors. Dev Biol 136: 372–380, 1989
5. Zhang J-M, Sieber-Blum M: Characterization of the norepinephrine uptake system and the role of norepinephrine in the expression of the adrenergic phenotype by quail neural crest cells in clonal culture. Brain Res 570: 251–258, 1992
6. Strudel G, Recasens M, Mandel P: Identification de catecholamines et de serotonine dans les chordes d'embryons de poulet. CR Acad Sci Paris 284: 967–969, 1977
7. Rothman TP, Gershon MD, Holtzer H: The relationship of cell division to the acquisition of adrenergic characteristics by developing sympathetic ganglion cell precursors. Dev Biol 65: 322–341, 1978
8. Ren ZG, Pörzgen P, Zhang J-M, Chen XR, Amara SG, Blakely RD, Sieber-Blum M: Autocrine regulation of NET expression. Submitted
9. Sieber-Blum M, Cohen AM: Clonal analysis of quail neural crest cells: They are pluripotent and differentiate *in vitro* in the absence of non-crest cells. Dev Biol 80: 96–106, 1980
10. Sieber-Blum M: The neural crest colony assay: assessing molecular influences on development in culture. In: The Neuron in tissue Culture, IBRO. John Wiley & Sons Ltd. 1999, pp 5–22
11. Hamburger V, Hamilton H: A series of normal stages in the development of the chick embryo. J Morphol 88: 49–92, 1951
12. Zhang J-M, Dix J, Langtimm-Sedlak C, Trusk T, Schroeder B, Strosberg AD, Winslow JW, Sieber-Blum M: Neurotrophin-3 and norepinephrine-mediated adrenergic differentiation and the inhibitory action of desipramine and cocaine. J Neurobiol 32: 262–280, 1997

13. Gaese F, Kolbeck R, Barde Y-A: Sensory ganglia require neurotrophin-3 early in development. Development 120: 1613–1619, 1994
14. Panabieres F, Piechaczyk M, Rainer B, Dani C, Fort P, Riaad S, Marti L, Imbach JL, Jeanteur P, Blanchard J-MM: Complete nucelotide sequence of the messenger RNA coding for chicken muscle glyceraldehyde-3-phosphate dehydrogenase. Biochem Biophys Res Commun 118: 767–773, 1984
15. Schroeter S, Apparsundaram S, Wiley RG, Miner LAH, Sesack SR, Blakely RD: Immunolocalization of the cocaine- and antidepressant-sensitive l-norepinephrine transporter. J Comp Neurol: 1999 (in press)
16. Sieber-Blum M: Commitment of neural crest cells to the sensory neuron lineage. Science 243: 1608–1611, 1989
17. Ren ZG, Pörzgen P, Schroeter S, Amara S, Blakely R, Sieber-Blum M: Embryonic NET expression in the nervous and cardiovascular systems and in muscle cells. Submitted
18. Guglielmone R, Panzica GC: Topographic, morphologic and developmental characterization of the nucleus *loci coerulei* in chicken. Cell Tissue Res 225: 95–110, 1982
19. Henion PD, Garner AS, Large TH, Weston JA: TrkC-mediated NT-3 signaling is required for the early development of a subpopulation of neurogenic neural crest cells. Dev Biol 172: 602–613, 1995
20. Merlio J-P, Ernfors P, Jaber M, Persson H: Molecular cloning of rat TrkC and identification of cells expressing mRNA for members of the trk family in the rat central nervous system. Neuroscience: 1992
21. Von Bartheld CS, Schober A, Kinoshita Y, Williams R, Ebendahl T, Bothwell T: Noradrenergic neurons in the locus ceruleus of birds express TrkA, transport, NGF, and respond to NGF. J Neurosci 15: 2225–2239, 1995
22. Shannon JR, Flattem NL, Jordan J, Jacob G, Black BK, Biaggioni I, Blakely RD, Robertson D: Orthostatic intolerance and tachycardia associated with norepinephrine transporter deficiency. New Engl J Med 432: 541–549, 2000
23. Amara SG, Kuhar MJ: Neurotransmitter transporters: Recent progress. Annu Rev Neurosci 16: 73–93, 1993
24. Lo L, Tiveron MC, Anderson DJ: MASH1 activates expression of the paired domain transcription factor Phx2a, and couples pan-neuronal and subtype-specific components of autonomic neuronal identity. Development 125: 609–620, 1998
25. Kim HS, Seo H, Yang C, Brunet JF, Kim KS: Noradrenergic-specific transcription of the dopamine beta-hydroxylase gene requires synergy of multiple cis-acting elements including at least two Phox2a-binding sites. J Neurosci 15: 8247–8260, 1998
26. Hurt H, Brodsky N, Betancourt L, Braitman LE, Malmud E, Giannetta J: Cocaine-exposed children: Follow-up through 30 months. J Dev Behav Ped 16: 29–35, 1995
27. Gingras JL, O'Donnell KJ, Hume RF: Maternal cocaine addiction and fetal behavioral state. I: A human model for the study of sudden infant death syndrome. Med Hypoth 33: 227–230, 1990
28. Gingrass JL, Weese-Mayer D: Maternal cocaine addiction. II: An animal model for the study of brainstem mechanisms operative in sudden infant death syndrome. Med Hypoth 33: 227–230, 1990
29. Gingrass JL, Weese-Mayer DE, Hume RF Jr, O'Donnell KJ: Cocaine and development: Mechanism of fetal toxicity and neonatal consequences of prenatal cocaine exposure. Early Hum Dev 31: 1–24, 1992
30. Hill RM, Tennyson LM: Maternal drug therapy: Effect on fetal and neonatal growth and neurobehavior. Neurol Toxicol 7: 121–140, 1986
31. Volpe JJ: Effect of cocaine use in the fetus. New Engl J Med 327: 399–407, 1992

PART II
CATECHOLAMINE AND ANGIOTENSIN II INFLUENCES ON GENE EXPRESSION

Molecular and Cellular Biochemistry **212:** 73–79, 2000.

Catecholamines and angiotensinogen gene expression in kidney proximal tubular cells

John S.D. Chan,[1] Tian-Tian Wang,[1] Shao-Ling Zhang,[1] Xing Chen[1] and Serge Carrière[2]

[1]*University of Montreal, Maisonneuve-Rosemont Hospital Research Center, Montreal, Quebec;* [2]*Servier Amérique, Laval, Quebec, Canada*

Abstract

To investigate the molecular mechanism(s) of action of catecholamines on the expression of the angiotensinogen (ANG) gene in kidney proximal tubular cells, we used opossum kidney (OK) cells with a fusion gene containing the 5′-flanking regulatory sequence of the rat ANG gene fused with a human growth hormone (hGH) gene as a reporter, pOGH (rANG N-1498/+18), permanently integrated into their genomes. The level of expression of the ANG-GH fusion gene was quantified by the amount of immunoreactive-hGH (IR-hGH) secreted into the medium. The addition of norepinephrine (NE), isoproterenol (a β_1/β_2-adrenergic receptor (AR) agonist) and iodoclonidine (an α_2-AR agonist) stimulated the expression of the ANG-GH fusion gene in a dose-dependent manner, whereas the addition of epinephrine and phenylephrine (α_1-AR agonist) had no effect. The stimulatory effect of NE was blocked by the presence of propranolol (β-AR blocker), atenolol (β_1-AR blocker), yohimbine (α_2-AR blocker), Rp-cAMP (an inhibitor of cAMP-dependent protein kinase AI & AII) and staurosporine (an inhibitor of protein kinase C), but was not blocked by ICI 118, 551 (β_2-AR blocker) and prazosin (α_1-AR blocker). The addition of a combination of isoproterenol and iodoclonidine or a combination of 8-Bromo-cAMP (8-Br-cAMP) and phorbol 12-myristate (PMA) synergistically stimulated the expression of the ANG-GH fusion gene as compared to the addition of isoproterenol, iodoclonidine, 8-Br-cAMP or PMA alone. Furthermore, the addition of NE, 8-Br-cAMP or PMA stimulated the expression of pOGH (rANG N-806/-779/-53/+18), a fusion gene containing the putative cAMP responsive element (CRE, ANG N-806/-779) upstream of the ANG promoter (ANG N-53/+18) in OK cells, but had no effect on the expression of fusion genes containing the mutant of the CRE. Gel mobility shift assays revealed that the ANG-CRE binds with the DNA-binding domain (bZIP 254-327) of the cAMP-responsive binding protein (CREB). The binding of the labeled ANG-CRE to CREB (bZIP254-327) was displaced by unlabeled ANG-CRE and the CRE of the somatostatin gene but not by the mutants of the ANG CRE. Finally, NE stimulated the phosphorylation of CREB in OK cells. These studies demonstrate that the molecular mechanism(s) of NE action on the expression of the ANG gene in OK cells may be mediated via both the PKA and PKC signalling pathways and via the phosphorylation of CREB. The phosphorylated CREB then interacts with the CRE in the 5′-flanking region of the ANG gene and subsequently stimulates the gene expression. (Mol Cell Biochem **212**: 73–79, 2000)

Key words: adrenergic receptor, renin-angiotensin system (RAS), gene expression, kidney

Abbreviations: AC – adenylyl cyclase; ANG – angiotensinogen; Ang I – angiotensin I; Ang II – angiotensin II; ACE – angiotensin converting enzyme; AR– adrenergic receptor; CRE – cyclic AMP responsive element; CREB – cyclic AMP responsive element binding protein; NE – norepinephrine; OK – opossum kidney; PKA – protein kinase A; PKC – protein kinase C; RAS – renin-angiotensin system

Address for offprints: J.S.D. Chan, University of Montreal, Maisonneuve-Rosemont Hospital Research Center, 5415 Boulevard de l'Assomption, Montreal, Quebec, Canada, H1T 2M4

Systemic and local intra-renal renin-angiotensin system (RAS)

Angiotensinogen (ANG) is a glycoprotein consisting of 452 amino acid residues (approximately 62,000 daltons in mol. wt.) with the angiotensin II (Ang II) sequence occurring at its amino terminal portion [1, 2]. Ang is predominantly expressed in the liver. The ANG gene is approximately 11.8 kilobase pairs long, consisting of 5 exons separated by 4 introns [3, 4]. The ANG gene is a single copy and has been mapped to chromosomes 19, 8 and 1 for the rat, mouse and human, respectively [5–7]. Genetic linkage studies in humans show that the ANG gene locus is associated with a predisposition to the development of hypertension in certain populations [8–15], whereas a weak or no genetic linkage is associated with the gene locus of renin [16], ACE [17–19] and Ang II-receptor [20, 21]. Studies have also shown that transgenic mice which carry an exogenous ANG gene and/or renin gene develop hypertension [22]. These studies demonstrate that there is a strong relationship between the expression of the ANG gene and the development of hypertension.

The systemic RAS has been well recognized and extensively studied for the past four decades. The existence of a local intrarenal RAS, however, has now generally been accepted [23, 24]. The mRNA and protein components of the RAS, including ANG, renin, ACE and Ang II receptors are expressed in murine renal proximal tubular cells [25–38]. We [39], as well as Ingelfinger *et al.* [40] have demonstrated that the ANG mRNA is expressed in opossum kidney (OK) cells. Moreover, we have recently reported that the ANG protein is secreted from rat immortalized renal proximal tubular cells (IRPTC) as measured by a specific radioimmunoassay for rat ANG (RIA-rANG) [41], providing evidence that the intrarenal Ang II is derived from the ANG that is synthesized within the RPTC. Physiological studies have also shown that the levels of proximal tubular luminal Ang I or Ang II are as high as 10^{-9} M [42–44]. Thus, the high levels of Ang I and Ang II found in the renal tubular fluid can not be accounted for from the circulation (i.e. 10^{-12} M).

Ang II has multiple effects on the kidney. Ang II inhibits renin release from cells of the juxtaglomerular apparatus [45] and facilitates sympathetic nerve transmission [46]. Ang II preferentially constricts the postglomerular efferent arterioles more than the preglomerular afferent arterioles. This action helps to maintain the glomerular filtration rate when renal blood flow falls [47, 48]. Ang II also modifies the filtration coefficient (K_f) via the contraction of the mesangial cells (125). Numerous studies have shown that Ang II at the physiological range of 10^{-12}–10^{-10} M increases sodium and fluid reabsorption by renal proximal tubules, whereas higher Ang II concentrations (10^{-8}–10^{-6} M) inhibit sodium transport [49–56]. The stimulatory effect of Ang II is abolished by micromolar concentrations of amiloride [56], indicating that the effect of Ang II is mediated via the stimulation of the Na^+/H^+ exchanger (i.e. NHE_3).

In summary, all of the above studies suggest that the local intra-renal RAS may play an important role in the modulation of the physiology of the renal proximal tubular cells (i.e. sodium and fluid reabsorption) and the over-expression of renal Ang II may play a role in the development of hypertension.

Catecholaminergic receptors in kidney proximal tubular cells

Renal proximal tubular cells are richly innervated by adrenergic nerves [57]. Denervation and activation of renal nerves have been shown to diminish and to increase sodium and fluid reabsorption by proximal tubules, respectively [58–63]. The mechanism(s) in which renal nerves exert their effects, however, is not well understood.

Norepinephrine (NE) is the major endogenous neurotransmitter of the renal sympathetic nerve [64, 65]. NE interacts with both the α- and β-adrenergic receptors (ARs) in the renal proximal tubules. The existence of functional β-Ars along the rat and rabbit renal proximal tubules has been reported [66, 67] and studies by Taniguchi *et al.* [68] have localized the β-AR mRNA in the early proximal convoluted tubule. Similarly, α-ARs have been detected on rat kidney proximal tubules with the dominance of the α_2-ARs [69–72]. It is estimated that the number of α_2-ARs is over 2-fold greater than α_1-ARs in rat kidney tubules [73]. A ratio of 2.9 (i.e. α_2-ARs vs. α_1-ARs) has been found in spontaneously hypertensive rats (SHR) and both renal α_1- and α_2-ARs densities were higher in SHR than in normotensive WKY rats at all ages [74], suggesting that the biologic effect of α_2-ARs may predominate over the effect of α_1-ARs in SHR. Studies have also shown that renal nerves, acting through α-ARs enhance proximal tubular sodium reabsorption (anti-natriuresis) in the kidney [72,75]. All of these studies suggest that NE, acting via β- and/or α-ARs, may contribute to the development of hypertension.

Molecular mechanism(s) of action of norepinephrine on angiotensinogen gene expression in kidney proximal tubular cells

A functional relationship between renal sympathetic nerve activity and the activation of renal RAS has been demonstrated in 1994 by the studies of Nakamura and Johns [76] and ourselves [77]. Studies by Nakamura and Johns [76] reported that low levels of renal nerve stimulation decrease sodium excretion and increase the levels of ANG mRNA but not renin mRNA in the rat kidney *in vivo*. Administration of atenolol (a β_1-AR blocker) blocks the effect of renal nerve stimulation on the increase of the ANG mRNA. Our *in vitro* studies [77] have shown that isoproterenol (a β_1/β_2-AR agonist) stimulates the ANG gene expression in opossum kidney (OK) proximal tubular cells. The OK proximal tubular cells were obtained from the American Tissue Culture Collection (ATCC) (Rockville, MD, USA). This cell line is derived from the kidney of a female American opossum and retains several properties of proximal tubular epithelial cells in culture [78, 79] and expresses a low level of ANG mRNA [39, 40]. Both β and α-ARs are also expressed in OK cells [80–82].

We have shown that OK cells express the fusion genes containing the 5′-flanking sequence of the ANG gene fused with either the hGH gene (i.e. pOGH (rANG N-1498/+18) or the chloramphenicol acetyltransferase (CAT) coding sequence (i.e. pOCAT (rANG N-1498/+18)) when these fusion genes are transiently transfected into OK cells [39, 83]. Moreover, the addition of thyroid hormones or dexamethasone stimulates the expression of these fusion genes in OK cells [39, 83]. These studies demonstrates that the fusion genes are functional and are hormone-responsive in OK cells. Furthermore, these studies suggest that OK cells, with the fusion gene stably integrated into their genomes and are able to express the fusion genes, would represent an ideal *in vitro* system to study the molecular mechanism(s) of action of catecholamines on ANG gene expression.

We have constructed fusion genes with various lengths of the 5′-flanking region of rat ANG gene fused with a hGH gene as reporter and stably transfected them into OK cells [84–88]. The level of expression of the fusion gene was quantified by the amount of immunoreactive-hGH (IR-hGH) secreted into the medium. We have shown that the addition of NE directly stimulates the expression of (pOGH (rANG N-1498/+18) in OK 27 cells in a dose-dependent manner [89]. This stimulatory effect of NE is inhibited in the presence of propranolol (a β_1/β_2-AR blocker), atenolol and yohimbine (an α_2-AR blocker). On the other hand, the addition of ICI 118,511

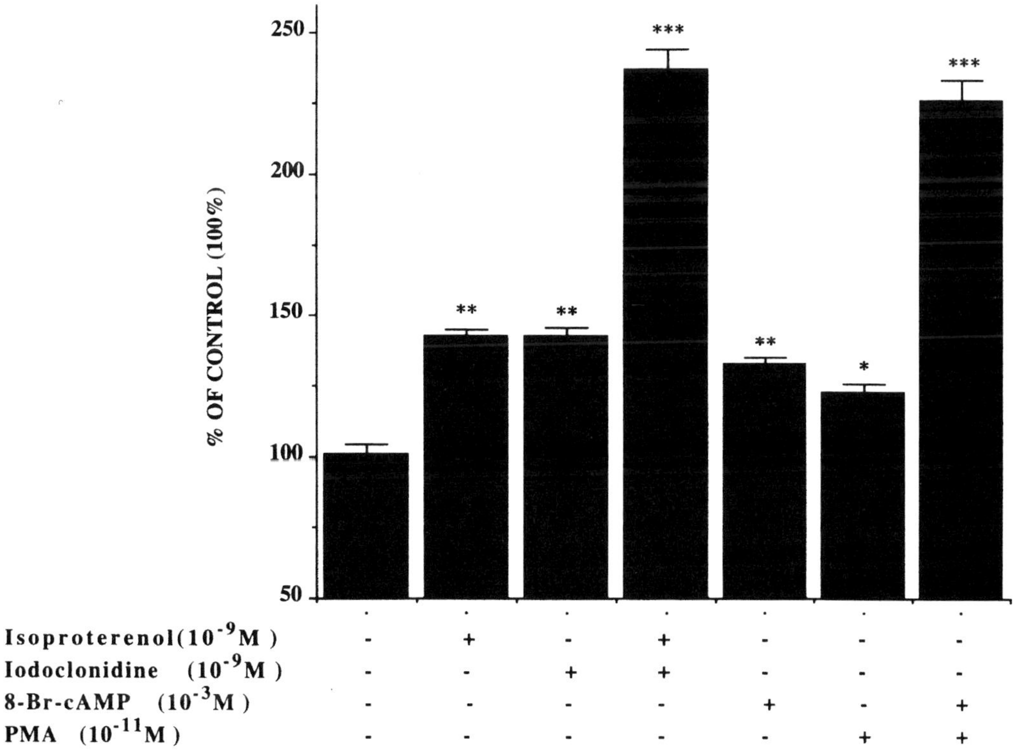

Fig. 1. Effect of iodoclonidine and isoproterenol or PMA and 8-Bromo-cAMP on the expression of pOGH (rANG N-1498/+18) in OK 27 cells. Cells were plated at a density of 2×10^5 cells/well in six-well plates for up to 24 h and incubated overnight in DMEM containing 10% FBS. The growth of the cells was arrested by incubation in serum-free DMEM for 24 h. Subsequently, isoproterenol (10^{-9} M) or iodoclonidine (10^{-9} M) or 8-Bromo-cAMP (10^{-3} M) or PMA (10^{-11} M) were added to the culture medium containing 1% depleted FBS and incubated for 24 h. Media were harvested after the incubation and assayed for immunoreactive human growth hormone (IR-hGH). Levels of IR-hGH in the medium in the absence of isoproterenol or iodoclonidine or 8-Bromo-cAMP or PMA are expressed as 100% (control). Each point represents the mean ± S.D. of at least 3 dishes (*$p \leq 0.05$; **$p \leq 0.01$; ***$p \leq 0.005$). The experiment has been done 3 times in triplicate.

(a β_2-AR blocker) and prazosin (an α_1-AR blocker) did not inhibit the stimulatory effect of NE on the expression of the pOGH (rANG N-1498/+18) in OK 27 cells. These studies demonstrate that the effect of NE is mediated via the β_1-AR and/or the α_2-AR. These studies also confirm our previous studies that the activation of the β_1-AR or α_2-AR alone also stimulates the expression of pOGH (ANG N-1498/+18) in OK 27 cells [85, 87, 90].

Furthermore, we have shown that the addition of Rp-cAMP (an inhibitor of PKA), staurosporine or H-7 (inhibitors of PKC) inhibit the stimulatory effect of NE [89], indicating that the stimulatory effect of NE may be mediated via either the PKA or PKC signal transduction pathway or via a 'cross-talk' of both the PKA and PKC signal transduction pathways. Indeed, our studies showed that the stimulatory effect of the addition of a combination of isoproterenol and iodoclonidine or a combination of 8-Br-cAMP and PMA on the expression of the pOGH (rANG N-1498/+18) is significantly higher than the addition of individual isoproterenol, iodoclonidine, 8-Br-cAMP or PMA ($p \leq 0.05$) as shown in Fig. 1, suggesting a synergistic effect of β_1- and α_2-adrenoceptors or PKA and PKC pathways on the ANG gene expression in OK cells.

Most interestingly, we have shown that the addition of NE stimulates the expression of the pOGH (rANG N-806/-779/-53/+18) in OK 95 cells [89]. OK 95 cell is a transformant with the fusion gene, pOGH (rANG N-806/-779/-53/+18) stably integrated into the genome. The DNA fragment, ANG N-806/-779 contains a CRE-motif (that is, N-795 to N-788, TGACGTAC), which is almost identical to the consensus CRE-motif of the somatostatin gene (that is TGACGTCA) [91]. Moreover, the stimulatory effect of NE is inhibited in the presence of propranolol or yohimbine [89], further demonstrating that the NE-responsive element is localized in the DNA fragment, ANG N-806/-779 of the rat ANG gene.

The addition of 8-Bromo-cAMP, forskolin or PMA alone also stimulates the expression of the pOGH (rANG N-806/-779/-53/+18) in OK 95 cells but not the expression of the mutants of pOGH (rANG N-806/-779/-53/+18) [89]. These studies suggest that the molecular mechanism(s) for the stimulatory effect of NE might be mediated via the phosphorylation of the nuclear cAMP-responsive element binding protein (CREB) at serine 133 via either PKA or PKC or a combination of both pathways, since studies have shown that CREB can be phosphorylated at serine 133 by cAMP-dependent protein kinase A [92], or by PKC [93, 94] and this phosphorylation increases the binding affinity of CREB to CRE [93, 94] which subsequently enhances gene expression. This possibility is supported by our studies which showed that the addition of isoproterenol enhances the stimulating effect of CREB on the expression of the pOGH (ANG N-1498/+18) in OK 27 cells [95] and the DNA fragment rat ANG N-806/-779 binds with the DNA-binding domain (bZIP 254-327) of the CREB as shown in Fig. 2. The binding of the labeled ANG

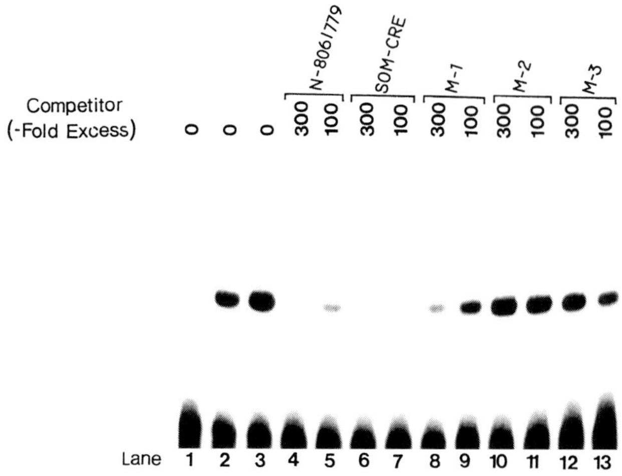

Fig. 2. Gel mobility shift assay of the radioactive-labeled DNA fragment. ANG-CRE with the purified DNA binding domain of CREB (CREB bZIP (254-327)). The labeled DNA probe (0.1 pmole) was incubated with BSA (10 μg) (lane 1) or CREB bZIP (254-327) (1 μg) (lanes 2–13 in the presence of 0.03 units of poly dI/dC. Competition with various amounts of unlabeled ANG-CRE (ANG N806/-779), SOM-CRE, mutant 1, mutant 2 and mutant 3 of ANG-CRE are shown in lanes (4 and 5), (6 and 7), 8 and 9), (10 and 11) and (12 and 13), respectively. Similar observations were obtained from 2 other experiments.

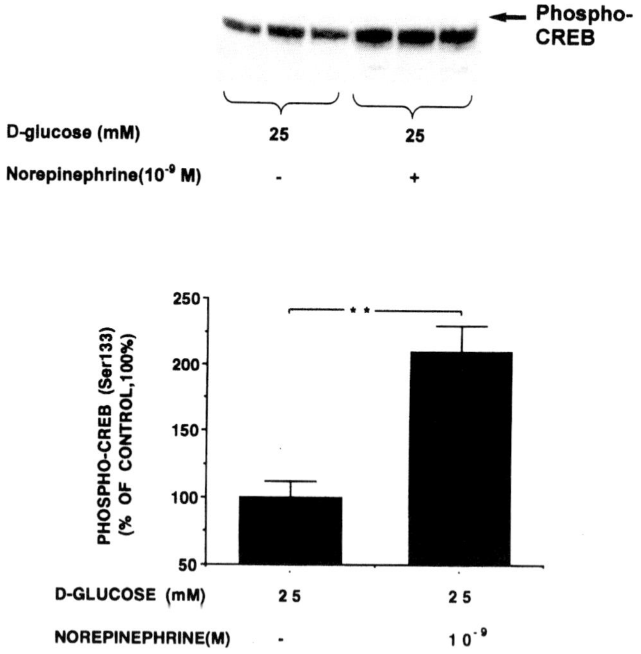

Fig. 3. Effect of norepinephrine on the phosphorylation of CREB in OK 96 cells. Cells were incubated in medium containing 1% depleted FBS in the absence or presence of NE (10^{-9} M) for 10 min. Then, cells were harvested assayed for the phosphorylated CREB by employing the Phospho Plus CREB antibody kit. The relative densities for the phosphorylated CREB in the cells incubated without NE added were considered as control (100%). Each point represents the mean ± S.D. of 3 dishes (*$p \leq 0.05$; **$p \leq 0.01$; ***$p \leq 0.005$). Similar results were obtained from another experiment.

N-806/-779 to CREB (bZIP 254-327) is displaced by unlabeled ANG N-806/-779 and the CRE of the somatostatin gene but not by the mutants of the ANG N-806/-779.

Most convincingly, our recent studies showed that the addition of NE stimulates the phosphorylation of CREB in OK 96 cells (an OK cell line with both pOGH (rANG N-806/-779/-53/+18) and pRSV/CREB stably co-transfected into the genome) [90] as shown in Fig. 3. These studies demonstrate that the stimulatory effect of NE on the expression of the ANG gene is mediated, at least in part, via the phosphorylation of CREB. This phosphorylated CREB then interacts with the CRE in the 5′-flanking region of the ANG gene and thereby enhances gene expression.

Conclusion

The molecular mechanism(s) of action of NE on the expression of the renal ANG gene is depicted in Fig. 4. In this model, we propose that NE interacts with both the α_2- and β_1-ARs in the renal proximal tubules and activates both PKC and PKA signal transduction pathways, respectively. The activation of PKC and/or PKA stimulates the phosphorylation of CREB in the nucleus. This phosphorylated CREB then binds to the CRE in the 5′-flanking of the ANG gene and subsequently stimulates the transcription of the ANG gene.

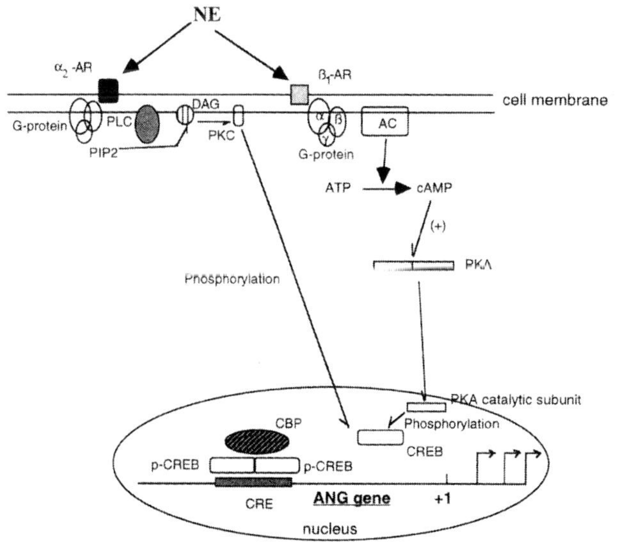

Fig. 4. Model of the molecular mechanism(s) of action of norepinephrine on the expression of the ANG gene in OK cells. Norepinephrine stimulates the expression of the ANG gene via both α_2- and β_1-ARs and then activates both PKA and PKC signalling pathways, respectively, resulting in phosphorylation of CREB. The phosphorylated CREB then interacts with the CRE in the 5′-flanking region of the ANG gene and subsequently triggers gene transcription.

Acknowledgement

This work was supported by grants from the Medical Research Council of Canada (MT-11568 to J.S.D.C. and S.C. and MT-13420 to J.S.D.C.). We thank Mrs. Ilona Schmidt for her secretarial assistance and Dr. Kenneth D. Roberts for his valuable comments.

References

1. Ohkubo H, Kageyama R, Ujihara M, Hirose T, Inayama S, Nakanishi S: Cloning and sequence analysis of cDNA for rat angiotensinogen. Proc Natl Acad Sci USA 80: 2196–3000, 1983
2. Kageyama R, Ohkubo H, Nakanishi S: Primary structure of human pre-angiotensinogen deduced from the cloned cDNA sequence. Biochemistry 23: 3603–3609, 1984
3. Gaillard I, Clauser E, Corvol P: Structure of human angiotensinogen gene. DNA 8: 87–89, 1989
4. Tanaka T, Ohkubo H, Nakanishi S: Common structural organization of the angiotensinogen and the a1-antitrypsin genes. J Biol Chem 259: 8063–8065, 1984
5. Mori M, Ishigaki K, Yamada T, Chen H, Sugiyama T, Serikawa T: Restriction fragment length polymorphisms of the angiotensinogen gene in inbred rat strains and mapping of the gene on chromosome 19q. Cytogenet Cell Genet 50: 42–45, 1989
6. Clouston WM, Fournier RE, Richard RI: The angiotensinogen gene is located on mouse chromosome 8. FEBS Lett 255: 419–422, 1989
7. Gaillard-Sanchez I, Mattei MG, Clauser E, Corvol P: Assignment by *in situ* hybridization of the angiotensinogen gene to chromosome band 1q4, the same region as the human renin gene. Hum Genet 84: 341–343, 1990
8. Jeunemaître X, Soubrier F, Kotelevtsev YV, Lifton RP, Williams CS, Charru A, Hunt SC, Hopkins PN, Williams RR, Lalouel JM, Corvol P: Molecular basis of human hypertension: role of angiotensinogen. Cell 71: 169–180, 1992
9. Hata A, Namikara C, Sasaki M, Sato K, Nakamura T, Tamura K, Lalouel JM: Angiotensinogen as a risk factor for essential hypertension in Japan. J Clin Invest 93: 1285–1287, 1994
10. Bloem LJ, Manatunga AK, Tewksbury DA, Howard Pratt J: The serum angiotensinogen concentration and variants of the angiotensinogen gene in white and black children. J Clin Invest 95: 948–953, 1995
11. Forrester T, McFarlane-Anderson N, Bennet F, Wilks R, Puras A, Cooper R, Rotimi C, Durazo R, Tewksbury D, Morrison L: Angiotensinogen and blood pressure among blacks: findings from a community survey in Jamaica. J Hypertens 14: 315–321, 1996
12. Rotimi C, Puras A, Cooper R, McFarlane-Anderson N, Forrester T, Ogunbiyi O, Morrison L, Ward R: Polymorphisms of renin-angiotensin genes among Nigerians, Jamaicans and African Americans. Hypertension 27: 558–563, 1996
13. Atwood LD, Kammerer CM, Samollow PB, Hinson JE, Shade RE, MaCluer JW: Linkage of essential hypertension to the angiotensinogen locus in Mexican Americans. Hypertension 30: 326–330, 1997
14. Chiang F-F, Hsu K-L, Tseng C-D, Hsiao W-H, Lo H-M, Chein T-H, Tseng Y-Z: Molecular variant M235T of the angiotensinogen gene is associated with essential hypertension in Taiwanese. J Hypertens 15: 607–611, 1997
15. Fardella CE, Claverie X, Vignolo P, Montero J, Villarroel L: T235 variant of the angiotensinogen gene and blood pressure in the Chilean population. J Hypertens 16: 829–833, 1998

16. Jeunemaître X, Charru A, Rigat B, Houot AM, Soubrier F, Corvol P: Sib-pair linkage analysis of renin gene haplotypes in human essential hypertension. Hum Genet 88: 301–306, 1992
17. Jeunemaître X, Lifton RP, Hunt SC, Williams RR, Lalouel JM: Absence of linkage between the angiotensin converting enzyme and human essential hypertension. Nature Gen 1: 72–75, 1992
18. Ueda S, Weir CJ, Inglis GC, Murray GD, Muir KW, Lees KR: Lack of association between angiotensin converting enzyme gene insertion/deletion polymorphism and stroke. J Hypertens 13: 1597–1601, 1995
19. O'Donnell CJ, Lindpaintner K, Larson MG, Rao VS, Ordovas JM, Schaefer EJ, Myers RH, Levy D: Evidence for association and genetic linkage of the angiotensin-converting enzyme locus with hypertension and blood pressure in man but not women in the Framingham heart study. Circulation 97: 1766–1772, 1998
20. Bonnardeaux A, Davies E, Jeunemaître X, Féry I, Charru A, Clauser E, Tiret L, Combien F, Corvol P, Soubrier F: Angiotensin II type 1 receptor gene polymorphism in human essential hypertension. Hypertension 24: 63–69, 1994
21. Schmidt S, Beige J, Walla-Friedel M, Michel MC, Sharma AM, Ritz E: A polymorphism in the gene for the angiotensin II type 1 receptor is not associated with hypertension. J Hypertens 15: 1385–1388, 1997
22. Fukamizu A, Sugimura K, Takimoto E, Sugiyama F, Seo MS, Takahashi S, Hatae T, Kajiwara N, Yagami K, Murakami K: Chimeric renin-angiotensin system demonstrates sustained increase in blood pressure of transgenic mice carrying both human renin and angiotensinogen genes. J Biol Chem 268: 11617–11621, 1993
23. Dzau VJ, Ingelfinger JR: Molecular biology and pathophysiology of the intrarenal renin-angiotensin system. J Hypertens 7(suppl 7): 53–58, 1989
24. Johnston CI, Fabris B., Jandeleit K: Intrarenal renin-angiotensin system in renal physiology and pathophysiology. Kidney Int 44(suppl 42): 559–563, 1993
25. Richou JP, Cordonnier JL, Bouhnik J, Clauser E, Corvol P, Menard P, Grignon G: Immunocytochemical localization of angiotensinogen in rat liver and kidney. Cell Tissue Res 122: 439–451, 1983
26. Ingelfinger JR, Zuo WM, Fon EA, Ellison KE, Dzau VJ: *In situ* hybridization evidence for angiotensinogen mRNA in the rat proximal tubule. A hypothesis for the intrarenal renin-angiotensin system. J Clin Invest 85: 417–423, 1990
27. Wolf G, Neilson EG: Angiotensin II as a hypertrophogenic cytokine for proximal tubular cells. Kidney Int 43(suppl 39): S100–S107, 1992
28. Terada Y, Tomita K, Nonoguch H, Marumo F: PCR localization of angiotensin II receptor and angiotensinogen mRNA in rat kidney. Kidney Int 43: 1251–1259, 1993
29. Tang SS, Jung F, Diamant D, Brown D, Bachinsky D, Hellman P, Ingelfinger J: Temperature-sensitive SV 40 immortalized rat proximal tubule cell line has functional renin-angiotensin system. Am J Physiol 268: F435–F446, 1995
30. Loghman-Adham M, Rohrwasser A, Helin C, Zhang S, Terreros D, Inoue I, Lalouel J-M: A conditionally immortalized cell line from murine proximal tubule. Kidney Int 52: 229–239, 1997
31. Chen M, Harris MP, Rose D, Smart A, He XR, Kretyler M, Briggs JP, Schnermann. Renin and renin mRNA in proximal tubule of the rat kidney. J Clin Invest 94: 237–243, 1994
32. Xie MH, Liu FY, Wong PC, Timmermans PBMWM, Cogan MG: Proximal nephron and renal effects of Dup 753, a nonpeptide angiotensin II receptor antagonist. Kidney Int 38: 473–479, 1990
33. Cogan MG, Liu FY, Wong PC, Timmermans PBMWM: Comparison of inhibitory potency by nonpeptide angiotensin II receptor antagonist PD 123177 and Dup 753 on proximal nephron and renal transport. J Pharmacol Exp Ther 259: 687–691, 1991
34. Burns KD, Inagami T, Harris RC: Cloning of a rabbit kidney cortex AT_1 angiotensin II receptor that is present in proximal tubules epithelium. Am J Physiol 264: (Renal Fluid Electrolyte Physiol 33): F645–F654, 1993
35. Harris RC, Becker BN, Cheng HF: Acute and chronic mechanisms for regulatory proximal tubule angiotensin II receptor expression. J Am Soc Nephrol 8: 306–313, 1997
36. Cheng HF, Wang J-L, Vinson GP, Harris RC: Young SHR express increased type 1 angiotensin II receptors in renal proximal tubule. Am J Physiol 274: (Renal Physiol 43): F10–F17, 1998
37. Kontogiannis J, Burns KD: Role of AT_1 angiotensin II receptors in renal ischemic injury. Am J Physiol 274 (Renal Physiol 43): F79–F90, 1998
38. Zhang CJ, Mayeux PR: Angiotensin II signalling activates the NO-cGMP pathway in rat proximal tubules. Life Sci 63: PL75, 1998
39. Chan JSD, Chan AHH, Nie Z-R, Sikstrom R, Lachance S, Hashimoto S, Carrière S: Thyroid hormone, L-T_3, stimulates the expression of the angiotensinogen gene in cultured opossum kidney (OK) cells. J Am Soc Nephrol 2: 1360–1367, 1992
40. Ingelfinger JR, Boubounes B, Jung FF, Tang S: High glucose downregulates expression of renin angiotensin system (RAS) in opossum kidney cells. Pediatr Nephrol 29: (abstr) 344A, 1991
41. Wang L, Lei C, Zhang S-L, Roberts KD, Tang S-S, Ingelfinger JR, Chan JSD, Synergistic effect of dexamethasone and isoproterenol on the expression of angiotensinogen in immortalized rat proximal tubular cells. Kidney Int 53: 287–295, 1998
42. Seikaly MG, Arant BS, Seney FD, Rutlege L and Green J: Endogenous angiotensin concentrations in specific intrarenal fluid compartments of the rat. J Clin Invest 86: 1352–1357, 1990
43. Braam B, Mitchell KD, Fox J and Navar LG: Proximal tubular secretion of the angiotensin II in rats. Am J Physiol 264: F891–F898, 1993
44. Navar LG, Lewis L, Hymel A, Braam B and Mitchell KD: Tubular fluid concentrations and kidney contents of angiotensin I and II in anesthesized rats. J Am Soc Nephrol 5: 1153–1158, 1994
45. Davis JO, Freeman RH: Mechanisms regulating renin release. Physiol Rev 56: 1–56, 1976
46. Ekboir AS, Enera MA: Pre- and post-junctional potentiation of the adrenergic neurotransmission by angiotensin II in the perfused rabbit kidney. Gen Pharmacol 11: 395, 1980
47. Gagnon JA, Keller HL, Kokotis W, Schrier RW: Analysis of role of renin-angiotensin system in autoregulation of glomerular filtration. Am J Physiol 219: 491–496, 1970
48. Hall JE, Guyton AC, Jackson TE, Coleman TG, Lohmeier TE, Trippodo NC: Control of glomerular filtration rate by renin-angiotensin system. Am J Physiol 233: F366–F372, 1977
49. Blantz RC, Konnen KS, Tucker BJ: Angiotensin II effects upon the glomerular microcirculation and ultrafiltration coefficient of the rat. J Clin Invest 57: 419–434, 1976
50. Schuster VL, Kokko JP, Jacobson HR: Angiotensin II directly stimulates sodium reabsorption by angiotensin II in the kidney. J Clin Invest 73: 507–515, 1984
51. Liu FY, Cogan MG: Angiotensin II stimulates early proximal bicarbonate absorption in the rat by decreasing cyclic adenosine monophosphate. J Clin Invest 84: 83–91, 1989
52. Harris PJ, Young JA: Dose-dependent stimulation and inhibition of proximal tubular sodium reabsorption by angiotensin II in rat kidney. Pflügers Arch 367: 1295–1297, 1977
53. Hall JE, Guyton AC, Trippodo NC, Lohmeier TE, McCrea RE, Cowley ALW Jr: Intrarenal control of electrolyte excretion by angiotensin II. Am J Physiol 232: F538–F544, 1977
54. Cogan MG: Angiotensin II: A powerful controller of sodium transport in the early proximal tubule. Hypertension 15: 451–458, 1990
55. Wang T, Chan YL: Mechanism of angiotensin II action on proximal tubular transport. J Pharmacol Exp Ther 252: 689–695, 1990
56. Schelling J, Singh H, Marzec R, Linas SL: Angiotensin II-dependent proximal tubule sodium transport is mediated by cAMP modulation of phospholipase C. Am J Physiol 267: C1239–C1245, 1994
57. Barajas L: Innervation of the renal cortex. Fed Proc 37: 1192–1201, 1978

58. Bello-Reusse E, Colindres RE, Pastoriza Munoz E, Müller RA, Gottschalk CW: Effects of acute unilateral renal denervation in the rat. J Clin Invest 56: 208–217, 1975
59. Pelayo JC, Zeigler MG, Jose PA, Blants RC: Renal denervation in the rat: Analysis of glomerular and proximal tubule function. Am J Physiol 244: F70–F77, 1983
60. DiBona GF: Neural control of renal tubular sodium reabsorption in the dog. Fed Proc 37: 1214–1217, 1978
61. Rudd AM, Grippo RS, Arendshorst WJ: Acute renal denervation produces a diuresis and natriuresis in young SHR but not WKY rats. Am J Physiol 251: F655–F661, 1986
62. Benscath PL, Asztalos B, Bzalay L, Takace L: Renal handling of sodium after chronic sympathectomy in the anesthetized rat. Am J Physiol 236: F513–F518, 1979
63. Liard JF: Renal denervation delays blood pressure increase in the spontaneously hypertensive rat. Experientia 33: 339–340, 1979
64. DiBona GF: The function of renal nerves. Rev Physiol Biochem Pharmacol 94: 75–81, 1982
65. DiBona GF: Neural control of renal function: Role of renal alpha adrenoceptors. J Cardiovasc Pharmacol 7(suppl 8): S12–S23, 1985
66. Jacobs WR, Chan YL: Evidence for the presence of functional beta-adrenoceptor along the proximal tubule of the rat kidney. Biochem Biophys Res Commun 141: 334–339, 1986
67. Kudo K, Kondo Y, Abe K, Igarashi Y, Tada K, Yoshinaga K: Evidence for presence of functional β-adrenoceptor in rabbit S-2 proximal straight tubules. Am J Physiol 261: F393–F399, 1991
68. Taniguchi S, Watanabe T, Nakas A, Seki G, Uwatoko S, Suzuki K, Kurokawa K: Distribution of β_2-adrenergic receptor mRNA expression along the hamster nephron segments. FEBS Lett 318: 65–70, 1993
69. Insel PA, Snavely MD, Healy DP, Munzel PA, Potenza CL, Nord EP: Radiologic binding and functional assays demonstrate post-synaptic α_2-receptors on proximal tubules of rat and rabbit kidney. J Cardiovasc Pharmacol 7(suppl 8): S1–S8, 1985
70. Pettinger WA, Unemura S, Smyth DD, Jeffries WB: Renal α_2-adrenoceptors and the adenylate cyclase-cAMP system: Biochemical and physiological interactions. Am J Physiol 252: F199–F208, 1987
71. Sundaresan PR, Barac-Nieto M, Stambo GW, Kelvie SL: Glomerular and tubular α_1- and α_2-adrenoceptors in the rat kidney: Distribution in basolateral and brush border membranes of tubular cells. J Cardiovasc Pharmacol 13: 16–24, 1989
72. Gesek FA, Stradhoy JW: Dual interactions between α_2-adrenoceptor agonists and the proximal Na^+-H^+ exchanger. Am J Physiol 258: F636–F642, 1990
73. Jensen RE, Berndt WO: Characterization of adrenergic receptors of proximal tubular basolateral membranes. Life Sci 43: 1473–1478, 1988
74. Sanchez A, Vidal MJ, Martinez-Sierra R, Saiz J: Ontogeny of renal alpha-1 and alpha-2 adrenoceptors in the spontaneously hypertensive rat. J Pharmacol Exp Ther 237: 972–979, 1986
75. Baines AD, Ho P: Specific α_1-, α_2- and β-response to norepinephrine in pyruvate-perfused rat kidneys. Am J Physiol 252: F170–F176, 1987
76. Wang TT, Lachance S, Delalandre A, Carrière S, Chan JSD: Alpha-adrenoceptors and angiotensinogen gene expression in opossum kidney cells. Kidney Int 48: 139–145, 1995
77. Wang TT, Chen M, Lachance S, Delalandre A, Carrière S, Chan JSD: Isoproterenol and 8-bromo-cyclic adenosine monophosphate stimulate the expression of the angiotensinogen gene in opossum kidney cells. Kidney Int 46: 703–710, 1994
78. Koyama H, Goodpasture C, Miller MM, Teplitz RL, Riggs AD: Establishment and characterization of a cell line from the American opossum (*Didelphys virginia*). *In Vitro* (Rockville) 14: 239–246, 1978
79. Pollock AS, Warnock DG, Strewler GJ: Parathyroid hormone inhibition of Na^+-H^+ antiporter activity in a cultured renal cell line. Am J Physiol 250: F217–F225, 1986
80. Murphy TJ, Bylund DB: Characterization of alpha^2 adrenergic receptors in the OK cell, an opossum kidney cell line. J Pharmacol Exp Ther 244: 457–478, 1988
81. Cheng L, Liang CT, Precht P, Sacktor B: Alpha$_2$ adrenergic modulation of the parathyroid hormone-inhibition on phosphate uptake in OK cells. Biophys Res Commun 155: 74–82, 1988
82. Cheng L, Precht P, Franck D, Liang CT: Dopamine stimulation of cAMP production in cultured opossum kidney cells. Am J Physiol 258: F877–F882, 1990
83. Chan JSD, Ming M, Nie Z-R, Sikstrom R, Lachance S, Carrière S: Hormonal regulation of expression of the angiotensinogen gene in cultured opossum kidney (OK) proximal tubular cells. J Am Soc Nephrol 2: 1516–1522, 1992
84. Chan JSD, Chan AHH, Jiang Q, Nie Z-R, Lachance S, Carrière S: Molecular cloning and expression of the rat angiotensinogen gene. Pediatr Nephrol 4: 429–435, 1990
85. Wang TT, Chen M, Lachance S, Delalandre A, Carrière S, Chan JSD: Isoproterenol and 8-bromo-cyclic adenosine monophosphate stimulate the expression of the angiotensinogen gene in opossum kidney cells. Kidney Int 46: 703–710, 1994
86. Ming M, Wang TT, Lachance S, Delalandre A, Carrière S, Chan JSD: Expression of the angiotensinogen gene is synergistically stimulated by 8-Bromo-cAMP and dexamethasone in opossum kidney cells. Am J Physiol 268: R105–R111, 1995
87. Wang TT, Lachance S, Delalandre A, Carrière S, Chan JSD: Alpha-adrenoceptors and angiotensinogen gene expression in opossum kidney cells. Kidney Int 48: 139–145, 1995
88. Wang TT, Lachance S, Delalandre A, Carrière S, Chan JSD: Dopaminergic receptors and angiotensinogen gene expression in opossum kidney cells. Am J Physiol R519–R527, 1996
89. Wang TT, Wu X-H, Zhang S-L, Chan JSD: The molecular mechanism(s) of action of norepinephrine on the expression of the angiotensinogen gene in opossum kidney cells. Kidney Int 54: 785–795, 1998
90. Wang TT, Chen X, Wu X-H, Zhang S-L, Chan JSD: Molecular mechanism(s) of action of isoproterenol on the expression of the angiotensinogen gene in opossum kidney proximal tubular cells. Kidney Int 55: 1713–1723, 1999
91. Montminy MR, Sevarino KA, Wagner JA, Mandel G, Goodman RH: Identification of a cyclic AMP responsive element within the rat somatostatin gene. Proc Natl Acad Sci USA 83: 6682–6686, 1986
92. Gonzalez GA, Montminy MR: Cyclic AMP stimulates somatostatin gene transcription by phosphorylation of CREB at serine 133. Cell 59: 675–680, 1989
93. Bundle P, Nakajima T, Montminy M: Multiple protein kinase A regulated events are required for transcriptional induction by cAMP. Proc Natl Acad Sci USA 92: 10521–10525, 1995
94. Yamamoto KR, Gonzalez GA, Biggs WH III, Montminy MR: Phosphorylation-induced binding and transcriptional efficacy of nuclear factor CREB. Nature 334: 394–398, 1988
95. Qian JF, Wang TT, Wu X-H, Wu J, Lachance S, Carrière S, Chan JSD: Angiotensinogen gene expression is stimulated by the cAMP-responsive element binding protein (CREB) in opossum kidney cells. J Am Soc Nephrol 8: 1072–1079, 1997

Molecular and Cellular Biochemistry 212: 81–90, 2000.

cAMP increases the expression of human angiotensinogen gene through a combination of cyclic AMP responsive element binding protein and a liver specific transcription factor

Chittampalli S. Narayanan, Yanning Cui, Shubhra Kumar and Ashok Kumar

Department of Pathology, New York Medical College, Valhalla, NY, USA

Abstract

Angiotensinogen is the glycoprotein precursor of one of the most potent vasoactive hormones angiotensin-II which plays an important role in the regulation of blood pressure. We show here that the promoter activity of reporter constructs containing human angiotensinogen promoter is increased by cAMP treatment on transient transfection in HepG2 cells. We have identified a composite cAMP responsive element, located around 840 bases upstream from the transcriptional initiation site, in the promoter of human angiotensinogen gene. This element is recognized by members of CREB/ATF as well as C/EBP family of transcription factors. Another C/EBP binding site that is not recognized by CREB is located 10 bases upstream from this site. We show that co-transfection of CREB increases the promoter activity of reporter constructs containing human angiotensinogen gene promoter attached to the CAT gene. We also show that co-transfection of DBP (which is a member of C/EBP family of transcription factors) increases promoter activity of these reporter constructs. (Mol Cell Biochem **212**: 81–90, 2000)

Key words: cAMP, CRE, CREB, DBP, C/EBP, angiotensinogen gene, transcriptional regulation, HepG2

Introduction

The renin-angiotensin system plays an important role in the regulation of blood pressure, fluid balance and electrolyte homeostasis [1, 2]. Angiotensin-II, which is one of the most potent vasoactive hormones, is obtained from its precursor molecule, angiotensinogen, by the successive proteolytic actions of renin and angiotensin converting enzyme. Prolonged mental stress increases the blood pressure through an increased activation of the sympathetic nervous system [3, 4]. It has been known for a long time that β-adrenergenic stimulation can increase renin synthesis through increased secretion of cAMP and a cAMP responsive element (CRE) has been identified in the promoter of human renin gene [5–7]. Although previous studies have suggested that cAMP treatment also increases angiotensinogen mRNA level in rat hepatocytes [8] and mouse hepatoma cells [9], the role of cAMP in the expression of the human angiotensinogen gene remains to be analyzed.

Cyclic AMP stimulates the target gene expression via a conserved CRE which consists of a palindromic sequence TGACGTCA [10, 11]. Cyclic AMP responsive element binding protein (CREB) was the first protein to be cloned by screening a λgt11 expression library using a CRE oligonucleotide probe [12], and shown to be involved in constitutive and inducible expression of a target gene [13]. Structural analysis of CREB indicated that it contains a consensus protein kinase-A phosphorylation site [14]. It was later shown that cAMP treatment results in the release and translocation of protein kinase-A (PK-A) to the nucleus which phosphorylates

Address for offprints: A. Kumar, Department of Pathology, New York Medical College, Valhalla, NY 10595, USA

CREB at serine-133. The phosphorylated CREB then binds to the CRE and stimulates the expression of a target gene [15–17]. Recent studies have shown that phosphorylated CREB also binds to specific amino acid sequence of CREB-binding protein (CBP) [18, 19] which has an intrinsic histone acetyl transferase activity and this interaction results in modification of the chromatin at this site. In addition, CBP interacts with various proteins involved in transcriptional initiation complex ultimately resulting in stabilization of the pre-initiation complex and transcriptional activation of the promoter [20]. Two additional CREB family members were subsequently identified: ATF-1 and cAMP response element modulator (CREM), and both of them contain a consensus PK-A binding site [21, 22]. In addition, tissue specific factors have been identified that bind to un-phosphorylated CREB and increase promoter activity through a CRE [23].

Angiotensinogen gene is primarily expressed in the liver and to a smaller extent in the brain, liver, kidney, and adipose tissue [24]. However, molecular mechanisms involved in tissue specific expression of the human angiotensinogen gene and its modulation by cAMP are not known. We have recently shown that liver specific transcription factor DBP binds to the proximal promoter of the human angiotensinogen gene and co-transfection of an expression vector containing coding sequence of DBP increases the expression of a reporter construct containing 1.2 Kb promoter of the human angiotensinogen gene attached to the CAT gene (pHAG1.2CAT) [25]. We have also shown that another liver enriched transcription factor HNF3 binds to the nucleotide sequence located in exon-1 of the human angiotensinogen gene and plays an important role in the liver specific expression of this gene [26]. Human angiotensinogen gene contains a polymorphic site C/A at –20 located between the TATA box and transcriptional initiation site [27]. We [28, 29] and others [30] have shown that reporter constructs containing human angiotensinogen gene promoter with nucleoside C at –20 have increased basal transcriptional activity on transient transfection in human liver derived HepG2 cells. An ubiquitous transcription factor MLTF binds to the nucleotide sequence located between the TATA box and transcriptional initiation site when nucleoside C is present at –20 and site specific mutation of this site drastically reduces the promoter activity of the human angiotensinogen gene. We have also shown that estrogen receptor-α (ERα) binds to this region of the promoter when nucleoside A is present at –20 and co-transfection of an expression vector containing human ERα increases the promoter activity of reporter constructs containing human angiotensinogen gene promoter with nucleoside A at –20 [29]. In addition, we have shown that orphan receptor Arp-1 binds to the nucleotide sequence located between the TATA box and transcriptional initiation site and co-transfection of an expression vector containing Arp-1 coding sequence reduces estrogen induced promoter activity [31]. In this paper, we show that cAMP treatment increases the promoter activity of the reporter construct containing human angiotensinogen gene promoter. We also show that human angiotensinogen gene contains a pseudo symmetrical cAMP responsive element located at 840 bases upstream from the transcriptional initiation site which binds to recombinant CREB. This element has only one mismatch with binding sites of AP1 and C/EBP family of transcription factors. Surprisingly, another cis-acting DNA element which binds to C/EBP family is located at 10 bp upstream from this CRE site in the human angiotensinogen gene promoter. Co-transfection of CREB increases the promoter activity of either pHAG1.2CAT or a reporter construct $(HAG\text{-}CRE)_3$ TK-CAT containing three copies of this CRE attached to the TK-CAT (80 bp of the herpes virus thymidine kinase promoter attached to the CAT gene) on transient transfection in HepG2 cells. We also show that liver specific transcription factor DBP (which is a member of C/EBP family of transcription factors) increases the promoter activity of these constructs.

Materials and methods

Materials

Expression plasmid pSVoCAT, pGem-T, and RSV-β-gal were obtained from Promega. The expression vector RSV-CREB (coding sequence of CREB attached to the Rous sarcoma virus promoter) was provided by Dr. Marc Montiminy (Joslin Diabetes Center, Boston, MA, USA). The expression vector CMV-DBP (DBP coding sequence attached to the Cytomegalo virus promoter) and DBP antibody were provided by Dr. Uri Schibler (University of Geneva, Switzerland). CREB antibody was purchased from Santa Cruz (Santa Cruz, CA, USA). Restriction enzymes were purchased from New England Biolab or Gibco/BRL. Plasmid kits were obtained from Qiagen (Chatsworth, CA, USA).

Plasmid construction

The reporter construct pHAG1.2-CAT was synthesized by attaching ~ 1.2 kb of 5′-flanking region of the human angiotensinogen gene [27, 28] in front of the chloramphenicol acetyl transferase (CAT) gene in the expression vector pSVoCAT. The ~ 1.2 kb region of the promoter contained 1223 bases of the 5′-flanking region and 44 bases of the first exon of the human angiotensinogen gene that were obtained by polymerase chain reaction of the human genomic DNA. The nucleotide sequence of the insert was confirmed by sequence analysis (the sequence in our clone had nucleoside A at –20 and at –6) and the promoter region was then subcloned in pSVoCAT.

Plasmid (HAG-CRE)$_3$-CAT was synthesized by attaching three copies of the oligonucleotide HAG-CRE in front of TK-CAT. Nucleotide sequence of different constructs were confirmed by restriction and sequence analysis. Plasmid DNAs for transfection were prepared by Qiagen column and quality of plasmid DNA was checked by gel electrophoresis.

Cell culture and transient transfection

Human hepatoma cells (HepG2) were grown as monolayers in Dulbecco's modified Eagle's medium (DMEM) supplemented with 10% fetal calf serum, 100 U/ml penicillin, and 100 μg/ml streptomycin in an atmosphere of 5% CO_2. Transient DNA transfections were performed by the calcium phosphate precipitation method using supercoiled plasmid DNA at 10–20 μg/plate and pRSV-β-gal (2 μg) as an internal control . The total weight of DNA was kept constant at 30 μg/plate. Cells were harvested after 48 h of transfection, total extract was prepared by three cycles of freezing and thawing in liquid nitrogen, an aliquot (5 λ) was used for the β-gal assay and rest of the extract was heated at 65°C for 5 min. After centrifugation, an aliquot of the extract (after normalization with the β-gal activity) was used to perform CAT assay using ^{14}C-chloramphenicol as a substrate followed by separation of acetylated products by thin layer chromatography using silica gel plates. After autoradiography, spots corresponding to ^{14}C-chloramphenicol and its acetylated derivatives were scraped from TLC plates and radioactivity in each spot was measured using a liquid scintillation counter. CAT activity was determined by dividing the counts in acetylated spots with the total number of counts (present in acetylated and non-acetylated spots). Transient transfections were performed at least three times using at least two different preparations of plasmid DNAs. HepG2 cells were treated with 8-Br-cAMP (0.5 mM), forskolin (10 μM) and 3-isobutyl-1-methyl xanthine (IBMX, 0.5 mM) for 6 h to study cAMP induced promoter activity.

Gel mobility shift assay

The probes for gel mobility shift analysis were chemically synthesized, annealed and radiolabeled at the 5′-ends by polynucleotide kinase and [γ-^{32}P] ATP. Radiolabeled oligonucleotides (20,000–50,000 cpm), 1–2 μg of poly(dI-dC), and 5–10 μg of the protein extract were incubated in a solution containing 10 mM HEPES (pH 7.5), 50 mM KCl, 5 mM $MgCl_2$, 0.5 mM EDTA, 1 mM DTT, 12.5% glycerol in ice for 30 min and separated on a 5–8% polyacrylamide gel at room temperature. After 2–3 h, the gel was dried under vacuum and protein-nucleic acid complexes were identified by autoradiography. For supershift with CREB and DBP antibodies, 1 μl of antibody was added to the reaction mixture which was then incubated for 30 min. Nuclear extracts for gel mobility shift assay were prepared by modification of a previously described method [32]. Recombinant CREB and DBP were synthesized by *in vitro* coupled transcription translation of respective vectors using Promega TNT kit following the conditions described by the manufacturer.

Oligonucleotides

The nucleotide sequence of forward and reverse primers used for the construction of pHAG 1.2CAT were 5′-GAGTCCCT-ATCTATAGGAACA and 5′-CATACCC TTCTGCTGTAG-TAC. The double stranded oligonucleotide HAG-CRE was obtained by annealing oligonucleotides 5′-ACCTGCCTC-ACCCACTGCGTCACTTGT and its complementary oligonucleotide This oligonucleotide corresponds to the nucleotide sequence –829 to –855 of the human angiotensinogen gene promoter. The double stranded oligonucleotides CRE1 and CRE2 were obtained by annealing CTAGCCACTGCGTCA-CTTG and CTAGCAACCTGCCTCACCC with their respective complementary oligonucleotides. The nucleotide sequence of oligonucleotides containing consensus C/EBP and DBP binding sites were CTAGACATTGCACAATCT and TGG-TATGATTTTGTAATGGG respectively. The nucleotide sequence of HNF4, NF-1, and mut CRE-1 oligonucleotides were CTAGTAGGCA AGCTTCATAT, CCTTTGGCATGCTGC-CAATATG, CTAGCCACTGTTTCACTTG respectively. Nucleotide sequence of only one strand is shown and the double stranded oligonucleotides were obtained by annealing the complimentary sequence.

Results

cAMP treatment increases the promoter activity of pHAG1.2CAT in HepG2 cells

To examine the effect of cAMP on the human angiotensinogen gene expression, we transiently transfected HepG2 cells with the reporter construct pHAG1.2CAT and its 5′-deletion mutant Sma1 (that contained only 350 bp of the 5′-flanking region of the human angiotensinogen gene attached to the CAT gene). The cells were then incubated in the absence and presence of cAMP, IBMX and forskolin, and the promoter activity was analyzed by the CAT assay. Results of this experiment, shown in Fig. 1, suggested that the promoter activity of pHAG1.2CAT was increased by 1.6 fold on cAMP treatment whereas the promoter activity of the deletion mutant Sma1 remained unchanged.

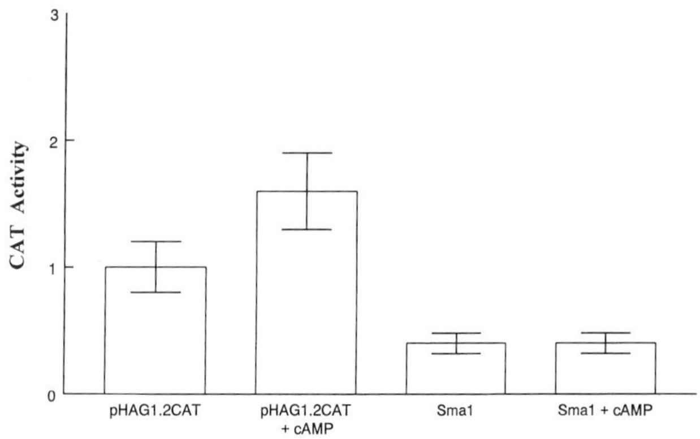

Fig. 1. cAMP treatment increases the promoter activity of a reporter construct containing 1.2 Kb of the human angiotensinogen gene promoter. Reporter constructs pHAG1.2CAT (containing 1.2 Kb of the 5′-flanking region) and Sma1 (containing 350 bp of the 5′-flanking region) of the human angiotensinogen gene were transiently transfected in HepG2 cells which were then treated in the absence and presence of cAMP, forskolin and IBMX for 6 h. Promoter activity was analyzed after 48 h of transfection after normalization with the β-gal activity. Results show mean value of three experiments assuming the promoter activity of pHAG1.2CAT as one.

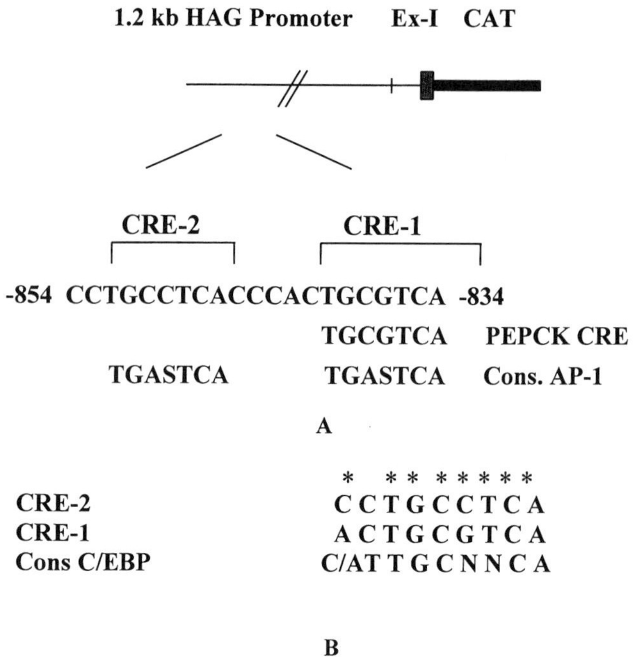

Fig. 2. Human angiotensinogen gene contains a cAMP responsive element around 840 bases upstream from the transcriptional initiation site. The top line shows construction of the reporter construct pHAG1.2CAT. (A) Nucleotide sequence of the human angiotensinogen gene promoter located between –834 and –854 and position of CRE-1 and CRE-2 sites, and sequence homology between PEPCK CRE and consensus AP-1 sites. (B) Sequence homology between CRE-1, CRE-2 and consensus C/EBP binding site. Identical bases are shown by asterisks.

Nucleotide sequence of the human angiotensinogen gene contains a cAMP responsive element around –850

An examination of the nucleotide sequence of 1.2 Kb region of the human angiotensinogen gene promoter revealed that nucleotide sequence around –840 contains a half CRE which is also present in the promoter of PEPCK gene. Moreover, this is a composite element and contains two related binding sites: CRE-1 and CRE-2. CRE-1 has sequence homology with CRE half site. On the other hand both CRE-1 and CRE-2 have only one mismatch with a consensus AP-1 binding site (Fig. 2A). In addition, both CRE-1 and CRE-2 have one mismatch with a consensus C/EBP binding site (Fig. 2B).

CRE-1 and CRE-2 bind to different transcription factors

To examine whether CRE-1 and CRE-2 bind to transcription factors present in liver cells, we performed a gel shift assay using radiolabeled oligonucleotide HAG-CRE (containing both CRE-1 and CRE-2 sites) in the presence of HepG2 nuclear extract. Results of this experiment (Fig. 3) suggested that this oligonucleotide formed two prominent complexes a and b (lane 1) which were competed out with a 100 fold excess of cold HAG-CRE. On the other hand, cold CRE-2 competed out only the faster moving complex b and cold CRE-1 competed out both the complexes. Taken together, results of this experiment suggested that CRE-1 and CRE-2 bind to different transcription factors present in HepG2 cells.

CRE-1 binds to CREB, AP-1, and C/EBP family of transcription factors

Sine nucleotide sequence of CRE-1 has homology with CRE and AP-1 binding sites, we next performed a gel shift assay with radiolabeled CRE-1 to examine whether this sequence binds to these transcription factors. Results of this experiment (Fig. 4A) showed that CRE-1 formed two complexes with HepG2 extract (lane 1) which were competed out with cold HAG-CRE (lane 2) and CRE-1 (lane 4). On the other hand cold CRE-2 competed out only the faster moving band (lane 3). We next performed these reactions in the presence of CREB and c-fos antibodies. Results of these experiments suggested that CREB antibody almost completely supershifted the slower moving complex (lane 5) and c-fos antibody partially supershifted the slower moving complex. We believe that the slower moving band consists of a complex of CREB and c-fos which is supershifted by both of these antibodies. Since CRE-1 also has sequence homology with C/EBP binding site, we next performed a gel shift assay in the presence of cold oligonucleotides containing C/EBP consensus binding site, non-specific HNF4 binding site, and DBP antibody. Results of this experiment (Fig. 4B) showed that the faster moving complex (shown by a solid arrow) was removed in the presence of cold oligonucleotides containing consensus DBP and C/EBP binding sites (lanes 2 and 3) but not in the presence of non-specific HNF4 binding site (lane 4). In addition, the faster moving complex was supershifted in the presence of increasing amount of DBP antibody (shown by a broken arrow) (lanes 5 and 6). Taken together, results of this experiment suggested that CRE-1 binds to CREB, c-fos, and C/EBP family of transcription factors.

Recombinant CREB and DBP bind to CRE-1

In order to further confirm that CREB and DBP bind to CRE-1, we performed a gel shift assay in the presence of recombinant CREB and DBP obtained by *in vitro* coupled transcription-translation using rabbit reticulocyte system. Results of this experiment showed that the major protein:DNA complex obtained with radiolabeled CRE-1 and r-CREB (Fig. 5A, lane 1) (shown by a solid arrow) was reduced in the presence of cold CRE-1 oligonucleotide (lane 2) but not in the presence of an oligonucleotide where CRE site was mutated (lane 3). In addition, CREB antibody supershifted the complex (lane 4) (broken arrow) whereas pre-immune serum had no effect (data not shown). Similarly, CRE-1 formed a complex with r-DBP (Fig. 5B, lane 1) which was partially supershifted in the presence of DBP antibody (lane 2). Results of these experiments suggested that r-CREB and r-DBP bind to CRE-1.

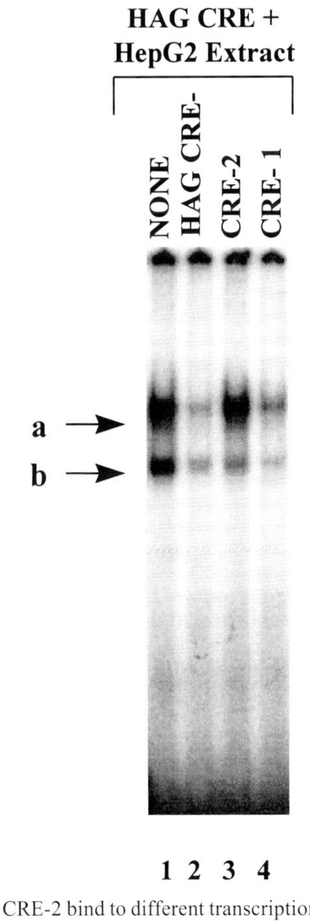

Fig. 3. CRE-1 and CRE-2 bind to different transcription factors present in HepG2 cells. A gel shift assay with radiolabelled HAG-CRE in the presence of HepG2 nuclear extract produced two complexes (a and b) (lane 1) which were competed out in the presence of a 100 fold excess of cold HAG-CRE (lane 2) or CRE-1 alone (lane 4). However, cold CRE-2 competed only the faster moving complex b (lane 3).

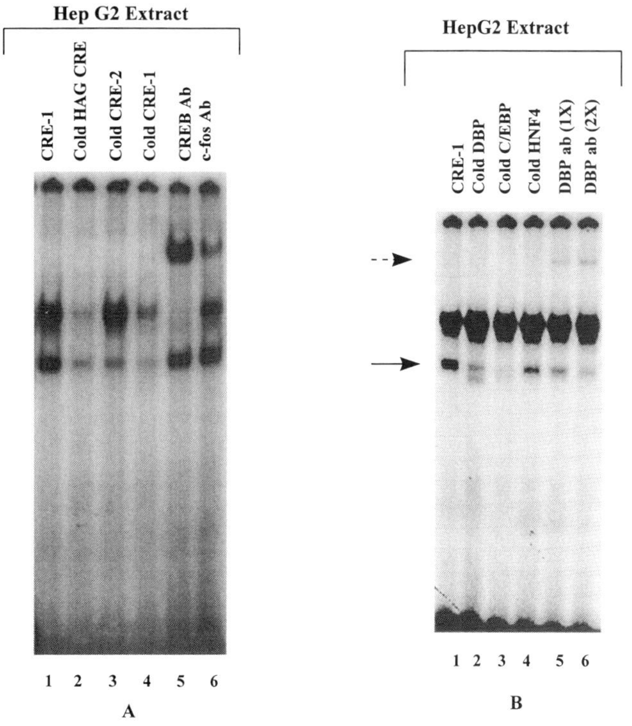

Fig. 4. CREB and C/EBP family of transcription factors present in HepG2 cells bind to CRE-1. (A) Radiolabelled CRE-1 produced two complexes in the presence of HepG2 nuclear extract (lane 1), which were competed out by a 100 fold excess of cold HAG-CRE (lane 2) and cold CRE-1 (lane 4) but only the faster moving complex was competed out by cold CRE-2 (lane 3). The slower moving complex was also supershifted in the presence of CREB and c-fos antibodies (lanes 5 and 6 respectively). (B) The faster moving complex obtained with CRE-1 and HepG2 extract (lane 1, solid arrow) was competed out by a 100 fold excess of cold oligonucleotides containing consensus DBP (lane 2) and C/EBP binding sites (lane 3) but not in the presence of a non-specific oligonucleotide containing consensus HNF4 binding site (lane 4). The faster moving complex was also partially supershifted in the presence of increasing amount of DBP antibody (lanes 5 and 6).

CRE-2 also binds to C/EBP family of transcription factors

In order to identify transcription factor(s) that bind to CRE-2, we first performed a gel shift assay with radiolabeled CRE-

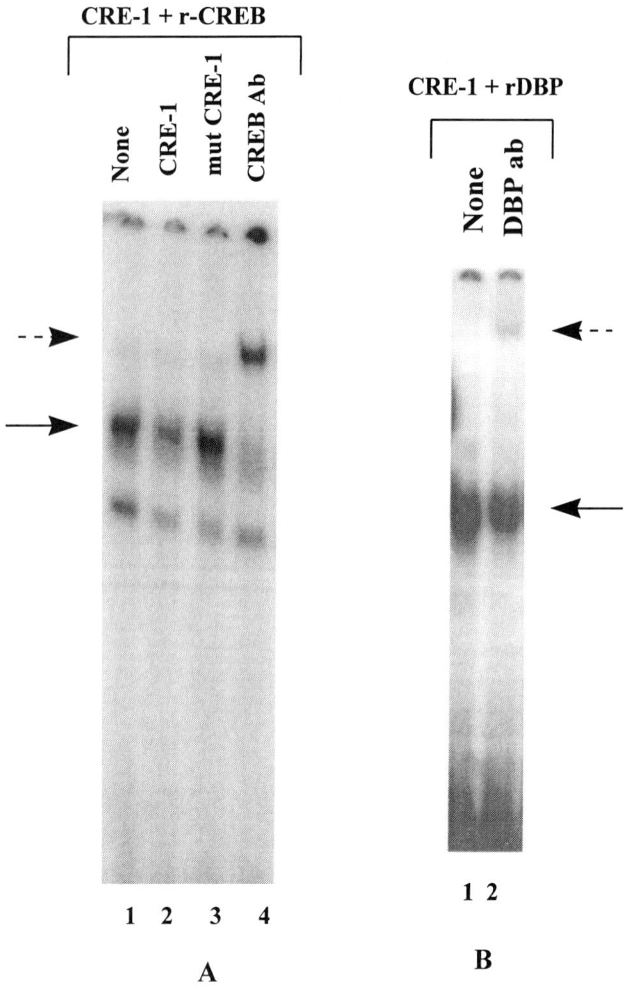

Fig. 5. Recombinant CREB and DBP bind to CRE-1. Radiolabelled CRE-1 was used in a gel shift assay in the presence of recombinant CREB and DBP obtained by *in vitro* coupled transcription translation system. (A) CRE-1 produced a major complex with r-CREB (lane 1, solid arrow) which was competed out with cold CRE-1 (lane 2) but not with the mutated CRE-1 (lane 3). The major complex was also supershifted in the presence of CREB antibody (lane 4, broken arrow). (B) CRE-1 also formed a complex with r-DBP (lane 1, solid arrow) which was partially supershifted in the presence of DBP antibody (lane 2, broken arrow).

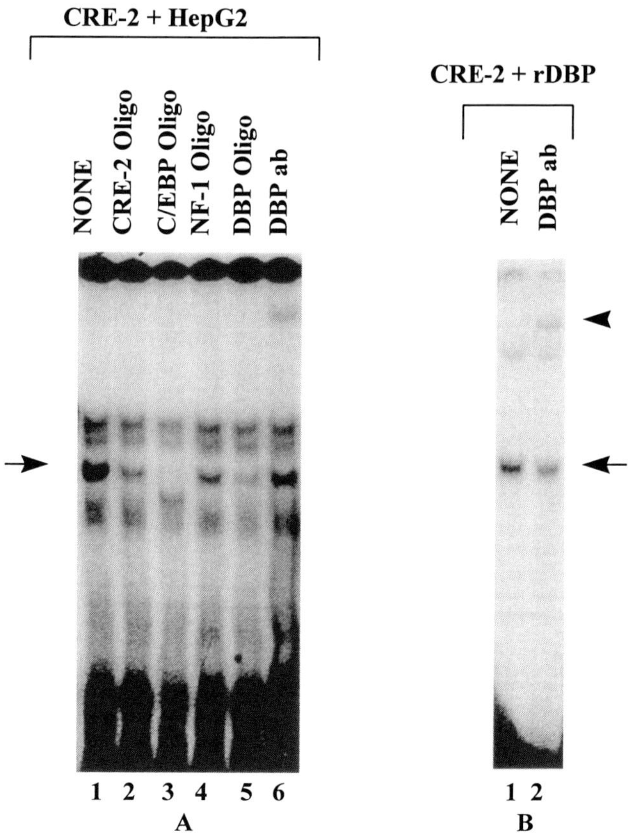

Fig. 6. CRE-2 binds to C/EBP family of transcription factors. (A) Radiolabelled CRE-2 produced one major and several minor complexes with HepG2 nuclear extract (lane 1, solid arrow) which were competed out with cold CRE-2 (lane 2), oligonucleotides containing consensus C/EBP (lane 3) and DBP (lane 5) binding sites but not the NF-1 binding site (lane 4). The major complex was also partially supershifted in the presence of DBP antibody (lane 6). (B) CRE-2 binds to recombinant DBP. Radiolabelled CRE-2 formed a complex with r-DBP obtained by in-vitro coupled transcription translation system (lane 1) and this complex was partially supershifted in the presence of DBP antibody (lane 2).

2 in the presence of HepG2 nuclear extract. Results of this experiment (Fig. 6A) suggested that CRE-2 produced one major and several minor complexes with HepG2 nuclear extract (lane 1). The major complex was reduced in the presence of a 100 fold excess of cold CRE-2 oligonucleotide (lane 2) , C/EBP oligonucleotide (lane 3), and DBP oligonucleotide (lane 5) but only partially in the presence of a non-specific NF-1 oligonucleotide (lane 4). In addition, DBP antibody reduced the intensity of the major complex and produced a supershift (lane 6). The presence of multiple bands in lane 1 can be explained by the fact that HepG2 cells contain multiple factors that belong to the C/EBP family which can all bind to a C/EBP consensus site. Radiolabelled CRE-2 also formed a complex with r-DBP (Fig. 6B, lane 1) which was partially supershifted in the presence of DBP antibody (lane 2).

Co-transfection of CREB and DBP increase the promoter activity of reporter construct containing three CRE binding sites attached to the TK-CAT gene

To understand the role of CRE binding site in the transcriptional regulation of the human angiotensinogen gene, we constructed a reporter construct where three copies of HAG-CRE were attached in front of the TK-CAT gene (Fig. 7A). This reporter construct was then transfected in HepG2 cells in the absence and presence of 1 and 2 μg of RSV-CREB and CMV-DBP. After two days of transfection, promoter activ-

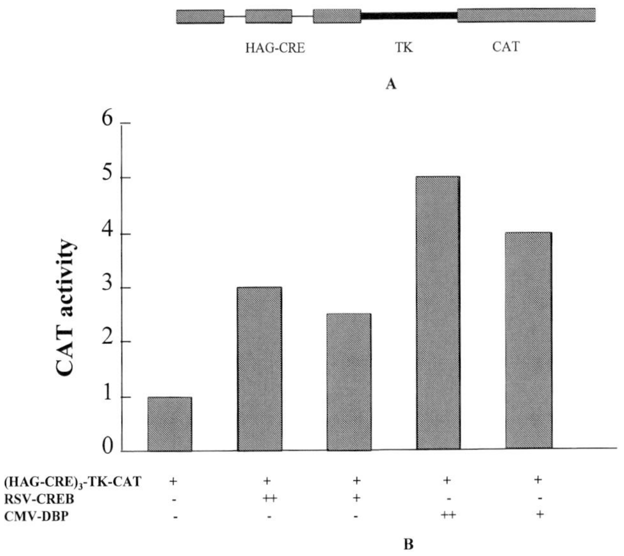

Fig. 7. The promoter activity of the reporter construct $(HAG\text{-}CRE)_3$ TK-CAT is increased on co-transfection of expression vectors RSV-CREB and CMV-DBP. (A) The reporter construct $(HAG\text{-}CRE)_3$ TK-CAT was prepared by attaching three copies of the oligonucleotide HAG-CRE in front of the reporter construct TK-CAT that contains 80 bp of Herpes virus thymidine kinase promoter attached to the CAT gene. (B) The reporter construct $(HAG\text{-}CRE)_3$TK-CAT (5 μg) and RSV-β gal (1 μg) were transfected in the absence or presence of RSV-CREB or CMV-DBP(1 and 2 μg each) in HepG2 cells and promoter activity was analyzed by the CAT assay after normalization with the β-gal assay. CAT activity was calculated by assuming the basal promoter activity of $(HAG\text{-}CRE)_3$ TK-CAT as one.

ity was analyzed by the CAT assay after normalization with the β-gal activity. Results of this experiment (Fig. 7B) showed that co-transfection of CREB and DBP increased the expression of reporter construct $(HAG\text{-}CRE)_3$ TK-CAT by 3 and 5 fold in HepG2 cells.

Co-transfection of CREB and DBP increases the promoter activity of pHAG1.2CAT

To understand the role of this CRE in transcriptional regulation in the context of human angiotensinogen gene promoter, we co-transfected different amounts of the expression vector RSV-CREB (1 and 2 μg) and a fixed amount of pHAG1.2CAT (10 μg) in HepG2 cells. After transfection, the promoter activity was analyzed by the CAT assay. Results of this experiment (Fig. 8) suggested that CREB increased the expression of pHAG1.2CAT by about 5 fold. Similarly, co-transfection of CMV-DBP increased the expression of pHAG1.2CAT in HepG2 cells by about 12 fold.

Discussion

In this paper we show for the first time that cAMP treatment increases the expression of a reporter construct pHAG1.2CAT

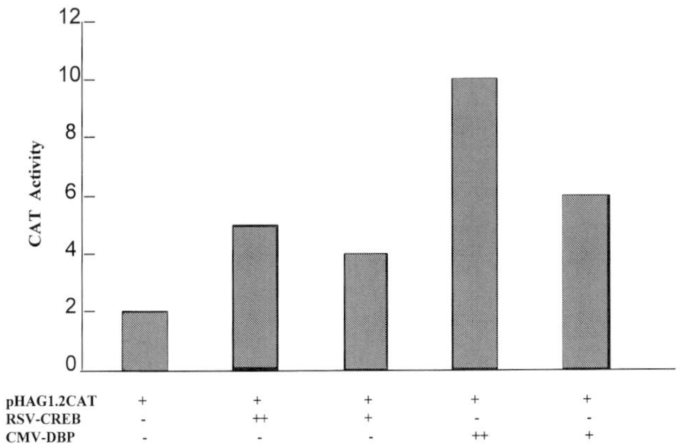

Fig. 8. The promoter activity of the reporter construct pHAG1.2CAT is increased on co-transfection of expression vectors RSV-CREB and CMV-DBP. The reporter construct pHAG1.2CAT (5 μg) and RSV-β gal (1 μg) were transfected in the absence or presence of RSV-CREB or CMV-DBP (1 and 2 μg each) in HepG2 cells and promoter activity was analyzed by the CAT assay after normalization with the β-gal assay. CAT activity was calculated by assuming the basal promoter activity of pHAG1.2CAT as one. Results show mean value of 3 experiments.

on transient transfection in HepG2 cells. We also show that a CRE-like element CTGCGTCA located between 834 and 840 bases upstream from the transcriptional initiation site of the human angiotensinogen gene binds to recombinant CREB and is involved in cAMP induced expression of this gene. In addition, this element has sequence homology with C/EBP binding site and liver enriched transcription factor DBP (which belongs to C/EBP family) binds to this sequence. We also show that co-transfection of DBP increases the promoter activity of reporter constructs containing human angiotensinogen gene promoter in HepG2 cells.

CREB/ATF proteins belong to bZIP family of transcription factors that also include AP-1, and C/EBP group of proteins. This family of transcription factors plays an important role in cAMP, TPA and cytokine induced signal transduction mechanism. CREB/ATF family of transcription factors normally bind to the palindromic sequence TGACGTCA; AP-1 family of transcription factors normally bind to the palindromic sequence TGAC/GTCA; and C/EBP family of transcription factors normally bind to the palindromic sequence TT/GNNGCAAT/G/ [33, 34]. bZIP family of transcription factors contain a basic region (a relatively high concentration of the positively charged amino acids, lysine and arginine) which resides next to a series of amino acids (leucine zipper) containing leucine at every seventh position (heptad repeats) and is flanked by transactivation domain [35]. As a result of leucine zipper, these transcription factors can form homo- as well as hetero-dimers with other members of bZIP family of transcription factors [36]. Radiolabeled peptides containing the bZIP domain of CREB/ATF family of transcription factors have been used to screen cDNA libraries to identify heterodimerizing partners. These techniques have so far resulted in the isolation of cDNA clones of a total of ten proteins. Binding studies with recombinant proteins have shown that different members of CREB/ATF family can selectively heterodimerize with other members of bZIP family of transcription factors and heterodimerization can affect their transactivation potential [37].

Although CREB/ATF family of transcription factors preferably bind to the symmetrical binding site TGACGTCA, they also bind and activate promoters that contain a pseudo-symmetrical binding site CGTCA. Thus cAMP inducible cis-acting DNA elements in promoters of phosphoenol pyruvate kinase [38], tyrosine aminotransferase [39, 40], prepro-enkephalin [41], and pro-interleukin-1β [42] genes contain CGTCA instead of TGACGTCA. Depending on the nucleotide sequence, these pseudo symmetrical cAMP responsive elements can bind to CREB/ATF as well as members of the AP-1 and C/EBP family of transcription factors. Thus the nucleotide sequence of CRE in the PEPCK gene promoter is CTTACGTCA which contains the CRE half site (CGTCA) and binds to the r-CREB. In addition, this sequence has homology with C/EBP binding site in the reverse orientation (TGACGTAAG) and also binds to C/EBP family of transcription factors. As a result, co-transfection of C/EBP-β increases CREB induced promoter activity of PEPCK gene on transient transfection in hepatoma cells. Similarly, the promoter of pro-interleukin-1β gene contains the nucleotide sequence GTTACGTCA which has a half CRE site in the forward orientation and a C/EBP binding site in the reverse orientation. As a result, this sequence is recognized by both the CREB and C/EBP-β. On the other hand, cAMP responsive element of the proenkephalin gene promoter contains the nucleotide sequence TG<u>C</u>GTCA which has a CRE half site (CGTCA) and has only one mismatch (underlined) with AP-1 binding site (TGAC/GTCA). As a result, proenkephalin gene promoter has been shown to bind to CREB and AP-1 family of transcription factors, and activated by co-transfection of expression vectors containing coding sequence of CREB/ATF as well as AP-1 family of transcription factors.

DNA binding studies with purified recombinant CREB protein have shown that Ser-133 phosphorylation has no effect on the affinity of CREB for a consensus CRE site. However, a two-three fold effect of Ser-133 phosphorylation was observed for the binding of pseudo-symmetrical CRE binding sites [43]. Since, the angiotensinogen production is increased in cultured hepatocytes by glucocorticoid, cAMP and IL6 treatment [8] and human angiotensinogen gene promoter contains a pseudo-symmetrical CRE, it is tempting to speculate that this sequence normally binds to C/EBP family of transcription factors in the absence of cAMP. However, in physiological conditions where cAMP levels are increased, the CREB gets phosphorylated and phosphorylated CREB then displaces C/EBP family of transcription factors resulting in cAMP induced transcription of this gene.

Previous studies have shown that cAMP or isoprenaline treatment increases the expression of reporter constructs containing rat angiotensinogen gene promoter in mouse hepatoma cells [9]. A CRE like sequence (TGACGTAC) is located between –795 and –788 of the rat angiotensinogen gene [44] and Chan *et al.* have shown that co-transfection of an expression vector containing CREB coding region increases the expression of rat angiotensinogen gene [45]. In a recent paper, this group has also shown that the 43 kDa CREB and a novel 52 kDa protein present in the mouse liver binds to this sequence [9]. The nucleotide sequence of this element (GTA<u>CGTCA</u> in reverse orientation) has homology with the HAG-CRE (CTG<u>CGTCA</u>). However, the 52 kDa liver specific protein that binds to the rat sequence remains to be identified and the role of C/EBP family of transcription factors in the expression of rat angiotensinogen gene remains to be investigated.

It has been shown previously that a single CRE is not sufficient for cAMP induced transcriptional activation and CREB has to co-operatively interact with other transcription factors for a robust cAMP induced promoter activity. Thus, gel shift and transient transfection studies have shown that although CREB binds to the CRE located at –85 in the promoter of PEPCK gene, the binding of liver specific transcription factors to at least three upstream sites is essential for cAMP induced promoter activity of this gene [46, 47]. Similarly, in the case of tyrosine amino transferase gene, it has been shown that CREB co-operatively interacts with liver enriched transcription factor HNF4 for optimum cAMP induced promoter activity [40]. It has been shown recently that cAMP increases C/EBP-β gene expression and therefore CREB and C/EBP-β may co-operatively increase cAMP induced promoter activity of a gene containing CRE and C/EBP binding sites [48]. In addition, CBP acts as a co-activator of CREB and co-transfection of CBP increases CREB induced promoter activity. It has been shown that CBP also binds to a number of transcription factors including C/EBP-β and HNF-4, and may co-ordinate cAMP induced promoter activity through co-operative interaction with these transcription factors. We have shown that human angiotensinogen gene promoter contains multiple C/EBP and HNF4 binding sites (unpublished data) and it is possible that these transcription factors co-operatively interact with CREB to increase cAMP induced promoter activity of the human angiotensinogen gene.

Acknowledgements

We thank Dr. Marc Montminy for CREB expression vector and Dr. Uri Schibler for DBP expression vector and DBP antibody. We also thank Jie Zhou for help in initial phase of this work. This work was supported by NIH grants HL 49884 and 59547 (to AK).

References

1. Corvol P, Jeunemaitre X, Charru A, Kotelevtsev Y, Soubrier F: Role of the renin-angiotensin system in blood pressure regulation and in human hypertension: new insights from molecular genetics. Rec Prog Horm Res 50: 287–308, 1995
2. Brasier AR, Li J: Mechanisms for inducible control of angiotensinogen gene transcription. Hypertension 27: 465–475, 1996
3. Mancia G, Grassi G, Giannattasio C, Seravalle G: Sympathetic activation in the pathogenesis of hypertension and progression of organ damage. Hypertension 34: 724–728, 1999
4. Mancia G: Bjorn Folkow Award Lecture. The sympathetic nervous system in hypertension. J Hypertens 15: 1553–1565, 1997
5. Ying L, Morris BJ, Sigmund CD: Transactivation of the human renin promoter by the cyclic AMP/protein kinase A pathway is mediated by both cAMP-responsive element binding protein-1 (CREB)-dependent and CREB-independent mechanisms in Calu-6 cells. J Biol Chem 272: 2412–2420, 1997
6. Germain S, Konoshita T, Philippe J, Corvol P, Pinet F: Transcriptional induction of the human renin gene by cyclic AMP requires cyclic AMP response element-binding protein (CREB) and a factor binding a pituitary-specific trans-acting factor (Pit-1) motif. Biochem J 316 (Part 1): 107–113, 1996
7. Tamura K, Umemura S, Yamaguchi S, Iwamoto T, Kobayashi S, Fukamizu A, Murakami K, Ishii M: Mechanism of cAMP regulation of renin gene transcription by proximal promoter. J Clin Invest 94: 1959–1967, 1994
8. Ohtani R, Yayama K, Takano M, Itoh N, Okamoto H: Stimulation of angiotensinogen production in primary cultures of rat hepatocytes by glucocorticoid, cyclic adenosine 3′,5′ monophosphate, and interleukin-6. Endocrinology 130: 1331–1338, 1992
9. Ming M, Wu J, Lachance S, Delalandre A, Carriere S, Chan JS: Beta-adrenergic receptors and angiotensinogen gene expression in mouse hepatoma cells *in vitro*. Hypertension 25: 105–109, 1995
10. Montminy M: Transcriptional regulation by cyclic AMP. Annu Rev Biochem 66: 807–822, 1997
11. Meyer TE, Habener JF: Cyclic adenosine 3′,5′-monophosphate response element binding protein (CREB) and related transcription-activating deoxyribonucleic acid-binding proteins. Endocr Rev 14: 269–290, 1993
12. Hoeffler JP, Meyer TE, Yun Y, Jameson JL, Habener JF: Cyclic AMP-responsive DNA-binding protein: Structure based on a cloned placental cDNA. Science 242: 1430–1433, 1988
13. Gonzalez GA, Yamamoto KK, Fischer WH, Karr D, Menzel P, Biggs W3, Vale WW, Montminy MR: A cluster of phosphorylation sites on the cyclic AMP-regulated nuclear factor CREB predicted by its sequence. Nature 337: 749–752, 1989
14. Gonzalez GA, Montminy MR: Cyclic AMP stimulates somatostatin gene transcription by phosphorylation of CREB at serine 133. Cell 59: 675–680, 1989
15. Lamph WW, Dwarki VJ, Ofir R, Montminy M, Verma IM: Negative and positive regulation by transcription factor cAMP response element-binding protein is modulated by phosphorylation. Proc Natl Acad Sci USA 87: 4320–4324, 1990
16. Yamamoto KK, Gonzalez GA, Menzel P, Rivier J, Montminy MR: Characterization of a bipartite activator domain in transcription factor CREB. Cell 60: 611–617, 1990
17. Dwarki VJ, Montminy M, Verma IM: Both the basic region and the 'leucine zipper' domain of the cyclic AMP response element binding (CREB) protein are essential for transcriptional activation. EMBO J 9: 225–232, 1990
18. Chrivia JC, Kwok RP, Lamb N, Hagiwara M, Montminy MR, Goodman RH: Phosphorylated CREB binds specifically to the nuclear protein CBP. Nature 365: 855–859, 1993
19. Parker D, Ferreri K, Nakajima T, LaMorte VJ, Evans R, Koerber SC, Hoeger C, Montminy MR: Phosphorylation of CREB at Ser-133 induces complex formation with CREB-binding protein via a direct mechanism. Mol Cell Biol 16: 694–703, 1996
20. Goldman PS, Tran VK, Goodman RH: The multifunctional role of the co-activator CBP in transcriptional regulation. Rec Prog Horm Res 52: 103–119, 1997
21. Hai TW, Liu F, Coukos WJ, Green MR: Transcription factor ATF cDNA clones: an extensive family of leucine zipper proteins able to selectively form DNA-binding heterodimers (published erratum appears in Genes Dev 4: 682, 1990). Genes Dev 3: 2083–2090, 1989
22. Foulkes NS, Borrelli E, Sassone Corsi P: CREM gene: Use of alternative DNA-binding domains generates multiple antagonists of cAMP-induced transcription. Cell 64: 739–749, 1991
23. Fimia GM, De Cesare D, Sassone-Corsi P: CBP-independent activation of CREM and CREB by the LIM-only protein ACT. Nature 398: 165–169, 1999
24. Campbell DJ, Habener JF: Angiotensinogen gene is expressed and differentially regulated in multiple tissues of the rat. J Clin Invest 78: 31–39, 1986
25. Narayanan CS, Cui Y, Kumar A: DBP binds to the proximal promoter and regulates liver-specific expression of the human angiotensinogen gene. Biochem Biophys Res Commun 251: 388–393, 1998
26. Cui Y, Narayanan CS, Zhou J, Kumar A: Exon-I is involved in positive as well as negative regulation of human angiotensinogen gene expression. Gene 224: 97–107, 1998
27. Fukamizu A, Takahashi S, Seo MS, Tada M, Tanimoto K, Uehara S, Murakami K: Structure and expression of the human angiotensinogen gene. Identification of a unique and highly active promoter. J Biol Chem 265: 7576–7582, 1990
28. Zhao YY, Zhou J, Kumar A. Molecular mechanism of human essential hypertension: A mutation in the proximal promoter of the angiotensinogen gene from hypertensive patients increases its transcription in human hepatoma cells. Circulation 94(suppl 1): (abstr) I-693, 1996
29. Zhao YY, Zhou J, Narayanan CS, Cui Y, Kumar A: Role of C/A polymorphism at –20 on the expression of human angiotensinogen gene. Hypertension 33: 108–115, 1999
30. Yanai K, Nibu Y, Murakami K, Fukamizu A: A cis-acting DNA element located between TATA box and transcription initiation site is critical in response to regulatory sequences in human angiotensinogen gene. J Biol Chem 271: 15981–15986, 1996
31. Narayanan CS, Cui Y, Zhao YY, Zhou J, Kumar A: Orphan receptor Arp-1 binds to the nucleotide sequence located between TATA box and transcriptional initiation site of the human angiotensinogen gene and reduces estrogen induced promoter activity. Mol Cell Endocrinol 148: 79–86, 1999
32. Dignam JD, Lebovitz RM, Roeder RG: Accurate transcription initiation by RNA polymerase II in a soluble extract from isolated mammalian nuclei. Nucl Acids Res 11: 1475–1489, 1983
33. Friedman AD, Landschulz WH, McKnight SL: CCAAT/enhancer binding protein activates the promoter of the serum albumin gene in cultured hepatoma cells. Genes Dev 3: 1314–1322, 1989
34. Friedman AD, McKnight SL: Identification of two polypeptide segments of CCAAT/enhancer- binding protein required for transcriptional activation of the serum albumin gene. Genes Dev 4: 1416–1426, 1990

35. Vinson CR, Sigler PB, McKnight SL: Scissors-grip model for DNA recognition by a family of leucine zipper proteins. Science 246: 911–916, 1989
36. Vallejo M, Ron D, Miller CP, Habener JF: C/ATF, a member of the activating transcription factor family of DNA-binding proteins, dimerizes with CAAT/enhancer-binding proteins and directs their binding to cAMP response elements. Proc Natl Acad Sci USA 90: 4679–4683, 1993
37. Hai T, Curran T: Cross-family dimerization of transcription factors Fos/Jun and ATF/CREB alters DNA binding specificity. Proc Natl Acad Sci USA 88: 3720–3724, 1991
38. Hanson RW, Reshef L: Regulation of phosphoenolpyruvate carboxykinase (GTP) gene expression. Annu Rev Biochem 66: 581–611, 1997
39. Weih F, Stewart AF, Boshart M, Nitsch D, Schutz G: *In vivo* monitoring of a cAMP-stimulated DNA-binding activity. Genes Dev 4: 1437–1449, 1990
40. Nitsch D, Boshart M, Schutz G: Activation of the tyrosine aminotransferase gene is dependent on synergy between liver-specific and hormone-responsive elements. Proc Natl Acad Sci USA 90: 5479–5483, 1993
41. Chu HM, Tan Y, Kobierski LA, Balsam LB, Comb MJ: Activating transcription factor-3 stimulates 3′,5′-cyclic adenosine monophosphate-dependent gene expression. Mol Endocrinol 8: 59–68, 1994
42. Tsukada J, Saito K, Waterman WR, Webb AC, Auron PE: Transcription factors NF-IL6 and CREB recognize a common essential site in the human prointerleukin 1 beta gene. Mol Cell Biol 14: 7285–7297, 1994
43. Nichols M, Weih F, Schmid W, DeVack C, Kowenz-Leutz E, Luckow B, Boshart M, Schutz G: Phosphorylation of CREB affects its binding to high and low affinity sites: Implications for cAMP induced gene transcription. EMBO J 11: 3337–3346, 1992
44. Chan JS, Chan AH, Jiang Q, Nie ZR, Lachance S, Carriere S: Molecular cloning and expression of the rat angiotensinogen gene. Pediatr Nephrol 4: 429–435, 1990
45. Qian JF, Wang TT, Wu XH, Wu J, Ge C, Lachance S, Carriere S, Chan JS: Angiotensinogen gene expression is stimulated by the cAMP- responsive element binding protein in opossum kidney cells. J Am Soc Nephrol 8: 1072–1079, 1997
46. Leahy P, Crawford DR, Grossman G, Gronostajski RM, Hanson RW: CREB binding protein coordinates the function of multiple transcription factors including nuclear factor I to regulate phosphoenolpyruvate carboxykinase (GTP) gene transcription. J Biol Chem 274: 8813–8822, 1999
47. Roesler WJ, Graham JG, Kolen R, Klemm DJ, McFie PJ: The cAMP response element binding protein synergizes with other transcription factors to mediate cAMP responsiveness. J Biol Chem 270: 8225–8232, 1995
48. Niehof M, Manns MP, Trautwein C: CREB controls LAP/C/EBP beta transcription. Mol Cell Biol 17: 3600–3613, 1997

Molecular and Cellular Biochemistry **212**: 91–98, 2000.

The role of Ca^{2+} mobilization and heterotrimeric G protein activation in mediating tyrosine phosphorylation signaling patterns in vascular smooth muscle cells

Peter P. Sayeski, M. Showkat Ali and Kenneth E. Bernstein

Department of Pathology and Laboratory Medicine, Emory University School of Medicine, Atlanta, GA, USA

Abstract

This work investigated the role of Ca^{2+} mobilization and heterotrimeric G protein activation in mediating angiotensin II-dependent tyrosine phosphorylation signaling patterns. We demonstrate that the predominant, angiotensin II-dependent, tyrosine phosphorylation signaling patterns seen in vascular smooth muscle cells are blocked by the intracellular Ca^{2+} chelator BAPTA-AM, but not by the Ca^{2+} channel blocker verapamil. Activation of heterotrimeric G proteins with NaF resulted in a divergent signaling effect; NaF treatment was sufficient to increase tyrosine phosphorylation levels of some proteins independent of angiotensin II treatment. In the same cells, NaF alone had no effect on other cellular proteins, but greatly potentiated the ability of angiotensin II to increase the tyrosine phosphorylation levels of these proteins. Two proteins identified in these studies were paxillin and Jak2. We found that NaF treatment alone, independent of angiotensin II stimulation, was sufficient to increase the tyrosine phosphorylation levels of paxillin. Furthermore, the ability of either NaF and/or angiotensin II to increase tyrosine phosphorylation levels of paxillin is critically dependent on intracellular Ca^{2+}. In contrast, angiotensin II-mediated Jak2 tyrosine phosphorylation was independent of intracellular Ca^{2+} mobilization and extracellular Ca^{2+} entry. Thus, our data suggest that angiotensin II-dependent tyrosine phosphorylation signaling cascades are mediated through a diverse set of signaling pathways that are partially dependent on Ca^{2+} mobilization and heterotrimeric G protein activation. (Mol Cell Biochem **212**: 91–98, 2000)

Key words: vascular smooth muscle, angiotensin II, tyrosine phosphorylation, intracellular Ca^{2+} and heterotrimeric G proteins

Introduction

Vascular smooth muscle cells (VSMC) are the principal component of the blood vessel wall and are responsible for maintaining vascular tone. They also express an abundant amount of the angiotensin II type AT_1 receptor. This G protein-coupled receptor mediates smooth muscle cell contraction through a transient increase in intracellular Ca^{2+} via the production of 1,4,5-IP_3. Although this class of receptors lacks intrinsic kinase activity, it is nonetheless able to initiate tyrosine phosphorylation signaling cascades through the activation of intracellular, non-receptor tyrosine kinases. In the case of angiotensin II, treatment of cells results in a ligand-dependent activation of the non-receptor tyrosine kinases Jak2, Tyk2, Fyn, c-Src, FAK, and Pyk2 [1–4]. Recent studies indicate that these tyrosine phosphorylation signaling cascades mediate many angiotensin II signaling events including cytoskeletal rearrangement, induction of the early response genes and proliferative growth responses [5–7]. To underscore the importance of this fact, a recent review identified nineteen different proteins that have increased tyrosine phosphorylation levels as a result of angiotensin II treatment [8].

Address for offprints: K.E. Bernstein, Department of Pathology and Laboratory Medicine, 1639 Pierce Drive, 7107 WMB, Emory University School of Medicine, Atlanta, GA 30322, USA

The exact mechanism(s) that couple AT_1 receptor activation to non-receptor tyrosine kinase activation are poorly understood. Based on our current understanding of second messenger molecules, several possibilities exist. First, Ca^{2+} mobilization may be a link coupling AT_1 receptor activation with tyrosine kinase activation; the tyrosine kinase Pyk2 is activated by numerous ligands, but all appear to require Ca^{2+} mobilization [9–11]. A second possibility is that heterotrimeric G protein activation mediates the activation of non-receptor tyrosine kinases such as c-Src [12]. Yet another possibility is that adaptor proteins serve as a bridge linking receptor activation with non-receptor tyrosine kinase activation. For instance, the adaptor protein SH2-Bβ has been implicated in the activation of Jak2 tyrosine kinase [13].

In the case of angiotensin II, we hypothesized that Ca^{2+} mobilization and heterotrimeric G protein activation were both critical mediators bridging AT_1 receptor activation to tyrosine kinase activation. Here, we utilized specific inhibitors of Ca^{2+} mobilization and an activator of heterotrimeric G proteins to delineate the mechanisms by which angiotensin II initiates tyrosine kinase phosphorylation signaling cascades. We conclude that angiotensin II-dependent tyrosine phosphorylation signaling cascades are mediated through a diverse set signaling pathways and that a subset of these are dependent on intracellular Ca^{2+} mobilization and G protein activation.

Materials and methods

Cell culture

VSMC were grown in DMEM + 10% FBS at 37°C in a 5% CO_2 humidified atmosphere and used between passages 8–22. One hundred mm dishes at approximately 80% confluence were growth arrested by incubation in serum free DMEM for 48 h before use. Cell culture reagents were obtained from Gibco/BRL. Inhibitors were purchased from Calbiochem. We previously demonstrated that the concentration and duration of inhibitor pretreatment used in these studies is sufficient to block the desired effect [14, 15]. All other reagents were purchased from Sigma Chemical.

Immunoprecipitation

To prepare cellular lysates, cells were washed with two volumes of ice-cold PBS containing 1 mM Na_3VO_4 and lysed in 1.0 ml RIPA buffer (20 mM Tris pH 7.5, 10% glycerol, 1% Triton X-100, 1% deoxycholic acid, 0.1% SDS, 2.5 mM EDTA, 50 mM NaF, 10 mM $Na_4P_2O_7$, 4 mM benzamidine, 1 mM AEBSF, 1 mM Na_3VO_4, and 10 μg/ml aprotinin). The samples were quickly sonicated and then incubated on ice for 30–60 min. The lysates were spun at 12,000 × g for 5 min at 4°C and the supernatants were quantitated using the D_c protein assay (Bio-Rad). Normalized protein lysates (approximately 400 μg/ml) were immunoprecipitated with 2 μg of anti-phosphotyrosine-monoclonal antibody (Transduction Laboratories; clone PY20) and 20 μl of a 50% slurry of Protein A/G Plus agarose beads (Santa Cruz Biotechnology) for 2–16 h at 4°C. Immune complexes were washed 3 times with Wash Buffer (25 mM Tris pH 7.5, 150 mM NaCl, and 0.1% Triton X-100) and resuspended in sample buffer. Proteins were separated by 8% SDS-PAGE and transferred onto nitrocellulose membranes (Schleicher and Scheull).

Western blotting

After blocking for 1 h in 5% milk/TBST (w/v) at 23°C, nitrocellulose membranes were probed with primary antibody for 1–2 h at 23°C in 5% milk/TBST. Blots were washed with TBST and proteins were visualized with ECL following the manufacturers instructions (NEN Life Sciences). Blotting antibodies purchased from Upstate Biotechnology were monoclonal anti-phosphotyrosine (clone 4G10) and polyclonal anti-Jak2. Monoclonal anti-paxillin was from Transduction Laboratories.

Results

Angiotensin II-dependent tyrosine phosphorylation signaling patterns require the release of intracellular Ca^{2+}

Treatment of cells with angiotensin II results in the activation of at least six non-receptor tyrosine kinases [1–4]. These activated kinases then initiate a series of tyrosine phosphorylation signaling cascades by phosphorylating intracellular protein substrates which in turn yield a specific cellular response. We hypothesized that Western blotting cellular protein lysates with an antibody directed against the phosphotyrosine moiety of proteins would serve as a good indicator of these tyrosine phosphorylation signaling patterns that move through VSMC after angiotensin II treatment. To test this, VSMC were treated with 10^{-7} M angiotensin II for the indicated times and soluble cell lysates were immunoprecipitated and Western blotted with an anti-phosphotyrosine monoclonal antibody [anti-Tyr(P)-mAb] (Fig. 1). We observed a predominant ligand-dependent increase in the tyrosine phosphorylation levels of several protein bands in the 66–80 kDa range and the 120–140 kDa range. Longer exposures of this blot identified additional bands which also had angiotensin II-dependent increased tyrosine phosphorylation levels (data not shown). We concluded that this method of immunoprecipitating and Western blotting VSMC lysates with anti-Tyr(P)-mAb provided a unique fingerprint of the tyrosine phosphorylation signaling patterns that move through VSMC after angiotensin II treatment.

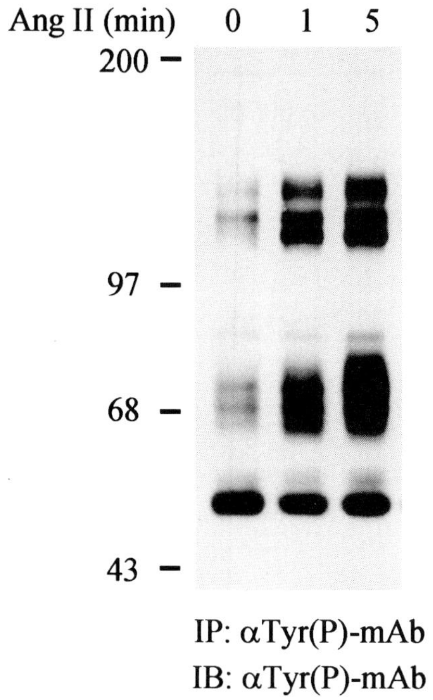

Fig. 1. Angiotensin II-induced tyrosine phosphorylation signaling patterns in VSMC. Quiescent VSMC were stimulated with 10^{-7} M angiotensin II for the indicated times. The resulting lysates were immunoprecipitated and Western blotted with anti-Tyr(P)-mAb. Shown is one of nine representative results.

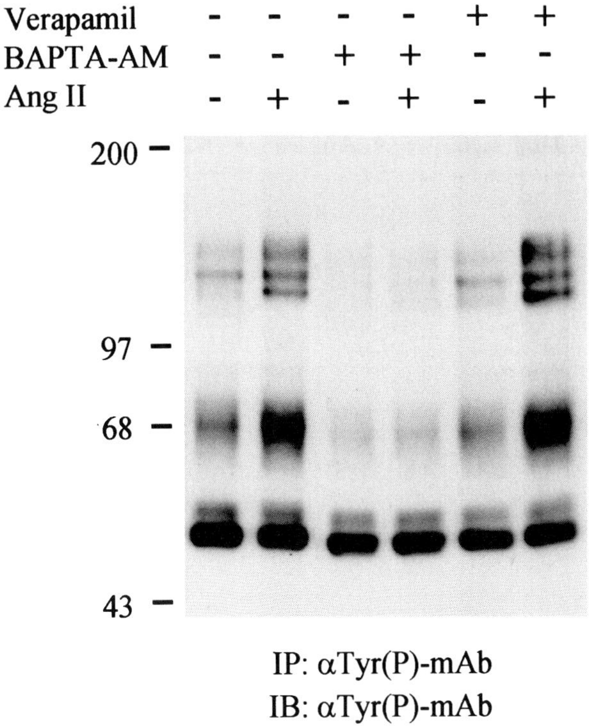

Fig. 2. The role of Ca^{2+} mobilization in mediating angiotensin II-dependent tyrosine phosphorylation signaling patterns in VSMC. Quiescent VSMC were pretreated as follows: 10 min with DMSO (lanes 1 and 2); 10 min with 50 μM BAPTA-AM (lanes 3 and 4); 0.5 min with 10 μM verapamil (lanes 5 and 6). The cells were then left untreated (–) or treated (+) with 10^{-7} M angiotensin II for 1 min. The resulting lysates were immunoprecipitated and Western blotted with anti-Tyr(P)-mAb. Shown is one of three representative results.

Next, we investigated the role of intracellular and extracellular Ca^{2+} in mediating these tyrosine phosphorylation signaling patterns. VSMC were pretreated with BAPTA-AM, a chelator of intracellular Ca^{2+}. Since voltage-dependent L-type Ca^{2+} channels are expressed on VSMC, we also used the Ca^{2+} channel blocker verapamil. The cells were then left untreated or treated for 1 min with 10^{-7} M angiotensin II and lysates were prepared. The soluble lysates were immunoprecipitated and Western blotted with anti-Tyr(P)-mAb to measure tyrosine phosphorylation signaling patterns (Fig. 2). We observed that pretreatment of cells with BAPTA-AM eliminated nearly all of the angiotensin II-mediated increase of tyrosine phosphorylation while verapamil was without effect. These results indicate that the predominant tyrosine phosphorylation signaling pattern resulting from angiotensin II treatment of VSMC is critically dependent on the mobilization of intracellular Ca^{2+}, but independent of extracellular Ca^{2+} entry.

Role of heterotrimeric G protein activation on tyrosine phosphorylation signaling patterns in VSMC

To investigate the role of heterotrimeric G protein activation in mediating this tyrosine phosphorylation fingerprint, VSMC were treated with the G protein activator NaF. The mechanism of activation is the formation of AlF_4^-, which in turn stabilizes the active GTP protein complex [16]. We successfully used this methodology to show that heterotrimeric G protein activation is sufficient to increase tyrosine phosphorylation levels of the adaptor protein, p130Cas [15]. In the current study, VSMC were either left untreated or pretreated with two doses of NaF (1 or 5 mM) for times ranging from 1–60 min. One group of cells was then treated with 10^{-7} M angiotensin II while the other cells were left untreated. Soluble lysates were prepared and proteins were immunoprecipitated and Western blotted with anti-Tyr(P)-mAb (Fig. 3A). For the protein band in the 66–80 kDa range, NaF treatment increased the tyrosine phosphorylation of this protein in both a dose and time-dependent manner. Furthermore, this was done independent of angiotensin II treatment, suggesting that G protein activation alone is sufficient to activate the tyrosine kinase(s) that phosphorylate this protein. For those proteins in the 120–140 kDa range, three distinct protein bands were observed (approximately 120, 130 and 140 kDa; arrowheads). For the 120 kDa protein band, NaF treatment alone was able to increase tyrosine phosphoryla-

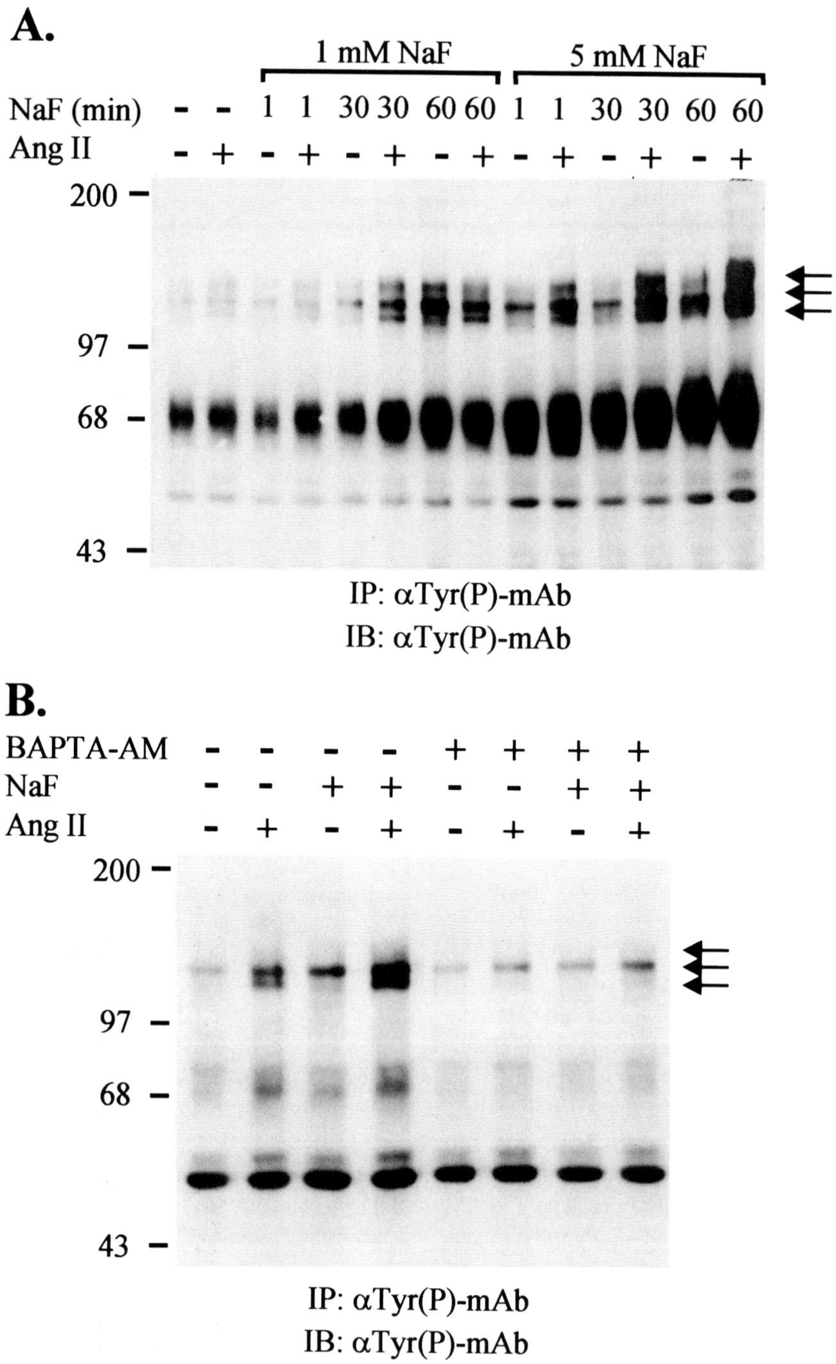

Fig. 3. The role of heterotrimeric G protein activation and intracellular Ca^{2+} mobilization on patterns of angiotensin II-dependent tyrosine phosphorylation in VSMC. (A) Quiescent VSMC were pretreated with either 1 or 5 mM NaF for the indicated times and then left untreated (–) or treated (+) with 10^{-7} M angiotensin II for 1 min. The resulting lysates were then immunoprecipitated and Western blotted with anti-Tyr(P)-mAb. (B) Quiescent VSMC were first pretreated with 50 μM BAPTA-AM (+) for 10 min and then treated with 5 mM NaF (+) for 30 min. The cells were then left untreated (–) or treated (+) with 10^{-7} M angiotensin II for 1 min. The resulting lysates were immunoprecipitated and Western blotted with anti-Tyr(P)-mAb. Shown is one of three representative results for each.

tion levels, but only at the highest doses and longest periods of NaF incubation. When angiotensin II was added to cells that had been pretreated with NaF, there was a marked synergy of the tyrosine phosphorylation levels of this protein; the combination of NaF plus angiotensin II was significantly greater than that of NaF alone. The 130 kDa band was unique in that NaF treatment alone increased its tyrosine phosphorylation levels suggesting that G protein activation is sufficient for activating the tyrosine kinase(s) that phosphorylate this protein. For the cells which received NaF plus angiotensin II, there was a marked synergy of the tyrosine phosphorylation levels when compared to cells that received only NaF. The 140 kDa band was characterized as having properties similar to that of the 120 kDa band; NaF treatment alone was mostly without effect, but that the combination of NaF plus angiotensin II had a marked synergy on the tyrosine phosphorylation levels when compared to cells that only received NaF.

Collectively, the data in Fig. 3A demonstrate a marked diversity in the ability of heterotrimeric G protein activation to mediate tyrosine phosphorylation signaling patterns in VSMC. In the case of the 66–80 kDa protein, NaF is sufficient to induce tyrosine phosphorylation and there is no statistically significant additive effect with angiotensin II treatment. For the 120 and 140 kDa bands, NaF alone is mostly without effect, but the combination of NaF plus angiotensin II has a marked potentiation of tyrosine phosphorylation levels when compared to treatment with NaF alone. Finally, the 130 kDa band has increased tyrosine phosphorylation with NaF, but unlike the 66–80 kDa band, there is a marked additive effect when cells receive both NaF and angiotensin II.

Role of heterotrimeric G protein activation and intracellular Ca^{2+} mobilization on tyrosine phosphorylation patterns in VSMC

In order to gain a better understanding of the interplay between intracellular Ca^{2+} mobilization and G protein activation in generating this tyrosine phosphorylation fingerprint, VSMC were treated with a combination of BAPTA-AM, NaF and angiotensin II. The cellular lysates were then immunoprecipitated and Western blotted with anti-Tyr(P)-mAb (Fig. 3B). In lane 2, angiotensin II treatment alone resulted in the expected tyrosine phosphorylation signaling pattern seen previously. Pretreatment with NaF alone (lane 3) again produced the effect of increasing the tyrosine phosphorylation levels of the 130 kDa protein band. Also, addition of angiotensin II after NaF pretreatment again resulted in a marked synergy of tyrosine phosphorylation levels when compared to cells that received only angiotensin II (lane 2 vs. 4). Pretreatment with BAPTA-AM eliminated virtually all the angiotensin II-dependent tyrosine phosphorylation increases with the exception of the 130 kDa band (lane 5 vs. 6). This result indicates that the activation of the kinase that is responsible for the angiotensin II-dependent increased tyrosine phosphorylation of this band, occurs independent of intracellular Ca^{2+} mobilization. Finally, the last set of cells first received BAPTA-AM (to chelate intracellular Ca^{2+}) and then received NaF (to activate heterotrimeric G proteins). The cells were then left untreated or treated with angiotensin II (lane 7 vs. 8). With the exception of the 130 kDa band, all protein bands had significantly reduced tyrosine phosphorylation signaling patterns. With respect to the 130 kDa band, the ability of NaF to increase tyrosine phosphorylation levels was reduced when the cells were treated with BAPTA-AM prior to NaF addition. However, the angiotensin II-dependent increase of this band persisted in the presence of BAPTA-AM, suggesting that NaF increases the phosphorylation of this protein via Ca^{2+}-dependent tyrosine kinases while angiotensin II increases the phosphorylation of this band via Ca^{2+}-independent tyrosine kinases.

Regulation of paxillin and Jak2 tyrosine phosphorylation

While the protocol in Figs 1, 2, and 3 is a good indicator of the tyrosine phosphorylation signaling patterns in VSMC, its shortcoming is that the identity of the specific protein bands is not known. A likely candidate for the 66–80 kDa band is the cytoskeletal protein, paxillin. Previous reports demonstrated that paxillin migrates at about 68–72 kDa, that it is tyrosine phosphorylated in response to angiotensin II and that this phosphorylation is blocked by BAPTA-AM [14, 17]. To determine whether this band is paxillin, VSMC were left untreated or pretreated with 5 mM NaF for times ranging from 1–60 min. The cells were then left untreated or treated for 1 minute with 10^{-7} M angiotensin II. Soluble lysates were immunoprecipitated with anti-Tyr(P)-mAb and then Western blotted with anti-paxillin-mAb (Fig. 4A). This under exposed blot clearly demonstrates that NaF treatment alone is sufficient to increase paxillin tyrosine phosphorylation in a manner identical to the 66–80 kDa protein in Fig. 3A. Although there appears to be an additive effect of angiotensin II plus NaF when compared to NaF alone, scanning densitometry of these bands from 3 independent experiments failed to yield a statistically significant additive effect. The absence of a significant additive effect of angiotensin II and NaF is consistent with a common signaling mechanism for these two activators. To determine the role of Ca^{2+} mobilization, G protein activation and angiotensin II treatment on the pattern of paxillin tyrosine phosphorylation, VSMC were treated exactly as described in Fig. 3B. This time however, the lysates were immunoprecipitated with anti-Tyr(P)-mAb and then Western blotted with anti-paxillin-mAb (Fig. 4B). The pattern of paxillin tyrosine phosphorylation was identical to the 66–80 kDa band seen in Fig. 3B. Independent treatment with either angiotensin II or NaF was able to increase the tyrosine phosphorylation levels of paxillin, but the combination of the two activators had no statistically significant additive effect when compared to NaF treatment alone. Furthermore, pretreatment with BAPTA-AM blocked both the angiotensin II and NaF mediated increases of paxillin tyrosine phosphorylation.

A second band of interest to us was the 130 kDa protein band. Unlike paxillin, its angiotensin II-dependent tyrosine phosphorylation persisted in the presence of BAPTA-AM. A likely candidate for this band was the tyrosine kinase, Jak2. This protein is 130 kDa in mass and is tyrosine phosphorylated in response to angiotensin II [1]. Previously, we showed that the angiotensin II-dependent tyrosine phosphorylation of Jak2 is mediated by the AT_1 receptor motif, [319]YIPP, located

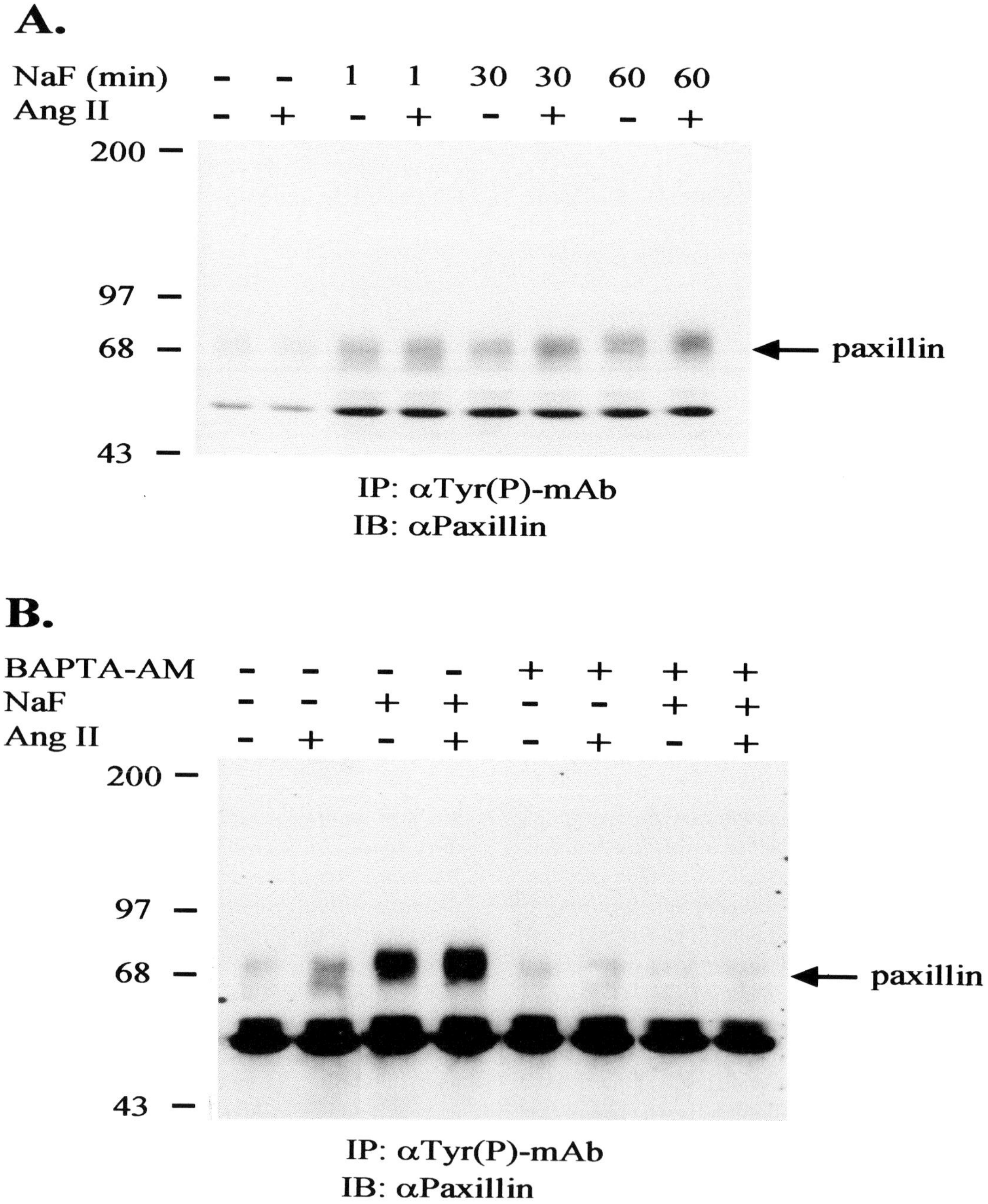

Fig. 4. The role of heterotrimeric G protein activation and Ca^{2+} mobilization on paxillin and Jak2 tyrosine phosphorylation. (A) Quiescent VSMC were pretreated with 5 mM NaF for the indicated times and then left untreated (–) or treated (+) with 10^{-7} M angiotensin II for 1 min. The resulting lysates were immunoprecipitated with anti-Tyr(P)-mAb and Western blotted with anti-paxillin-mAb. (B) Quiescent VSMC were treated exactly as described in Fig. 3B. The resulting lysates were immunoprecipitated with anti-Tyr(P)-mAb and Western blotted with anti-paxillin-mAb. (C) Quiescent VSMC were pretreated with inhibitors as described in Fig. 2. The cells were then treated with 10^{-7} M angiotensin II for the indicated times. The resulting lysates were immunoprecipitated with anti-Tyr(P)-mAb and Western blotted with anti-Jak2-pAb. Shown is one of three representative results for (A) and (B) and one of two representative results for (C).

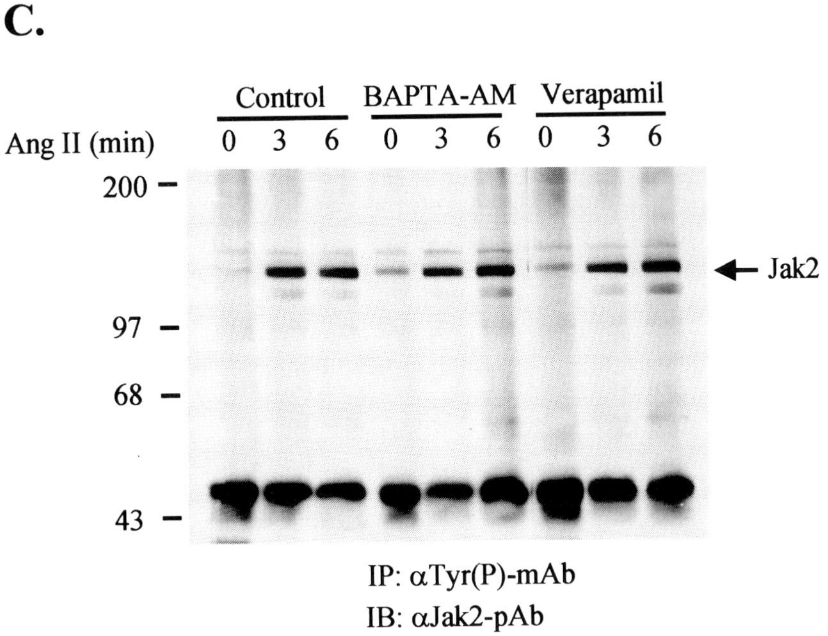

Fig. 4. Continued.

on the intracellular carboxyl terminal tail of the receptor [18]. While this protein motif is required for Jak2 activation, it has never been implicated in any Ca^{2+}-dependent signaling events. To determine whether the 130 kDa band was Jak2, VSMC were pretreated with either DMSO, BAPTA-AM or verapamil and then treated with 10^{-7} M angiotensin II for the indicated times. The resulting lysates were immunoprecipitated with anti-Tyr(P)-mAb and Western blotted with anti-Jak2-pAb (Fig. 4C). We observed that the angiotensin II-dependent tyrosine phosphorylation of Jak2 was nearly identical across all three conditions, indicating that the angiotensin II-dependent tyrosine phosphorylation of Jak2 is independent of intracellular Ca^{2+} mobilization and of extracellular Ca^{2+} entry. Consistent with these results is the observation that the angiotensin II-dependent tyrosine phosphorylation of Jak2 persists in the presence thapsigargin, a sesquiterpene which blocks intracellular Ca^{2+} mobilization by inhibiting the endoplasmic reticular Ca^{2+}-ATPase (data not shown). Thus, these data suggest that the 130 kDa protein band is most likely Jak2 tyrosine kinase and that the angiotensin II-dependent tyrosine phosphorylation of Jak2 is independent of Ca^{2+} mobilization.

In summary, NaF treatment alone, independent of angiotensin II stimulation, is sufficient to increase the tyrosine phosphorylation levels of paxillin. The ability of either NaF and/or angiotensin II to increase the tyrosine phosphorylation levels of paxillin is critically dependent on intracellular Ca^{2+}. In contrast, the angiotensin II-dependent tyrosine phosphorylation of Jak2 is completely independent of any Ca^{2+} mobilization.

Discussion

Angiotensin II is the effector molecule of the renin-angiotensin system. It is vital for maintaining a wide variety of physiological responses including salt and water balance, blood pressure and vascular tone. These effects are transduced through the G protein-coupled AT_1 receptor. Although this receptor lacks intrinsic tyrosine kinase activity, activation of the receptor by angiotensin II leads to the activation of at least six different non-receptor tyrosine kinases and the subsequent tyrosine phosphorylation of at least nineteen different proteins [1–4, 8]. These tyrosine phosphorylation signaling pathways alter specific cellular responses that allow the cell to adapt to its ever changing external milieu. However, the mechanism(s) by which the activated AT_1 receptor initiates these tyrosine kinase signaling pathways is poorly understood.

In this report, we tested the hypothesis that Ca^{2+} mobilization and heterotrimeric G protein activation are critical intermediates by which the AT_1 receptor activates tyrosine kinases. Using specific inhibitors of Ca^{2+} mobilization and an activator of heterotrimeric G proteins, we conclude that the predominant, angiotensin II-dependent tyrosine phosphorylation fingerprint, is mediated through a diverse set of signaling pathways, and that a subset of these are dependent on intracellular Ca^{2+} mobilization and heterotrimeric G protein activation. In the case of the cytoskeletal protein, paxillin, we demonstrate that heterotrimeric G protein activation is sufficient to increase its tyrosine phosphorylation levels. Moreover, the ability of either NaF and/or angiotensin

II to increase the tyrosine phosphorylation levels of paxillin is critically dependent on intracellular Ca^{2+}. In contrast, the mechanism by which the AT_1 receptor activates Jak2 appears to be completely independent of Ca^{2+} mobilization.

In summary, these results provide insight into the mechanisms by which a catalytically inactive receptor activates tyrosine kinase signaling pathways. Our results suggest that the mechanisms by which the AT_1 receptor activates intracellular tyrosine kinase signaling pathways are characterized by a marked level of signaling diversity that are both dependent and independent of Ca^{2+} mobilization and G protein activation.

Acknowledgements

We thank Kim Hawks for outstanding technical assistance. This work was supported by grants from the National Institutes of Health DK39777, DK44280, DK45215, DK51445 and HL47035. Dr. Sayeski has been supported by grants T32-DK07298 and F32-HL09678. Dr. Ali has been supported by National Institutes of Health grant HL61710.

References

1. Marrero MB, Scheiffer B, Paxton WG, Heerdt L, Berk BC, Delafontaine P, Bernstein KE: Direct stimulation of Jak/STAT pathway by the angiotensin II AT_1 receptor. Nature 375: 247–250, 1995
2. Ishida M, Marrero MB, Schieffer B, Bernstein KE, Berk BC: Angiotensin II activates pp60$^{c\text{-}src}$ in vascular smooth muscle cells. Circ Res 77: 1053–1059, 1995
3. Sadoshima J, Izumo S: The heterotrimeric G_q protein-coupled angiotensin II receptor activates p21^{ras} via the tyrosine kinase-Shc-Grb2-Sos pathway in cardiac myocytes. EMBO J 15: 775–787, 1996
4. Li X, Earp HS: Paxillin is tyrosine-phosphorylated by and preferentially associates with the calcium-dependent tyrosine kinase in rat liver epithelial cells. J Biol Chem 272: 14341–14348, 1997
5. Turner CE, Pietras KM, Taylor DS, Molloy CJ: Angiotensin II stimulation of rapid paxillin tyrosine phosphorylation correlates with the formation of focal adhesions in rat aortic smooth muscle cells. J Cell Science 108: 333–342, 1995
6. Sayeski PP, Ali MS, Safavi A, Lyles M, Kim SO, Frank SJ, Bernstein KE: A catalytically active Jak2 is required for the angiotensin II-dependent activation of Fyn. J Biol Chem 274: 33131–33142, 1999
7. Teutsch B, Bihoreau C, Monnot C, Bernstein KE, Murphy TJ, Alexander RW, Corvol P, Clauser E: A recombinant rat vascular AT_1 receptor confers growth properties to angiotensin II in Chinese hamster ovary cells. Biochem Biophys Res Comm 187: 1381–1388, 1992
8. Sayeski PP, Ali MS, Semeniuk DJ, Doan TN, Bernstein KE: Angiotensin II signal transduction pathways. Reg Peptides 78: 19–29, (review) 1998
9. Sasaki H, Nagura K, Ishino M, Tobioka H, Sasaki T: Cloning and characterization of cell adhesion kinase β, a novel protein tyrosine kinase of the focal adhesion kinase subfamily. J Biol Chem 270: 21206–21219, 1995
10. Lev S, Moreno H, Martinez R, Canoll P, Peles E, Musacchio JM, Plowman GD, Rudy B, Schlessinger J: Protein tyrosine kinase Pyk2 involved in Ca^{2+}-induced regulation of ion channel and MAP kinase function. Nature 376: 737–745, 1995
11. Yu H, Li X, Marchetto GS, Dy R, Hunter D, Calvo B, Dawson TL, Wilm M, Anderegg RJ, Graves LM, Earp HS: Activation of a novel calcium-dependent protein tyrosine kinase. Correlation with c-jun N-terminal kinase but not mitogen-activated protein kinase activation. J Biol Chem 271: 29993–29998, 1996
12. Luttrell LM, Hawes BE, van Biesen T, Luttrell DK, Lansing J, Lefkowitz RJ: Role of c-Src tyrosine kinase in G protein-coupled receptor- and Gβγ subunit-mediated activation of mitogen-activated protein kinases. J Biol Chem 271: 19443–19450, 1996
13. Rui L, Carter-Su C: Identification of SH2-Bβ as a potent cytoplasmic activator of the tyrosine kinase Janus kinase 2. Proc Natl Acad Sci (USA) 96: 7172–7177, 1999
14. Harp JB, Sayeski PP, Scanlon M, Bernstein KE, Marrero MB: Role of intracellular calcium in the angiotensin II-mediated tyrosine phosphorylation and dephosphorylation of PLC-γ1. Biochem Biophys Res Commun 232: 540–544, 1997
15. Sayeski PP, Ali MS, Harp JB, Marrero MB, Bernstein KE: Phosphorylation of p130Cas by angiotensin II is dependent on c-Src, intracellular Ca^{2+}, and protein kinase C. Circ Res 81: 1279–1288, 1998
16. Blackmore PF, Bocckino SB, Waynick LE, Exton JH: Role of guanine nucleotide-binding regulatory protein in the hydrolysis of hepatocyte phosphatidylinositol 4,5-bisphosphate by calcium mobilizing hormones and the control of cell calcium. Studies utilizing aluminum fluoride. J Biol Chem 260: 14477–14488, 1985
17. Leduc L, Meloche S: Angiotensin II stimulates tyrosine phosphorylation of the focal adhesion-associated protein paxillin in aortic smooth muscle cells. J Biol Chem 270: 4401–4404, 1995
18. Ali MS, Sayeski PP, Hayzer DJ, Dirksen LB, Marrero MB, Bernstein KE: Dependence of the motif YIPP for the physical association of Jak2 kinase with the intracellular carboxyl tail of the angiotensin II AT_1 receptor. J Biol Chem 272: 23382–23388, 1997

Molecular and Cellular Biochemistry **212**: 99–109, 2000.

Functional cross-talk between the cyclic AMP and Jak/STAT signaling pathways in vascular smooth muscle cells

Sylvain Meloche, Stéphane Pelletier and Marc J. Servant
Research Centre, Centre hospitalier de l'Université de Montréal and Department of Pharmacology, Université de Montréal, Montreal, Quebec, Canada

Abstract

Angiotensin II (Ang II), the primary effector of the renin-angiotensin system, is a multifunctional hormone that plays an important role in vascular function. In addition to its classical vasoconstrictor action, more recent studies demonstrated that Ang II stimulates the growth of a number of cell types, including vascular smooth muscle cells (SMC) (reviewed in [1–3]). *In vivo* studies have shown that chronic infusion of Ang II leads to the development of vascular hypertrophy in rats, whereas administration of angiotensin-converting enzyme (ACE) inhibitors or Ang II receptor antagonists prevents or regresses vascular hypertrophy in models of genetic and experimental hypertension [4]. Consistent with *in vivo* data, several laboratories have shown that Ang II stimulates protein synthesis and induces cellular hypertrophy, but not cell proliferation, in cultured aortic SMC [5–9]. Ang II also induces directed migration (chemotaxis) of vascular SMC [10, 11], although its effect is less prominent than that of platelet-derived growth factor (PDGF). The cellular mechanisms underlying these diverse actions of Ang II are not clearly understood but are likely to involve the activation of distinct signaling pathways. (Mol Cell Biochem **212**: 99–109, 2000)

Key words: cell growth, angiotensin II, signal transduction, cyclic AMP, protein kinase A, Janus kinases

Signal transduction pathways of the AT_1 receptor: An overview

Ang II exerts its physiological actions by interacting with specific membrane receptors in target cells. Two pharmacologically distinct subtypes of receptors, designated AT_1 and AT_2, have been described based on their affinity for the selective receptor antagonists losartan (AT_1) and PD123319 (AT_2) (reviewed in [12]). Pharmacological studies have indicated that the AT_1 receptor is the predominant subtype expressed in most adult tissues and is mediating the known biological effects of Ang II [13, 14]. The biological function of the AT_2 receptor remains controversial but recent studies suggest that it may counteract AT_1 receptor-mediated responses [15, 16]. The signaling pathways activated by the AT_2 receptor are still poorly understood and will not be reviewed here. Molecular cloning studies have revealed that the AT_1 and AT_2 receptor subtypes belong to the superfamily of seven transmembrane domain receptors [17–20].

Binding of Ang II to the AT_1 receptor stimulates heterotrimeric G proteins which couple the receptor to multiple effector enzymes in the plasma membrane. The best documented signaling pathway of the AT_1 receptor is the stimulation of phospholipase C (PLC), which generates the two second messengers diacylglycerol and inositol 1, 4, 5-trisphosphate (IP_3) (reviewed in [21, 22]). Diacylglycerol is a physiological activator of protein kinase C (PKC) isoforms, while IP_3 binds to specific receptor-channels to release Ca^{2+} from intracellular stores. The activation of PLC by Ang II is most likely mediated by a member of the Gq family of G

Present address: M.J. Servant, Lady Davis Institute for Medical Research, Sir Mortimer B. Davis Jewish General Hospital, Montreal, Quebec, Canada H3T 1E2
Address for offprints: S. Meloche, Research Centre, Centre hospitalier de l'Université de Montréal, Hôtel-Dieu Campus, 3840 St. Urbain Street, Montreal, Quebec, Canada, H2W 1T8

proteins [23, 24] which activate PLC-β isoenzymes in a pertussis toxin (PTX)-insensitive manner [25]. Multiple evidence plead in favor of a critical role for Ca^{2+} in the vasoconstrictor and hypertrophic action of Ang II (unpublished data). In addition to PLC, Ang II rapidly activates class IA phosphoinositide 3-kinase (PI3-kinase) to generate the lipid second messengers phosphatidylinositol (PtdIns) 3, 4-P_2 and PtdIns 3, 4, 5-P_3 ([26]; unpublished data). Activation of the PI3-kinase pathway is required for the stimulation of mRNA translation and protein synthesis by Ang II in vascular SMC ([26]; Voisin L, Giasson E, Meloche S, manuscript in preparation). In several of its target tissues, Ang II inhibits adenylyl cyclase via a PTX-sensitive G_i protein [27]. However, there is no evidence that Ang II modulates the activity of adenylyl cyclase in vascular SMC and PTX has no effect on Ang II-induced protein synthesis (unpublished data). Ang II has also been shown to activate phospholipase D, which hydrolyses phosphatidylcholine into phosphatidic acid and choline [28]. Phosphatidic acid is further metabolized to diacylglycerol by phosphatidate phosphohydrolase and may thus prolong diacylglycerol accumulation. The exact role of phospholipase D in the vascular effects of Ang II remains to be established. Finally, recent works have shown that Ang II stimulates the production of reactive oxygen species (ROS) in vascular SMC via a membrane-associated NADH/NAPDH oxidase system similar to the neutrophil NADPH oxidase [29, 30]. There is evidence that ROS are integral part of the signaling pathways which mediate the hypertrophic [30] and vasoconstrictor [31] actions of the hormone.

The immediate consequence of these early signals is the activation of a complex network of protein kinases which integrate and propagate the signals to the appropriate cellular compartments. Ang II has been shown to stimulate the activity of multiple serine/threonine kinases in vascular SMC, including PKC isoforms [32–34], the MAP kinases ERK1/ERK2 [35–37], JNK1/JNK2 ([38]; unpublished data) and p38 [31, 38, 39], p70 S6 kinase [9], Akt ([40]; unpublished data), Ca^{2+}/calmodulin-dependent protein kinase II [41] and casein kinase I [42]. We have demonstrated that activation of ERK1/ERK2 [37] and p70 S6 kinase [9] is necessary for the stimulation of global protein synthesis by Ang II. More recently, we showed that the p38 MAP kinase pathway selectively modulates the vasoconstrictor action of Ang II in vascular smooth muscle [31]. In common with cytokine receptors, AT_1 receptor activation was also found to increase tyrosine phosphorylation of multiple substrates in vascular SMC (reviewed in [43]).

Importance of tyrosine phosphorylation in AT_1 receptor signaling

The reversible phosphorylation of proteins on tyrosine residues has been recognized for a long time as the primary mechanism by which receptor tyrosine kinases and cytokine receptors regulate the proliferation and differentiation of cells (reviewed in [44, 45]). It is only recently, however, that the importance of this signaling pathway has been appreciated in the case of G protein-coupled receptors. Initial studies by several independent groups revealed that agonists of G protein-coupled receptors such as thrombin, bombesin, vasopressin, endothelin, or lysophosphatidic acid (LPA) stimulate tyrosine phosphorylation of multiple substrates in their target cells [46–50]. Ang II was also reported to induce rapid tyrosine phosphorylation in liver epithelial cells [51], glomerular mesangial cells [47], vascular SMC [52–54], and cardiac fibroblasts [55]. The observation that tyrosine kinase inhibitors can block the stimulation of DNA synthesis induced by thrombin [56], endothelin [57], and bombesin [58] suggested for the first time that tyrosine phosphorylation plays an important role in the growth response to this class of receptors. Using a similar pharmacological approach, we have shown that selective inhibition of tyrosine phosphorylation with genistein and herbimycin A completely abolishes Ang II-stimulated protein synthesis in rat aortic SMC [54]. Interestingly, tyrosine kinase inhibitors were also found to block Ang II-mediated contraction of gastric smooth muscle strips [59], thereby suggesting that tyrosine phosphorylation plays a central role in both the vasoconstrictive and hypertrophic actions of Ang II.

In vascular SMC, treatment with Ang II increases the phosphotyrosine content of numerous proteins with apparent molecular masses ranging from 180 to 42 kDa. We have identified the major tyrosine phosphorylated substrate in vascular SMC as paxillin, a 68 kDa cytoskeletal protein localized to focal adhesions [60]. Other tyrosine phosphorylated proteins which have been identified in Ang II-stimulated vascular SMC include p125 focal-adhesion kinase ($p125^{FAK}$) [61–63], the adaptor protein Shc [63, 64], PLCγ1 [65], Src [66], STATs [67], insulin receptor substrate 1 (IRS-1) [68], SHP-2 [68], the focal adhesion-associated kinase Pyk2 [63, 69], the focal adhesion proteins $p130^{cas}$ [70–72] and tensin [72], and the receptor for epidermal growth factor (EGF) ([73]; Foisy S, Giasson E, Voisin L, Meloche S, manuscript in preparation). It is interesting to note that most of these signaling proteins are traditionally phosphorylated in response to the activation of receptor tyrosine kinases or cytokine receptors. Two mechanisms for activation of tyrosine kinases by the AT_1 receptor have been proposed. The first mechanism is by transactivation of a tyrosine kinase receptor, such as the EGF receptor, which then functions as a scaffold for assembly of other signaling molecules. Such a mechanism was first demonstrated in an elegant study showing that the EGF receptor is rapidly tyrosine-phosphorylated upon stimulation of Rat1 cells with endothelin, LPA and thrombin, and that inhibition of EGF receptor function suppresses MAP kinase activation and c-*fos* gene induction by these agonists [74]. More recently, Ang II was also shown to transactivate the EGF receptor tyrosine kinase in vascular SMC ([73]; Foisy S, Giasson E, Voisin L, Meloche S, manuscript in preparation). The second mechanism involves direct recruitment of tyrosine kinases to the AT_1 receptor. It has been reported that Ang II binding to the AT_1 receptor promotes the association of the tyrosine kinase Jak2 with the receptor (see below).

One family of tyrosine kinases that has emerged as an integral aspect of AT_1 receptor signaling in recent years is the Janus kinase (Jak) family. Two groups initially reported the unexpected observation that the G protein-coupled receptor agonist Ang II stimulates the Jak/STAT pathway in vascular SMC [67] and cardiac myocytes [75]. These findings have now been reproduced in many other studies ([63, 76–81]; Fig. 1). Intriguingly, Bernstein and colleagues presented evidence that Jak2 forms a stable complex with the intracellular carboxyl tail of the AT_1 receptor [67, 76]. Various lines of evidence suggest that the Jak/STAT pathway may play an important role in mediating some of the biological actions of the AT_1 receptor. For instance, treatment of vascular SMC with the tyrphostin AG490 or AG9, two potent inhibitors of Jak2 and Tyk2, significantly inhibits the growth promoting effect of Ang II (our unpublished observations). In a recent study, Pan *et al.* [80] reported that acute pressure overload activates the Janus kinases Jak1, Jak2 and Tyk2 in the rat heart *in vivo*. The activation of Jak2 and Tyk2 was blocked by the AT_1 receptor antagonist E4177. It is worth mentioning that activation of the Jak/STAT pathway is not the exclusive privilege of the AT_1 receptor, as we have found that other agonists of G protein-coupled receptors stimulate tyrosine phosphorylation of Tyk2 in vascular SMC (Fig. 2).

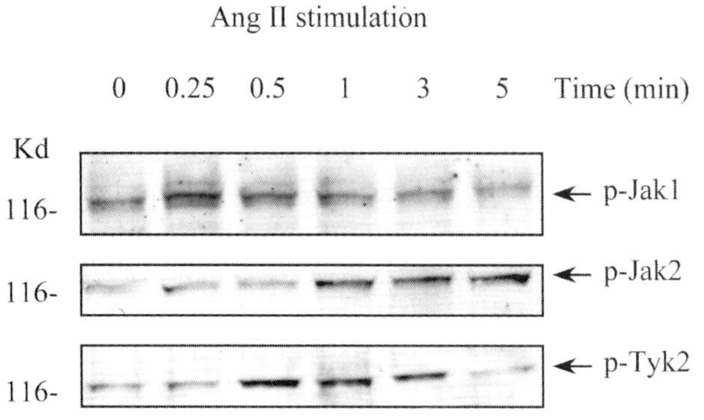

Fig. 1. Ang II stimulates the activation of the Janus kinases Jak1, Jak2 and Tyk2 in vascular SMC. Quiescent rat aortic SMC were stimulated for the indicated times with 10 nM Ang II. Cellular lysates were prepared and analyzed by immunoblotting with phospho-specific antibodies selective for the activating tyrosine residues of Jak1, Jak2 and Tyk2.

The Jak/STAT pathway

Signaling by cytokine receptors depends upon their association with Janus kinases, which couple ligand binding to tyrosine phosphorylation of signaling proteins recruited to the receptor complex (reviewed in [45, 82, 83]). Among these are the signal transducers and activators of transcription (STATs), a family of transcription factors that contribute to the diversity of cytokine responses. Janus kinases form a small family of intracellular tyrosine kinases which is comprised of four members: Jak1, Jak2, Jak3, and Tyk2 (reviewed in [84, 85]). With the exception of Jak3, which is primarily expressed in hematopoietic cells, Janus kinases are ubiquitously expressed. The general mode of cytokine signaling is the following: ligand binding induces receptor oligomerization, resulting in trans-phosphorylation and activation of one or more Janus kinases. Activated Janus kinases in turn, phosphorylate STAT proteins (which are recruited to the tyrosine phosphorylated receptor through their SH2 domain) on a single tyrosine residue, inducing the homo or hetero-dimerization of the STATs and their translocation to the nucleus [86]. The activation of STATs has also been reported to occur through receptor tyrosine kinases [87] and, as mentioned previously, G protein-coupled receptors ([67, 75]; Fig. 2).

The functional importance of the Janus kinases was first revealed by studies in cell lines defective in interferon signaling and confirmed by the development of knockout mouse strains. Jak3-deficient mice exhibit severe defects in lymphoid cell development [88–90]. Jak2 deficiency is embryonic lethal [91, 92], and Jak1 is perinatally lethal [93]. Embryonic fibroblasts derived from Jak1 and Jak2 deficient mice fail to respond to specific subsets of cytokine receptors. Conversely, overexpression or hyperactivation of Janus kinases has been associated with various forms of leukemias (see [84] for review).

Cyclic AMP antagonizes AT_1 receptor biological responses in vascular SMC

Cyclic AMP (cAMP) is a ubiquitous second messenger that is synthesized in cells by adenylyl cyclases in response to

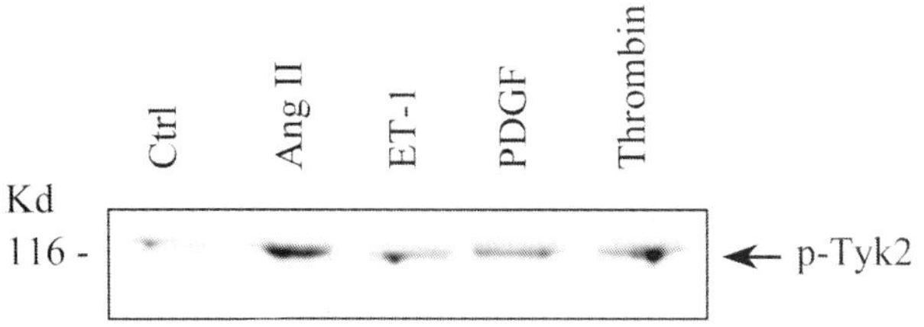

Fig. 2. Activation of Tyk2 by agonists of G protein-coupled receptors and PDGF in aortic SMC. Quiescent rat aortic SMC were stimulated for 3 min with the following agonists: Medium (Ctrl), Ang II (10 nM), endothelin-1 (ET-1, 100 nM), thrombin (1 U/ml), and PDGF-BB (PDGF, 50 ng/ml). Cellular lysates were prepared and analyzed by immunoblotting with a phospho-Tyk2 antibody (Tyr 1054/Tyr 1055).

many extracellular stimuli. With few exceptions, the cellular effects of cAMP are mediated trough activation of cAMP-dependent protein kinase (PKA), a multifunctional serine/threonine kinase [94]. In the absence of cAMP, PKA exists as an inactive tetrameric complex containing two regulatory subunits and two catalytic subunits. Binding of cAMP to the regulatory subunit decreases its affinity for the catalytic subunit and promotes dissociation into a dimer of regulatory subunits and two active catalytic subunits. Activated PKA is then able to phosphorylate target proteins in all cellular compartments. The magnitude and duration of the cAMP response is determined by the rate of cAMP hydrolysis by cyclic nucleotide phosphodiesterases [95].

cAMP signaling regulates a vast array of cellular processes. Notably, cAMP has been shown to modulate growth factor-induced responses by interfering with different signaling pathways. The best documented example is the ability of cAMP to inhibit or stimulate cell proliferation depending on the cell type. Elevation of intracellular cAMP stimulates the proliferation of thyrocytes, keratinocytes and Swiss 3T3 fibroblasts; in contrast, cAMP inhibits proliferation of NIH 3T3 fibroblasts, macrophages, lymphoid cells and many tumor cells (reviewed in [96–98]). Activation of the cAMP/PKA pathway also markedly inhibits the proliferation of vascular SMC both *in vitro* and *in vivo* ([99, 100]; unpublished data). Several mechanisms have been put forward to explain the anti-mitogenic action of cAMP. A number of studies have proposed that cAMP inhibits cell proliferation by interfering with Ras-dependent activation of the mitogen-activated protein (MAP) kinases extracellular signal-regulated kinase 1 (ERK1)/ERK2 [101–105]. Biochemical analysis of the various intermediates in the signaling cascade suggested that cAMP inhibits signal transmission at the level of Raf-1 [101, 102, 105]. However, treatment of other cell types with cAMP-raising agents was found to block growth factor-induced proliferation without preventing ERK1/ERK2 activation [106–108]. Similarly, elevation of cAMP levels was shown to inhibit the activation of p70 S6 kinase in certain cell types [107, 109], but not in others [106]. The cAMP/PKA signaling pathway also targets regulators of the cell cycle machinery. Treatment of macrophages with analogs of cAMP was found to raise the levels of the cyclin-dependent kinase (Cdk) inhibitor $p27^{Kip1}$, thereby increasing its association with cyclin D-Cdk4 and preventing the activation of Cdk4 [110]. In fibroblast cells, the cell cycle arrest imposed by cAMP is associated with the inhibition of growth factor-dependent cyclin D1 induction and $p27^{Kip1}$ degradation [111]. cAMP was also shown to downregulate c-*myc* mRNA expression in several cell types [112–114]. However, the significance of this observation is not known because constitutive expression of c-*myc* was insufficient to override the anti-proliferative effect of cAMP in macrophages [114]. It is likely that the targets of the cAMP/PKA pathway will vary depending on cell type origin and the nature of the mitogenic signals.

In addition to its well defined modulatory effect on cell multiplication, cAMP also exerts negative regulatory effects on the stimulation of protein synthesis by hypertrophic factors [63, 115, 116]. We have observed that a variety of agents known to elevate the intracellular concentration of cAMP potently inhibit the stimulatory effect of Ang II on protein synthesis in rat aortic SMC [63]. The inhibition was dose-dependent and was observed at all concentrations of Ang II. Other important physiological effects of Ang II are also subject to the modulatory action of cAMP in vascular SMC. Agents that raise cAMP levels cause a relaxation of arterial smooth muscle and antagonize the vasoconstrictor effect of Ang II [117]. cAMP also inhibits the chemotactic response of arterial SMC to Ang II *in vitro* [10]. Thus, the cAMP signaling system impacts on several important cellular processes in vascular smooth muscle.

Cross-talk between cAMP and the Jak/STAT pathway

The cross-talk potential of the cAMP signaling system is known for more than two decades but the mechanisms by which cAMP exerts its inhibitory action on various cellular processes remain poorly understood. We have found that elevation of intracellular cAMP levels inhibits Ang II-stimulated protein synthesis in rat aortic SMC, without affecting the basal rate of protein synthesis [63]. To understand the molecular basis of this cross-talk, we have examined the effect of agents that raise cAMP levels on various AT_1 receptor-mediated signaling events [63]. Short-term cAMP treatment has no effect on the agonist binding properties of the AT_1 receptor, a finding consistent with the lack of optimal PKA consensus site in the sequence of the rat AT_{1A} and AT_{1B} receptors. We also found that cAMP elevation does not interfere with the activation of PLC or the induction of the early growth response genes c-*fos*, c-*jun,* and *egr-1* by Ang II. As mentioned previously, PKA activation has been shown to inhibit the ERK MAP kinase pathway in many cell types, including human arterial SMC [103]. Activation of the ERK pathway plays an important role in the regulation of mRNA translation and protein synthesis, by inducing the phosphorylation of the translation initiation factor eIF4E through activation of the protein kinase Mnk1 [118]. However, in contrast to the above studies, we [63] and others [116] did not observe any significant effect of cAMP-raising agents on Ang II-dependent activation of ERK1 and ERK2 in rat aortic SMC. Similarly, elevation of cAMP levels strongly inhibits thrombin-stimulated DNA synthesis in these cells without compromising the effect of thrombin on ERK1/ERK2 acti-

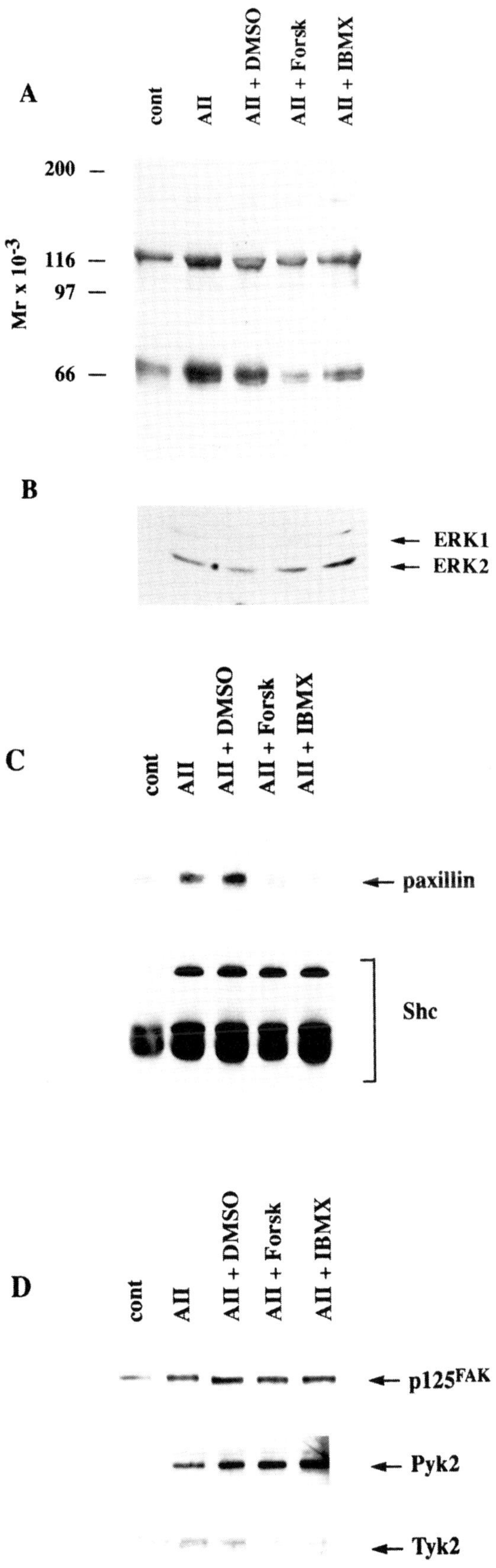

vation [108]. These observations clearly highlight the importance of the cell type (and species) in determining the overall influence of cAMP on other signaling systems, but also argue against the idea that the ERK MAP kinase pathway is a major target of the growth inhibitory action of cAMP. The p70 S6 kinase is another important regulator of translation [119]. Activation of p70 S6 kinase leads to increased phosphorylation of 40S ribosomal protein S6, an event which is believed to up-regulate the translation of a family of mRNAs that include ribosomal proteins and elongation factors [120]. We found that raising cAMP levels does not interfere with the activation of p70 S6 kinase nor with the phosphorylation of the translational repressor 4E-binding protein 1 (4E-BP1), another downstream target of mTOR (mammalian target of rapamycin), in Ang II-stimulated rat aortic SMC ([63]; unpublished data). These findings also suggest that the growth inhibitory action of cAMP in these cells is not at the level of PI3-kinase, since full activation of p70 S6 kinase is dependent on phosphorylation of Thr 229 by PDK1 [121], a downstream effector of PI3-kinase [122].

Because of the importance of tyrosine phosphorylation signaling in the hypertrophic action of Ang II [54], we also examined the effect of cAMP-raising agents on tyrosine kinase-regulated pathways. We showed that cAMP elevation completely abolishes Ang II-dependent tyrosine phosphorylation of the focal adhesion-associated protein paxillin in aortic SMC (Fig. 3). In contrast, cAMP does not prevent tyrosine phosphorylation of the adaptor protein Shc, another target of AT_1 receptor signaling. Most importantly, we found that cAMP elevation completely blocks Ang II-induced tyrosine phosphorylation of the Janus kinase Tyk2 in aortic SMC, without interfering with phosphorylation of the tyrosine kinases $p125^{FAK}$ and Pyk2. These results revealed for the first time the existence of a functional cross-talk between the cAMP signaling system and the Jak/STAT pathway in vascular smooth muscle. Using a recently developed phosphospecific antibody to Tyk2, we have extended our original findings and demonstrated that cAMP blocks phosphoryla-

Fig. 3. Elevation of cAMP inhibits Ang II-stimulated tyrosine phosphorylation in aortic SMC. Quiescent rat aortic SMC were pretreated for 30 min in the absence or presence of 10 μM forskolin (Forsk) or 0.1 mM 3-isobutyl-1-methylxanthine (IBMX). The cells were then stimulated or not (cont) with 10 nM Ang II for either 1 min (Shc, $p125^{FAK}$, Pyk2), 3 min (panel A, Tyk2), or 5 min (panel B, paxillin). (A) Protein tyrosine phosphorylation. Cellular lysates were analyzed by anti-phosphotyrosine immunoblotting. (B) Tyrosine phosphorylation of the MAP kinases ERK1/ERK2. Cell lysates were analyzed as in panel A. (C) Tyrosine phosphorylation of paxillin and Shc. Cell lysates were subjected to immunoprecipitation with anti-paxillin or anti-Shc antibody and analyzed by immunoblotting with anti-phosphotyrosine. (D) Tyrosine phosphorylation of cytosolic tyrosine kinases. Cell lysates were subjected to immunoprecipitation with a specific antibody to $p125^{FAK}$, Pyk2, or Tyk2 and analyzed by immunoblotting with anti-phosphotyrosine. (Reproduced from J Biol Chem 272: 26879–26886, 1997).

tion of Tyk2 on the activating Tyr 1054/Tyr 1055 residues (which directly reflects enzymatic activation) (Fig. 4). Because of the pivotal role played by Janus kinases in the regulation of normal cell growth by cytokines, it is tempting to speculate that Tyk2 may represent a critical target of the growth inhibitory action of cAMP in vascular SMC. In agreement with this idea, we have observed in preliminary studies that pharmacological inhibition of Janus kinases markedly reduces the stimulatory effect of Ang II on global protein synthesis (unpublished observations). Janus kinases are also good candidates for the inhibitory action of cAMP on vascular SMC chemotaxis. It has been recently reported that urokinase-type plasminogen activator (uPA) stimulates tyrosine phosphorylation of Jak1 and Tyk2 in vascular SMC, and that the two kinases co-localize with uPA receptor at the leading edge of migrating cells after wounding [123]. In addition, genetic studies in zebrafish have shown that Jak1 activation is necessary for cell migration and anterior specification of the embryo [124]. All these observations suggest that Janus kinases are important regulators of cell migration. Interestingly, cAMP-elevating agents were also shown to inhibit interleukin-6-induced STAT activation in monocytes [125], and to inhibit interferon β-stimulated tyrosine phosphorylation of Jak1, Tyk2, STAT1 and STAT2 in the myeloma cell line U266 [126]. Thus, modulation of the Jak/STAT pathway by cAMP is likely to play an important role in the regulation of growth, immune and inflammatory responses.

The mechanisms whereby cAMP inhibits the activation of Tyk2 in vascular SMC is not known. Pharmacological evidence suggest that the effect of cAMP is mediated at least in part by PKA, since pre-incubation of aortic SMC with the PKA inhibitors Rp-cAMPS and H-89 restored Ang II-dependent activation of Tyk2 in cells exposed to prostaglandin E_1 (Fig. 5). One intriguing possibility is that PKA directly phosphorylates the Tyk2 kinase. Examination of the amino acid sequence of Janus kinases reveals that Tyk2, but not Jak1 or Jak2, contains a possible PKA recognition site at position 209. Phosphorylation of this site may induce a conformational change of the Tyk2 protein, preventing its subsequent

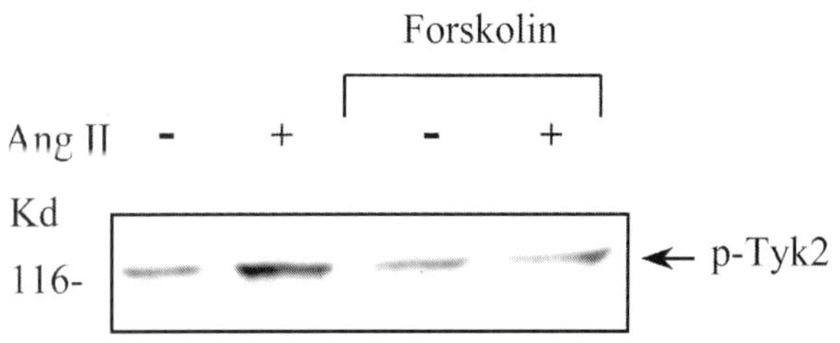

Fig. 4. Elevation of cAMP inhibits Ang II-dependent activation of Tyk2 in aortic SMC. Quiescent rat aortic SMC were pretreated for 45 min in the absence or presence of 10 μM forskolin and then stimulated or not for 3 min with 10 nM Ang II. Cellular lysates were prepared and analyzed by immunoblotting with a phospho-Tyk2 antibody (Tyr 1054/Tyr 1055).

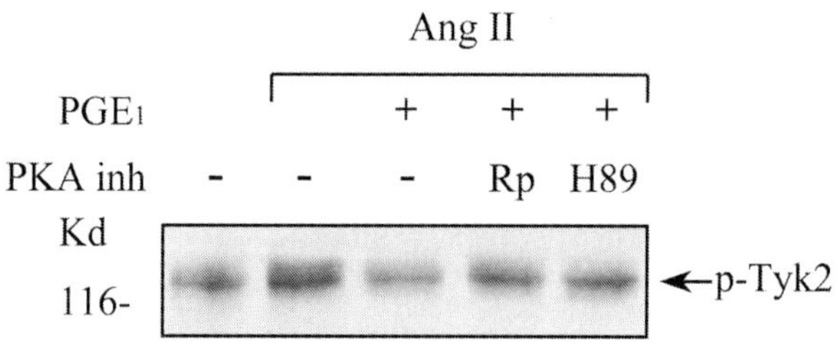

Fig. 5. Incubation with PKA inhibitors reverses the inhibitory effect of cAMP on Ang II-stimulated Tyk2 activity in aortic SMC. Quiescent rat aortic SMC were pre-incubated for 60 min with the PKA inhibitors Rp-cAMPS (Rp, 100 μM) or H-89 (1 μM) prior to treatment with 1 μM PGE_1 for 30 min. The cells were then stimulated or not for 3 min with 10 nM Ang II. Cellular lysates were prepared and analyzed by immunoblotting with a phospho-Tyk2 antibody (Tyr 1054/Tyr 1055).

phosphorylation on activating tyrosine residues. The mode of activation of Janus kinases by G protein-coupled receptors remains obscure but, by analogy with cytokine receptors, may require the oligomerization and trans-phosphorylation of Janus kinase molecules. If this is the case, then inactivation of Tyk2 by PKA could also interfere with the activation of other Janus kinase family members (such as Jak2) oligomerized with Tyk2. In this regard, it is noteworthy that recent findings suggest that G protein-coupled receptors also exist as dimeric structures in cell membranes [127]. A second possibility is that PKA directly or indirectly stimulates the activity of a protein tyrosine phosphatase which inactivates Janus kinases. For example, the tyrosine phosphatase SHP-1 binds to specific tyrosine phosphorylated residues in the cytoplasmic domain of the erythropoietin receptor via its SH2 domain and dephosphorylates Jak2 [128]. SHP-1 also binds directly to Jak2 and Tyk2 in intact cells, causing their dephosphorylation [129, 130]. The related phosphatase SHP-2 was also reported to associate with Jak1 in monocytes [125]. SHP-$2^{-/-}$ fibroblasts show excessive activation of STATs by interferons, a response which is attenuated by reintroduction of SHP-2 protein [131]. It was also reported that phorbol ester treatment causes the tyrosine dephosphorylation of Tyk2 but not of Jak1 in interferon-stimulated monocytes [132]. Finally, a third possibility is that cAMP elevation may downregulate Janus kinase protein levels in vascular SMC. While such a mechanism would not explain the rapid inactivation of Tyk2 observed after short-term treatment with cAMP-raising agents, it may however contribute to the inhibition of cellular responses after long-term or chronic exposure. For example, a recent study showed that prostaglandin E_2 and agents that increase cAMP levels selectively decrease the abundance of Jak3 protein in T lymphocytes [133]. The suppression of Jak3 expression was correlated with impaired interleukin-2-dependent signaling and proliferation. Additional studies are clearly necessary to examine these various possibilities and others.

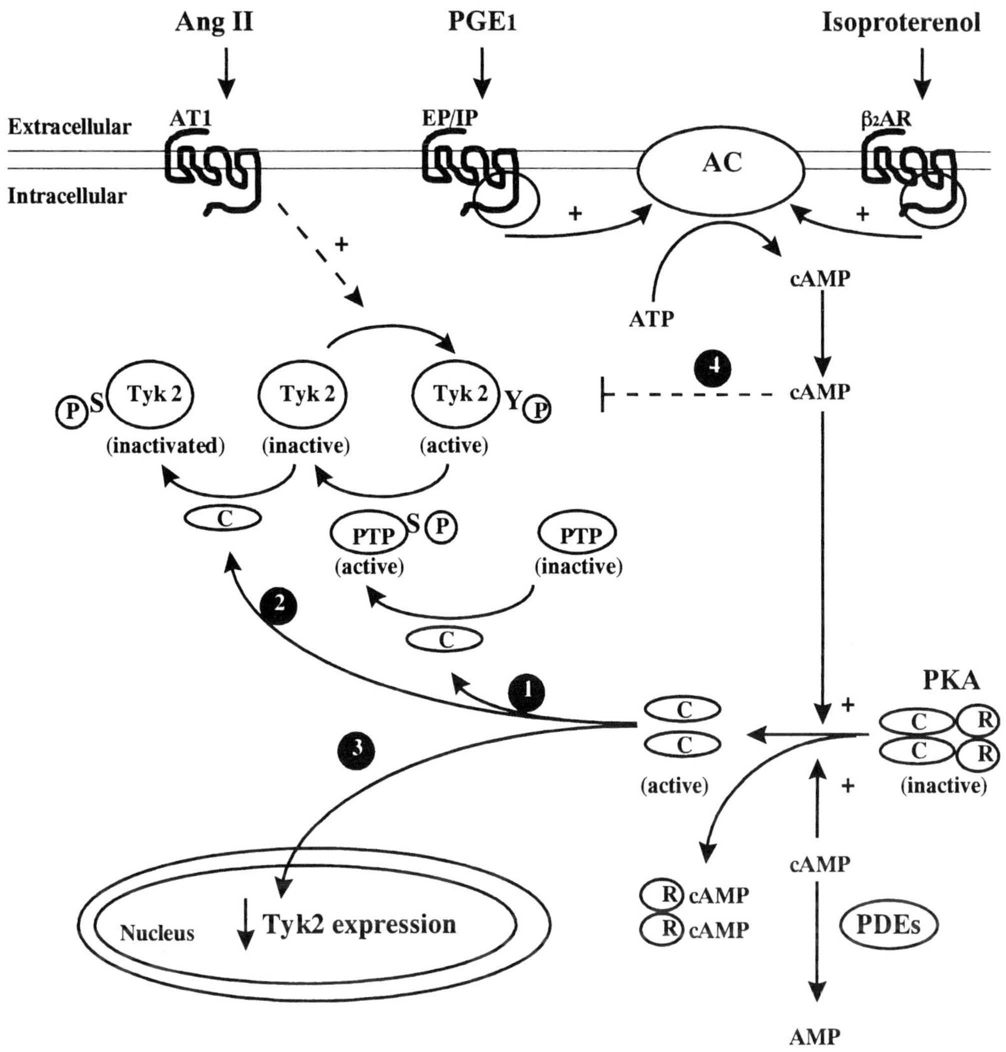

Fig. 6. Proposed model of the cross-talk between cAMP and the Jak/STAT pathway in vascular SMC. This diagram summarizes various possibilities whereby elevation of intracellular cAMP levels may interfere with the activation of Tyk2 kinase by the Ang II AT_1 receptor. (1) PKA stimulates the activity of a protein tyrosine phosphatase which dephosphorylates and inactivates Janus kinases. (2) PKA directly phosphorylates Tyk2, which prevents its subsequent phosphorylation on activating tyrosine residues. (3) PKA downregulates the expression of Tyk2 protein. (4) cAMP inhibits Tyk2 activation by a PKA-independent mechanism. These mechanisms are not mutually exclusive. Abbreviations: PGE_1 – prostaglandin E_1; EP/IP – prostaglandin receptor; AR – adrenergic receptor; AC – adenylyl cyclase; PDE – cyclic nucleotide phosphodiesterase; PTP – protein tyrosine phosphatase.

Conclusions

It is clear that the cAMP/PKA signaling pathway can modulate or antagonize responses of vascular SMC to Ang II and other vasoconstrictor agents. Although the importance of this cross-talk in the physiological setting is less well characterized, a number of observations suggest that cAMP plays a major role in regulating vascular growth. For example, Indolfi *et al.* [100] reported that local administration of cAMP and phosphodiesterase-inhibitor drugs (aminophylline and amrinone) to rats markedly inhibits neointima formation after balloon injury *in vivo*. Another recent study showed that overexpression of prostacyclin synthase, the enzyme which catalyzes the synthesis of prostacyclin (PGI_2), in lung epithelial cells protects against the development of pulmonary hypertension in mice exposed to chronic hypoxia by preventing vasoconstriction and vessel wall hypertrophy [134]. A detailed understanding of how cAMP modulates vascular responses is clearly an important issue, and may prove useful for the identification of novel cellular targets for the treatment of vascular proliferative disorders.

Acknowledgements

Work in the author's laboratory was supported by grants from the Heart and Stroke Foundation of Canada and the Medical Research Council of Canada. S. Meloche is a Scientist of the

Medical Research Council of Canada. M.J. Servant was recipient of a studentship from the Heart and Stroke Foundation of Canada.

References

1. Schelling P, Fischer H, Ganten D: Angiotensin and cell growth: A link to cardiovascular hypertrophy? J Hypertens 9: 3–15, 1991
2. Rosendorff C: The renin-angiotensin system and vascular hypertrophy. J Am Coll Cardiol 28: 803–812, 1996
3. Pratt RE: Angiotensin II and the control of cardiovascular structure. J Am Soc Nephrol 10(suppl 11): S120–S128, 1999
4. Metsärinne KP, Stoll M, Falkenhahn M, Gohlke P, Unger T: Inhibiting the effects of angiotensin on cardiovascular hypertrophy. In: J.M. Saavedra, P.B.M.W.M. Timmermans (eds). Angiotensin Receptors. Plenum Press, New York, 1994, pp 235–253
5. Geisterfer AAT, Peach MJ, Owens GK: Angiotensin II induces hypertrophy, not hyperplasia, of cultured rat aortic smooth muscle cells. Circ Res 62: 749–756, 1988
6. Berk BC, Vekshtein V, Gordon HM, Tsuda T: Angiotensin II-stimulated protein synthesis in cultured vascular smooth muscle cells. Hypertension 13: 305–314, 1989
7. Itoh H, Pratt RE, Dzau VJ: Atrial natriuretic polypeptide inhibits hypertrophy of vascular smooth muscle cells. J Clin Invest 86: 1690–1697, 1990
8. Grainger DJ, Witchell CM, Weissberg PL, Metcalfe JC: Mitogens for adult rat aortic vascular smooth muscle cells in serum-free primary culture. Cardiovasc Res 28: 1238–1242, 1994
9. Giasson E, Meloche S: Role of p70 S6 protein kinase in angiotensin II-induced protein synthesis in vascular smooth muscle cells. J Biol Chem 270: 5225–5231, 1995
10. Dubey RK, Jackson EK, Luscher TF: Nitric oxide inhibits angiotensin II-induced migration of rat aortic smooth muscle cell. Role of cyclic-nucleotides and angiotensin 1 receptors. J Clin Invest 96: 141–149, 1995
11. Xi XP, Graf K, Goetze S, Fleck E, Hsueh WA, Law RE: Central role of the MAPK pathway in ang II-mediated DNA synthesis and migration in rat vascular smooth muscle cells. Arterioscler Thromb Vasc Biol 19: 73–82, 1999
12. Timmermans PBMWM, Wong PC, Chiu AT, Herblin WF, Benfield P, Carini DJ, Lee RJ, Wexler RR, Saye JAM, Smith RD: Angiotensin II receptors and angiotensin II receptor antagonists. Pharmacol Rev 45: 205–251, 1993
13. Smith RD, Chiu AT, Wong PC, Herblin WF, Timmermans PBMWM: Pharmacology of nonpeptide angiotensin II receptor antagonists. Annu Rev Pharmacol Toxicol 32: 135–165, 1992
14. Brooks DP, Ruffolo RRJ. Functions mediated by peripheral angiotensin II receptors. In: R.R. Ruffolo Jr (ed). Angiotensin II Receptors Volume 1: Molecular Biology, Biochemistry, Pharmacology, and Clinical Perspectives, 1st edn. CRC Press, Boca Raton, 1994, pp 71–102
15. Horiuchi M, Akishita M, Dzau VJ: Recent progress in angiotensin II type 2 receptor research in the cardiovascular system. Hypertension 33: 613–621, 1999
16. de Gasparo M, Siragy HM: The AT2 receptor: Fact, fancy and fantasy. Reg Pep 81: 11–24, 1999
17. Murphy TJ, Alexander RW, Griendling KK, Runge MS, Bernstein KE: Isolation of a cDNA encoding the vascular type-1 angiotensin II receptor. Nature 351: 233–236, 1991
18. Sasaki K, Yamano Y, Bardhan S, Iwai N, Murray JJ, Hasegawa M, Matsuda Y, Inagami T: Cloning and expression of a complementary DNA encoding a bovine adrenal angiotensin II type-1 receptor. Nature 351: 230–233, 1991
19. Kambayashi Y, Bardhan S, Takahashi K, Tsuzuki S, Inui H, Hamakubo T, Inagami T: Molecular cloning of a novel angiotensin II receptor isoform involved in phosphotyrosine phosphatase inhibition. J Biol Chem 268: 24543–24546, 1993
20. Mukoyama M, Nakajima M, Horiuchi M, Sasamura H, Pratt RE, Dzau VJ: Expression cloning of type 2 angiotensin II receptor reveals a unique class of seven-transmembrane receptors . J Biol Chem 268: 24539–24542, 1993
21. Catt KJ, Sandberg K, Balla T. Angiotensin II receptors and signal transduction mechanisms. In: M.K. Raizada, M.I. Phillips, C. Sumners (eds). Cellular and Molecular Biology of the Renin-Angiotensin System, 1st edn. CRC Press, Boca Raton, 1993, pp 307–356.
22. Griendling KK, Ushio-Fukai M, Lassègue B, Alexander RW: Angiotensin II signaling in vascular smooth muscle: New concepts. Hypertension 29: 366–373, 1997
23. Schelling JR, Nkemere N, Konieczkowski M, Martin KA, Dubyak GR: Angiotensin II activates the beta 1 isoform of phospholipase C in vascular smooth muscle cells. Am J Physiol 272: C1558–C1566, 1997
24. Ushio-Fukai M, Griendling KK, Akers M, Lyons PR, Alexander RW: Temporal dispersion of activation of phospholipase C-beta 1 and -gamma isoforms by angiotensin II in vascular smooth muscle cells. Role of alphaq/11, alpha 12, and beta gamma G protein subunits. J Biol Chem 273: 19772–19777, 1998
25. Exton JH: Phosphoinositide phospholipases and G proteins in hormone action. Annu Rev Physiol 56: 349–369, 1994
26. Saward L, Zahradka P: Angiotensin II activates phosphatidylinositol 3-kinase in vascular smooth muscle cells. Circ Res 81: 249–257, 1997
27. Edwards RM, Ruffolo RR Jr: Angiotensin II receptor signal transduction mechanisms. In: R.R. Ruffolo Jr (ed). Angiotensin II Receptors Volume 1: Molecular Biology, Biochemistry, Pharmacology, and Clinical Perspectives, 1st edn. CRC Press, Boca Raton, 1994, pp 53–70.
28. Lassègue B, Alexander RW, Clark M: Phosphatidylchonine is a major source of phosphatidic acid and diacylglycerol in angiotensin II-stimulated vascular smooth muscle cells. Biochem J 292: 509–517, 1993
29. Griendling KK, Minieri CA, Ollerenshaw JD, Alexander RW: Angiotensin II stimulates NADH and NADPH oxidase activity in cultured vascular smooth muscle cells. Circ Res 74: 1141–1148, 1994
30. Zafari AM, Ushio-Fukai M, Akers M, Yin Q, Shah A, Harrison DG, Taylor WR, Griendling KK: Role of NADH/NADPH oxidase-derived H_2O_2 in angiotensin II-induced vascular hypertrophy. Hypertension 32: 488–495, 1998
31. Meloche S, Landry J, Huot J, Houle F, Marceau F, Giasson E: The p38 mitogen-activated protein kinase pathway regulates angiotensin II-induced contraction of rat vascular smooth muscle. Am J Physiol (in press)
32. Lang U, Vallotton MB: Effects of angiotensin II and of phorbol ester on protein kinase C activity and on prostacyclin production in cultured rat aortic smooth-muscle cells. Biochem J 259: 477–483, 1989
33. Tsuda T, Alexander RW: Angiotensin II stimulates phosphorylation of nuclear lamins via a protein kinase C-dependent mechanism in cultured vascular smooth muscle cells. J Biol Chem 265: 1165–1170, 1990
34. Damron DS, Nadim HS, Hong SJ, Darvish A, Murray PA: Intracellular translocation of PKC isoforms in canine pulmonary artery smooth muscle cells by ANG II. Am J Physiol 274: L278–L288, 1998
35. Duff JL, Berk BC, Corson MA: Angiotensin II stimulates the pp44 and pp42 mitogen-activated protein kinases in cultured rat aortic smooth muscle cells. Biochem Biophys Res Commun 188: 257–264, 1992

36. Tsuda T, Kawahara Y, Ishida Y, Koide M, Shii K, Yokoyama M: Angiotensin II stimulates two myelin basic protein/microtubule-associated protein 2 kinases in cultured vascular smooth muscle cells. Circ Res 71: 620–630, 1992
37. Servant MJ, Giasson E, Meloche S: Inhibition of growth factor-induced protein synthesis by a selective MEK inhibitor in aortic smooth muscle cells. J Biol Chem 271: 16047–16052, 1996
38. Kusuhara M, Takahashi E, Peterson TE, Abe J, Ishida M, Han J, Ulevitch R, Berk BC: p38 Kinase is a negative regulator of angiotensin II signal transduction in vascular smooth muscle cells: Effects on Na^+/H^+ exchange and ERK1/2. Circ Res 83: 824–831, 1998
39. Ushio-Fukai M, Alexander RW, Akers M, Griendling KK: p38 Mitogen-activated protein kinase is a critical component of the redox-sensitive signaling pathways activated by angiotensin II. Role in vascular smooth muscle cell hypertrophy. J Biol Chem 273: 15022–15029, 1998
40. Ushio-Fukai M, Alexander RW, Akers M, Yin Q, Fujio Y, Walsh K, Griendling KK: Reactive oxygen species mediate the activation of Akt/protein kinase B by angiotensin II in vascular smooth muscle cells. J Biol Chem 274: 22699–22704, 1999
41. Abraham ST, Benscoter H, Schworer CM, Singer HA: In situ Ca^{2+} dependence for activation of Ca^{2+}/calmodulin-dependent protein kinase II in vascular smooth muscle cells. J Biol Chem 271: 2506–2513, 1996
42. Voisin L, Larose L, Meloche S: Angiotensin II stimulates serine phosphorylation of the adaptor protein nck: Physical association with the serine/threonine kinases pak1 and casein kinase I. Biochem J 341: 217–223, 1999
43. Berk BC, Corson MA: Angiotensin II signal transduction in vascular smooth muscle: Role of tyrosine kinases. Circ Res 80: 607–616, 1997
44. Ullrich A, Schlessinger J: Signal transduction by receptors with tyrosine kinase activity. Cell 61: 203–212, 1990
45. Taniguchi T: Cytokine signaling through nonreceptor protein tyrosine kinases. Science 268: 251–255, 1995
46. Golden A, Brugge JS: Thrombin treatment induces rapid changes in tyrosine phosphorylation in platelets. Proc Natl Acad Sci USA 86: 901–905, 1989
47. Force T, Kyriakis JM, Avruch J, Bonventre JV: Endothelin, vasopressin, and angiotensin II enhance tyrosine phosphorylation by protein kinase C-dependent and -independent pathways in glomerular mesangial cells. J Biol Chem 266: 6650–6656, 1991
48. Leeb-Lundberg LM, Song XH: Bradykinin and bombesin rapidly stimulate tyrosine phosphorylation of a 120-kDa group of proteins in Swiss 3T3 cells. J Biol Chem 266: 7746–7749, 1991
49. Zachary I, Gil J, Lehmann W, Sinnett-Smith J, Rozengurt E: Bombesin, vasopressin, and endothelin rapidly stimulate tyrosine phosphorylation in intact Swiss 3T3 cells. Proc Natl Acad Sci USA 88: 4577–4581, 1991
50. Hordijk PL, Verlaan I, van Corven EJ, Moolenaar WH: Protein tyrosine phosphorylation induced by lysophosphatidic acid in Rat-1 fibroblasts. Evidence that phosphorylation of map kinase is mediated by the Gi-p21ras pathway. J Biol Chem 269: 645–651, 1994
51. Huckle WR, Prokop CA, Dy RC, Herman B, Earp S: Angiotensin II stimulates protein-tyrosine phosphorylation in a calcium-dependent manner. Mol Cell Biol 10: 6290–6298, 1990
52. Tsuda T, Kawahara Y, Shii K, Koide M, Ishida Y, Yokoyama M: Vasoconstrictor-induced protein-tyrosine phosphorylation in cultured vascular smooth muscle cells. FEBS Lett 285: 44–48, 1991
53. Molloy CJ, Taylor DS, Weber H: Angiotensin II stimulation of rapid protein tyrosine phosphorylation and protein kinase activation in rat aortic smooth muscle cells. J Biol Chem 268: 7338–7345, 1993
54. Leduc I, Haddad P, Giasson E, Meloche S: Involvement of a tyrosine kinase pathway in the growth-promoting effects of angiotensin II on aortic smooth muscle cells. Mol Pharmacol 48: 582–592, 1995
55. Schorb W, Peeler TC, Madigan NN, Conrad KM, Baker KM: Angiotensin II-induced protein tyrosine phosphorylation in neonatal rat cardiac fibroblasts. J Biol Chem 269: 19626–19632, 1994
56. Weiss RH, Niccittelli R: Inhibition of tyrosine phosphorylation prevents thrombin-induced mitogenesis, but not intracellular free calcium release, in vascular smooth muscle cells. J Biol Chem 267: 5608–5613, 1992
57. Simonson MS, Herman WH: Protein kinase C and protein tyrosine kinase activity contribute to mitogenic signaling by endothelin-1. J Biol Chem 268: 9347–9357, 1993
58. Seckl M, Rozengurt E: Tyrphostin inhibits bombesin stimulation of tyrosine phosphorylation, c-*fos* expression, and DNA synthesis in Swiss 3T3 cells. J Biol Chem 268: 9548–9554, 1993
59. Hollenberg MD: Tyrosine kinase-mediated signal transduction pathways and the actions of polypeptide growth factors and G-protein-coupled agonists in smooth muscle. Mol Cell Biochem 149/150: 77–85, 1995
60. Leduc I, Meloche S: Angiotensin II stimulates tyrosine phosphorylation of the focal adhesion-associated protein paxillin in aortic smooth muscle cells. J Biol Chem 270: 4401–4404, 1995
61. Polte TR, Naftilan AJ, Hanks SK: Focal adhesion kinase is abundant in developing blood vessels and elevation of its phosphotyrosine content in vascular smooth muscle cells is a rapid response to angiotensin II. J Cell Biochem 55: 106–119, 1994
62. Okuda M, Kawahara Y, Nakayama I, Hoshijima M, Yokoyama M: Angiotensin II transduces its signal to focal adhesions via angiotensin II type 1 receptors in vascular smooth muscle cells. FEBS Lett 368: 343–347, 1995
63. Giasson E, Servant MJ, Meloche S: Cyclic AMP-mediated inhibition of angiotensin II-induced protein synthesis is associated with suppression of tyrosine phosphorylation signaling in vascular smooth muscle cells. J Biol Chem 272: 26879–26886, 1997
64. Linseman DA, Benjamin CW, Jones DA: Convergence of angiotensin II and platelet-derived growth factor receptor signaling cascades in vascular smooth muscle cells. J Biol Chem 270: 12563–12568, 1995
65. Marrero MB, Paxton WG, Duff JL, Berk BC, Bernstein KE: Angiotensin II stimulates tyrosine phosphorylation of phospholipase C-gamma 1 in vascular smooth muscle cells. J Biol Chem 269: 10935–10939, 1994
66. Ishida M, Marrero MB, Schieffer B, Ishida T, Berstein KE, Berk BC: Angiotensin II activates $pp60^{c\text{-}src}$ in vascular smooth muscle cells. Circ Res 77: 1053–1059, 1995
67. Marrero MB, Schieffer B, Paxton WG, Heerdt L, Berk BC, Delafontaine P, Bernstein KE: Direct stimulation of Jak/STAT pathway by the angiotensin II AT_1 receptor. Nature 375: 247–250, 1995
68. Ali MS, Schieffer B, Delafontaine P, Bernstein KE, Ling BN, Marrero MB: Angiotensin II stimulates tyrosine phosphorylation and activation of insulin receptor substrate 1 and protein-tyrosine phosphatase 1D in vascular smooth muscle cells. J Biol Chem 272: 12373–12379, 1997
69. Brinson AE, Harding T, Diliberto PA, He Y, Li X, Hunter D, Herman B, Earp HS, Graves LM: Regulation of a calcium-dependent tyrosine kinase in vascular smooth muscle cells by angiotensin II and platelet-derived growth factor. Dependence on calcium and the actin cytoskeleton. J Biol Chem 273: 1711–1718, 1998
70. Sayeski PP, Ali MS, Harp JB, Marrero MB, Bernstein KE: Phosphorylation of p130Cas by angiotensin II is dependent on c-Src, intracellular Ca^{2+}, and protein kinase C. Circ Res 82: 1279–1288, 1998
71. Takahashi T, Kawahara Y, Taniguchi T, Yokoyama M: Tyrosine phosphorylation and association of p130Cas and c-Crk II by ANG II in vascular smooth muscle cells. Am J Physiol 274: H1059–H1065, 1998
72. Ishida T, Ishida M, Suero J, Takahashi M, Berk BC: Agonist-stimulated cytoskeletal reorganization and signal transduction at focal adhesions in vascular smooth muscle cells require c-Src. J Clin Invest 103: 789–797, 1999

73. Eguchi S, Numaguchi K, Iwasaki H, Matsumoto T, Yamakawa T, Utsunomiya H, Motley ED, Kawakatsu H, Owada KM, Hirata Y, Marumo F, Inagami T: Calcium-dependent epidermal growth factor receptor transactivation mediates the angiotensin II-induced mitogen-activated protein kinase activation in vascular smooth muscle cells. J Biol Chem 273: 8890–8896, 1998
74. Daub H, Weiss FU, Wallasch C, Ullrich A: Role of transactivation of the EGF receptor in signalling by G-protein-coupled receptors. Nature 379: 557–560, 1996
75. Bhat GJ, Thekkumkara TJ, Thomas WG, Conrad KM, Baker KM: Angiotensin II stimulates *sis*-inducing factor-like DNA binding activity. J Biol Chem 269: 31443–31449, 1994
76. Ali MS, Sayeski PP, Dirksen LB, Hayzer DJ, Marrero MB, Bernstein KE: Dependence on the motif YIPP for the physical association of Jak2 kinase with the intracellular carboxyl tail of the angiotensin II AT1 receptor. J Biol Chem 272: 23382–23388, 1997
77. Bhat GJ, Baker KM: Angiotensin II stimulates rapid serine phosphorylation of transcription factor Stat3. Mol Cell Biochem 170: 171–176, 1997
78. Marrero MB, Schieffer B, Li B, Sun J, Harp JB, Ling BN: Role of Janus kinase/signal transducer and activator of transcription and mitogen-activated protein kinase cascades in angiotensin II- and platelet-derived growth factor-induced vascular smooth muscle cell proliferation. J Biol Chem 272: 24684–24690, 1997
79. McWhinney CD, Hunt RA, Conrad KM, Dostal DE, Baker KM: The type I angiotensin II receptor couples to Stat1 and Stat3 activation through Jak2 kinase in neonatal rat cardiac myocytes. J Mol Cell Cardiol 29: 2513–2524, 1997
80. Pan J, Fukuda K, Kodama H, Makino S, Takahashi T, Sano M, Hori S, Ogawa S: Role of angiotensin II in activation of the JAK/STAT pathway induced by acute pressure overload in the rat heart. Circ Res 81: 611–617, 1997
81. Kodama H, Fukuda K, Pan J, Makino S, Sano M, Takahashi T, Hori S, Ogawa S: Biphasic activation of the JAK/STAT pathway by angiotensin II in rat cardiomyocytes. Circ Res 82: 244–250, 1998
82. Ihle JN: Cytokine receptor signalling. Nature 377: 591–594, 1995
83. Stark GR, Kerr IM, Williams BR, Silverman RH, Schreiber RD: How cells respond to interferons. Annu Rev Biochem 67: 227–264, 1998
84. Aringer M, Cheng A, Nelson JW, Chen M, Sudarshan C, Zhou YJ, O'Shea JJ: Janus kinases and their role in growth and disease. Life Sci 64: 2173–2186, 1999
85. Yeh TC, Pellegrini S: The Janus kinase family of protein tyrosine kinases and their role in signaling. Cell Mol Life Sci 55: 1523–1534, 1999
86. Darnell JE Jr.: STATs and gene regulation. Science 277: 1630–1635, 1997
87. Leaman DW, Leung S, Li X, Stark GR: Regulation of STAT-dependent pathways by growth factors and cytokines. FASEB J 10: 1578–1588, 1996
88. Park SY, Saijo K, Takahashi T, Osawa M, Arase H, Hirayama N, Miyake K, Nakauchi H, Shirasawa T, Saito T: Developmental defects of lymphoid cells in Jak3 kinase-deficient mice. Immunity 3: 771–782, 1995
89. Nosaka T, van Deursen JM, Tripp RA, Thierfelder WE, Witthuhn BA, McMickle AP, Doherty PC, Grosveld GC, Ihle JN: Defective lymphoid development in mice lacking Jak3. Science 270: 800–802, 1995
90. Thomis DC, Gurniak CB, Tivol E, Sharpe AH, Berg LJ: Defects in B lymphocyte maturation and T lymphocyte activation in mice lacking Jak3. Science 270: 794–797, 1995
91. Neubauer H, Cumano A, Muller M, Wu H, Huffstadt U, Pfeffer K: Jak2 deficiency defines an essential developmental checkpoint in definitive hematopoiesis. Cell 93: 397–409, 1998
92. Parganas E, Wang D, Stravopodis D, Topham DJ, Marine JC, Teglund S, Vanin EF, Bodner S, Colamonici OR, van Deursen JM, Grosveld G, Ihle JN: Jak2 is essential for signaling through a variety of cytokine receptors. Cell 93: 385–395, 1998
93. Rodig SJ, Meraz MA, White JM, Lampe PA, Riley JK, Arthur CD, King KL, Sheehan KC, Yin L, Pennica D, Johnson EM Jr, Schreiber RD: Disruption of the Jak1 gene demonstrates obligatory and non-redundant roles of the Jaks in cytokine-induced biologic responses. Cell 93: 373–383, 1998
94. Taylor SS, Buechler JA, Yonemoto W: cAMP-dependent protein kinase: Framework for a diverse family of regulatory enzymes. Annu Rev Biochem 59: 971–1005, 1990
95. Houslay MD, Milligan G: Tailoring cAMP-signalling responses through isoform multiplicity. Trends Biochem. Sci 22: 217–224, 1997
96. Pastan IH, Johnson GS, Anderson WB: Role of cyclic nucleotides in growth control. Annu Rev Biochem: 491–523, 1975
97. Boynton AL, Whitfield JF. The role of cyclic AMP in cell proliferation: A critical assessment of the evidence. In: P. Greengard, G.A. Robison (eds). Advances in Cyclic Nucleotide Research. Raven Press, New York, 1983, pp 193–294
98. Dumont JE, Jauniaux JC, Roger PP: The cyclic AMP-mediated stimulation of cell proliferation. Trends Biochem Sci 14: 67–71, 1989
99. Assender JW, Southgate KM, Hallett MB, Newby AC: Inhibition of proliferation, but not of Ca^{2+} mobilization, by cyclic AMP and GMP in rabbit aortic smooth-muscle cells. Biochem J 288: 527–532, 1992
100. Indolfi C, Avvedimento EV, Di Lorenzo E, Esposito G, Rapacciulo A, Giuliano P, Grieco D, Cavuto L, Stingone AM, Ciullo I, Condorelli G, Chiariello M: Activation of cAMP-PKA signaling *in vivo* inhibits smooth muscle cell proliferation induced by vascular injury. Nat Med 3: 775–779, 1997
101. Burgering BMT, Pronk GJ, van Weeren PC, Charding P, Bos JL: cAMP antagonizes $p21^{ras}$-directed activation of extracellular signal-regulated kinase 2 and phosphorylation of mSos nucleotide exchange factor. EMBO J 12: 4211–4220, 1993
102. Cook SJ, McCormick F: Inhibition by cAMP of Ras-dependent activation of Raf. Science 262: 1069–1072, 1993
103. Graves LM, Bornfeldt KE, Raines EW, Potts BC, MacDonald SG, Ross R, Krebs EG: Protein kinase A antagonizes platelet-derived growth factor-induced signaling by mitogen-activated protein kinase in human arterial smooth muscle cells. Proc Natl Acad Sci USA 90: 10300–10304, 1993
104. Sevetson BR, Kong X, Lawrence JC Jr.: Increasing cAMP attenuates activation of mitogen-activated protein kinase. Proc Natl Acad Sci USA 90: 10305–10309, 1993
105. Wu J, Dent P, Jelinek T, Wolfman A, Weber MJ, Sturgill TW: Inhibition of the EGF-activated MAP kinase signaling pathway by adenosine 3′, 5′-monophosphate. Science 262: 1065–1069, 1993
106. Kahan C, Seuwen K, Meloche S, Pouyssegur J: Coordinate, biphasic activation of p44 mitogen-activated protein kinase and S6 kinase by growth factors in hamster fibroblasts. Evidence for thrombin-induced signals different from phosphoinositide turnover and adenylylcyclase inhibition. J Biol Chem 267: 13369–13375, 1992
107. Monfar M, Lemon KP, Grammer TC, Cheatham L, Chung J, Vlahos CJ, Blenis J: Activation of pp70/85 S6 kinases in interleukin-2-responsive lymphoid cells is mediated by phosphatidylinositol 3-kinase and inhibited by cyclic AMP. Mol Cell Biol 15: 326–337, 1995
108. Rao GN, Runge MS: Cyclic AMP inhibition of thrombin-induced growth in vascular smooth muscle cells correlates with decreased JNK1 activity and c-Jun expression. J Biol Chem 271: 20805–20810, 1996
109. Graves LM, Bornfeldt KE, Argast GM, Krebs EG, Kong X, Lin TA, Lawrence JC Jr.: cAMP- and rapamycin-sensitive regulation of the

association of eukaryotic initiation factor 4E and the translational regulator PHAS-I in aortic smooth muscle cells. Proc Natl Acad Sci USA 92: 7222–7226, 1995

110. Kato JY, Matsuoka M, Polyak K, Massague J, Sherr CJ: Cyclic AMP-induced G1 phase arrest mediated by an inhibitor (p27Kip1) of cyclin-dependent kinase 4 activation. Cell 79: 487–496, 1994

111. L'Allemain G, Lavoie JN, Rivard N, Baldin V, Pouyssegur J: Cyclin D1 expression is a major target of the cAMP-induced inhibition of cell cycle entry in fibroblasts. Oncogene 14: 1981–1990, 1997

112. Trepel JB, Colamonici OR, Kelly K, Schwab G, Watt RA, Sausville EA, Jaffe ES, Neckers LM: Transcriptional inactivation of c-*myc* and the transferin receptor in dibutyryl cyclic AMP-treated HL-60 cells. Mol Cell Biol 7: 2644–2648, 1987

113. Heldin NR, Paulsson Y, Forsberg K, Heldin CH, Westermark B: Induction of cyclic AMP synthesis by forskolin is followed by a reduction in the expression of c-*myc* messenger RNA and inhibition of ^{3}H-thymidine incorporation in human fibroblasts. J Cell Physiol 138: 17–23, 1989

114. Rock CO, Cleveland JL, Jackowski S: Macrophage growth arrest by cyclic AMP defines a distinct checkpoint in the mid-G_1 stage of the cell cycle and overrides constitutive c-*myc* expression. Mol Cell Biol 12: 2351–2358, 1992

115. Wolf G, Killen PD, Neilson EG: Intracellular signaling of transcription and secretion of type IV collagen after angiotensin II-induced cellular hypertrophy in cultured proximal tubular cells. Cell Reg 2: 219–227, 1991

116. Takahashi T, Kawahara Y, Okuda M, Yokoyama M: Increasing cAMP antagonizes hypertrophic response to angiotensin II without affecting Ras and MAP kinase activation in vascular smooth muscle cells. FEBS Lett 397: 89–92, 1996

117. Randall VA, MacLennan SJ, Martin GR, Wilson VG: The effect of forskolin on 5-HT1-like and angiotensin II-induced vasoconstriction and cyclic AMP content of the rabbit isolated femoral artery. Br J Pharmacol 118: 627–634, 1996

118. Sonenberg N, Gingras AC: The mRNA 5′ cap-binding protein eIF4E and control of cell growth. Curr Opin Cell Biol 10: 268–275, 1998

119. Jefferies HBJ, Thomas G. Ribosomal protein S6 phosphorylation and signal transduction. In: J.W.B. Hershey, M.B. Mathews, N. Sonenberg (eds). Translational Control. Cold Spring Harbor Laboratory Press, Cold Spring Harbor, 1996, pp 389–409.

120. Jefferies HB, Fumagalli S, Dennis PB, Reinhard C, Pearson RB, Thomas G: Rapamycin suppresses 5′TOP mRNA translation through inhibition of p70s6k. EMBO J 16: 3693–3704, 1997

121. Pullen N, Dennis PB, Andjelkovic M, Dufner A, Kozma SC, Hemmings BA, Thomas G: Phosphorylation and activation of p70s6k by PDK1. Science 279: 707–710, 1998

122. Alessi DR, James SR, Downes CP, Holmes AB, Gaffney PR, Reese CB, Cohen P: Characterization of a 3-phosphoinositide-dependent protein kinase which phosphorylates and activates protein kinase B alpha. Curr Biol 7: 261–269, 1997

123. Dumler I, Weis A, Mayboroda OA, Maasch C, Jerke U, Haller H, Gulba DC: The Jak/Stat pathway and urokinase receptor signaling in human aortic vascular smooth muscle cells. J Biol Chem 273: 315–321, 1998

124. Conway G, Margoliath A, Wong-Madden S, Roberts RJ, Gilbert W: Jak1 kinase is required for cell migrations and anterior specification in zebrafish embryos. Proc Natl Acad Sci USA 94: 3082–3087, 1997

125. Sengupta TK, Schmitt EM, Ivashkiv LB: Inhibition of cytokines and JAK-STAT activation by distinct signaling pathways. Proc Natl Acad Sci USA 93: 9499–9504, 1996

126. David M, Petricoin E III, Larner AC: Activation of protein kinase A inhibits interferon induction of the Jak/Stat pathway in U266 cells. J Biol Chem 271: 4585–4588, 1996

127. Salahpour A, Angers S, Bouvier M: Functional significance of oligomerization of G protein-coupled receptors. Trends Endocr Metab (in press)

128. Klingmuller U, Lorenz U, Cantley LC, Neel BG, Lodish HF: Specific recruitment of SH-PTP1 to the erythropoietin receptor causes inactivation of JAK2 and termination of proliferative signals. Cell 80: 729–738, 1995

129. Yetter A, Uddin S, Krolewski JJ, Jiao H, Yi T, Platanias LC: Association of the interferon-dependent tyrosine kinase Tyk-2 with the hematopoietic cell phosphatase. J Biol Chem 270: 18179–18182, 1995

130. Jiao H, Berrada K, Yang W, Tabrizi M, Platanias LC, Yi T: Direct association with and dephosphorylation of Jak2 kinase by the SH2-domain-containing protein tyrosine phosphatase SHP-1. Mol Cell Biol 16: 6985–6992, 1996

131. You M, Yu DH, Feng GS: Shp-2 tyrosine phosphatase functions as a negative regulator of the interferon-stimulated Jak/STAT pathway. Mol Cell Biol 19: 2416–2424, 1999

132. Petricoin E III, David M, Igarashi K, Benjamin C, Ling L, Goelz S, Finbloom DS, Larner AC: Inhibition of alpha interferon but not gamma interferon signal transduction by phorbol esters is mediated by a tyrosine phosphatase. Mol Cell Biol 16: 1419–1424, 1996

133. Kolenko V, Rayman P, Roy B, Cathcart MK, O'Shea J, Tubbs R, Rybicki L, Bukowski R, Finke J: Downregulation of JAK3 protein levels in T lymphocytes by prostaglandin E2 and other cyclic adenosine monophosphate-elevating agents: impact on interleukin-2 receptor signaling pathway. Blood 93: 2308–2318, 1999

134. Geraci MW, Gao B, Shepherd DC, Moore MD, Westcott JY, Fagan KA, Alger LA, Tuder RM, Voelkel NF: Pulmonary prostacyclin synthase overexpression in transgenic mice protects against development of hypoxic pulmonary hypertension. J Clin Invest 103: 1509–1515, 1999

Molecular and Cellular Biochemistry **212**: 111–119, 2000.

The inducible cAMP early repressor ICERIIγ inhibits CREB and AP-1 transcription but not AT_1 receptor gene expression in vascular smooth muscle cells

Xiaofei Wang and T.J. Murphy

Department of Pharmacology, Emory University School of Medicine, and the Graduate Program in Molecular Therapeutics and Toxicology, Graduate Division of Biological and Biomedical Sciences, Emory University, Atlanta, GA, USA

Abstract

Growth factors and agonists of both Gαs and Gαq-coupled receptor agonists each down regulate angiotensin AT_1 receptor gene expression in vascular smooth muscle cells. To test whether the transcription factor CREB is in the pathway involved in modulation of AT_1-receptor gene expression, AT_1 receptor mRNA levels were measured after stimulation with forskolin, angiotensin II and PDGF-BB in VSMC expressing the inducible cAMP early repressor protein, ICERIIγ, fused to the green fluorescent protein (eGFP). This fusion protein inhibits luciferase induction from promoters driven by either CREB response elements (CRE) or, unexpectedly, by tetradecanoylphorbol acetate response elements (TRE), which bind activator protein (AP-1). EGFP-ICERIIγ expression inhibits occupancy of CRE and TRE elements by endogenous VSMC nuclear proteins. Nevertheless, agonist-induced AT_1-receptor mRNA down regulation is unaffected by eGFP-ICERIIγ expression. Although CREB transcription is activated by multiple classes of agonists in VSMC, and ICER can inhibit agonist-induced transcription mediated by both CREB and AP-1 enhancers, these transcription factors are not obligate components of the pathways regulating AT_1-R gene expression in rat VSMC. (Mol Cell Biochem **212**: 111–119, 2000)

Key words: angiotensin, down regulation, luciferase, retroviral gene transfer, nuclear repressors

Introduction

The participation of angiotensin II in blood volume and pressure homeostasis occurs through activation of Gαq-coupled AT_1-receptors (AT_1-R) [1]. AT_1-R are located on the surface of several types of cells within diverse tissues such as the kidney, adrenal gland, cardiac muscle, neurons and in blood vessels (for reviews see refs [2, 3]). Recent animal studies have confirmed long held notions that AT_1-R gene expression is regulated dynamically in all of these tissues under various physiological conditions [4–7]. Nevertheless, the cellular and molecular mechanisms associated with these responses remains poorly understood [8]. Since AT_1-R levels will have a major effect on end organ sensitivity to angiotensin II, defects in the pathways involved in modulation of AT_1-R gene expression may contribute to the development and progression of cardiovascular diseases.

Although cell surface AT_1-R are internalized by angiotensin II, sustained decreases in AT_1-R protein expression in VSMC only occur when mRNA levels are reduced [9]. Down regulation of AT_1-R mRNA occurs in VSMC following activation by several growth factors [10, 11], by various Gαq-coupled receptor agonists [12, 13] and also by cAMP-elevating agents [14, 15]. The effects of growth factor and Gαq-coupled re-

Address for offprints: T.J. Murphy, Department of Pharmacology, Emory University School of Medicine, 5031 O.W. Rollins Research Building, Atlanta, GA 30322, USA ceptor activation results from suppression of AT_1-R gene transcription [10, 12, 13]. In contrast, cAMP-elevating agents do not suppress AT_1-R gene transcription and down regulate the mRNA solely by a protein kinase A (PKA)-dependent post-transcriptional mechanism [15, 16].

Previous observations are consistent with the possibility that regulation of AT_1-R mRNA stability may involve *de novo* gene induction. First, general transcriptional inhibitors reduce the rapidity and degree by which agonist-mediated destabilization of the AT_1-R mRNA occurs. Second, reductions in AT_1-R mRNA levels are usually evident only after a lag period following agonist treatment, suggesting the hypothesis that a modulator might be induced first [10, 12, 13, 15]. Since the cAMP-responsive binding protein (CREB) is a major mediator of PKA-dependent transcriptional responses [17], but can also be activated by engagement of other signaling pathways [17, 18], this study was designed to determine if activation of CREB mediated transcription contributes to AT_1-R down regulation.

The cAMP-responsive modulator (CREM) gene produces several alternatively spliced transcripts whose proteins are more related to CREB than to other members of the basic leucine zipper transcription factor superfamily. Several CREM gene products can inhibit CREB mediated transcription including the immediate cAMP early repressor protein ICERIIγ [19]. ICERIIγ is composed of a leucine zipper region and the second of two CREM gene DNA binding domains, but since it lacks kinase inducible transactivation domains it can serve as a nuclear antagonist of CREB-mediated transcription [20]. Other products of the CREM gene have been shown to also bind to tetradecanolylphorbol acetate responsive elements (TRE) *in vitro* and to inhibit AP-1 mediate transcription in choriocarcinoma cells [21]. Because of this, we examined the ability of various agonists to stimulate both CREB and AP-1 mediated transcription in smooth muscle cells and how ICERIIγ might interfere with these processes. Using a novel approach that leads to high yield expression of recombinant genes in VSMC, a fusion protein composed of eGFP and ICERIIγ is shown to inhibit transcription directed by both of these elements. However, ICERIIγ expression does not influence down regulation of AT_1-R gene expression in VSMC.

Materials and methods

Cell culture

Rat thoracic aorta VSMC are kept as a continuous cell line and were kindly provided by Dr. R.W. Alexander (Emory University). They are cultured in high glucose Dulbecco's modified Eagle's medium (DMEM) with 3.7 mg/ml of $NaHCO_3$, 100 unit/ml penicillin, 100 μg/ml streptomycin, and 10% heat-inactivated calf serum in a humidified incubator with 5% CO_2 at 37°C. After reaching confluence, the cells were growth-arrested before experiments by culturing in serum-free media for 24 h. For the experiments reported herein, cells were used between passages 10–25 after primary explant. The retroviral producer cell line Bing CAK8 (ATCC CRL 11554) [22] was cultured under identical conditions except that 10% fetal bovine serum was used in place of calf serum.

Plasmid construction

EGFP and eGFP/ICERIIγ expression vectors

A Hind III-Not I restriction fragment of pEGFP-N1 (Clontech Inc., Palo Alto, CA, USA) was subcloned into pTJM2, a derivative of the expression vector pCDM8 (Invitrogen Inc., Carlsbad, CA, USA), in which the supF gene is replaced by a beta-lactamase cassette, creating the intermediate expression plasmid pTJM2-eGFP. The human ICERIIγ cDNA was amplified by reverse transcriptase-polymerase chain reaction (RT-PCR) using primers 5′-TATGTACAAGCCCAACATGGCTG-3′ and 5′-CTAATCTGTTTTGGGAGAGCAAATG-3′. To create pTJM2-eGFP/ICERIIγ, in which ICERIIγ is a carboxyl extension of eGFP, the PCR product was digested with BsrG I and cloned into pTJM2-eGFP that was prepared first with NotI cleavage and Klenow to blunt before cutting with BsrG I. Subsequently, a blunt ended EcoRI-XbaI fragment was excised from this latter vector, ligated with BstXI adapters and subcloned into the BstXI sites of the retroviral vector pTJM9 [23], a derivative of the retroviral vector pLNCX [24]. This places expression of eGFP/ICERIIγ under the control of an internal CMV promoter. A control retroviral vector that expresses eGFP without the ICERIIγ sequences was created by subcloning a Hind III-Not I fragment of pEGFP-N1 into pTJM9.

CREB and AP1 retroviral luciferase reporter vectors

CRE and AP1 responsive elements were cloned into the retroviral reporter vector pKA9 as the sole upstream enhancers in a minimal human IL-2 gene promoter (–72 to +47) [23]. The CRE elements were designed from the CREB enhancers within the CREM intragenic ICER promoter [20]. The oligonucleotides (5′-GATCAGGCTGG<u>TGACGTCA</u>CTG<u>TGATGTCA</u>GTGCTC-3′ and 5′-GATCGAGCACTGACATCACAGTGACGTCACCAGCCT-3′) were annealed and concatenated 4 × and inserted into a unique BamHI site upstream of the IL-2 promoter creating an enhancer with 8 CRE repeats. The TRE sequences are derived from the human collagenase gene and were formed by annealing and 2 × concatenation of the oligonucleotides 5′-TCGAA<u>TGAGTCA</u>GCTCTA<u>TGAGTCA</u>GATCCA<u>TGAGTCA</u>GCTCTA<u>TGAGTCA</u>G-3′ and 5′-GATCTGACTCATAGAGCTGACTCATGGATCTGACTCATAGAGCTGACTCAT-3′. These were subcloned into SalI and BamHI sites of the reporter vector creating an enhancer with 8 TRE repeats. The promoter/enhancer regions of these vectors were confirmed by sequencing reactions.

Retroviral production and infection of VSMC

Bing Cak-8 retroviral producer cells were transfected transiently with retroviral plasmids to produce helper virus free infectious supernatant which was then used to infect VSMC as described in detail previously [23]. For reporter studies, the cells were first infected with reporter virus and stable populations of infected cells were obtained after selection for 7–10 days with 150 μg/ml G418. These cell lines were subsequently infected with virus prepared from the eGFP or eGFP/ICERIIγ expression vectors.

eGFP detection and cell sorting

The cells expressing either eGFP or eGFP/ICERIIγ were therefore enriched by fluorescence activated cell sorting (FACS) as described previously [25]. To prepare cells for FACS, VSMC were dispersed by treatment with trypsin and passed through a 70 mm cellulose acetate filter to remove cell clusters. Cells were collected at the flow cytometry facility at the Emory University Hospital using a FACS Vantage flow cytometry system (Becton Dickinson Inc., San Jose, CA, USA).

DNA binding assays

After treatments, VSMC were collected in cold PBS, resuspended in 4 ml of buffer A (10 mM HEPES, 1.5 mM $MgCl_2$, 10 mM KCl, and 0.5 mM DTT, pH 7.9), and incubated 10 min on ice. After homogenization by 10 strokes in glass dounce homogenizer, nuclei were collected by centrifugation in a Beckman GS-6 rotor at 3,750 rpm for 15 min at 4°C. The nuclear pellet was resuspended in 500 μl buffer C (20 mM HEPES, 25% glycerol, 0.42 M NaCl, 1.5 mM $MgCl_2$, and 0.2 mM EDTA, pH 7.9), homogenized again, and stirred slowly at 4°C for 30 min. Following centrifugation at 19,000 rpm for 30 min at 4°C in a Beckman TLS55 rotor, the supernatant was recovered and dialyzed overnight at 4°C against 300 ml of buffer D (20 mM HEPES, 20% glycerol, 0.1 M KCl, 0.2 mM EDTA, 0.5 mM DTT, 0.5 mM PMSF, pH 7.9). The nuclear extract was then centrifuged at 19,000 rpm for 30 min at 4°C, before freezing supernatant aliquots at –70°C until used. CRE and TRE oligonucleotides used in binding assays were the same as those described above for construction of the reporter vectors. Mutated CRE oligos were: 5′-GATCAGGCTGGAGATGTTACTGAGATGTTAGTGCTC-3′ and 5′-GATCGAGCACTAACATCTCAGTAACATCTCCAGCCT-3′, which were designed according to Foulkes *et al.* [19]. Binding probes (100 pmole) were labeled by the Klenow fill-in method with [α-^{32}P]dATP (800 Ci/mmol) and radioactivity was determined after passing the samples through gel exclusion columns. For the DNA binding reactions, 2×10^4 cpm (~ 1 nM) of radiolabeled oligonucleotides were combined with 2 μg of poly dI-dC and the indicated molar excess of competitive oligonucleotides. Nuclear protein (1–10 μg) was added and the samples were incubated at 37°C for 10 min before loading onto non-denaturing 4% polyacrylamide gels, which were subsequently developed on phosphorimage cassettes.

Luciferase assay

Cells were grown to confluence in 12 × 20 mm-diameter multiwell culture plates. The luciferase assay was performed as described previously [26] and is a modification of the method reported by de Wet *et al.* [27].

Northern hybridization analysis

After treating cells with agonists for the indicated periods of time, RNA was extracted and Northern blot hybridization assays were done using a [^{32}P]-dCTP labeled cDNA probe as described previously [15]. The hybridization signals were quantified from phosphorimages using a volume integration protocol in the Molecular Dynamics Inc., ImageQuant software package.

Western blot

Western blot detection of CREB was performed by using the PhosphoPlus™ CREB (Ser133) antibody kit (New England Biolab, Natick, MA, USA). VSMC cultured in 35 mm-diameter dishes were treated with agonists for the indicated times before washing with PBS, and lysis in 100 μl SDS sample buffer (62.5 mM Tris-Cl, pH 6.8, 2% SDS, 10% glycerol, 50 mM DTT, 0.1% bromophenol blue). Protein samples were sonicated for 2 sec, heated to 95°C for 5 min, and cooled on ice. After a brief spin, 10 μl of samples were separated on precast 10% SDS-PAGE gels and transferred to Immobilon P membranes. After blotting with 0.5% non-fat dry milk, the membranes were incubated overnight with 1:1000 dilution of either anti-CREB antibody or anti-phospho-CREB antibody at 4°C. The blots were developed using a chemiluminescent detection method and anti-rabbit secondary antibody conjugated to horseradish peroxidase purchased from New England Biolabs.

Results

Various agonists that down regulate AT_1-R gene expression in VSMC were tested for their ability to activate CREB first by western blot analysis using a anti-CREB antibody, which recognizes a phosphorylated epitope flanking the essential serine at position 133 [28]. As shown in Fig. 1, treatment of VSMC for 10 min with forskolin (10 μM), angiotensin II (100 nM) and PDGF-BB (50 ng/ml) increased phosphorylated CREB levels after 10 and 30 min compared to non-stimulated cells. The total amount of CREB in the samples was comparable, as detected by an anti-CREB antibody that does not discriminate between the phosphorylated and non-phosphorylated forms of CREB.

Retroviral vectors were used to deliver luciferase reporter genes because these VSMC are difficult to transfect with plasmid DNA yet are efficiently transduced with retroviruses. Previous studies with this reporter system have established that the quiescent minimal human IL-2 gene promoter is induced faithfully only when specific heterologous enhancer cassettes are inserted in its 5′ region. The CREB responsive vector pCRE-IL2-Luc contains CRE sequences that direct cAMP-dependent induction of ICER expression from within the CREM gene (Fig. 2) [20]. The tetradecanoylphorbol acetate response element (TRE) in the pTRE-IL2-Luc vector is based upon a well-characterized AP-1 responsive enhancer within the human collagenase gene [29]. A retroviral construct without enhancer was used as negative control (pΦIL2-Luc). Cultured VSMC infected with either reporter retrovirus or enhancerless control were treated with either forskolin, angiotensin II, PDGF-BB or PMA for various times before luciferase levels were measured. As shown in Fig. 3, none of the agonists stimulated luciferase expression in cells infected with retrovirus made from enhancerless pΦIL2-Luc vector. Comparison of forskolin and PMA responses provide the clearest evidence that the activation properties of the CRE and TRE enhancers differ in VSMC. PMA, but not forskolin stimulates TRE-mediated luciferase expression whereas both drugs effectively stimulate CRE-mediated transcription. Furthermore, angiotensin II modestly stimulates TRE-mediated transcription (4.7 ± 1.4 fold at 4 h; n = 5), but is routinely a more effective activator of CRE-mediated transcription (37 ± 7 fold at 4 h; n = 5). Lastly, the temporal characteristics of TRE- and CRE-mediated luciferase expression differ, such that the latter is induced earlier by angiotensin II, PDGF-BB and PMA than is the latter. These findings suggest the two elements respond to distinct classes of transcription factors.

The results of electrophoretic mobility shift assays are consistent with this interpretation. [^{32}P]-Labeled CRE oligonucleotides are bound nuclear proteins in VSMC that are selectively inhibited by an unlabeled oligonucleotide containing a CRE, but not by an oligonucleotide bearing mutations within the CRE or by an unlabeled TRE oligonucleotide (Fig. 4A). When [^{32}P]-labeled TRE oligonucleotides are used as probe, a single protein binding complex is also formed by VSMC nuclear extracts (Fig. 4B). This complex was competed by unlabeled TRE, and although inhibited with 100-

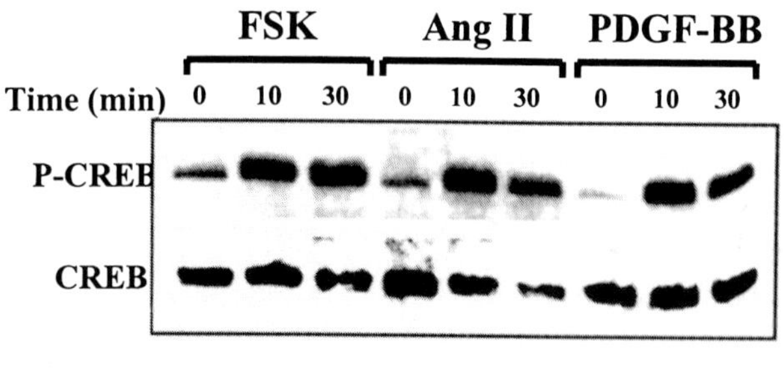

Fig. 1. Agonist-mediated CREB phosphorylation. VSMC received no treatment (0 min) or were treated for 10 or 30 min with 10 μM forskolin (FSK), 100 nM angiotensin II (Ang II), or 50 ng/ml PDGF-BB (PDGF-BB) before cell extracts were prepared and analyzed by Western blot. Phosphorylated CREB was detected using an antibody recognizing phosphorylated serine 133 (P-CREB). Non phosphorylated CREB was detected after stripping the same blot using an anti-CREB antibody that does not discriminate phosphorylated and unphosphorylated CREB (CREB). The results are representatives of 3 experiments.

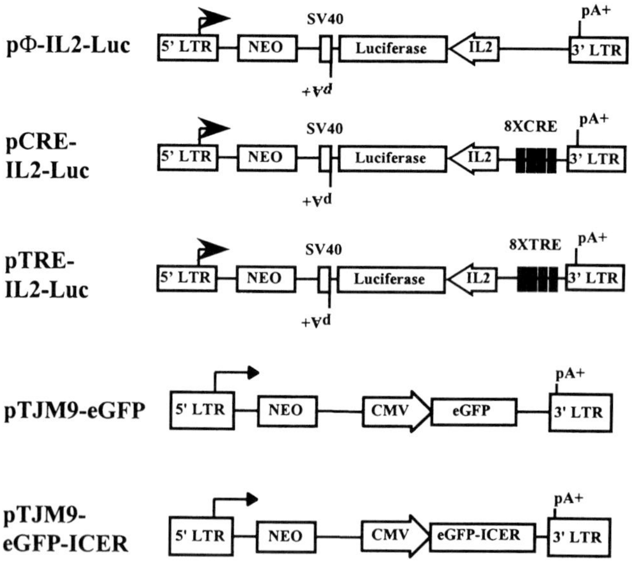

Fig. 2. Retroviral vectors. For clarity, only the retroviral components of the circular plasmid vectors are shown. The neomycin resistance gene (NEO) is expressed from the 5′ long terminal repeat (LTR) promoter in each vector but G418 (150 μg/ml) was only used to select for cells infected with luciferase reporter viruses. The luciferase reporter vectors pΦ-IL2-Luc, pCRE-IL2-Luc and pTRE-IL2-Luc, which are respectively enhancerless, CREB responsive or AP-1 responsive, utilize a minimal promoter from the human IL-2 gene [46] to drive luciferase expression from the strand opposite of the 5′ LTR promoter. Polyadenylation signals (pA) from the SV40 virus are inserted to stop luciferase transcription. Either eGFP or eGFP-ICERIIγ are expressed from an internal CMV promoter and FACS, rather than G418 was used to select for cells expressing these.

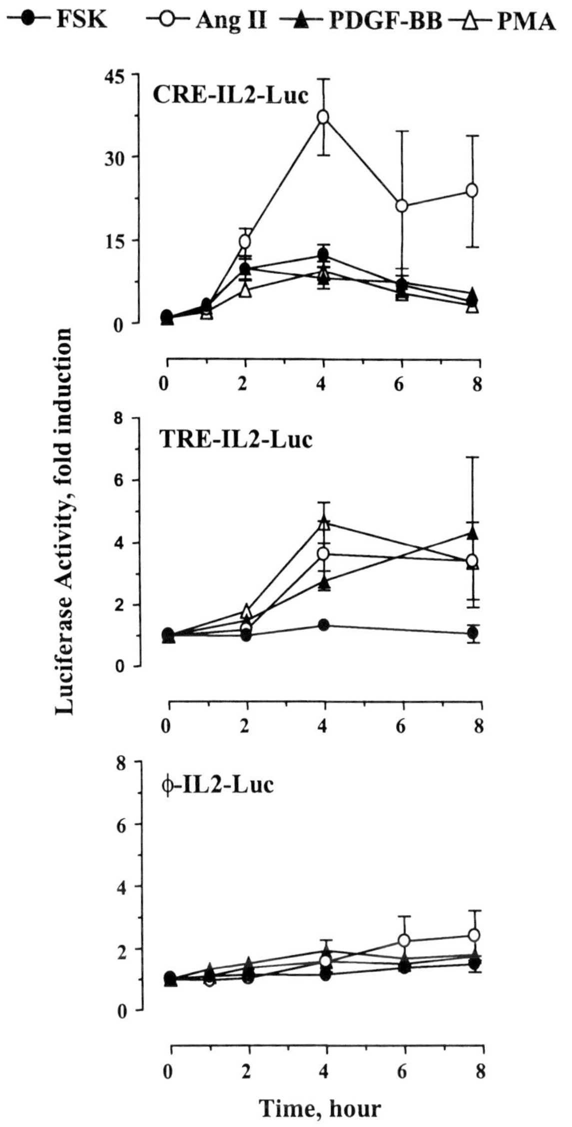

Fig. 3. Luciferase reporter activities. VSMC were infected with retroviruses responsive to CREB (top panel), AP-1 (middle panel) or enhancerless (bottom panel) and treated with either 10 μM forskolin (●), 100 nM angiotensin II (O), 50 ng/ml PDGF-BB (▲) or with 100 nM PMA (Δ) for the indicated times before measuring luciferase levels. Agonist stimulated luciferase values are normalized to the level of luciferase in vehicle treated cells and expressed as fold over basal. Each data point is mean ± S.E.M. of 3–5 independent experiments performed either in duplicate or triplicate.

fold molar excess of CRE, lower concentrations of unlabeled CRE are less effective at inhibiting TRE than CRE binding activity (Fig. 4B). Although the extracts used in these binding experiments were prepared from untreated cells, these findings were no different from those using nuclear extracts prepared from VSMC treated for 30 min with 10 μM forskolin, 100 nM Ang II, and 50 ng/ml PDGF-BB (data not shown). These observations indicate that distinct groups of VSMC nuclear factors bind to either the CRE or TRE sequences, and are most likely CREB and AP-1 family members, respectively.

A cDNA encoding the ICERIIγ isoform of CREM was fused in-frame to the 3′ end of enhanced green fluorescent protein (eGFP) cDNA, and expressed from a retroviral CMV promoter in either CRE or TRE luciferase reporter VSMC. Separate control cell lines were prepared in which eGFP

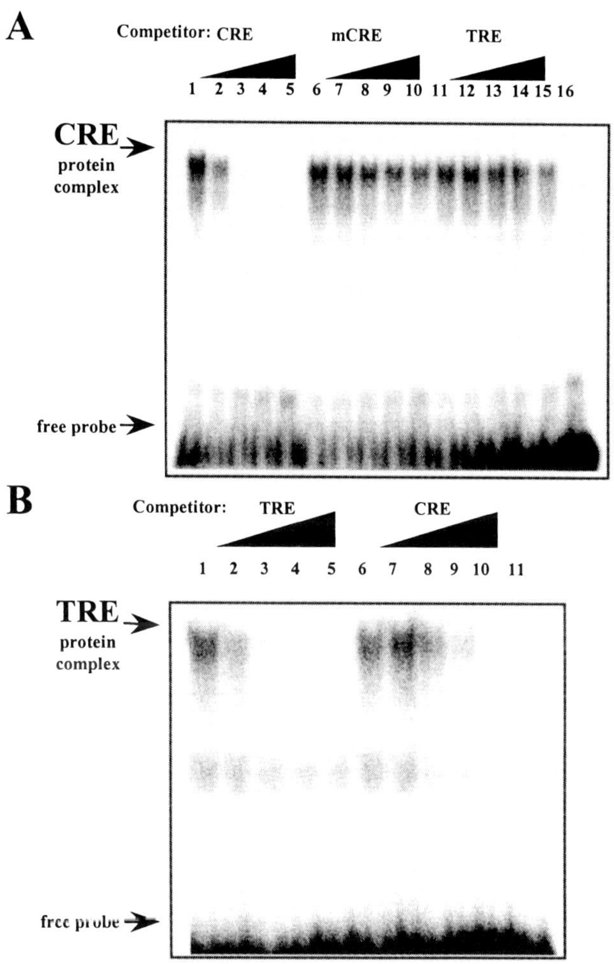

Fig. 3. Distinct nuclear protein complexes form on CRE and TRE enhancers in VSMC. *Panel A: Protein binding to the CRE.* A [^{32}P]-labeled CRE oligonucleotide (~ 1 nM final concentration) was incubated with 5 μg of nuclear extract prepared from VSMC in the absence (lanes 1, 6 and 11) or presence of unlabeled competitor oligonucleotides. Lanes 2–5 contain 3, 10, 30, and 100-fold molar excess of an unlabeled CRE competitor. Lanes 7–10 show lack of competition using the same relative concentrations of a mutant CRE oligonucleotide. Lanes 12–15 shows lack of competition with the same relative concentrations of a TRE oligonucleotide. Lane 16 shows radiolabeled CRE probe mobility in the absence of nuclear protein. *Panel B: Protein binding to the TRE.* A [^{32}P]-labeled TRE oligonucleotide (~ 1 nM final concentration) was incubated with 5 μg of nuclear extract in the absence (lanes 1 and 6) or presence of oligonucleotide competitors including a TRE oligonucleotide (lanes 2–5) or a CRE oligonucleotide (lanes 7–10). Lane 11 shows radiolabeled TRE probe mobility in the absence of nuclear protein. The data shown are representative of 3 experiments.

without the ICERIIγ sequence was expressed instead. Cells expressing this or the fusion protein were then enriched by fluorescence activated cell sorting (FACS) in a manner analogous to our recent study on the effects of a PKA inhibitor/eGFP fusion protein [25]. The following series of experiments were conducted using three different culture preparations in which quantitative FACS analysis indicated > 90% of the cells express the indicated recombinant proteins. Microscopic inspection indicated that expression of the eGFP-ICERIIγ fusion protein is restricted to the nucleus, whereas the control eGFP protein was expressed in all cell compartments (data not shown).

The cells were then stimulated for 4 h with either vehicle, 10 μM forskolin, 100 nM angiotensin II, or 50 ng/ml PDGF-BB before luciferase activity was measured. As shown in Fig. 5, expression of eGFP-ICERIIγ fully inhibits luciferase expression from both the CRE and TRE reporters relative to that seen in control cells expressing eGFP alone. Nuclear protein extracts were prepared from VSMC expressing either eGFP or EGFP-ICERIIγ and DNA binding assays were performed to determine if eGFP-ICERIIγ occupies both CRE and TRE sites. As shown in Fig. 6, slowly migrating complexes bound on CRE and TRE oligonucleotides were observed with increasing amounts of protein derived from eGFP control cell extracts. These bands likely represent CREB and AP-1 protein complexes, respectively. These slower complexes were absent in nuclear extracts prepared from VSMC expressing the eGFP-ICERIIγ protein. Instead, a faster migrating complex is observed in binding assays using these latter extracts, particularly when relatively higher concentrations of nuclear protein are used in the assay. Taken together, the simplest interpretation of these observations is that eGFP-ICERIIγ occupies the CRE and TRE elements thereby blocking the ability of CREB and AP-1 to form DNA binding complexes.

The effect of eGFP-ICERIIγ expression on agonist induced down regulation of AT_1-R gene expression was assessed by Northern hybridization analysis. Figure 7A displays a representative phosphorimage of AT_1-R mRNA hybridization signals within total RNA samples at various times after forskolin stimulation. The Northern protocol detects specific signals from a major 2.2 kb transcript and a minor 3.2 kb alternately spliced transcript [30], which are reduced clearly and progressively following a single bolus dose of the drug. These findings are consistent with had is observed in non-recombinant VSMC [15]. Expression of eGFP-ICERIIγ does not affect basal levels of AT_1-R mRNA abundance, because the hybridization signals in the zero hour time point are comparable in two cell lines. As is evident, eGFP-ICERIIγ expression does not affect forskolin-stimulated down regulation of the mRNA either. The result of quantitative phosphorimage analysis from three independent experiments using forskolin, angiotensin II and PDGF-BB are compiled in Fig. 7B. EGFP-ICERIIγ expression does not

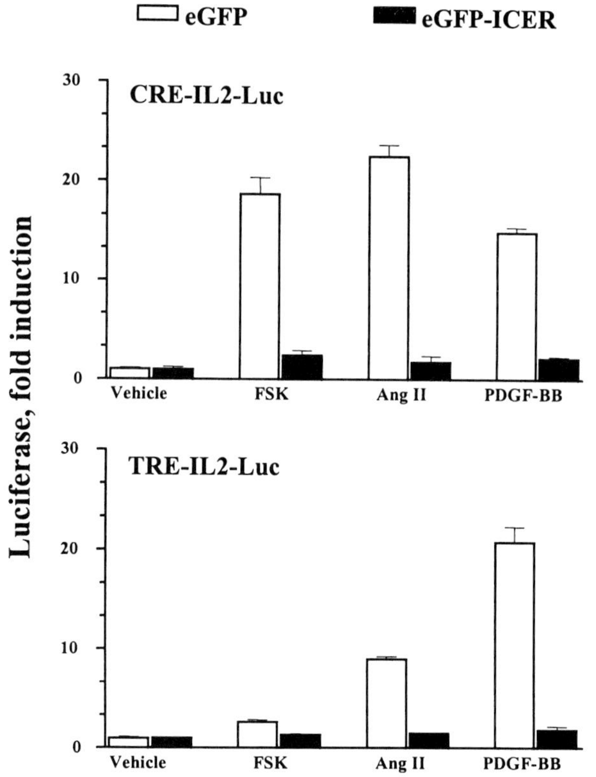

Fig. 5. Expression of eGFP-ICERIIγ inhibits CREB and AP-1 dependent transcription. Luciferase Reporter VSMC infected with either CRE (top panel) or TRE (lower panel) reporter retroviruses were infected again with a retrovirus expressing either eGFP (open bars) for control or a eGFP-ICERIIγ retrovirus (shaded bars). After FACS enrichment of expressing cells, the cells were stimulated for 4 h with either 10 μM forskolin, 100 nM angiotensin II, or 50 ng/ml PDGF-BB before luciferase levels were measured. Bar heights are mean ± S.E.M. of triplicate determinants from a single experiment. Similar differences between the effects of eGFP and eGFP-ICERIIγ were observed from 2 other experiments.

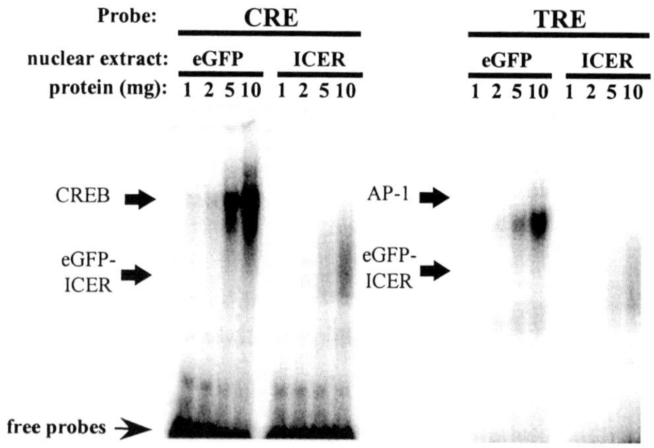

Fig. 6. eGFP-ICERIIγ expression inhibits nuclear protein binding on CRE and TRE enhancers in VSMC. Nuclear extract (1, 2, 5, and 10 μg of protein) prepared from cells expressing eGFP-ICERIIγ (ICER) or eGFP as control were incubated with either radiolabeled CRE (left panel) or TRE (right panel) probes and subjected to electrophoretic mobility analysis. The results are representative of 3 experiments.

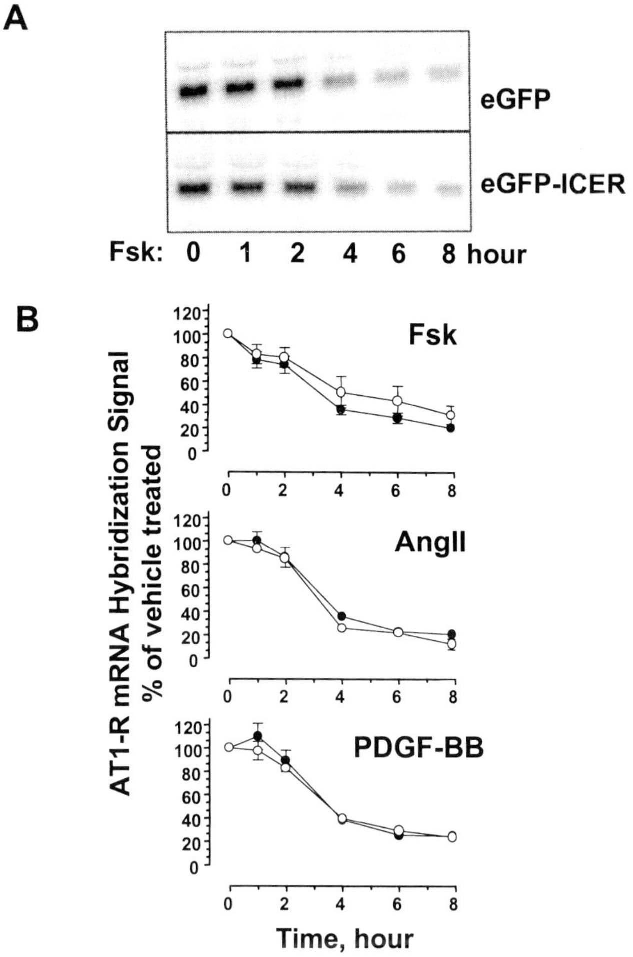

Fig. 7. eGFP-ICERIIγ expression does not inhibit agonist mediated down regulation of AT_1-R mRNA levels. VSMC expressing either ICERIIγ or eGFP as control were stimulated with a single bolus dose of either 10 μM forskolin, 100 nM angiotensin II or 50 ng/ml PDGF-BB and incubated for the indicated times before extracting RNA and performing Northern hybridization analysis. *Panel A*: AT_1-R mRNA hybridization signals obtained in a representative experiment following treatment with forskolin. *Panel B*: Data quantified from phosphorimage analysis where each point represents the mean ± S.E.M. from 3 independent experiments.

interfere with down regulation of AT_1-R mRNA expression induced by any of these agonists.

Discussion

The only direct evidence published to date that documents CREB-mediated gene transcription in VSMC is a study of homocysteine induced cyclin A gene expression [31]. Although activation of various signaling pathways in VSMC by homocysteine is consistent with mediation through cell surface receptors [32, 33], this possibility remains to be clarified. To our knowledge, the present study thus provides the first direct evidence of receptor-mediated induction of CREB-dependent transcription in VSMC. The importance of this finding rests in the data showing that CREB-mediated transcription in VSMC can be stimulated not only by an activator of adenylyl cyclase, but also through activation of growth factor and Gαq-coupled receptors. Although CREB was discovered as a nuclear target of PKA, this has led to a general perception that CREB mediated transcription is controlled dominantly by cAMP-dependent signaling [34–36]. However, subsequent studies clearly indicate that CREB can be phosphorylated and CREB-mediated transcription can be induced by activation of PKA-independent signaling cascades. For example CREB is a substrate for calcium regulated kinases [37] and also for various mitogen activated kinases [38–42]. Thus, CREB can function as a nuclear integrator of information from multiple signaling cascades. Given the great interest in the regulation of VSMC by large numbers of diverse extracellular signaling molecules, it is surprising that a general role for CREB in regulation of VSMC gene expression has been largely overlooked. Because it can be regulated by so many agonists, CREB may contribute more significantly to specific gene expression responses triggered by a wider variety of VSMC agonists than is currently presumed. A great deal more work will be necessary to understand how under what conditions PKA-independent CREB kinases function in VSMC and how CREB participates in VSMC gene expression.

A recent study has shown that PKA activation is necessary for AT_1-R mRNA down regulation by the adenylyl cyclase activator forskolin [25]. That study has also demonstrated a partial role for PKA in angiotensin II, but not PDGF-BB, induced AT_1-R mRNA down regulation. Those results imply that signals from Gαq-coupled receptors and growth factor receptors can bypass PKA to regulate AT_1-R gene expression. This raises an intriguing notion that down regulation of AT_1-R gene expression could reflect convergence of information from various receptor classes upon some process more directly involved in supervising control of AT_1-R mRNA down regulation in VSMC. CREB was considered an attractive potential target for this control point in part because general transcriptional inhibitors consistently attenuate agonist-mediated vascular AT_1-R down regulation [10, 13, 15]. Such findings raise the possibility that induction of an as yet unknown gene or genes may be necessary to down regulate AT_1-R gene expression.

The present study was designed to test of this notion, by inhibiting CREB function at its site of action through exploiting the known ability of CREM gene products to bind to CREB enhancers [19]. The effect of eGFP-ICERIIγ expression on TRE-mediated transcription was also assessed because of previous observations which suggested that CREM gene products can also interfere with AP-1 mediated transcription [21] (see below). Although we show that expression of the eGFP-ICERIIγ protein is a very effective inhibitor of both CREB and AP-1 dependent transcription in VSMC, it has no effect on AT_1-R gene expression. Given that eGFP-ICERIIγ has no affect on AT_1-R gene expression, other classes of transcription factors might induce such a gene and explain this putative pathway for down regulation of AT_1-R gene expression. The possibility that the attenuation of agonist-mediated AT_1-R mRNA decay by general transcriptional inhibitors reflects an artifact associated the wide spectrum of side effects these drugs cannot be dismissed [16].

Due to alternative splicing, no less than four ICER proteins can be generated *in vivo* by after induction of an intragenic cAMP-inducible promoter within the CREM gene [20]. These and the other CREM-derived repressors utilize one or both of the DNA binding domains found within the CREM gene [17]. The ability of ICER and the larger CREM-α, -β and -γ gene products to repress cAMP-dependent transcriptional responses by occupying CREB enhancer elements has been recognized for some time [17, 19]. These CREM-derived repressors function as feedback controllers that participate in a growing number of physiological processes [43], although there is no evidence yet that they might be induced in VSMC. The sequences of the two basic leucine zipper DNA binding domains of CREM are more similar to those in the CREB proteins than to those in AP-1 protein family members [17]. On this basis, one would expect that the spectrum of DNA binding activity by ICER and other CREM products would be more similar to CREB than to AP-1. Since endogenous VSMC CREB-related proteins do not appear to bind efficiently to TRE elements, or to functionally activate TRE-reporter gene expression (as evidenced by a lack of effect of forskolin on TRE-mediated luciferase expression), it is notable that the eGFP-ICERIIγ fusion protein can inhibit transcription from both elements. The simplest explanation for this is that eGFP-ICERIIγ occupies both elements equally well in VSMC and thereby inhibits the ability of endogenous proteins to bind to these sites (Fig. 6). Alternately, eGFP-ICERIIγ may form weakly binding heterodimers with CREB and AP-1 proteins.

To our knowledge, the ability of ICER to bind to a consensus AP-1 enhancer element and inhibit transcription has not been specifically documented previously. The binding activities of the larger CREMα and CREMβ proteins when synthesized in bacteria, which each have the same DNA binding domain found in ICERIIγ but have more extensive upstream sequences, have been examined [21]. They can bind very effectively to CRE elements but only bind a consensus TRE sequence if the reactions are supplemented with mammalian cell extracts [21]. This suggests that CREM binding to TRE sequences involves some additional co-factor and the present findings raises the possibility that this putative co-factor can interact with CREM and ICER in their common domains. A more recent study has shown that ICER can inhibit nuclear factor of activated T cell (NFAT)-mediated transcription in lymphocytes, in part by blocking AP-1 binding upon composite NFAT enhancer elements [44]. On these enhancers, NFAT functions to facilitate AP-1 binding over DNA sequences that are only distantly related to consensus AP-1 sites, and these sequences tend to have only weak affinity for AP-1 in the absence of NFAT [45]. Together with the present observations, ICER and the other CREM repressors may have a broader spectrum of activity than thought previously and may interfere equally well with both CREB and AP-1 dependent transcriptional processes. If so, ICERIIγ may prove to be a broad spectrum antagonist of both CREB and AP-1 mediated transcription not only on consensus elements for these factors, but also on elements that diverge significantly from this consensus. ICERIIγ can thus have considerable utility for understanding how these protein classes contribute to the regulation of specific genes in VSMC. The highly efficient retroviral expression system described in this study should prove quite useful for accelerating the pace of discovery along these lines.

Acknowledgements

This work was supported by grants HL56107 and HL52810 from the National Heart Lung and Blood Institute. TJM is an Established Investigator of the American Heart Association.

References

1. Griendling KK, Lassegue B, Murphy TJ, Alexander RW: Angiotensin II receptor pharmacology. Adv Pharmacol 28: 269–306, 1994
2. Clauser E, Curnow KM, Davies E, S, Teutsch B, Vianello B, Monnot C, Corvol P: Angiotensin II receptors: Protein and gene structures, expression and potential pathological involvements. Eur J Endocrinol 134: 403–411, 1996
3. Matsusaka T, Ichikawa I: Biological functions of angiotensin and its receptors. Annu Rev Physiol 59: 395–412, 1997
4. Du Y, Guo DF, Inagami T, Speth RC, Wang DH: Regulation of ANG II-receptor subtype and its gene expression in adrenal gland. Am J Physiol 271: H440–H446, 1996
5. Wang DH, Du Y, Yao A, Hu Z: Regulation of type 1 angiotensin II receptor and its subtype gene expression in kidney by sodium loading and angiotensin II infusion. J Hypertens 14: 1409–1415, 1996
6. Wang DH, Du Y, Yao A: Regulation of the gene-encoding angiotensin II receptor in vascular tissue. Microcirculation 3: 237–239, 1996
7. Wang DH, Yao A, Zhao H, DiPette DJ: Distinct mechanisms of modulation of angiotensin II type I receptor gene expression in heart and aorta. Hypertension 29: 1104–1108, 1997
8. Regitz-Zagrosek V, Neuss M, Holzmeister J, Warnecke C, Fleck E: Molecular biology of angiotensin receptors and their role in human cardiovascular disease. J Mol Med 74: 233–251, 1996
9. Adams B, Obertone TS, Wang X, Murphy TJ: Relationship between internalization and messenger RNA decay in down-regulation of recombinant type-1 angiotensin II receptor (AT1) expression in smooth muscle cells. Mol Pharmacol 55: 1028–1036, 1999

10. Nickenig G, Murphy TJ: Down-regulation by growth factors of vascular smooth muscle angiotensin receptor gene expression. Mol Pharmacol 46: 653–659, 1994
11. Ullian MF, Raymond JR, Willingham MC, Paul RV: Regulation of vascular angiotensin II receptors by EGF. Am J Physiol 273: C1241–C1249, 1997
12. Lassegue B, Alexander RW, Nickenig G, Clark M, Murphy TJ, Griendling KK: Angiotensin II down-regulates the vascular smooth muscle AT1 receptor by transcriptional and post-transcriptional mechanisms. Evidence of homologous and heterologous regulation. Mol Pharmacol 48: 601–609, 1995
13. Nickenig G, Murphy TJ: Enhanced angiotensin receptor type 1 mRNA degradation and induction of polyribosomal mRNA binding proteins by angiotensin II in vascular smooth muscle cells. Mol Pharmacol 50: 743–751, 1996
14. Chen X, Nishimura J, Hasna J, Kobayashi S, Shikasho T, Kanaide H: Protein kinase C and protein kinase A regulate the expression of angiotensin II receptor mRNA in smooth muscle cells. Eur J Pharmacol 267: 175–183, 1994
15. Wang X, Nickenig G, Murphy TJ: The vascular smooth muscle type I angiotensin II receptor mRNA is destabilized by cAMP-elevating agents. Mol Pharmacol 52: 781–787, 1997
16. Xu K, Murphy TJ: Reconstitution of angiotensin receptor mRNA down regulation in vascular smooth muscle: Post-transcriptional control by protein kinase A but not mitogenic signalling directed by the 5′-untranslated region. J Biol Chem 275: 7604–7611, 2000
17. Sassone-Corsi P: Transcription factors responsive to cAMP. Ann Rev Cell Dev Biol 11: 355–377, 1995
18. Brindle PK, Montminy MR: The CREB family of transcription activators. Curr Opin Genet Dev 2: 199–204, 1992
19. Foulkes NS, Borrelli E, Sassone-Corsi P: CREM gene: Use of alternative DNA-binding domains generates multiple antagonists of cAMP-induced transcription. Cell 64: 739–749, 1991
20. Molina CA, Foulkes NS, Lalli E, Sassone-Corsi P: Inducibility and negative autoregulation of CREM: An alternative promoter directs the expression of ICER, an early response repressor. Cell 75: 875–886, 1993
21. Masquilier D, Sassone-Corsi P: Transcriptional cross-talk: Nuclear factors CREM and CREB bind to AP-1 sites and inhibit activation by Jun. J Biol Chem 267: 22460–22466, 1992
22. Pear WS, Nolan GP, Scott ML, Baltimore D: Production of high-titer helper-free retroviruses by transient transfection. Proc Natl Acad Sci USA 90: 8392–8396, 1993
23. Boss V, Abbott KL, Wang XF, Pavlath GK, Murphy TJ: The cyclosporin A-sensitive nuclear factor of activated T cells (NFAT) proteins are expressed in vascular smooth muscle cells. Differential localization of NFAT isoforms and induction of NFAT-mediated transcription by phospholipase c-coupled cell surface receptors. J Biol Chem 273: 19664–19671, 1998
24. Miller AD, Rosman GJ: Improved retroviral vectors for gene transfer and expression. Biotechniques 7: 980–990, 1989
25. Wang X, Murphy TJ: Inhibition of cyclic AMP-dependent kinase by expression of a protein kinase inhibitor/enhanced green fluorescent fusion protein attenuates angiotensin II-induced type 1 AT1 receptor mRNA down-regulation in vascular smooth muscle cells. Mol Pharmacol 54: 514–524, 1998
26. Takeuchi K, Alexander RW, Nakamura Y, Tsujino T, Murphy TJ: Molecular structure and transcriptional function of the vascular AT1a angiotensin receptor gene. Circ Res 73: 612–621, 1993
27. De Wet JR, Wood KV, DeLuca M, Helinski DR, Subramani S: Firefly luciferase gene: Structure and expression in mammalian cells. Mol Cell Biol 7: 725–737, 1987
28. Gonzalez GA, Montminy MR: Cyclic AMP stimulates somatostatin gene transcription by phosphorylation of CREB at serine 133. Cell 59: 675–680, 1989
29. Angel P, Imagawa M, Chiu R, Stein B, Imbra RJ, Rahmsdorf HJ, Jonat C, Herrlich P, Karin M: Phorbol ester-inducible genes contain a common cis element recognized by a TPA-modulated trans-acting factor. Cell 49: 729–739, 1987
30. Murphy TJ, Alexander RW, Griendling KK, Runge MS, Bernstein KE: Isolation of a cDNA encoding the vascular type-1 angiotensin II receptor. Nature 351: 233–236, 1991
31. Tsai JC, Wang H, Perrella MA, Yoshizumi M, Sibinga NE, Tan LC, Haber E, Chang TH, Schlegel R, Lee ME: Induction of cyclin A gene expression by homocysteine in vascular smooth muscle cells. J Clin Invest 97: 146–153, 1996
32. Dalton ML, Gadson PF Jr, Wrenn RW, Rosenquist TH: Homocysteine signal cascade: Production of phospholipids, activation of protein kinase C, and the induction of c-fos and c-myb in smooth muscle cells. Faseb J 11: 703–711, 1997
33. Brown JC, Rosenquist TH, Monaghan DT: ERK2 activation by homocysteine in vascular smooth muscle cells [in process citation]. Biochem Biophys Res Commun 251: 669–676, 1998
34. Montminy MR, Bilezikjian LM: Binding of a nuclear protein to the cyclic-AMP response element of the somatostatin gene. Nature 328: 175–178, 1987
35. Gonzalez GA, Yamamoto KK, Fischer WH, Karr D, Menzel P, Biggs WD, Vale WW, Montminy MR: A cluster of phosphorylation sites on the cyclic AMP-regulated nuclear factor CREB predicted by its sequence. Nature 337: 749–752, 1989
36. Lee KAW, Masson N: Transcriptional regulation by CREB and its relatives. Biochim Biophys Acta 1174: 221–233, 1993
37. Sheng M, Thompson MA, Greenberg ME: CREB: A Ca(2+)-regulated transcription factor phosphorylated by calmodulin-dependent kinases. Science 252: 1427–1430, 1991
38. Ginty DD, Bonni A, Greenberg ME: Nerve growth factor activates a Ras-dependent protein kinase that stimulates c-fos transcription via phosphorylation of CREB. Cell 77: 713–725, 1994
39. Xing J, Ginty DD, Greenberg ME: Coupling of the RAS-MAPK pathway to gene activation by RSK2, a growth factor-regulated CREB kinase. Science 273: 959–963, 1996
40. Pierrat B, Correia JS, Mary JL, Tomas-Zuber M, Lesslauer W: RSK-B, a novel ribosomal S6 kinase family member, is a CREB kinase under dominant control of p38alpha mitogen-activated protein kinase (p38alphaMAPK). J Biol Chem 273: 29661–29671, 1998
41. Xing J, Kornhauser JM, Xia Z, Thiele EA, Greenberg ME: Nerve growth factor activates extracellular signal-regulated kinase and p38 mitogen-activated protein kinase pathways to stimulate CREB serine 133 phosphorylation. Mol Cell Biol 18: 1946–1955, 1998
42. Deak M, Clifton AD, Lucocq LM, Alessi DR: Mitogen- and stress-activated protein kinase-1 (MSK1) is directly activated by MAPK and SAPK2/p38, and may mediate activation of CREB. Embo J 17: 4426–4441, 1998
43. Della Fazia MA, Servillo G, Sassone-Corsi P: Cyclic AMP signalling and cellular proliferation: Regulation of CREB and CREM. FEBS Lett 410: 22–24, 1997
44. Bodor J, Habener JF: Role of transcriptional repressor ICER in cyclic AMP-mediated attenuation of cytokine gene expression in human thymocytes. J Biol Chem 273: 9544–9551, 1998
45. Rao A, Luo C, Hogan PG: Transcription factors of the NFAT family: Regulation and function. Annu Rev Immunol 15: 707–747, 1997
46. Durand DB, Shaw J-P, Bush MR, Replogle RE, Belagaje R, Crabtree GR: Characterization of antigen receptor response elements within the interleukin-2 enhancer. Mol Cell Biol 8: 1715–1724, 1988

Molecular and Cellular Biochemistry **212**: 121–129, 2000.

Angiotensin II-induced changes in G-protein expression and resistance of renal microvessels in young genetically hypertensive rats

Subhash J. Vyas,[1] Christopher M. Blaschak,[1] Mala R. Chinoy[3] and Edwin K. Jackson[2]

Departments of [1]Medicine, [2]Pharmacology, Center for Clinical Pharmacology, University of Pittsburgh School of Medicine, Pittsburgh, PA; [3]Department of Surgery, The Pennsylvania State University College of Medicine, The Milton S. Hershey Medical Center, Hershey, PA, USA

Abstract

Altered regulation of cAMP may contribute to enhanced renal reactivity to angiotensin II (Ang II) in spontaneously hypertensive rats (SHR). Such a phenomenon may occur in renal preglomerular arterioles and may involve changes in expression of GTP-binding regulatory proteins. We have examined the effects of Ang II on steady state levels of $G_{\alpha i\text{-}1,2}$, $G_{\alpha i\text{-}3}$, $G_{\alpha s}$ and $G_{\alpha q}$ in preglomerular arterioles from young marginally hypertensive SHR and on mean arterial pressure (MAP), renal vascular resistance (RVR) and renal cAMP excretion (UcAMP.V). Young (5–6 week old) SHR and Wistar Kyoto (WKY) rats received Ang II (35 ng/kg/min, s.c.) or vehicle for 7 days via osmotic minipumps. Urine was collected over the last 24 h. On day seven, MAP and renal blood flow were measured in anesthetized rats and RVR was determined. Preglomerular arterioles were isolated by perfusing the kidneys with iron oxide and using a series of mechanical steps coupled with the use of a magnet to retain iron-laden vessels. Membranes were prepared and the expressions of $G_{\alpha i\text{-}1,2}$, $G_{\alpha i\text{-}3}$, $G_{\alpha s}$ and $G_{\alpha q}$ were evaluated by Western immunoblotting. Baseline MAP (124 ± 6 mmHg) was only marginally ($p > 0.05$) higher in SHR when compared with WKY rats (110 ± 4 mmHg). RBF (3.04 ± 0.16 mL/min) was significantly lower and RVR (41.10 ± 1.37 mmHg.min/mL) was significantly higher in SHR when compared to age-matched WKY rats (4.36 ± 0.30 mL/min and 25.79 ± 1.58 mmHg.min/mL, respectively). Ang II significantly increased MAP in SHR (17 mmHg) but not in WKY rats. These increases in MAP were accompanied by significant increases in RVR in SHR (48% over control) but not in WKY rats. Compared to WKY rats, preglomerular arterioles from SHR exhibited significantly higher basal expression of $G_{\alpha i\text{-}1,2}$ (11- fold), $G_{\alpha i\text{-}3}$ (13-fold) and $G_{\alpha s}$ (3-fold). Chronic infusion of Ang II, however, downregulated the expression of $G_{\alpha s}$ (by 53%; $p < 0.05$), $G_{\alpha i\text{-}1,2}$ (by 72%; $p < 0.05$) and $G_{\alpha i\text{-}3}$ (by 35%; $p > 0.05$) in SHR preglomerular arterioles but significantly upregulated the expression of these proteins in WKY by 3-, 8- and 15-fold, respectively. Basal levels of $G_{\alpha q}$ were not different in preglomerular arterioles from the two strains but were downregulated by Ang II in both WKY (74% of basal) and SHR (52% of control). Baseline UcAMP.V was significantly lower in SHR (31.22 ± 6.51 nmol/24 h) compared with WKY rats (65.33 ± 3.60 nmol/24 h). Chronic Ang II infusion significantly increased UcAMP.V in SHR as well as WKY rats. These data clearly demonstrate that expressions of G_i isoforms as well as G_s in renal microvessels are elevated during early stages of hypertension and suggest that the elevated levels of G_i proteins may be directly associated with a blunted adenylyl cyclase-cAMP cascade in the renal microvasculature. Furthermore, Ang II appears to directly downregulate the expression of G_s in young SHR but not in young WKY renal microvessels. Such diversity in its effect on G-protein expression may be important for enhanced renal sensitivity to Ang II in SHR. (Mol Cell Biochem **212**: 121–129, 2000)

Key words: angiotensin II, SHR, preglomerular arterioles, G-proteins, cAMP, hypertension

Abbreviations: SHR – spontaneously hypertensive rat; WKY – Wistar-Kyoto; cAMP – adenosine 3′, 5′-cyclic monophosphate; Ang II – angiotensin II; MAP – mean arterial pressure; RBF – renal blood flow; RVR – renal vascular resistance; GFR – glomerular filtration rate; UcAMP.V – urinary cAMP excretion

Address for offprints: S.J. Vyas, Center for Clinical Pharmacology, Scaife Hall, Room 623, University of Pittsburgh Medical Center, 200 Lothrop Street, Pittsburgh, PA 15213-2582, USA

Introduction

Several studies document an enhanced renal responsiveness to angiotensin II (Ang II) in adult spontaneously hypertensive rats (SHR) [1–3] as well as young, marginally hypertensive SHR [4, 5]. A similar renal hyper-responsiveness to Ang II exists in human subjects with essential hypertension [6] and in normotensive human subjects with positive family histories of hypertension [7]. Such a genetically determined and kidney selective hyper-responsiveness to Ang II may substantially contribute to the development and/or maintenance of hypertension in both SHR and humans.

The increased renovascular sensitivity to Ang II may be due to an alteration in signaling pathways that regulate cAMP in the afferent microvessels [5]. In the isolated perfused kidney, the ability of Ang II to inhibit stimulus-induced release of cAMP is enhanced in the SHR [8]. Moreover, pertussis toxin normalizes the renovascular sensitivity to Ang II in the SHR [9], and in cultured preglomerular arteriolar smooth muscle cells, pertussis toxin increases total cAMP release in SHR cells but not in WKY cells [10]. These studies suggest an altered regulation of G-proteins in SHR preglomerular microvessels and, thus, of cAMP. These alterations in regulation of cAMP may significantly contribute to elevated renal vascular resistance in genetic hypertension.

Although several studies demonstrate altered G-protein mRNA [11] and expression of G-proteins in heart and aortic tissue in young [12] as well as adult [13, 14] SHR, the profile for expression of G-proteins in SHR renal resistance microvasculature during the early stages of hypertension is unknown. Furthermore, how Ang II influences the profile of G-proteins in this segment of vasculature from young SHR is not known. Consequently, we examined the effects of Ang II on the expressions of G_i and G_s in preglomerular arterioles isolated from young marginally hypertensive SHR and correlated these alterations to changes in renal vascular resistance, mean arterial pressure and renal cyclic AMP excretion. Because adenylyl cyclase is indirectly regulated by ligands that activate phospholipase C [15], such as Ang II, we also evaluated the effects of this peptide on the expressions of G_q in preglomerular microvessels of young SHR.

Materials and methods

Male SHR and WKY rats (5–6 weeks of age) were ordered from Taconic Farms (Germantown, NY, USA) and were housed at the University of Pittsburgh Central Animal Facility with controlled temperature, relative humidity and light cycle (22°C, 55% and 7 a.m. to 7 p.m., respectively). Animals were treated in accordance with institutional guidelines. The rats were maintained on Prolab Isopro RMH 3000 formula chow (PMI Nutrition International, Richmond, IN, USA). The studies were conducted with prior approval from the Institutional Animal Care and Use Committee.

On the day of experiment (day 1), SHR and WKY rats were anesthetized with halothane and an osmotic minipump (model 2002, 14-day pump, mean delivery rate 0.5 μL/h; Alza, Palo Alto, CA, USA) was implanted subcutaneously. These rats were randomized to receive osmotic minipumps delivering either angiotensin II (Ang II, 35 ng/kg/min, s.c.) or its vehicle (control). Rats were housed individually in metabolic cages from days 5–7 with free access to drinking water and rat chow. During the last 24 h, urine was collected into receptacles containing 50 μL of an antibiotic/antimycotic solution (10,000 units penicillin, 10 mg streptomycin and 25 μg amphotericin B per mL, Sigma Chemical Co., St. Louis, MO, USA). Urine volume was recorded and the urine samples were frozen at –70°C until assayed for cyclic AMP, creatinine and electrolytes.

Following the 24-h urine collection period and 7 days after implantation of minipumps, rats were anesthetized with Inactin® (5-sec-butyl-5-ethyl-2-thiobarbituric acid; 100 mg/kg, i.p., RBI Biochemicals International, Natick, MA, USA). A PE-50 catheter was quickly inserted into the left carotid artery and a blood sample (1 mL) was immediately collected from this catheter for analysis of plasma electrolytes and creatinine. A PE-240 catheter was inserted into trachea for facilitating spontaneous respiration, and these rats were prepared for measurement of blood pressure, heart rate and renal blood flow. Briefly, rats were placed on pre-heated Deltaphase Isothermal pad (Braintree Scientific Inc., Braintree, MA, USA), and body temperature was maintained at 37 ± 0.5°C by positioning a heat lamp above the animal at an appropriate distance and was continuously monitored with a digital rectal microprobe thermometer (model BAT-12, Physitemp Instruments Inc., Clifton, NJ, USA). The PE-50 catheter in the left carotid artery was connected to a digital blood pressure analyzer (Micro-Med Inc., Louisville, KY, USA) for continuous measurement of mean arterial blood pressure (MAP) and heart rate (HR). The abdominal cavity was exposed through a midline incision, and a transit-time blood flow probe (model #1RB493, Transonic Systems Inc., Ithaca, NY, USA) was placed around the left renal artery and was connected to a 2-channel, small animal, digital, transit-time blood flow meter with a digital display (model T206; Transonic Systems Inc.). The animals were allowed to stabilize for 60 min. After stabilization, baseline MAP, HR and renal blood flow (RBF) were recorded over a 30 min period. MAP and HR were time-averaged each 10-min period, and RBFs were recorded every 5 min and the values were averaged to provide means for the entire period. Renal vascular resistance (RVR) was calculated by dividing MAP by RBF. Urine (24 h) and plasma creatinine and electrolytes were measured using a Creatinine Analyzer (Beckman, Brea, CA,

USA) and a flame photometer (model IL943, Instrumentation Laboratories Inc., Lexington, MA, USA), respectively. Glomerular filtration rate was estimated as the renal clearance of creatinine. Urine cAMP concentrations were measured using a method described later in this section.

Isolation of renal preglomerular arterioles

After recording baseline MAP, HR and RBF, rats were prepared for isolation of preglomerular arterioles by slight modification of a technique recently used by us [10]. Briefly, a PE-190 catheter was placed in abdominal aorta, and 10 mL of saline were infused into the abdominal aorta rostral to the left renal artery after ligating the mesenteric artery and abdominal aorta proximal to the heart to flush blood out of the renal circulation. Thereafter, a 5% w/v suspension of iron oxide particles (ferroso-ferric oxide, Aldrich Chemical Co., Milwaukee, WI, USA) in saline was infused into the kidneys. The kidneys, now loaded with the iron oxide particles, were removed and placed in ice-cold phosphate buffered saline (DPBS, pH 7.4). All further procedures were carried out at 4°C. Kidneys from 3–4 rats from each treatment group were pooled to provide enough arteriolar tissue for analysis of G-proteins. Kidneys were decapsulated, the cortex separated carefully and chopped on ice-cold DPBS to very fine fragments. The fine dispersion of this cortical tissue was poured into a 20 mL syringe and passed twice sequentially through 16, 18, 20, 21 and 22 gauge blunt hypodermic needles and then ten times through a 23 gauge needle. This procedure shears off glomeruli and removes blood vessels from arcuate and interlobular segments. The iron laden microvessels were separated by holding a magnet against the side of the container while decanting the non-vascular tissue. The preparation was washed 4–5 times with DPBS, repeating the separation procedure until the supernatant was free of any tissue particles. A clear suspension containing the preglomerular arterioles was transferred onto a 149 μm sieve (Small Parts Inc., Miami, FL, USA) and washed with cold DPBS. The arterioles retained on the mesh were resuspended in PBS and purity of the arterioles as well as absence of glomeruli in the preparation was established by light microscopy. Only microvessels free of glomeruli and tubules and less than 100 μm in diameter were used for the experiments. Arterioles were centrifuged at 14,000 rpm for 1 min to remove DPBS, and the arteriolar pellet was frozen at –80°C.

Preparation of plasma membranes from preglomerular arterioles

Arterioles were thawed, washed once with PBS and homogenized on ice in 200 μL of tris-EDTA (TE, 10 mM, pH 7.4) buffer containing protease inhibitors (antipain, 2 μg/mL; aprotinin, 1 μg/mL; leupeptin, 2 μg/mL). The homogenate was then passed through a 25 gauge needle and centrifuged at 2,000 rpm at 4°C on Beckman J2-MI centrifuge for 15 min. The supernatants, containing the membrane fraction, were transferred to fresh microfuge tubes and centrifuged at 17,000 rpm at 4°C for 15 min. The pellets containing plasma membranes were suspended in 200 μL of TE buffer, and the suspensions were centrifuged at 17,000 rpm at 4°C for 15 min. The supernatants were discarded and the pellets, containing the membrane fraction, were resuspended in 100 μL of TE buffer for approximately 30 min at 4°C. These suspensions were then heated at 85°C for 2 min on a heat block. The protein content in the membranes were determined using the Bio-Rad Dye (Bio-Rad Labs, Hercules, CA, USA), Bradford Assay [16] as described by Blewett *et al.* [17]. The membranes were then suspended in a lysis buffer to provide a final concentration of 1 μg protein/mL.

Western blot analyses

Twenty μg of the membrane proteins from renal preglomerular arterioles from control as well as Ang II-treated SHR and WKY rats were resolved on 10% SDS-PAGE [18]. The proteins were then electroblotted onto a PVDF membrane (Millipore, Bedford, MA, USA). Immunoblotting was performed using the methods recently described by us [10]. Briefly, the membranes were blocked with 5% milk for 1 h and incubated for 3 h at room temperature with $G_{\alpha i\text{-}1,2}$, $G_{\alpha i\text{-}3}$, $G_{\alpha s}$ and $G_{\alpha q}$ antibodies (Calbiochem, San Diego, CA, USA) at a dilution of 1:1000 each in 2% milk in tris-buffered saline containing Triton X 100 (TBST). The membranes were then washed in TBST and incubated with horseradish peroxidase-conjugated donkey anti-rabbit IgG secondary antibody (Amersham, Arlington Heights, IL, USA) at 1:2500 dilution. The antigen-antibody complexes were detected using an enhanced chemiluminescence (ECL) kit for Western blot detection (Amersham). Densitometric analyses were done on a 100A Molecular Dynamics Densitometer (Molecular Dynamics, Sunnyvale, CA, USA) using the Protein Data Basis Incorporation software.

Measurement of cAMP

cAMP in urine was measured by a high performance liquid chromatographic-fluorometric assay by slightly modifying a method described by us recently [10]. Fifty μL of thawed urine sample were added to 20 μL of 10 μM adenine 9-β-D-arabinoside (internal standard) and 1 mL of 0.5 mol/L ammonium sulphate buffer (pH 9.3) in a 1.5 mL polypropylene microfuge tube and vortexed. Cyclic AMP in each of these samples was extracted by loading 1 mL volume of this mixture onto a C_{18} reverse-phase Sep-Pak cartridge (Part #WAT-051910, Waters Corporation, Milford, MA, USA). Three mL of 5 mM ammonium sulphate buffer (pH 9.3) was then added to the column. cAMP retained by the column was then eluted with 2 mL of 10 mM phosphoric acid containing 10% methanol. The first 0.5 mL of the eluate was discarded, and the remaining was collected and used for analysis. Prior to their use, the Sep-Pak cartridges were activated by eluting 2 mL methanol and then 2 mL of deionized water through the columns. To the 1.5 mL of the eluate, 40 μL of 0.5 M acetate buffer and 40 μL of 50% chloroacetaldehyde (aqueous solution) were added. The tubes were capped, vortexed and incubated at 80°C for 1 h for complete derivatization of cAMP. Eighty μL of the derivatized sample were injected into an ISCO (Lincoln, NE, USA) HPLC system (pump model 2350, gradient programmer model 2360, 4.6 × 250 mm C_{18} reverse-phase column with 5 μm particle size; ChemResearch Data Management System). The fluorometric detection was achieved at an excitation wavelength of 275 nm and an emission wavelength of 420 nm using a Waters 470 fluorescence detector. The mobile phase consisted of 10 mmol/L citrate-buffer with 4% acetonitrile (pH 4.0) and was run isocratically at 1 mL/min. A standard curve for cAMP was constructed using the ratio of areas of cAMP and the internal standard.

Statistical analysis

Data are presented as mean ± S.E.M. Statistical analyses were performed on a personal computer with Number Cruncher Statistical System (NCSS) software package (Kaysville, UT, USA). Data were analyzed using analysis of variance followed by Duncan's multiple range test for multiple comparisons. $p < 0.05$ denotes statistical significance.

Results

The influence of a chronic (1 week) infusion of a moderately low-dose of Ang II (35 ng/kg/min, s.c.) on MAP, RVR, UcAMP.V and on steady-state levels of GTP-binding proteins including $G_{\alpha i}$, $G_{\alpha s}$ and $G_{\alpha q}$ in preglomerular arterioles from young prehypertensive SHR was examined in the present study. Baseline (control) MAP was only marginally ($p > 0.05$) higher in SHR when compared with WKY rats (Fig. 1). Ang II (35 ng/kg/min, s.c., 7 days) significantly increased MAP (17 mmHg) in SHR but not in the WKY rats (Fig. 1). Baseline RBF was significantly lower in the SHR but Ang II did not significantly alter RBF in either strain (Table 1). However, post-Ang II RBF in SHR was still significantly lower as compared with RBF in WKY rats after similar treatment. Correspondingly, RVR in the young SHR was observed to

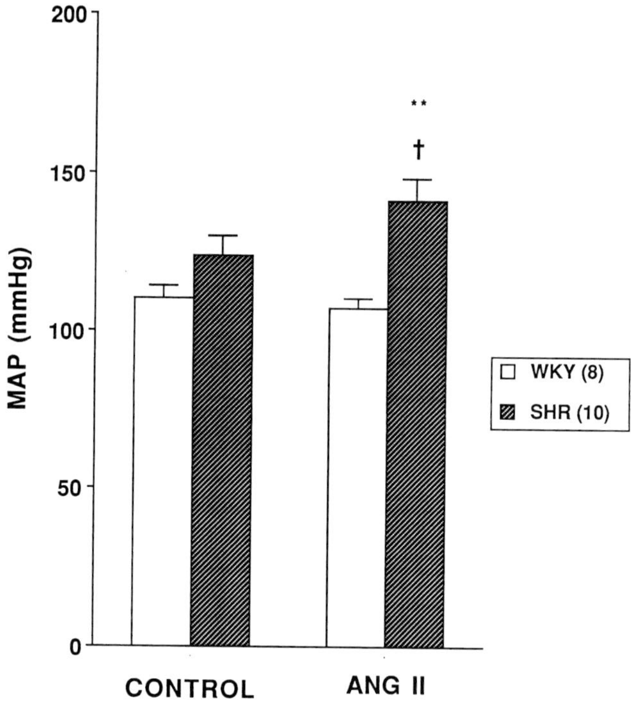

Fig. 1. Effects of angiotensin II (Ang II; 35 ng/kg/min, s.c.; 7 days) on mean arterial pressure (MAP) in young (6–7 weeks of age) SHR and WKY rats. Data are presented as mean ± S.E.M. Numbers in parentheses represent number of rats per group. Data were analyzed using analysis of variance followed by Duncan's *post-hoc* test. †$p < 0.05$ vs. WKY (within treatment); **$p < 0.05$ vs. strain control.

Table 1. Effects of angiotensin II (Ang II; 35 ng/kg/min, s.c.; 7 days) on renal blood flow (RBF), glomerular filtration rate (GFR), urine volume (UV), urinary sodium and potassium excretions (UNaV, UKV) and fractional excretions of sodium and potassium (FENa, FEK) in young prehypertensive SHR and WKY rats

	Control		Ang II	
	WKY	SHR	WKY	SHR
RBF (mL/min)	4.36 ± 0.30	3.04 ± 0.16*	4.93 ± 0.37	2.75 ± 0.33*
GFR (mL/min)	1.14 ± 0.17	0.45 ± 0.07*	2.81 ± 0.27**	0.67 ± 0.08*
UV (mL/24 h)	13.07 ± 1.25	7.00 ± 1.41*	19.10 ± 2.30**	8.67 ± 0.84*
UNaV (mmol/24 h)	1.58 ± 0.03	0.79 ± 0.04*	1.57 ± 0.22	0.88 ± 0.11*
UKV (mmol/24 h)	4.19 ± 0.11	2.19 ± 0.09*	4.25 ± 0.58	2.40 ± 0.30*
FENa (%)	1.10 ± 0.16	1.37 ± 0.20	0.97 ± 0.40	1.01 ± 0.16
FEK (%)	2.90 ± 0.40	3.73 ± 0.59	2.45 ± 1.01	2.71 ± 0.42

Data are presented as mean ± S.E.M. (n = 6–8). Data were analyzed by analysis of variance followed by Duncan's *post-hoc* test. *$p < 0.05$ vs. WKY from same treatment; **$p < 0.05$ vs. strain control.

be significantly higher when compared with WKY rats (Fig. 2). As shown in Figs 1 and 2, Ang II-induced increases in MAP in SHR were accompanied by significant increases in RVR (48% increase over control). Ang II did not affect either MAP (Fig. 1) or RVR (Fig. 2) in young WKY rats.

The effects of Ang II on glomerular filtration rate and urinary volume, sodium and potassium excretions as well as fractional excretions of sodium and potassium are depicted in Table 1. Resting GFR in SHR was significantly lower and amounted to approximately 40% of that in WKY rats. Ang II infusion caused large and significant increases in GFR (146% over control) in WKY rats but only produced a relatively smaller and nonsignificant increase (49% over control) in GFR in SHR (Table 1). Basal urine volume in SHR was also significantly lower (54% of WKY rats). Ang II infusion did not significantly change urine volume in SHR but actually increased (46% over control) the same in WKY rats. Baseline urinary excretion of sodium and potassium were significantly lower in SHR and these parameters were not significantly affected by chronic infusion of Ang II in either strain. Fractional excretion of sodium and potassium were not different between SHR and WKY rats, nor were these variables altered by Ang II treatment.

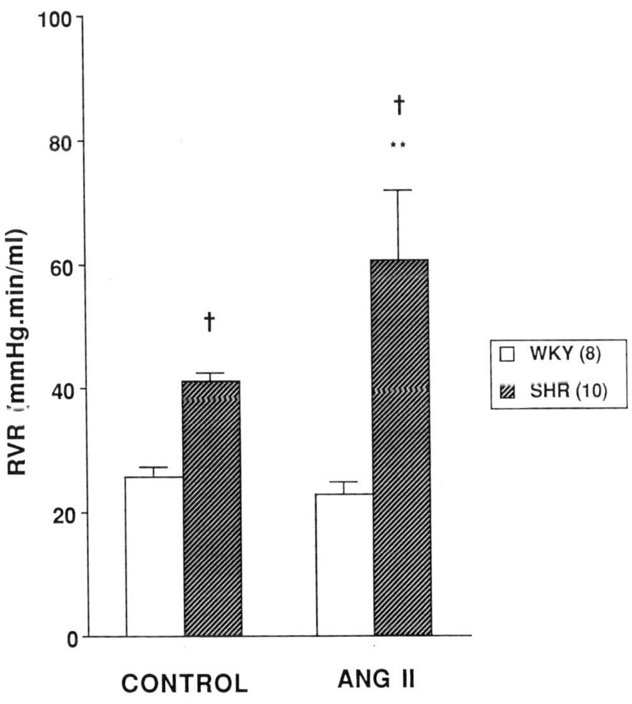

Fig. 2. Effects of angiotensin II (Ang II; 35 ng/kg/min, s.c.; 7 days) on renal vascular resistance (RVR) in young (6–7 weeks of age) SHR and WKY rats. Data are presented as mean ± S.E.M. Numbers in parentheses represent number of rats per group. Data were analyzed using analysis of variance followed by Duncan's *post-hoc* test. †$p < 0.05$ vs. WKY (within treatment); **$p < 0.05$ vs. strain control.

The effects of Ang II on steady state expressions of $G_{\alpha i}$, $G_{\alpha s}$ and $G_{\alpha q}$ are shown in Figs 3–5. As shown, compared with WKY rats, renal preglomerular arterioles from SHR exhibited much higher basal expression of $G_{\alpha i\text{-}1,2}$ (11-fold), $G_{\alpha i\text{-}3}$ (13-fold) and $G_{\alpha s}$ (3-fold). Chronic infusion of Ang II, however, significantly downregulated the expression of $G_{\alpha s}$ (by 53%), $G_{\alpha i\text{-}1,2}$ (by 72%) and tended to decrease $G_{\alpha i\text{-}3}$ (by 35%; $p > 0.05$) in preglomerular arterioles from young SHR while upregulating the expression of all of these proteins in WKY rats by 3-, 8- and 15-fold, respectively (Figs 4 and 5). Basal levels of $G_{\alpha q}$ were not different in preglomerular arterioles from the two strains and were downregulated by Ang II in both SHR (52% of control) and WKY rats (74% of basal) (Figs 3 and 5). Baseline UcAMP.V was significantly lower in SHR compared to WKY rats (Fig. 6). Chronic Ang II infusion significantly increased UcAMP.V in SHR as well as WKY rats (Fig. 6), but more so in WKY rats.

Discussion

In the present study we evaluate the hypothesis that altered G-protein-mediated regulation of adenylyl cyclase may contribute to elevated renal vascular resistance in genetically hypertensive rats and, thus, to the development of hypertension. The results from this study clearly demonstrate that, even at an early stage of development of hypertension, there is an enhanced expression of the inhibitory GTP-binding regulatory proteins in the renal microvessels of spontaneously hypertensive rats. Furthermore, elevated expression of G_i proteins in young genetically hypertensive rats is accompanied by significant increases in the renal vascular resistance and decreases in renal cAMP production. These results support the hypothesis that increased expression of inhibitory G-proteins at an early stage may participate in the development of hypertension in SHR.

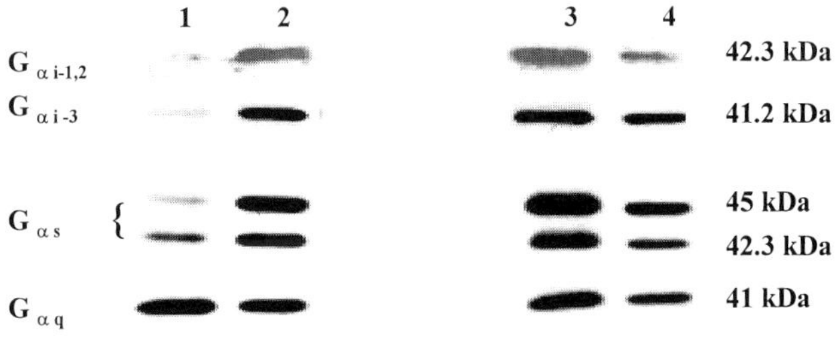

Fig. 3. Western immunoblot for $G_{\alpha i\text{-}1,\text{-}2}$, $G_{\alpha i\text{-}3}$, $G_{\alpha s}$ and $G_{\alpha q}$ in preglomerular arterioles from young (6–7 weeks of age) SHR and age-matched WKY rats. Kidneys from 3–4 rats from each treatment group were pooled and preglomerular arterioles were isolated as described in the 'Materials and methods' section. Twenty microgram of membrane protein from each group were resolved using 10% SDS-PAGE. Lanes (left to right) are as follows: 1 – WKY control; 2 – WKY Ang II; 3 – SHR control; 4 – SHR Ang II.

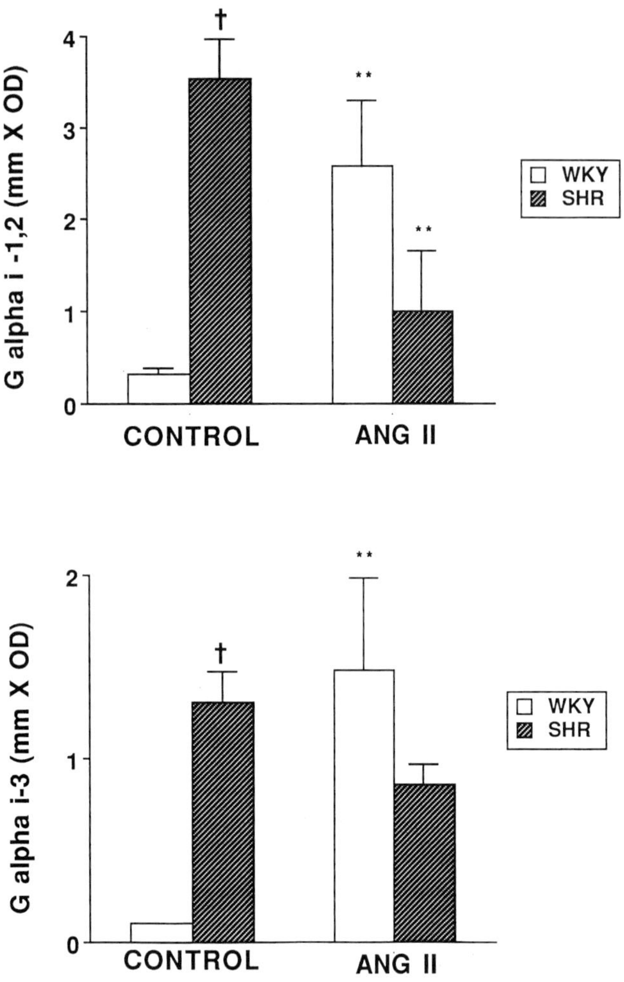

Fig. 4. Effects of angiotensin II (Ang II; 35 ng/kg/min, s.c.; 7 days) on steady-state levels of $G_{\alpha i\text{-}1,\text{-}2}$ (top) and $G_{\alpha i\text{-}3}$ (bottom) in renal preglomerular arterioles from young (6–7 weeks of age) SHR and WKY rats. Kidneys from 3–4 rats from each treatment group were pooled and preglomerular arterioles were isolated as described in the 'Materials and methods' section. Data are presented as mean ± S.E.M. (2–3 separate determinations) of units (mm × OD) obtained from densitometric analysis of immunoblots. Data were analyzed using analysis of variance followed by Duncan's *post-hoc* test. $^{\dagger}p < 0.05$ vs. WKY (within treatment); $^{**}p < 0.05$ vs. strain control.

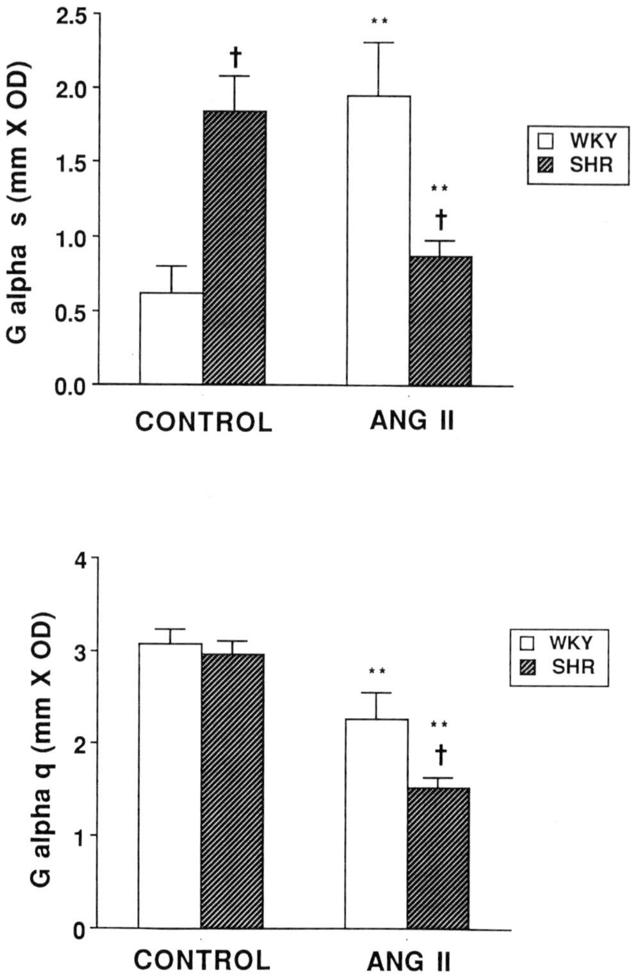

Fig. 5. Effects of angiotensin II (Ang II; 35 ng/kg/min, s.c.; 7 days) on steady-state levels of $G_{\alpha s}$ (top) and $G_{\alpha q}$ (bottom) preglomerular arterioles from young (6–7 weeks of age) SHR and WKY rats. Kidneys from 3–4 rats from each treatment group were pooled and preglomerular arterioles were isolated as described in the 'Materials and methods' section. Data are presented as mean ± S.E.M. (2–3 separate determinations) of units (mm × OD) obtained from densitometric analysis of immunoblots. Data were analyzed using analysis of variance followed by Duncan's *post-hoc* test. $^{\dagger}p < 0.05$ vs. WKY (within treatment); $^{**}p < 0.05$ vs. strain control.

Young SHR, i.e. 5–6 weeks of age, are generally identified with the onset of the developmental phase of hypertension. The difference in blood pressure between SHR and their normotensive controls (WKY rats) in the present study is modest and comparable to earlier reports from our laboratory [5] as well as from others [4, 19]. By using SHR that exhibited only a modest, statistically non-significant increase in blood pressure, we circumvent the confounding effects of established hypertension. Indeed, it is important to note that even at this age, RBF is significantly lower and RVR is significantly higher in SHR. These observations are entirely consistent with earlier findings [20–22] and the concept that the kidney plays an important role in the development of hypertension in SHR. In the present study, although Ang II does increase MAP in SHR, the Ang II-induced increase in MAP in SHR is modest and it is unlikely that such a narrow pressure spread between WKY rats and SHR accounts for the differences in G-protein expression in response to Ang II. That a low dose infusion of Ang II increases MAP as well as RVR in young SHR, but not in WKY rats, is consistent with, and further supports, the idea of enhanced renovascular sensitivity to Ang II in young SHR [4, 5].

We observe that renal preglomerular microvessels from young SHR exhibit significantly higher levels of not only $G_{\alpha i}$

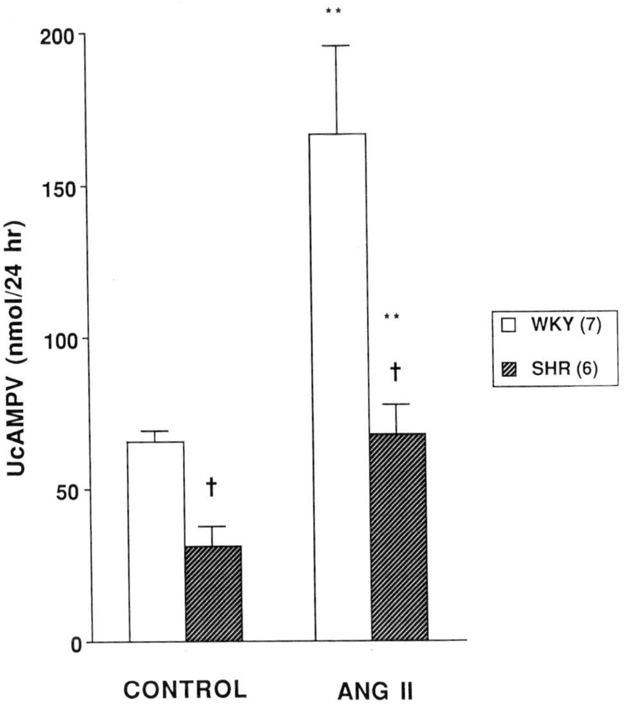

Fig. 6. Effects of angiotensin II (Ang II; 35 ng/kg/min, s.c.; 7 days) on mean urinary cyclic AMP excretion (UcAMPV) in young (6–7 weeks of age) SHR and WKY rats. Data are presented as mean ± S.E.M. Numbers in parentheses represent number of rats per group. Data were analyzed using analysis of variance followed by Duncan's *post-hoc* test. †$p < 0.05$ vs. WKY (within treatment); **$p < 0.05$ vs. strain control.

isoforms but also $G_{\alpha s}$. The ratios of levels of these G proteins in SHR renal arterioles relative to those in WKY arterioles are 11, 13 and 3 for $G_{\alpha i\text{-}1,2}$, $G_{\alpha i\text{-}3}$ and $G_{\alpha s}$, respectively. Guanine nucleotide regulatory proteins (G-proteins) play a significant role in modulation of vascular tone and compliance via several signaling mechanisms including the adenylyl cyclase-cAMP cascade (for review, see [23]). A defective regulation of the adenylyl cyclase-cAMP system in heart and aorta is implicated in the development of hypertension in SHR as well as other models of hypertension including DOCA-salt hypertension, and altered expressions of G_i and G_s exist in these models of hypertension [12, 24]. In cultured preglomerular arteriolar cells, basal cAMP formation in renal vascular smooth muscle cells obtained from SHR is significantly lower when compared with cells from WKY rats [10]. This deficit in cAMP in the SHR cells is, however, completely eliminated when the cells are treated with pertussis toxin, strongly implying an enhanced G_i-mediated modulation of adenylyl cyclase in the SHR renal microvasculature [10]. The findings of the present study that preglomerular arterioles from young SHR exhibit increased expression of inhibitory G-protein isoforms, coupled with the facts that these rats show decreased renal cAMP levels and higher renal vascular resistance, lend full support to the hypothesis that increased G_i-mediated modulation of adenylyl cyclase is associated with early stages of hypertension. The fact that expression of $G_{\alpha i\text{-}1,2}$ and $G_{\alpha i\text{-}3}$ in preglomerular arterioles from adult, fully hypertensive SHR is not higher than in WKY rats [25] emphasizes the possibility that higher G_i-protein expression is involved in the development of hypertension in SHR and, once hypertension is established the G_i-protein levels in SHR arterioles are normalized. The observation that G_i-protein expression is not different in adult SHR is supported by our earlier findings that G_i mRNA levels in preglomerular arterioles are also not different between adult SHR and WKY rats [26].

Importantly, in addition to $G_{\alpha i}$, $G_{\alpha s}$ levels are also higher (3-fold) in SHR yet renal cAMP production is lower in SHR compared with WKY. Although several groups report increases in G_i in heart and aorta in SHR, G_s is generally either decreased or not altered. The higher expression of G_s in SHR arterioles in the present study may be a characteristic of the resistance vasculature in general and/or may be a manifestation of the chronic influence of high G_i levels to upregulate G_s levels. Because adenylyl cyclase is under the dual control of G_s (stimulation) and G_i (inhibition) proteins [23, 27, 28], it is quite likely that the predominance of G_i isoforms in the SHR mask the effects of relatively moderate elevations in G_s on adenylyl cyclase resulting in lower renal cAMP levels in SHR. Basal expression of G_q is similar in preglomerular microvessels from young SHR and WKY rats. Thus, under resting conditions, it appears that the phospholipase C-inositol trisphosphate signaling pathway may modulate vascular tone in a similar fashion in both the young SHR and WKY rats, although 'downstream' signaling components of this pathway may be altered.

Chronic treatment with Ang II leads to diminished expression of G_s in the renal arterioles from young SHR. Ang II-induced decreases in G_s levels in SHR are accompanied by significant increases in RVR and MAP. This effect of Ang II on renal vascular resistance in SHR is consistent with our earlier observations [5]. However, the effects of Ang II on renal cAMP levels in the present study do not agree with our earlier observation that Ang II actually decreased renal cAMP excretion in SHR [5]. These differences are most likely due to the differences in dose and length of duration of Ang II treatment in these studies. Interestingly, Ang II, in addition to downregulating G_s levels, also downregulates the expressions of G_i isoforms as well as G_q in SHR preglomerular microvessels suggesting that such regulation of these G-proteins could be an adaptive phenomenon whereby vascular smooth muscle cells tend to maintain vascular tone. Contrary to decreases in G_s in SHR, in the WKY preglomerular arterioles, the expression of this protein is upregulated in response to chronic treatment with Ang II. Increases in arteriolar G_s, coupled with large increases in renal cAMP excretion in response to Ang II in WKY rats indicate that a much more efficient cellular adaptive mechanism exists in the normotensive rats which is able to compensate for the increases in expression of inhibitory G-proteins.

An important finding of the present study is that glomerular filtration is significantly diminished in the SHR when compared to their normotensive controls. This observation is consistent with available literature [21]. What is interesting here is that Ang II causes large (146% over control) and significant increases in glomerular filtration in the WKY rats but only produces a relatively small and statistically non-significant increase in GFR in SHR. Normally, Ang II produces greater percentage reductions in renal plasma flow than GFR. This response is believed to be due to a more dominant effect of Ang II on efferent arterioles as compared to afferent arterioles. The observation that Ang II produces large increases in GFR in WKY but not so in SHR is consistent with and corroborates the idea of enhanced sensitivity of afferent, not efferent, arterioles to Ang II in SHR [5]. Urine volume and urinary excretions of sodium and potassium are also suppressed in the SHR. These observations imply that not only renal hemodynamics but renal tubular function is also significantly altered in early hypertension. Fractional excretions of these electrolytes are neither different nor are they affected by Ang II. Furthermore, Ang II does not effect any changes in these variables in either strain. Taken together, these observations are indicative of the fact that at an early stage of hypertension, vascular and glomerular changes are more pronounced in SHR than the tubular changes.

In conclusion, expression of inhibitory G-proteins (G_i) is upregulated in renal preglomerular microvessels in SHR even at an early stage of development of hypertension. Such an increase in regulatory proteins is associated with elevated renal vascular resistance in these rats. Furthermore, Ang II-induced depression of stimulatory protein (G_s) in renal microvasculature in young SHR is associated with further increments in renal vascular resistance. These alterations in regulatory G-proteins may play an integral role in enhanced renal sensitivity to Ang II as well as in the development of hypertension in SHR.

Acknowledgements

The authors wish to thank Xiaoli Chi, M.D. for her expert technical assistance with Western Blot analyses. This work was supported by a grant from the Competitive Medical Research Fund, University of Pittsburgh, Pittsburgh, PA, USA.

References

1. Li P, Jackson EK: Enhanced slow-pressor response to angiotensin II in spontaneously hypertensive rats. J Pharmacol Exp Ther 251: 909–921, 1989
2. Chatziantoniou C, Arendshorst WJ: Angiotensin and thromboxane in genetically hypertensive rats: Renal blood flow and receptor studies. Am J Physiol 261: F238–F247, 1991
3. Kost CK Jr, Jackson EK: Enhanced renal angiotensin II subtype 1 receptor responses in the spontaneously hypertensive rat. Hypertension 21: 420–431, 1993
4. Chatziantoniou C, Daniels FH, Arendshorst WJ: Exaggerated renal vascular reactivity to angiotensin and thromboxane in young genetically hypertensive rats. Am J Physiol 259: F372–F382, 1990
5. Vyas SJ, Jackson EK: Angiotensin II: Enhanced renal responsiveness in young genetically hypertensive rats. J Pharmacol Exp Ther 273:768–777, 1995
6. Ljungman S, Aurell M, Hartford M, Wikstrand J, Berglund G: Effects of subpressor doses of angiotensin II on renal hemodynamics in relation to blood pressure. Hypertension 5: 368–374, 1983
7. Widgren BR, Herlitz H, Aurell M, Berglund G, Wikstrand J, Andersson OK: Increased systemic and renal vascular sensitivity to angiotensin II in normotensive men with positive family histories of hypertension. Am J Hypertens 5: 167–174, 1992
8. Vyas SJ, Mi Z, Jackson EK: The inhibitory effect of angiotensin II on stimulus-induced release of cAMP is augmented in the genetically hypertensive rat kidney. J Pharmacol Exp Ther 279: 114–119, 1996
9. Jackson EK: Pertussis toxin normalizes enhanced renovascular responses to angiotensin II in spontaneously hypertensive rats. Life Sci 54: 445–450, 1994
10. Vyas SJ, Mokkapatti R, Dubey RK, Chinoy MR, Jackson EK: Guanine nucleotide -binding inhibitory protein-mediated inhibition of adenylyl cyclase is enhanced in spontaneously hypertensive rat preglomerular arteriolar smooth muscle cells. J Pharmacol Exp Ther 285: 828–834, 1998
11. Thibault C, Anand-Srivastava MB: Altered expression of G-protein mRNA in spontaneously hypertensive rats. FEBS Lett 313: 160–164, 1992
12. Anand-Srivastava MB: G-proteins and adenylyl cyclase signalling in hypertension. Mol Cell Biochem 157: 163–170, 1996
13. Anand-Srivastava MB: Enhanced expression of inhibitory guanine nucleotide regulatory protein in spontaneously hypertensive rats. Relationship to adenylate cyclase inhibition. Biochem J 288: 79–85, 1992
14. Bohm M, Gierschik P, Knorr A, Larisch K, Weismann K, Erdmann E: Role of altered G-protein expression in the regulation of myocardial adenylyl cyclase activity and force of contraction in spontaneous hypertensive cardiomyopathy in rats. J Hypertens 10: 1115–1128, 1992
15. Lustig KD, Conklin BR, Herzmark P, Taussig R, Bourne HR: Type II adenylyl cyclase integrates coincident signals from Gs, Gi, and Gq. J Biol Chem 268: 13900–13905, 1993
16. Bradford MM: A rapid and sensitive method for the quantitation of microgram quantities of protein utilizing the principle of protein-dye binding. Anal Biochem 72: 248–254, 1976
17. Blewett CJ, Zgleszewski SE, Chinoy MR, Krummel TM, Cilley RE: Bronchial ligation enhances murine fetal lung development in whole-organ culture. J Pediatr Surg 31: 869–877, 1996
18. Laemmli UK: Cleavage of structural proteins during the assembly of the head of the bacteriophage T_4. Nature 227: 680–685, 1970
19. Beierwaltes WH, Arendshorst WA, Klemmer PJ: Electrolyte and water balance in young spontaneously hypertensive rats. Hypertension 4: 908–915, 1982
20. Dilley JR, Arendshorst WJ: Enhanced tubuloglomerular feedback activity in rats developing spontaneous hypertension. Am J Physiol 247: F672–F679, 1984
21. Dilley JR, Stier CT Jr, Arendshorst WJ: Abnormalities in glomerular function in rats developing spontaneous hypertension. Am J Physiol 246: F12–F20, 1984
22. Harrap SB, Doyle AE: Renal hemodynamics and total body sodium in immature spontaneously hypertensive and Wistar-Kyoto rats. J

Hypertens 4: S249–S252, 1986
23. Weiss RH: G protein-coupled receptor signalling in the kidney. Cell Signal 10: 313–320, 1998
24. Marcil J, Champlain J, Anand-Srivastava MB: Overexpression of Gi-proteins precedes the development of DOCA-salt-induced hypertension: Relationship with adenylyl cyclase. Cardiovasc Res 39: 492–505, 1998
25. Vyas SJ, Blaschak CM, Chinoy MR, Jackson EK: Angiotensin II-induced alterations in GTP-binding regulatory proteins in renal preglomerular arterioles from genetically hypertensive rats. Am J Hypertens 12: 102A, 1999
26. Mokkapatti R, Vyas SJ, Jackson EK: G protein mRNA expression in renal microvessels from spontaneously hypertensive and Wistar-Kyoto rats. Am J Physiol 273: F877–F882, 1997
27. Gilman AG: G proteins and dual control of adenylylate cyclase. Cell 36: 577–579, 1984
28. Sunahara RK, Dessauer CW, Gilman AG: Complexity and diversity of mammalian adenylyl cyclases. Annu Rev Pharmacol Toxicol 36: 461–480, 1996

Molecular and Cellular Biochemistry **212**: 131–134, 2000.

Angiotensin receptor II is present in dopaminergic cell line of rat substantia nigra and it is down regulated by aminochrome

Alexies Dagnino-Subiabre,[1] Katerine Marcelain,[1] Christian Arriagada,[2] Irmgard Paris,[1] Pablo Caviedes,[2] Raul Caviedes[2] and Juan Segura-Aguilar[1]

[1]*Program of Molecular and Clinical Pharmacology, ICBM, Faculty of Medicine;* [2]*Program of Morphology, ICBM, Faculty of Medicine, University of Chile, Santiago, Chile*

Abstract

Angiotensin receptor II mRNA was found to be expressed in dopaminergic neuronal cell line RCSN3 of rat substantia nigra using RT-PCR reaction. Aminochrome (150 μM), a metabolite of the dopamine oxidative pathway, was found to down regulate the expression of angiotensin receptor mRNA in RCSN3 cells by 83% ($p < 0.05$). (Mol Cell Biochem **212**: 131–134, 2000)

Key words: dopamine, aminochrome, angiotensin receptor II, substantia nigra, quinone

Introduction

The existence of a renin-angiotensin system in mammalian brain is supported by several investigations [1–3]. Angiotensin II (AGT II) receptors subtypes and their distribution in the central nervous system (CNS) have been reported [4–7]. There is evidence supporting the idea that AGT II is involved in cognitive processes. It has been shown that intracerebroventricular administration of AGT II facilitates the acquisition in active conditioning trial and retention in passive avoidance models [8–11], improving recognition, but not spatial memory in rats [12]. The facilitatory effect of AGT II on memory appears to be mediated, in part, by the dopaminergic system since it is abolished by pimozide, a dopaminergic receptor antagonist [13]. In agreement, a bilateral 6-OH-dopamine induced lesion of the dopaminergic projection from A10 ventral tegmental neurons to the central amygdala abolished the facilitatory effects of AGT II [14, 15] and recognition memory [16].

Dopamine has been reported to be oxidized enzymatically and nonenzymatically to *o*-quinone aminochrome [17–24]. The one-electron reduction of aminochrome to a leukoaminochrome *o*-semiquinone radical has been suggested to be responsible for the degeneration of dopaminergic system in Parkinson Disease [25, 26]. We have speculated that the AGT II system may be involved in the loss of memory in Parkinson's patients with dementia, therefore we decided to study the expression of AGT II receptor mRNA in RCSN-3 cells, to investigate a possible modulation by aminochrome of regulation of AGT II receptor. For comparison, the effect of aminochrome on the expression of DT-diaphorase mRNA was studied.

Materials and methods

Chemicals

Dopamine, was purchased from Sigma Chemical Co. (St Louis, MO, USA). Thermoscript RT-PCR system and Taq DNA polymerase were from Life Technologies (California,

Address for offprints: J. Segura-Aguilar, Programme of Molecular and Clinical Pharmacology, ICBM, Faculty of Medicine, University of Chile, Avd. Independencia 1027, Casilla 70000 Santiago-7, Chile

USA). RNeasy midi system was from QIAGEN (Hilden, Germany). Aminochrome was prepared by oxidizing dopamine with Manganese^{3+}-pyrophosphate complex according to Segura-Aguilar and Lind [18].

Cell culture

A RCSN-3 cell line was derived from the substantia nigra of a 4 month old Fisher 344 normal rat. The cell material used to establish primary cultures was transformed to a permanent cell line by exposing them to media conditioned by UCHT1 cells, a process that induces transformation in cell cultures [27]. The RCSN-3 cell line grows on monolayers, with a doubling time of 52 h, a plating efficiency of 21% and a saturation density of 410.000 cells/cm^2. RCSN-3 cells possess receptors for tetanus toxin, and immunohistochemical analysis has demonstrated the presence of neuronal markers such as microtubular associated protein-2 (MAP-2), neuronal specific enolase (NSE), parvalbumin, and tyrosine hydroxylase. Conversely, glial markers glial fibrillary acidic protein (GFAP) and S-100 have not been observed [28]. Also RCSN-3 cells exhibited intracellular fluorescence with paraformaldehyde-glioxilate and are positive for melanin staining, indicating the presence of catecholamines [28]. The cultures were kept in an incubator at 37°C with 100% humidity and an atmosphere of 10% CO_2. The cells were treated with 150 μM aminochrome.

Expression of AGT II receptor

The expression of AGT II receptor in a RCSN-3 cell line was studied by using the RT-PCR technique. The total RNA was isolated by using RNeasy Midi kit (QIAGEN). Five μg of the total RNA was used for the synthesis of a single strand DNA with the reverse transcriptase (RT) reaction. The RT-reaction was performed by using a Thermoscript RT-PCR system (Life Technologies) with Oligo $(dT)_{20}$ as primers. The amplification of ssDNA of AGT II was performed by PCR reaction using the following primers 5′- GATGCTGGTAGCCAAAGTCACC-3′ (upstream) and 5′- GATAAGGAAAGGGAACACGAAGC-3′ (downstream) designed from the cDNA sequence of AGT II receptor [29]. The PCR reaction was performed in three steps: (i) 95°C for 5 min; (ii) 20, 25–30, 35 and 40 cycles at 95°C for 40 sec, 65°C for 40 sec, 72°C for 40 sec; (iii) one cycle at 72°C for 10 min. The PCR incubation contained 6 ml of RT-incubation, 0.4 mM dNTP each, 3 mM $MgCl_2$, 2,5 μM primers, 5 μl 10 × PCR-buffer (GibcoBRL), 29 ml H_2O and 2 U Taq polymerase (GibcoBRL). The region amplified by PCR was between the bases 561–582 and 768–746, which resulted in a fragment of 208 bp [29]. PCR at different number of cycles was performed to estimate the differences in the level of expression of AGT II receptor mRNA between control and aminochrome-treated RCSN-3 cells. The PCR products were electrophoresed on 2% agarose gels, stained with ethidium bromide and photographed. The photographs were scanned and computer analyzed by using a Scion Image software (NIH, USA) in order to estimate the differences in the expression of AGT II mRNA. The number of pixels estimated by the software was plotted against the number of cycles. DT-diaphorase was used as control protein to demonstrate that the aminochrome effect on AGT II mRNA expression was specific.

Expression of DT-diaphorase

The PCR reaction for DT-diaphorase was performed as described by Arriagada *et al.* [30] using the following primers 5′-CAGAAACGACATCACAGGGGAG-3′ (upstream) and 5′-CAAGCACTCTCTCAAACCAGCC-3′ (downstream). The region amplified by PCR was between the bases 230 and 438, which resulted in a fragment of 209 bp [30].

Results

AGT II receptor mRNA was constitutively expressed in a dopaminergic neuronal cell line RCSN-3 of rat substantia nigra as assessed by RT-PCR technique. The effect of aminochrome on the expression of AGT II receptor was studied by

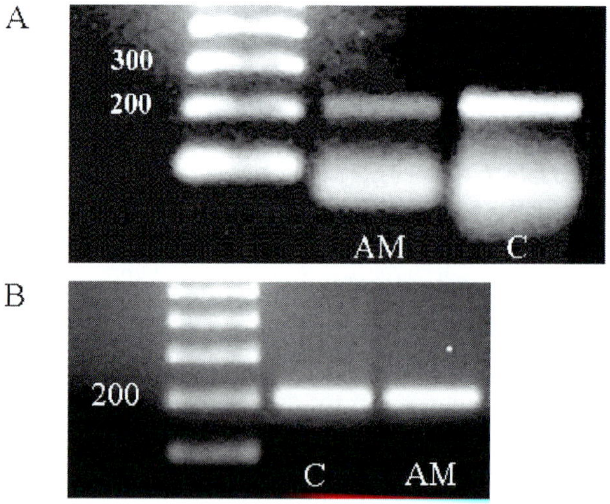

Fig. 1. Expression of angiotensin II receptor mRNA in RCSN3 cells of rat substantia nigra and the effect of aminochrome on expression of its mRNA. The expression of AGT II receptor (A) and DT-diaphorase (B) mRNA was measured using RT-PCR technique described under Materials and methods. The cells were incubated in the presence of 150 μM aminochrome (AM) during 2 h before extraction of total RNA. The AGT II band is observed at 208 kb. The control (C) cells were incubated under the same conditions in the absence of aminochrome.

incubating the cells with 150 μM of aminochrome during 2 h before extraction of the total RNA. Aminochrome was found to down regulate the expression of AGT II mRNA in RCSN3 cells (Fig. 1A). The specificity of this down regulation of AGT II mRNA by aminochrome was supported by the lack of effect of aminochrome on DT-diaphorase mRNA expression (Fig. 1B). PCR reactions at different cycle numbers was performed to estimate the down regulation of AGT II receptor exerted by aminochrome (Fig. 2). The down regulation was assessed as 83% at 40 cycles ($p < 0.05$).

Discussion

The brain renin-AGT II system has been suggested to exert facilitatory effect on memory [17], which appears to be mediated, in part, by the dopaminergic systems [18–20]. The finding that AGT II receptors are present in dopaminergic neurons of substantia nigra and that aminochrome down regulates the expression of AGT II mRNA receptor raises a question on the role of aminochrome in the regulation of brain AGT II receptor. Aminochrome can be formed *in vivo* by oxidation of dopamine by oxygen, transitions metals (iron, manganese or iodine) [17–20] and peroxynitrite radical [21]. Dopamine is also oxidized to aminochrome by several enzymes, such as prostaglandin H synthase, xanthine oxidase, several forms of cytochrome P450 and, specially, by CYP 1A2 [22–24]. The possible formation of aminochrome *in vivo* is supported by the finding that cysteinyl adducts, such as 5-cysteinyl-dopamine and quinone adducts, are present in rat, guinea pig and human brain [31–33]. In addition, aminochrome is the precursor of neuromelanin. The reductive metabolism of aminochrome to leukoaminochrome *o*-semiquinone has been proposed to be the reaction responsible for the degenerative process characterizing Parkinson's disease [25, 26]. Leucoaminochrome *o*-semiquinone is very reactive with oxygen, resulting in the reduction of dioxygen to superoxide radicals. Superoxide radicals, enzymatically or non-enzymatically, dismutate resulting in the formation of hydrogen peroxide and dioxygen. Hydrogen peroxide is the precursor of one of the most harmful free radicals, hydroxyl radical (·OH), in the presence of metals such as Fe^{2+} and Cu^{+}. Furthermore, leukoaminochrome *o*-semiquinone is a radical that can react by itself with nucleophile molecules, such as RNA, DNA, GSH, or induce lipid peroxidation and deactivation of enzymes by oxidation of essential thiol groups. Paradoxically, the antioxidant enzymes, superoxide dismutase and catalase, have been reported to play a prooxidante role during one-electron reduction of aminochrome by increasing the autoxidation rate of leukoaminochrome *o*-semiquinone radical [25]. However, whether AGT II receptor down regulation observed by aminochrome may in part explain the loss of memory in Parkinson's patients with dementia remains to be further investigated.

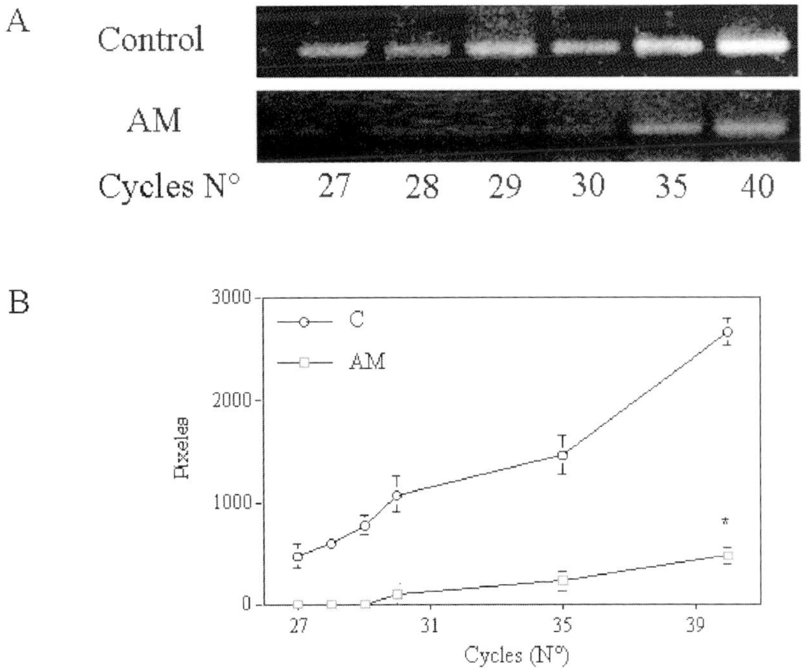

Fig. 2. Estimation of down regulation of AGT II receptor mRNA by using PCR. (A) PCR reactions at different cycle numbers in the presence and absence of 150 μM aminochrome. (B) Estimation of differences in the expression of AGT II mRNA was performed by scanning the gel showed in A. The values are the mean ± S.D. (n = 3). $p < 0.05$ vs. control was determined by using the unpaired Student's *t*-test.

Acknowledgements

This work was supported by grants from FONDECYT (No. 1990622), (No. 1980906) and DID of University of Chile.

References

1. Fisher-Ferraro C, Nahmod VE, Goldstein DJ, Finkielman S: Angiotensin and renin in the rat and dog brain. J Exp Med 133: 353–361, 1971
2. Ganten D, Minnich JE, Granger P, Hayduk K, Brecht HM, Barbeau A, Boucher R, Genest J: Angiotensin-forming enzyme in brain tissue. Science 173: 64–65, 1971
3. Ganten D, Hermann K, Bayer C, Unger T, Lang RE: Angiotensin synthesis in the brain and increased turnover in hypertensive rats. Science 221: 869–871, 1983
4. Wright JW, Miller-Wing AV, Shaffer MJ, Higginson C, Wright DE, Hanesworth JM, Harding JW: Angiotensin II 3-8 ANG IV hippocampal binding: Potential role in the facilitation of memory. Brain Res Bull 32: 497–502, 1993
5. Wright JW, Harding JW: Brain angiotensin receptor subtypes in the control of physiological and behavioral responses. Neurosci Biobehav Rev 18: 21–53, 1994
6. Hohle S, Blume A, Lebrun Ch, Culman J, Unger T: Angiotensin receptors in the brain. Pharmacol Toxicol 77: 306–315, 1995
7. Wright JW, Harding JW: Brain angiotensin receptor subtypes AT, AT, and AT and their functions. Reg Pep 59: 269–295, 1995
8. Braszko JJ, Wisniewski K, Kupryszewski G, Witczuk, B: Psychotropic effects of angiotensin II and III in rats: Locomotor and exploratory vs cognitive behaviour. Behav Brain Res 25: 195–203, 1987
9. Braszko JJ, Kupryszewski G, Witczuk B, Wisniewski, K: Angiotensin II-3-8-hexapeptide affects motor activity, performance of passive avoidance and conditioned avoidance response in rats. Neuroscience 3: 777–783, 1988
10. Braszko JJ, Wisniewski K: Effect of angiotensin II and saralasin on motor activity and the passive avoidance behavior of rats. Peptides 9: 475–479, 1988
11. Braszko JJ, Wlasienko J, Koziolkiewicz W, Janecka A, Wisniewski K: The 3–7 fragment of angiotensin II is probably responsible for its psychoactive properties. Brain Res 542: 49–54, 1991
12. Braszko JJ, Kulakowka A, Wisniewski K: Angiotensin II and its 3–7 fragment improve recognition but not spatial memory in rats. Brain Res Bull 37: 627–631, 1995
13. Wisniewski K, Braszko JJ: The significance of central monoamine synthesis in the angiotensin II AII improvement of learning. Clin Exp Hypertens A6: 2127–2131, 1984
14. Winnicka MM, Boguszewicz J, Braszko J: Lesion to central amygdala abolishes angiotensin II improvement of recall in passive avoidance situation. Pol J Pharmacol Pharm 40: 129–134, 1988
15. Winnicka MM, Braszko JJ, Wisniewski K: 6-OHDA lesions to amygdala and hippocampus attenuate memory enhancing effect of the 3–7 fragment of angiotensin II. Gen Pharmacol 30: 801–805, 1998
16. Winnicka MM, Braszko JJ: 6-OHDA lesions to the central amygdala abolish angiotensins facilitation of object recognition in rats. Gen Pharmacol 29: 239–243, 1997
17. Senoh S, Creveling CR, Udenfriend S and Witkof B: Chemical, enzymatic and metabolic studies on the mechanism of oxidation of dopamine. J Am Chem Soc 81: 6231–6240, 1959
18. Segura-Aguilar J, Lind C: On the mechanism of Mn^{3+} induce neurotoxicity of dopamine: Prevention of quinone derived oxygen toxicity by DT-diaphorase and superoxide dismutase. Chem Biol Interact 72: 309–324, 1989
19. Hawley MD, Tatawawadi SV, Piekarski S, Adams RN: Electrochemical studies of the oxidation pathways of catecholamines. J Am Chem Soc 89: 447–450, 1967
20. Harrison WH, Whisler WW, Hill BJ: Catecholamine oxidation and ionization properties indicated from the H^- release, tritium exchange, and spectral changes which occur during ferricyanide oxidation. Biochemistry 7: 3089–3093, 1968
21. Daveu C, Servy C, Dendane M, Marin P, Ducrocq C: Oxidation and nitration of catecholamines by nitrogen oxides derived from nitric oxide. Nitric Oxide 3: 234–243, 1997
22. Hastings TG: Enzymatic oxidation of dopamine: The role of prostaglandin H synthase. J Neurochem 64: 919– 924, 1995
23. Segura-Aguilar J: Peroxidase activity of liver microsomal vitamin D 25 hydroxylase catalyzes 25-hydroxylation of vitamin D_3 and oxidation of dopamine to aminochrome. Biochem Mol Med 58: 122–129, 1996
24. Foppoli C, Coccia R, Cini C, Rosei MA: Catecholamines oxidation by xanthine oxidase. Biochim Biophys Acta 1334: 200–206, 1997
25. Baez S, Linderson Y, Segura-Aguilar J: Superoxide dismutase and catalase enhance autoxidation during one-electron reduction of aminochrome by NADPH-cytochrome P-450 reductase. Biochem Mol Med 54: 12–18, 1995
26. Segura-Aguilar J, Metodiewa D, Welch CJ: Metabolic activation of dopamine *o*-quinones to *o*-semiquinones by NADPH cytochrome P450 reductase may play an important role in oxidative stress and apoptotic effects. Biochim Biophys Acta 1381: 1–6, 1998
27. Caviedes P, Olivares E, Salas K, Caviedes R, Jaimovich E: Calcium fluxes, ions currents and dihydropyridine receptors in a new immortal cell line from rat heart muscle. J Mol Cell Cardiol 25: 829–845, 1993
28. Arriagada C, Caviedes R, Caviedes P: Establishment and characterization of a neuronal cell line derived from the substantia nigra of the adult rat. Soc Neurosci 24: (abstr) 93.13, 1998
29. Kakar SS, Sellers JC, Devor DC, Musgrove LC, Neill JD: Angiotensin II type-1 receptor subtype cDNAs: Differential tissue expression and hormonal regulation. Biochem Biophys Res Commun 183: 1090–1096, 1992
30. Arriagada C, Dagnino-Subiabre A., Caviedes P, Armero JM, Caviedes R, Segura-Aguilar J: Studies of aminochrome toxicity in a mouse derived neuronal cell line: Is this toxicity mediated via glutamate transmission? Amino Acids 18: 363–373, 2000
31. Carlsson A, Fornstedt B: Catechol metabolites in the cerebrospinal fluid as possible markers in the early diagnosis of Parkinson's disease. Neurology 41: 50–51, 1991
32. La Voie MJ, Hastings TG: Dopamine quinone formation and protein modification associated with the striatal neurotoxicity of methamphetamine: Evidence against a role for extracellular dopamine. J Neurosci 19: 1484–1491, 1999
33. Hastings TG, Lewis DA, Zigmond MJ: Role of oxidation in the neurotoxic effects of intrastriatal dopamine injections. Proc Natl Acad Sci USA 93: 1956–1961, 1996

Molecular and Cellular Biochemistry **212**: 135–142, 2000.

Control of cardiomyocyte gene expression as drug target

Heinz Rupp, Martin Benkel and Bernhard Maisch
Molecular Cardiology Laboratory, Department of Internal Medicine and Cardiology, Philipps University of Marburg, Marburg, Germany

Abstract

Pressure overload of the heart is associated with a perturbed gene expression of the cardiomyocyte leading to an impaired pump function. The ensuing neuro-endocrine activation results in disordered influences of angiotensin II and catecholamines on gene expression. To assess whether angiotensin II type 1 receptor inhibition can also counteract a raised sympathetic nervous system activity, spontaneously hypertensive rats fed a hypercaloric diet were treated with eprosartan (daily 90 mg/kg body wt) and cardiovascular parameters were monitored with implanted radiotelemetry pressure transducers. Both, blood pressure and heart rate were increased ($p < 0.05$) by the hypercaloric diet. Although eprosartan reduced ($p < 0.05$) the raised systolic and diastolic blood pressure, the diet-induced rise in heart rate was blunted only partially. In addition to drugs interfering with the enhanced catecholamine influence, compounds should be considered that selectively affect cardiomyocyte gene expression via 'metabolic' signals. (Mol Cell Biochem **212**: 135–142, 2000)

Key words: gene expression, catecholamines, angiotensin II, heart failure, myosin, hypertension, eprosartan

Introduction

Despite recent progress in the treatment of heart diseases, cardiovascular mortality and morbidity remains high. Current therapeutic targets represent mostly surrogate end points and do not provide a causal approach. As exemplified by primary arterial hypertension, the treatment is monitored by the reduction of the pressure load imposed on vasculature and heart while underlying causes are not necessarily taken into account. Hypertension precedes heart failure in 90% of all cases, and the prevalence of heart failure has markedly increased over the last decades [1]. The question arises, therefore, whether the therapeutic efficacy could be increased by identifying additional drug targets. In the present study, the hypothesis is addressed that a disordered gene expression occurs in various cardiovascular diseases due to chronically altered influences of catecholamines, angiotensin II and so-called 'metabolic' signals. For the example of the myosin heavy chain (MHC) and sarcoplasmic reticulum (SR) Ca^{2+} ATPase (SERCA2) gene, disorders in gene expression are identified and the potential of drugs is examined to affect gene transcription. The hypothesis is also addressed that during progression of heart failure an imbalance of catecholamine and angiotensin II influences affects particularly fibroblast activity whereas the cardiomyocyte can be altered via so-called 'metabolic' signals.

Functional genomics of myosin heavy chain in normal heart

The myosin heavy chain gene is affected by diverse functional loads imposed on the heart. Some of these alterations and the associated changes in myosin ATPase activities have been discussed controversially ranging from detrimental to adaptive. Only few studies have, however, been carried out to assess the consequences of drug-induced alterations tar-

Address for offprints: H. Rupp, Molecular Cardiology Laboratory, Department of Internal Medicine and Cardiology, Philipps University of Marburg, Karl-von-Frisch-Strasse 1, 35033 Marburg, Germany geted specifically at cardiomyocyte gene expression. We examined, therefore, functional determinants of an altered MHC expression and attempted to delineate it's consequences for heart performance. Most of these experiments have been carried out in rats that exhibit a high proportion of alpha-MHC and their importance for human heart failure has thus been questioned. However, recent data show that even in human ventricles, alpha-MHC mRNA abundance ranges up to 35% [2] and alpha-MHC protein up to 7% [3] and it's disappearance has been inferred as an early process in the deterioration of heart performance [4]. We consider an altered MHC expression as a quantitative marker of changes in cardiomyocyte gene expression that permits to compare efficacies of transcriptional modulators.

Among neuro-endocrine interventions affecting MHC expression, thyroid hormones appear to be most potent. Since thyroid hormones exhibit a tachycardic action and increase oxygen consumption, they are not suitable for modulating cardiac gene expression on a long-term basis. Using thyroid hormone analogues with a reduced action on oxygen consumption but preserved effects on thyroid-linked gene expression [5], it has been shown that the function of infarcted rabbit hearts can be improved [6] and clinical trials with the lead compound DITPA are ongoing [7]. In addition to thyroid-linked signals, catecholamines and 'metabolic' signals appear to be useful for modulating MHC expression.

Catecholamines

In cultured cardiomyocytes, dibutyryl cyclic AMP increased the proportion of myosin V1 (alpha,alpha-MHC) by 47% which represents to our knowledge the most potent stimulus for alpha-MHC expression with the exception of thyroid hormones [8] (Fig. 1A). Forskolin and the beta-adrenergic receptor agonist isoproterenol exhibited a less pronounced increase. The action of isoproterenol was, however, not significantly greater than that of the alpha-adrenergic receptor agonist phenylephrine or BayK 8644 and ouabain (Fig. 1A). The action of isoproterenol was observed also in cardiomyocytes that had been arrested by KCl or verapamil indicating that an increased mechanical activity did not mediate the change in myosin expression. The effect of catecholamines appeared also not be linked to an increased glucose uptake of cardiomyocytes [8]. However, exotoxin A selectively prevented the effect of isoproterenol [9]. The increase in alpha-MHC mRNA abundance was associated with an antithetic decrease of beta-MHC mRNA and was transcriptionally regulated [10]. A 13 bp E-box motif has been identified which conferred a cyclic AMP-inducible expression of the alpha-MHC gene [11].

There is increasing evidence that catecholamines alter MHC expression also *in vivo*. Treatment with the beta-adrenergic receptor blocker atenolol and the catecholamine depleting agents guancthidine and reserpine reduced the proportion of myosin V1 [12] (Fig. 1B). Vice versa, repetitive isoproterenol and dobutamine injections increased the myosin V1 proportion [8]. Furthermore, the effect of intermittent swimming appears to be linked to a raised adrenergic activity since atenolol counteracted the increase in myosin V1 proportion [12] (Fig. 1B). The effect of atenolol was significantly greater in swimming rats than in sedentary rats of presumably lower adrenergic activity [12] . Since guanethidine did not prevent the increase in myosin V1 of swimming rats, it appears that circulating catecholamines from the adrenal gland have a major influence [13]. The effect of swimming was observed also in intermittently fasted rats which exhibit a reduced myosin V1 proportion [12] (Fig. 1C).

'Metabolic signals'

Functional studies indicate that signals in addition to thyroid hormones and catecholamines have to be inferred for MHC expression. In intermittently fasted rats, the proportion of myosin V1 was lower which cannot be accounted for solely by a reduced thyroid influence [12, 14]. The effect of fasting was partially prevented by 0.8% sucrose containing drinking water, whereby circulating thyroid hormones were not altered [14]. A 0.8% sucrose solution increased also the myosin V1 proportion of Wistar/WU rats [15]. Since during fasting insulin influences and associated reactions such as glucose utilization are reduced, insulin-linked signals appear to have a role in the altered MHC expression. Also the effect of 0.8% sucrose containing drinking water could be attributed to enhanced insulin influences arising from gastro-intestinal incretin factors [16]. Intermittent fasting had, however, no effect in rats with pressure overloaded left ventricles with a reduced myosin V1 proportion [17] indicating that intermittent fasting and pressure overload recruit similar signals (Fig. 1C). A 0.8% sucrose containing drinking water increased myosin V1 to a greater extent in intermittently fasted rats with normal than in pressure overloaded left ventricles [17] (Fig. 1C).

Myosin heavy chain expression in pressure overloaded heart

There is increasing evidence that chronic pressure overload of the heart results in cardiomyocyte hypertrophy with an inherent expression of genes that are not optimal for the imposed load. We examined the hypothesis that drugs targeted specifically at cardiomyocyte gene expression can prevent progression of heart failure. Of particular importance

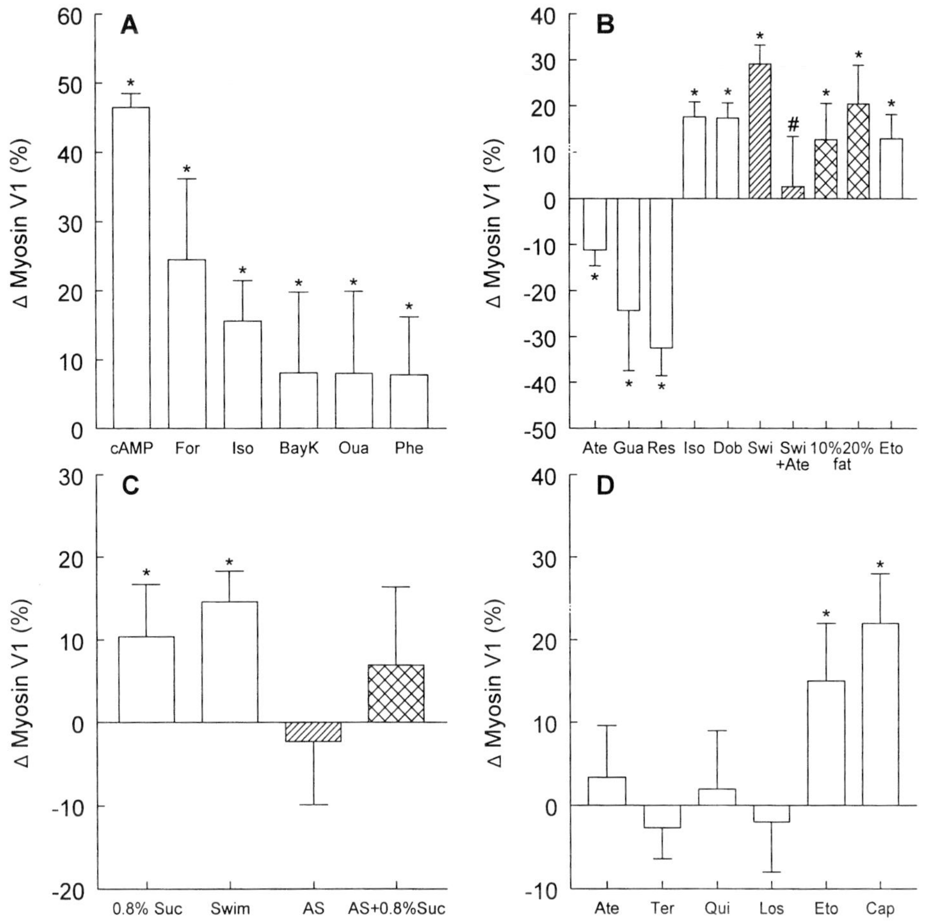

Fig. 1. Selected determinants of myosin isozyme V1 proportion of the rat left ventricle. (A) cultured cardiomyocytes, (B) normal rats, (C) intermittently fasted rats with reduced myosin V1 proportion and (D) rats with pressure overloaded hearts due to ascending aortic constriction with reduced myosin V1 proportion. Values represent differences in percentage points referred to the respective control experiments. Abbreviations are (A) [8]: cAMP – dibutyryl cyclic AMP; For – forskolin; Iso – isoproterenol; BayK – BayK 8644; Oua – ouabain; Phe – phenylephrine. (B) Ate – atenolol treated [12]; Gua – guanethidine treated [12]; Res – reserpine treated [12]; Iso – isoproterenol treated [8]; Dob – dobutamine treated [8]; Swi – swimming [12]; Swi+Ate – swimming rats treated with atenolol [12]; 10% and 20% fat diets [30]; Eto – etomoxir treated [30]. (C) 0.8% Suc – drinking water containing 0.8% sucrose [14]; Swim – swimming rats [12]; AS – rats with constriction of the abdominal aorta [17]; AS+0.8% Suc – corresponding AS rats given 0.8% sucrose containing drinking water [17]. (D) Ate – atenolol treated rats with ascending aortic constriction [18]; Ter – terazosine treated AS rats [18], Qui – quinapril treated AS rats [19]; Los – losartan treated AS rats [20]; Eto – etomoxir treated AS rats [22]; Cap – captopril treated AS rats [25]. *$p < 0.05$ vs. respective control cardiomyocytes or rats; #$p < 0.05$ vs. swimming rats. Values are means ± S.D.

was the question whether drugs can be identified that alter MHC expression although the pressure overload remained unaltered. For this purpose, rats with constriction of the ascending aorta were used. In contrast to experimental myocardial infarction, any changes in gene expression and associated alterations in function had to arise from direct effects on the left ventricular myocardium and a possible ventricular unloading can be excluded.

To examine whether the sympathetic system has any influence on MHC expression in overloaded non-failing heart, rats with ascending aortic constriction were treated with atenolol or the alpha-adrenergic receptor blocking drug terazosine. Treatment with atenolol or terazosine (Fig. 1D) did, however, not prevent the reduced myosin V1 proportion that is characteristic of pressure overloaded hearts [18]. The angiotensin II converting enzyme (ACE) inhibitor quinapril [19] and the angiotensin II type 1 receptor antagonist losartan [20] did also not prevent the reduction in myosin V1 proportion (Fig. 1D). Extending the observation that 0.8% sucrose containing drinking water increased myosin V1 proportion in Wistar/WU hearts [15] we examined the effect of the hypoglycemic drug etomoxir which inhibits mitochondrial carnitine palmitoyltransferase 1 and activates peroxisome proliferator-activated receptor (PPAR)-alpha [21]. The etomoxir treatment did not increase further left ventricular hypertrophy but partially prevented the reduction in myosin

V1 proportion [22] (Fig. 1D) and the initial stage of left ventricular dilatation [23]. In rats with abdominal aortic constriction, 0.8% sucrose containing drinking water also increased the myosin V1 proportion [24].

Unexpected was the observation that the ACE inhibitor captopril increased myosin V1 not only in rats with ascending aortic constriction but also in normal rats that occurred independently of bradykinin B_2 receptors [25] (Fig. 1D). Since captopril differed from other ACE inhibitors and improved insulin sensitivity [26], it's action on myosin expression could be attributed to metabolic signals related to an enhanced glucose utilization. Captopril could affect also intracellular redox potential [27] and thus various transcription factors. The complexity of the underlying signals is demonstrated by the finding that treatment of rats with abdominal aortic constriction with the hypoglycemic compound BM13.907 reduced the myosin V1 proportion in parallel with the amplified hypertrophy [17]. BM13.907 has been developed as a compound that increases glucose transporter translocation without having additional effects on insulin signaling [28].

MHC vs. SERCA2 gene expression for monitoring drug effects

In contrast to MHC expression, functional parameters of SERCA2a can less reproducibly be quantitated and do not permit a long-term comparison of drug efficacies. There is, however, increasing evidence that an increased myosin V1 proportion is not always associated with a higher mRNA abundance of SERCA2a or it's pump activity. Thus, in swimming rats [29] (unpublished) and normal rats fed 10–20% fat diets [30] a statistically significant increase in rate of SR Ca^{2+} uptake was not observed although the myosin V1 proportion was increased. Only in the case of 'metabolic' signals of intermittently fasted rats or rats with pressure overloaded left ventricles, it appears that an increased myosin V1 proportion is associated with a higher SR Ca^{2+} uptake rate. Thus, providing intermittently fasted rats 0.8% sucrose drinking water [14] or a 24% fat/32% sucrose diet (unpublished) partially prevented the reduction of SR Ca^{2+} uptake rate. Furthermore, etomoxir treatment of rats with abdominal aortic constriction or providing 0.8% sucrose containing drinking water raised the depressed SERCA2a mRNA abundance [31] and SR Ca^{2+} ATPase activity [24]. Etomoxir treatment also increased the rate of cardiac relaxation and contraction of rats with ascending aortic constriction [22, 23]. The concept of involvement of metabolically regulated signals for SERCA2 expression is strengthened by the finding that the SERCA2 promoter exhibits candidate binding sequences [32] (Fig. 2). The proximal promoter region of the SERCA2 genes has in addition to a thyroid response element, 9 potential Sp1-binding sites (5′-GGGCGG-3′, 5′-CCGCCC-3′ and 5′-GGGAGG-3′), and an E-box motif (5′-CACATG-3′), which may function as glucose response elements. This region also has 2 putative fatty-acid response elements (5′-GGGGGA-3′). It was proposed that the observed effects of etomoxir on SERCA2 gene expression arises from a shift in fuel metabolism involving glucose response elements and/or peroxisome proliferator-activated receptors [32]. Taken together, it appears that an increased number of active SERCA2a pumps is associated with a higher myosin V1 proportion. However, the opposite appears not necessarily to be the case.

Disordered cardiomyocyte gene expression as drug target during progression of heart failure

Progression of heart failure has been associated with an adverse remodeling of the extracellular matrix that is characterized by fibrosis [33]. Although an excessive collagen deposition is deleterious and contributes to deterioration of heart performance, there is accumulating evidence that gene expression of the cardiomyocyte is particularly critical during the early stages of heart failure. Since the proportion of myosin V1 is reduced in parallel with heart weight and since the cardiac SERCA2a activity appears to be reduced proportionally to heart weight (Rupp and Vetter, unpublished), the function of the cardiomyocyte becomes depressed in parallel with the hypertrophy process. The depressed function is expected to activate the sympathetic nervous system thereby favouring the expression of genes influenced by cyclic AMP signaling [34]. Since the sympathetic nervous system is of major importance for the release of renin, influences of angiotensin II are also enhanced. Angiotensin II stimulates fibroblast proliferation and collagen synthesis [35] and reduces cardiomyocyte alpha-MHC expression [36].

The occurrence of pulmonary congestion as a consequence of left ventricular failure of rats with ascending aortic constriction is associated with a further reduction in the myosin V1 proportion. Both, the reduced alpha-MHC expression and the associated fibrosis could be attributed to a greater angiotensin II influence. However, since neither quinapril nor losartan prevented these changes, additional effects have to be inferred such as a reduced catecholamine influence when congestive heart failure occurs. In accordance, hearts with this characteristic reduction in myosin V1 proportion had depleted norepinephrine stores [37].

In accordance with this contention is also the finding that downregulation of beta-adrenergic receptors occurs in the failing heart [38] resulting in an impaired signaling of cyclic AMP. Since various cyclic AMP dependent pathways coun-

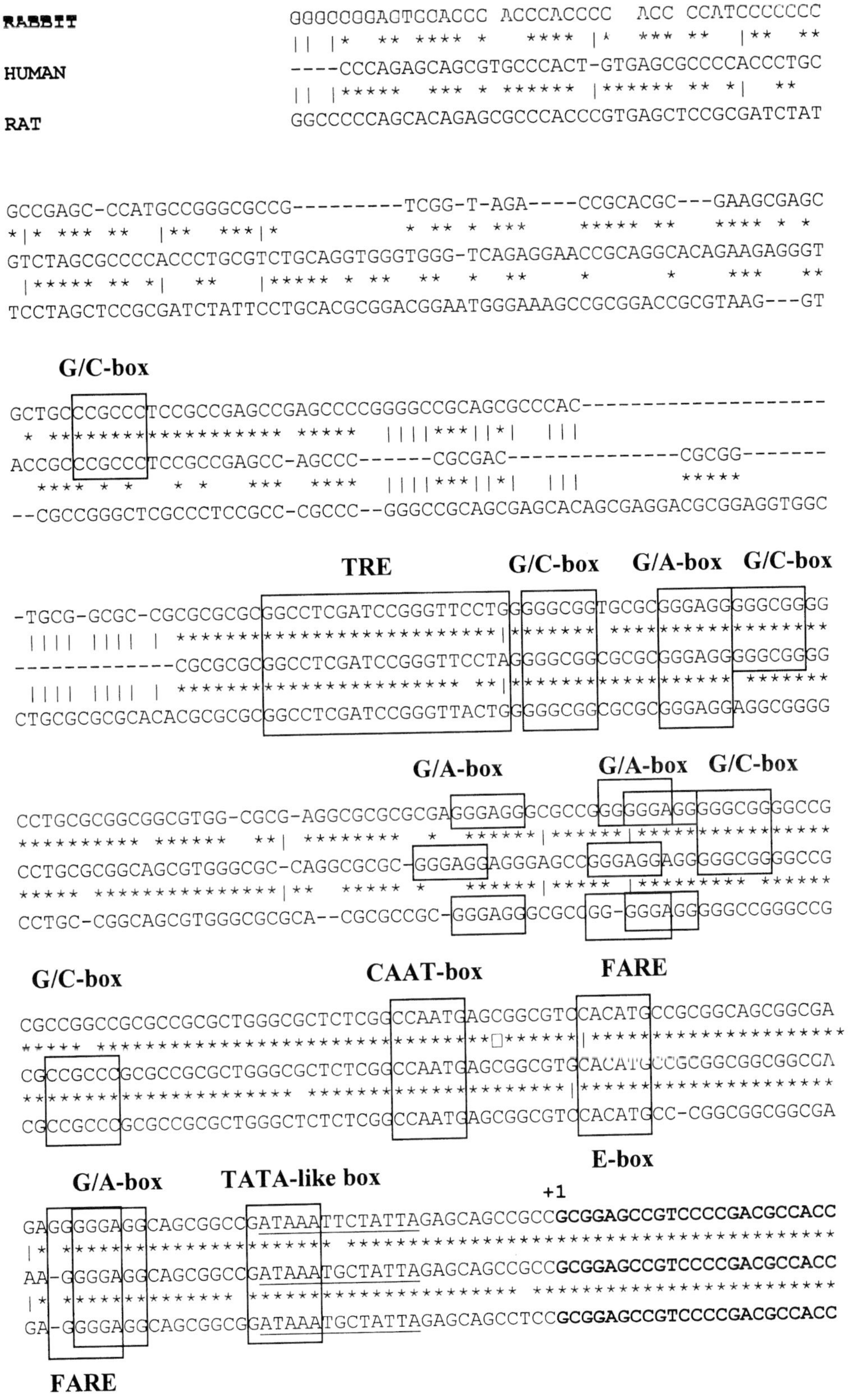

Fig. 2. Nucleotide sequence alignment of human, rabbit, and rat sarco(endo)plasmic reticulum Ca^{2+}-ATPase (SERCA2) genes ([32] with permission from Excerpta Medica Inc.) CAAT box, G-C boxes, G-A boxes, fatty acid response elements (FARE), and thyroid-responsive element (TRE) are boxed. The A-T-rich region is underlined. The 5′-non-translated sequence is in bold letters.

teract signaling by angiotensin II [39], an exaggerated angiotensin II influence is expected to ensue. The question arises, therefore, whether current drug interventions targeted at neuroendocrine activation can counteract such an imbalance in signaling for cardiomyocyte and fibroblast gene expression.

Centrally acting antisympathotonic drugs could reduce effects of an overactivity of the sympathetic nervous system and the consecutive renin release by the kidney [40, 41]. However, the local tissue production of renin and non-ACE pathways require additional drugs such as mast cell chymase inhibitors [42]. Of particular interest could also be AT_1 antagonists which might be expected to reduce not only angiotensin II but also catecholamine influences.

AT_1 antagonists- effects on sympathetic activity

Angiotensin II increases norepinephrine release from sympathetic nerve terminals [43, 44]. The AT_1 antagonist losartan prevented the potentiating action of angiotensin II in the vasculature while the AT_2 antagonist PD 123177 had no effect [45]. For the atria of the pithed rat it was shown that AT_1 antagonists reduce the angiotensin II induced potentiation of sympathetic nerve stimulation [46, 47]. It appears, however, that not all organs are affected in a similar manner. Thus, no effect of AT_1 antagonists was observed for the mouse brain [48]. The inhibitory effect of AT_1 antagonists appears to be more pronounced in the case of eprosartan [49, 50]. Eprosartan differs structurally from other known sartans and inhibited the pressor response induced by spinal cord stimulation of pithed rats while equivalent doses of other non-peptide angiotensin II receptor antagonists had no effect [50]. The question was, therefore, addressed whether eprosartan affects both the raised blood pressure and heart rate of rats fed a hypercaloric diet [51].

The present radiotelemetric data show that during the hypercaloric feeding, heart rate and blood pressure were significantly increased (Fig. 3) that is indicative of a raised sympathetic nervous system activity. A daily dose of 30 mg/kg body wt transiently reduced ($p < 0.05$) both systolic (Fig. 3) and diastolic (not shown) blood pressure to the level before the hypercaloric feeding was started. A higher dose (90 mg/kg) reduced blood consistently (Fig. 3). The increased heart rate was significantly reduced after administration of eprosartan (90 mg/kg). However, eprosartan did not blunt the rise of heart rate due to hypercaloric diet intake. Although an increased sympathetic nervous system activity is expected to raise angiotensin II, it is not known to what extent angiotensin potentiates sympathetic nervous system activity under these conditions. It can, therefore, not be decided whether the effect of eprosartan reflects an inhibitory action on an angi-

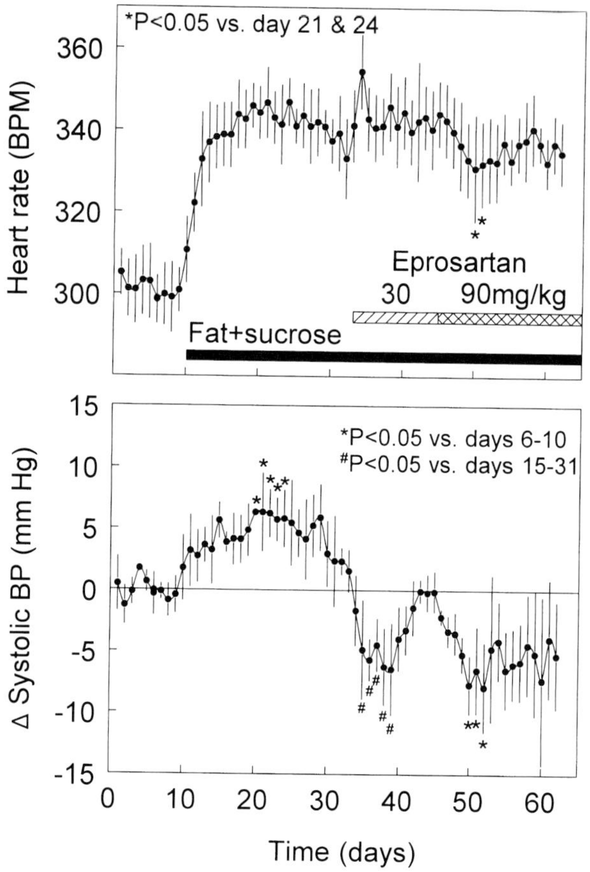

Fig. 3. Radiotelemetrically monitored (A) heart rate and (B) systolic blood pressure averaged (288 30-sec measurements) over 24 h. Rats were fed a 24% coconut fat/32% sucrose diet and were treated daily with 30 mg/kg body wt or 90 mg/kg eprosartan. The radiotelemetry system consisted of TA11PA-C40 pressure transmitters and RA 1010 receivers (Data Sciences, St. Paul, MN, USA). Statistical comparisons were made by ANOVA with repeated measures using 4 spontaneously hypertensive rats. Values are means ± S.D.

otensin II induced potentiation of sympathetic nervous system activity or a direct effect on the sympathetic nervous system activity. It nonetheless appears that in addition to eprosartan antisympathotonic drugs are required for normalizing the high heart rate and it's associated unfavorable reactions.

Taken together, the drug target of preserving a normal gene expression of the cardiomyocyte appears as an important goal which has, however, not been achieved yet. Accumulating evidence suggests that particularly during progression of heart failure, 'metabolic' signals could be recruited for preventing a protein phenotype of the cardiomyocyte that is associated with a depressed function leading to neuro-endocrine activation. Efforts should be made to identify markers of gene expression that are affected both by catecholamines and angiotensin II. During progression of heart failure, such markers could be used to derive a rational combination therapy targeted at the prevention of disorders in gene expression.

Acknowledgements

The experimental studies were supported by the German Research Foundation (Ru 245/7-1).

References

1. Lenfant C, Roccella EJ: A call to action for more aggressive treatment of hypertension. J Hypertens 17(suppl): S3–S7, 1999
2. Nakao K, Minobe W, Roden R, Bristow MR, Leinwand LA: Myosin heavy chain gene expression in human heart failure. J Clin Invest 100: 2362–2370, 1997
3. Miyata S, Minobe W, Bristow MR, Leinwand LA: Myosin heavy chain isoform expression in the failing and non-failing human heart. Circ Res 86: 386–390, 2000
4. Lowes BD, Minobe W, Abraham WT, Rizeq MN, Bohlmeyer TJ, Quaife RA, Roden RL, Dutcher DL, Robertson AD, Voelkel NF, Badesch DB, Groves BM, Gilbert EM, Bristow MR: Changes in gene expression in the intact human heart. Downregulation of alpha-myosin heavy chain in hypertrophied, failing ventricular myocardium. J Clin Invest 100: 2315–2324, 1997
5. Morkin E, Pennock GD, Raya TE, Bahl JJ, Goldman S: Development of a thyroid hormone analogue for the treatment of congestive heart failure. Thyroid 6: 521–526, 1996
6. Mahaffey KW, Raya TE, Pennock GD, Morkin E, Goldman S: Left ventricular performance and remodeling in rabbits after myocardial infarction. Effects of a thyroid hormone analogue. Circulation 91: 794–801, 1995
7. Spooner PH, Morkin E, Goldman S: Thyroid hormone and thyroid hormone analogues in the treatment of heart failure. Coron Artery Dis 10: 395–399, 1999
8. Rupp H, Berger HJ, Pfeifer A, Werdan K: Effect of positive inotropic agents on myosin isozyme population and mechanical activity of cultured rat heart myocytes. Circ Res 68: 1164–1173, 1991
9. Muller-Werdan U, Pfeifer A, Hubner G, Seliger C, Reithmann C, Rupp H, Werdan K: Partial inhibition of protein synthesis by Pseudomonas exotoxin A deranges catecholamine sensitivity of cultured rat heart myocytes. J Mol Cell Cardiol 29: 799–811, 1997
10. Gupta MP, Gupta M, Zak R: An E-box/M-CAT hybrid motif and cognate binding protein(s) regulate the basal muscle-specific and cAMP-inducible expression of the rat cardiac alpha-myosin heavy chain gene. J Biol Chem 269: 29677–29687, 1994
11. Gupta MP, Gupta M, Dizon E, Zak R: Sympathetic control of cardiac myosin heavy chain gene expression. Mol Cell Biochem 157: 117–124, 1996
12. Rupp H, Wahl R: Influence of thyroid hormones and catecholamines on myosin of swim- exercised rats. J Appl Physiol 68: 973–978, 1990
13. Ohkubo T, Jacob R, Rupp H: Swimming changes vascular fatty acid composition and prostanoid generation of rats. Am J Physiol 262: R464–R471, 1992
14. Rupp H, Elimban V, Dhalla NS: Diabetes-like action of intermittent fasting on sarcoplasmic reticulum Ca^{2+}-pump ATPase and myosin isoenzymes can be prevented by sucrose. Biochem Biophys Res Commun 164: 319–325, 1989
15. Rupp H, Wahl R, Jacob R: Remodelling of the myocyte at a molecular level – relationship between myosin isozyme population and sarcoplasmic reticulum. In: N.S. Dhalla, G.N. Pierce, R.E. Beamish (eds). Heart Function and Metabolism. Martinus Nijhoff Publishing, Boston, 1987, p. 307
16. Fukase N, Takahashi H, Manaka H, Igarashi M, Yamatani K, Daimon M, Sugiyama K, Tominaga M, Sasaki H: Differences in glucagon-like peptide-1 and GIP responses following sucrose ingestion. Diabetes Res Clin Pract 15: 187–195, 1992
17. Rupp H, Elimban V, Dhalla NS: Differential influence of fasting and BM13.907 treatment on growth and phenotype of pressure overloaded rat heart. Mol Cell Biochem 188: 209–215, 1998
18. Jacob R, Brandle M, Dierberger B, Rupp H: Functional consequences of cardiac hypertrophy and dilatation. Basic Res Cardiol 86(suppl 1): 113–130, 1991
19. Turcani M, Rupp H: Heart failure development in rats with ascending aortic constriction and angiotensin-converting enzyme inhibition. Br J Pharmacol (in press)
20. Turcani M, Rupp H: Development of pressure overload induced cardiac hypertrophy is unaffected by long-term treatment with losartan. Mol Cell Biochem 188: 225–233, 1998
21. Brandt JM, Djouadi F, Kelly DP: Fatty acids activate transcription of the muscle carnitine palmitoyltransferase I gene in cardiac myocytes via the peroxisome proliferator-activated receptor alpha. J Biol Chem 273: 23786–23792, 1998
22. Turcani M, Rupp H: Etomoxir improves left ventricular performance of pressure-overloaded rat heart. Circulation 96: 3681–3686, 1997
23. Turcani M, Rupp H: Modification of left ventricular hypertrophy by chronic etomixir treatment. Br J Pharmacol 126: 501–507, 1999
24. Rupp H, Elimban V, Dhalla NS: Sucrose feeding prevents changes in myosin isoenzymes and sarcoplasmic reticulum Ca^{2+}-pump ATPase in pressure-loaded rat heart. Biochem Biophys Res Commun 156: 917–923, 1988
25. Turcani M, Rupp H: Bradykinin (B2) independent effect of captopril on the development of pressure overload cardiac hypertrophy. Mol Cell Biochem 212: 219–225, 2000
26. Lithell HO: Insulin resistance and diabetes in the context of treatment of hypertension. Blood Press 3(suppl): 28–31, 1998
27. Anderson B, Khaper N, Dhalla AK, Singal PK: Anti-free radical mechanisms in captopril protection against reperfusion injury in isolated rat hearts. Can J Cardiol 12: 1099–1104, 1996
28. Obermaier-Kusser B, Muhlbacher C, Mushack J, Seffer E, Ermel B, Machicao F, Schmidt F, Haring HU: Further evidence for a two-step model of glucose-transport regulation. Inositol phosphate-oligosaccharides regulate glucose-carrier activity. Biochem J 261: 699–705, 1989
29. Pagani ED, Solaro RJ: Coordination of cardiac myofibrillar and sarcotubular activities in rats exercised by swimming. Am J Physiol 247: H909–H915, 1984
30. Rupp H, Wahl R, Hansen M: Influence of diet and carnitine palmitoyltransferase I inhibition on myosin and sarcoplasmic reticulum. J Appl Physiol 72: 352–360, 1992
31. Zarain-Herzberg A, Rupp H, Elimban V, Dhalla NS: Modification of sarcoplasmic reticulum gene expression in pressure overload cardiac hypertrophy by etomoxir. FASEB J 10: 1303–1309, 1996
32. Zarain-Herzberg A, Rupp H: Transcriptional modulators targeted at fuel metabolism of hypertrophied heart. Am J Cardiol 83: 31H–37H, 1999
33. Weber KT, Brilla CG: Pathological hypertrophy and cardiac interstitium. Fibrosis and renin- angiotensin-aldosterone system. Circulation 83: 1849–1865, 1991
34. Muller FU, Boknik P, Knapp J, Neumann J, Vahlensieck U, Oetjen E, Scheld HH, Schmitz W: Identification and expression of a novel isoform of cAMP response element modulator in the human heart. FASEB J 12: 1191–1199, 1998

35. Brilla CG, Zhou G, Matsubara L, Weber KT: Collagen metabolism in cultured adult rat cardiac fibroblasts: Response to angiotensin II and aldosterone. J Mol Cell Cardiol 26: 809–820, 1994
36. Yamazaki T, Shiojima I, Komuro I, Nagai R, Yazaki Y: Involvement of the renin-angiotensin system in the development of left ventricular hypertrophy and dysfunction. J Hypertens 12(suppl): S23–S27, 1994
37. Jacob R, Kissling G, Rupp H, Vogt M: Functional significance of contractile proteins in cardiac hypertrophy and failure. J Cardiovasc Pharmacol 10(suppl 6): S2–S12, 1987
38. Asano K, Dutcher DL, Port JD, Minobe WA, Tremmel KD, Roden RL, Bohlmeyer TJ, Bush EW, Jenkin MJ, Abraham WT, Raynolds MV, Zisman LS, Perryman MB, Bristow MR: Selective downregulation of the angiotensin II AT1-receptor subtype in failing human ventricular myocardium. Circulation 95: 1193–1200, 1997
39. Giasson E, Servant MJ, Meloche S: Cyclic AMP-mediated inhibition of angiotensin II-induced protein synthesis is associated with suppression of tyrosine phosphorylation signaling in vascular smooth muscle cells. J Biol Chem 272: 26879–26886, 1997
40. Ziegler D, Haxhiu MA, Kaan EC, Papp JG, Ernsberger P: Pharmacology of moxonidine, an I1-imidazoline receptor agonist. J Cardiovasc Pharmacol 27(suppl 3): S26–S37, 1996
41. Rupp H: Excess catecholamine syndrome. Pathophysiology and therapy. Ann NY Acad Sci 881: 430–444, 1999
42. Hayashi Y, Iijima K, Katada J, Kiso Y: Structure-activity relationship studies of chloromethyl ketone derivatives for selective human chymase inhibitors. Bioorg Med Chem Lett 10:199–201, 2000
43. Zimmerman BG, Kraft E: Blockade by saralasin of adrenergic potentiation induced by renin-angiotensin system. J Pharmacol Exp Ther 210: 101–105, 1979
44. Kaufman LJ, Vollmer RR: Endogenous angiotensin II facilitates sympathetically mediated hemodynamic responses in pithed rats. J Pharmacol Exp Ther 235: 128–134, 1985
45. Wong PC, Bernard R, Timmermans PB: Effect of blocking angiotensin II receptor subtype on rat sympathetic nerve function. Hypertension 19: 663–667, 1992
46. Brasch H, Sieroslawski L, Dominiak P: Angiotensin II increases norepinephrine release from atria by acting on angiotensin subtype 1 receptors. Hypertension 22: 699–704, 1993
47. Dendorfer A, Raasch W, Tempel K, Dominiak P: Interactions between the renin-angiotensin system (RAS) and the sympathetic system. Basic Res Cardiol 93(suppl 2): 24–29, 1998
48. Cox SL, Trendelenburg AU, Starke K: Prejunctional angiotensin receptors involved in the facilitation of noradrenaline release in mouse tissues. Br J Pharmacol 127: 1256–1262, 1999
49. Ohlstein EH, Gellai M, Brooks DP, Vickery L, Jugus J, Sulpizio A, Ruffolo RR Jr, Weinstock J, Edwards RM: The antihypertensive effect of the angiotensin II receptor antagonist DuP 753 may not be due solely to angiotensin II receptor antagonism. J Pharmacol Exp Ther 262: 595–601, 1992
50. Ohlstein EH, Brooks DP, Feuerstein GZ, Ruffolo RR, Jr: Inhibition of sympathetic outflow by the angiotensin II receptor antagonist, eprosartan, but not by losartan, valsartan or irbesartan: Relationship to differences in prejunctional angiotensin II receptor blockade. Pharmacology 55: 244–251, 1997
51. Rupp H, Maisch B: Radiotelemetric characterization of overweight-associated rises in blood pressure and heart rate. Am J Physiol 277: H1540–H1545, 1999

PART III
ANGIOTENSIN II INFLUENCES ON GENE EXPRESSION

Molecular and Cellular Biochemistry **212**: 145–153, 2000.

Designing antisense to inhibit the renin-angiotensin system

Dagmara Mohuczy and M. Ian Phillips
Department of Physiology, College of Medicine, University of Florida, USA

Abstract

Overactive renin-angiotensin system has been indicated in numerous pathological situations. Current treatment is based on pharmaceutical compounds, which work on the proteins level. Undisputedly helpful, it is not, however, flawless. Some of the drawbacks include adverse effects and non-compliance problem, since in many cases medicine has to be taken at least once a day for a long time. Therefore it seems logical to try a different approach, for instance to correct the disease at the gene expression level, possibly having a choice of shorter or longer-lasting effects. This current review combines results, relevant to the angiotensin system, with the antisense approach, which decreases amount of target protein by interfering at the mRNA level. Dependent on the tool used – oligodeoxynucleotide, plasmid or viral vector, the antisense effect lasts from few days to months. (Mol Cell Biochem **212**: 145–153, 2000)

Key words: angiotensin, antisense, oligodeoxynucleotide, plasmid, adeno-associated virus, hypertension

Introduction

Antisense nucleic acid is a sequence complementary to a sense, coding sequence.

Antisense inhibition has been developed, starting from cell culture applications through studies on animals, and reaching the point where it is being tested in clinical trials for HIV and cancer [1, 2]. The advantages and drawbacks of its use have been reviewed elsewhere [3]. Before 1992, however, antisense (AS) had not been applied *in vivo* with any success. There was much concern about the efficiency of cellular uptake of oligodeoxynucleotides (ODNs). In 1992–1993, few laboratories simultaneously and independently designed and tested AS-ODNs in the brain [4–8]. Cellular uptake was not a limiting factor in the central nervous system.

AS-ODNs have many potential attractive features as a new class of therapeutic agents to inhibit the renin-angiotensin system (RAS) [8]. Role of the components of RAS in a disease state has been described in many previous publications [9–14]. The first demonstration of antisense for reducing hypertension was the use of antisense to angiotensinogen mRNA and angiotensin II type 1 receptor (AT_1-R) mRNA in the brain [15]. To prolong the effect of antisense inhibition for weeks or months, DNA (partial or full-length) can be inserted in the antisense orientation in viral vectors.

Antisense oligonucleotides

Ideal antisense molecules

Table 1 lists the characteristics of AS-ODNs that need to be incorporated into their design and use.

AS-ODNs have several potential sites of action. AS-ODNs inhibit translation by hybridizing to the specific mRNA that they are designed for and the hybridization prevents either ribosomal assembly or ribosomal sliding along the mRNA [16]. This kind of action assumes that AS-ODNs are acting in the cytosol and do not affect measurable mRNA levels. Indeed, there are several articles reporting antisense effects without detectable change in target mRNA levels [6].

The general principles we follow in the design of the antisense are described in the following protocol.

Address for offprints: D. Mohuczy, Department of Physiology, College of Medicine, Box 100274, University of Florida, FL 32610-2452, USA

Table 1. Characteristics of ideal antisense ODNs

1.	The DNA sequence is specific and unique
2.	Uptake into cells is efficient
3.	The effect in cells is stable (for long-term treatment) or transient (for short-term treatment)
4.	There is no non-specific binding to protein
5.	Hybridization of the ODN is specific for the target DNA
6.	The targeted protein and/or mRNA level is reduced
7.	The ODN is not toxic
8.	No inflammatory or immune response is induced
9.	The ODN is more effective than appropriate sense and mismatch ODN controls

Protocol for design of AS-ODN

Check GenBank for mRNA sequence of the target protein in the species to be studied
If there is more than one lab cloning the same protein, compare the homology of different reports. These sequences will point out the controversial bases, which can be due to natural mutations in variant strains of the same species or sequencing errors. Try to avoid these debatable regions. Target the ODN to the identical regions, which ensures reliability of the sequence being used. The best targets for designing effective AS-ODNs are the 5′ cap region, the AUG translation initiation codon and the 3′ untranslated region of the mRNA [17–19].

Choose the length of the antisense
The AS-ODNs we are using are 15–20 bases long, but longer or full-length cDNA in the antisense direction is used in plasmid or viral vectors. When designing antisense molecules, one has to consider two antagonistic factors: the affinity of oligonucleotide to its target sequence, which is dependent on the number and composition of complementary bases, and the availability of the target sequence, which is dependent on the folding of the mRNA molecule [20–22].

Check if antisense sequence is unique using Blast Search
Antisense sequence should be unique and specific for the target mRNA.

Avoid antisense with complicated secondary structure and self-dimerizing
Complicated secondary structures like loops and hairpins in the antisense sequence and self-dimerization prevent degradation, but also make hybridization with target mRNA more difficult. We are using phosphorothioated antisense which is much more resistant to degradation than natural phosphodiester, therefore more important is to facilitate interaction with target mRNA. To check the presence of the secondary structure we use program for designing PCR primers.

Stability of ODNs

Oligonucleotides in their natural form as phosphodiesters are subject to rapid degradation in the blood, intracellular fluid or cerebrospinal fluid by exo- and endonucleases.

The most widely used modified ODNs are phosphorothioates, where one of the oxygen atoms in the phosphodiester bond between nucleotides is replaced with a sulfur atom. These phosphorothioate ODNs have greater stability in biological fluids than normal oligos. The half-life of a 15-mer phosphorothioate ODN is 9 h in human serum, 14 h in tissue culture medium and 19 h in cerebrospinal fluid [23].

Cellular uptake of ODNs

In order to hybridize with the target mRNA, AS-ODNs have to cross the cell membrane. Saturable uptake of ODNs reaches a plateau within hours, occurs rapidly and, depending on the cells, can be efficient [24–26]. Uptake is faster for shorter ODNs than for longer ones.

Pharmacology of AS-ODNs

Antisense inhibition can be considered pharmacologically as a drug-receptor interaction, where the oligonucleotide is the drug and the target sequence is the receptor. For binding between the two affinity is provided by hydrogen bonding between the Watson-Crick base pairs and base stacking in the double helix that is formed. In order to achieve pharmacological activity, a minimum number of 12–15 bases can provide the minimum level of affinity [26].

Increasing the length of the ODN should result in higher level of specificity, but it also decreases its uptake into cells [23]. With viral vectors, however, the length problem is overcome because the virus enters cells by binding to viral receptors on cell membranes. Therefore, in a viral vector a full-length DNA-antisense sequence can be used. The mechanism of action of antisense DNA differs from that of the antisense ODN. The antisense DNA produces an antisense mRNA that competes negatively with cell native mRNA.

Toxicity of AS-ODNs

AS-ODNs can inhibit protein synthesis in cultured cells in nmol/L doses. The therapeutic window for AS-ODNs is rather narrow: when testing for the optimal dose, small increments in the high nmol/L range should be tested [26, 27]. High concentrations may produce non-specific binding to cytosolic proteins and give misleading results.

Phosphodiester ODNs are degraded to their naturally occurring nucleotide building blocks relatively quickly, therefore no toxic reaction is expected from even high doses of phosphodiester. Studies on phosphorothioated ODNs rats show that following intravenous injection, phosphorothioated ODNs are taken up from plasma mainly by the liver, fat and muscle tissues. Phosphorothioate ODNs are excreted through the urine in 3 days, mainly in their original form. An apparently mild increase in plasma lactate dehydrogenase (LDH) and, to a lesser extent, indicators of a possible transient liver toxicity with very high doses of phosphorothioated ODN was observed [28]. Whole new classes of ODN backbone modification are being developed to avoid the possible liver toxicity in humans with phosphorothioates [1–7, 27].

Delivery of AS as a naked ODNs

Direct injection of the AS-ODN to inhibit renin-angiotensin system has been used in the experiments described below [29–31]. For injections into the brain, ODNs appear to be very successful. In a number of studies, using different antisense ODNs, there has been efficient uptake and effective reduction in protein as well as inhibition of the physiological parameters studied. Uptake is so efficient that one difficulty with intracerebroventricular injections is that the ODN tends to be taken up close to the site of injection and not to spread to other parts of the brain. While this has little impact for hypertension therapy, it is an important consideration in antisense strategies for the treatment of brain diseases, such as Parkinson's disease, thalamic pain, Alzheimer's disease and gliomas.

ODNs and liposomes

Liposomes which are self-assembling particles of bilipid layers have been used for encapsulating antisense ODN for delivery in blood and cell culture [32–35]. Antisense, directed to angiotensinogen mRNA in liposomes, has had successful results.

Tomita *et al.* [34] used liposome encapsulation of angiotensinogen antisense and a Sendai virus injected into the portal vein of spontaneously hypertensive rats (SHR). Blood pressure decreased for several days. However, they did not compare their results with the effects of naked ODN. Wielbo *et al.* [35] compared liposome-encapsulated antisense and naked ODN given intra-arterially. They found that liposome encapsulation was effective, whereas naked ODN was not, under the same conditions. Twenty-four h after injection of 50 μg of liposome-encapsulated antisense ODN, blood pressure decreased by 25 mm Hg. Neither empty liposome nor liposome-encapsulated scrambled ODN nor AS-ODN alone had a significant effect on blood pressure. Confocal microscopy of rat liver tissue 1 h after intra-arterial injection of 50 μg of not encapsulated fluorescein isothiocyanate (FITC) AS, or liposome-encapsulated FITC-conjugated AS, showed intense fluorescence in liver tissue sinusoids with the liposome-encapsulated ODN. Levels of protein (angiotensinogen and angiotensin II in the plasma) were significantly reduced in the liposome-encapsulated ODN group. Antisense alone, lipids alone, and scrambled ODN in liposomes had no effect on protein levels [35]. The short, single-stranded AS-ODNs are not in fact encapsulated but complexed with bilamellar vesicles by electrostatic interactions. Liposome development with cationic lipids also allows high transfection efficiency of plasmid DNA.

Plasmid vectors

Partial or full-length cDNA of the gene of interest can be subcloned in antisense orientation under chosen promoter in a plasmid vector. Antisense mRNA is then expressed and competes with host sense mRNA for translation machinery. Plasmid vectors cause a relatively small immune response and can deliver long DNA sequences. Current limitations are mainly a low efficiency of gene transfer *in vivo* and poor long-term expression.

We constructed plasmid vector containing part of the AT_1-R sequence in antisense orientation under the CMV promoter. This plasmid was tested for AT_1 receptor inhibition *in vitro*, using NG108-15 [36] and vascular smooth muscle cells (VSMC) [37]. The NG108-15 cells had a significant ($p < 0.01$) decrease in angiotensin II AT_1 receptors number comparing with the control cells. No effect was seen on the AT_2 receptors. Rat VSMC transfected with the plasmid showed decrease in AT_1-R mRNA and reduced calcium response to angiotensin II stimulation [37].

Also, we subcloned full-length angiotensinogen (AGT) cDNA in antisense orientation under the CMV promoter. This plasmid vector was used for transient transfection of hepatoma cells H4-IIE. Antisense expression was detected from 2 h after transfection and caused 50% reduction in AGT level secreted by the cells [38]. AGT-AS plasmid injected i.v. (3 mg/kg body wt) to adult SHRs caused a decrease in the blood pressure to a maximum of –22 mm Hg, as compared to sense or saline-injected animals, lasting up to 6 days [39].

For the local effect the most frequently used route of administration for plasmid DNA is probably intramuscular injection.

Generally, use of plasmid DNA is considered to be safer than viral vectors but expression is less efficient and shorter lasting, therefore, is useful when the transient effect is sufficient.

Viral vectors for antisense DNA delivery

Perfect viral vector

There are several viruses which have been tested for gene delivery, and each has its advantages, but does not fit perfectly to the description of the 'ideal viral vector'. To be the perfect vector, a virus should fulfill all of the following criteria listed in Table 2.

Retroviruses

These have been used primarily because of their high efficiency in delivering genes to dividing cells [40]. Retroviruses permit insertion and stable integration of single copy genes. Although effective in cell culture systems, they randomly integrate into the genome, which raises concerns about their safety for practical use *in vivo*. Because retroviruses can only act in dividing cells, they are ideal for tumor therapy but less desirable where other cells are dividing that need to be protected.

In hypertension research, retroviruses are being investigated in treating developing SHR. Our colleagues at the University of Florida delivered retrovirus vector (LNSV) containing an antisense DNA to AT_{1-B} receptor mRNA [41]. Injections in the heart of 5 day old SHR resulted in effective, long-term inhibition of AT_1 receptor mRNA and significant inhibition of the development of hypertension. Several measures indicated that the treatment reduced responsiveness of AT_1-R in vessels to angiotensin II stimulation [41]. Problem could arise, however from incorporating the virus into the germ line.

Adenoviruses

Adenovirus vectors have been tested successfully in their natural host cells, the respiratory endothelia, as well as other tissues such as vascular smooth and striated muscle, and brain [42–44].

Adenovirus is a double-stranded DNA with 2700 distinct adenoviral gene products. It infects many mammalian cell types because most cells have membrane receptors. The viruses enter the cell by receptor-induced endocytosis and translocate to the nucleus. Most adeno-virus vectors in their current form are episomal, that is, they do not integrate into the host DNA. They provide high levels of expression, but the episomal DNA will invariably become inactive after some time. In some species, e.g. mice, this time may be long, relative to their life span, but in humans it is a limitation of the virus as a vector. Repeated infections result in an inflammatory response with consequent tissue damage. This is because the adenovirus expresses genes that lead to immune cell attacks. This further limitation makes current recombinant adenovirus unsuitable for long-term treatment and several gene therapy trials using adenovirus vectors have failed to produce acceptable results. Preliminary studies with adenovirus vectors for delivery of AT_1-R mRNA antisense have been tested in rats and reduced developing hypertension in SHR [45]. The adenovirus is easy to produce and therefore useful for animal studies of mechanisms. However, the adenovirus as a vector has too many limitations at present to be successful in human gene therapy. Further engineering of the adenovirus may eventually avoid these limitations. Promising results have been obtained by Burcin *et al.* [46] who used adenovirus devoid of all viral coding sequences [47].

Adeno-associated virus (AAV)

The AAV has been gaining attention because its safety and efficiency [48]. It has been successfully used for delivering antisense RNA against alpha-globin [49] and HIV LTR [50], and it is our vector of choice for delivering antisense targeted to the renin-angiotensin system in hypertensive animal models.

AAV is a parvovirus, discovered as a contamination of adenoviral stocks. It is widespread (it is estimated that antibodies are present in 85% of people in the USA) and has not been linked to any disease. Its replication is dependent on the presence of a helper virus, such as adenovirus or herpes virus. Five serotypes have been isolated, of which AAV-2 is the best characterized. AAV has a single-stranded linear DNA which is encapsidated into capsid proteins VP_1, VP_2 and VP_3 to form an icosahedral virion of 20–24 nm in diameter [48].

The AAV DNA is approximately 4.7 kb long. It contains two open reading frames and is flanked by two inverted terminal repeats (TRs). There are two major genes in the AAV genome: *rep* and *cap*. *Rep* codes for proteins responsible for viral replications, whereas *cap* codes for capsid proteins VP_{1-3}. Each TR forms a T-shaped hairpin structure. These ter-

Table 2. Features of the perfect viral vector

1.	Is safe, so if it is known to cause disease, it must be re-engineered to be harmless
2.	Does not elicit an immune or inflammatory response
3.	Does not integrate into the genome randomly, as this would risk disrupting other cellular genes and mutagenesis
4.	Is replication-deficient so that it will not spread to other tissues or infect other individuals
5.	It delivers a defined gene copy number into each infected cell
6.	Is efficiently taken up by the target tissue, therefore, infects the target cells with high efficiency
7.	Has high capacity so it can accommodate the gene of interest, along with its regulatory sequences
8.	The recombinant DNA is packaged with high efficiency into the viral capsid
9.	Is easy to manipulate and produce in pure form

minal repeats are the only *cis*-components of the AAV necessary for packaging. Therefore, the AAV can be used as a vector with all viral coding sequences removed and replaced by the cassette of genes for delivery. Three viral promoters have been identified and named p5, p19 and p40, according to their map position. Transcription from p5 and p19 results in production of *rep* proteins, while transcription from p40 produces the capsid proteins [48].

Upon infection of a human cell, the wild-type AAV integrates into the q-arm of chromosome 19 [51, 52]. Although chromosomal integration requires the terminal repeats, the viral components responsible for site-specific integration have been recently targeted to the *rep* proteins [53]. With no helper virus present, AAV infection remains latent indefinitely. Upon super-infection of the cell with helper virus, the AAV genome is excised, replicated, packaged into virions and released to the extracellular fluid. This fact is the basis of recombinant AAV production for research.

Several factors prompted researchers to study the possibility of using recombinant AAV as an expression vector. Benefits and drawbacks of using AAV are listed in Table 3.

The advantages, particularly its safety, make AAV appear to be one of the best candidates for delivery of genes for long-term therapy. For more powerful expression native promoter can be replaced by a stronger one, like the cytomegalovirus (CMV) promoter. For expression in a particular tissue, there is an option of local injection or use of tissue-specific promoters. Recently, Flotte *et al.* [55] have established gene therapy Phase I trials for cystic fibrosis using AAV gene delivery in patients.

The general concept for antisense gene delivery in the AAV vector and the steps involved are shown in Fig. 1. To illustrate these steps, a brief description is given that is applicable to the renin-angiotensin system. Further details are presented in the reference [55].

The first step toward obtaining the recombinant AAV, is to construct a plasmid containing AAV terminal repeats (pAAV). Example below refers to AT1R-AS.

Method for preparation of pAAV-AT$_1$R-AS

The 749 bp fragment of the AT$_1$-R cDNA (–183 to 566) was amplified using polymerase chain reaction (PCR) and ligated to an AAV-derived vector (Fig. 2) in the antisense orientation,

Table 3. Features of AAV as a vector

Benefits	Drawbacks
1. Non-pathogenic	1. Capacity up to 4.4 kb
2. Very broad host range [54]	2. Complex packaging requires helper virus and helper plasmids with *rep* and *cap* genes
3. Virus sequences limited to ITRs (290 bp), therefore, does not evolve an inflammatory response	3. Multistep purification method
4. Long-lasting expression	

in place of *gfp*. The resulting plasmid vector (pAAV-AT$_1$R-AS) contained adeno-associated virus terminal repeats (TR), a cytomegalovirus promoter (Pcmv), the DNA encoding AT$_1$ receptor mRNA in the antisense orientation and a *neo*mycin resistance gene (*neo*^{r}) (Fig. 3). *Escherichia coli* bacteria (Sure II, Stratagene) were transformed with the plasmid according to the manufacturer protocol. The plasmid DNA was purified on CsCl gradient.

We constructed plasmids for both AT$_1$ receptor antisense and angiotensinogen antisense in the AAV-derived expression vector. Initially we used a plasmid containing AAV genome and 750 bp cDNA inserted into the AAV in the antisense direction downstream from the AAV promoter. The NG108-15 cells or hepatoma H4 cells were transfected with AT$_1$-R-AS or AGT-AS plasmid, respectively, using lipofectamine [56, 57]. In both cases, there were significant reductions in the appropriate proteins, namely AT$_1$-R and angiotensinogen. To test that the cells expressed AAV we used the *rep* gene product as a marker. Immunocytochemical staining with a rep protein antibody showed that the majority of cells in culture expressed the vector. A further development of the AAV was the insertion of more powerful and specific promoters than the p40 promoter. AAV with CMV promoter and *neo*mycin resistance (*neo*^{r}) gene as a selectable marker is now being used in our current experiments. The AAV cassette contained a marker, either the *gfp* gene which encodes the green fluorescent protein (Fig. 2, ref. [58]) or *lac*Z gene [59]. NG108-15 cells transfected with AAV plasmid containing the *gfp* and *neo*^{r} genes were selected by antibiotic, G418 (600 μg/ml), and the selected clones viewed for GFP expression. Very few cells died during selection. Two weeks after transfection all of the cells were expressing GFP. The transfection efficiency of this pAAV-*gfp* construct in different cell lines, including ATt20 (mouse pituitary cells), L929 (mouse fibroblasts), HEK293 (human embryonic kidney cells) and NG108-15 was over 50% [60].

Packaging rAAV

To prepare recombinant AAV, Human Embryonic Kidney (HEK293) cells are transfected with plasmid vector containing gene of interest in antisense orientation and AAV terminal repeats (pAAV-AS), together with helper plasmid delivering *rep* and *cap* genes (necessary for AAV replication) in *trans* using calcium phosphate method. Eight h after transfection, helper virus is added at a multiplicity of infection (MOI) of 5. Some methods use plasmids-only packaging, without the helper virus. There are few methods used to package and purify the rAAV. Details are described in refs [61, 62]. The titer of the virus is assayed using HEK293 cells, wild type AAV and adenovirus.

In our studies, recombinant AAV-AT$_1$R-AS was tested for AT$_1$ receptor inhibition *in vitro*, using vascular smooth muscle cells [37]. Transduced cells, without G418 selection, expressed

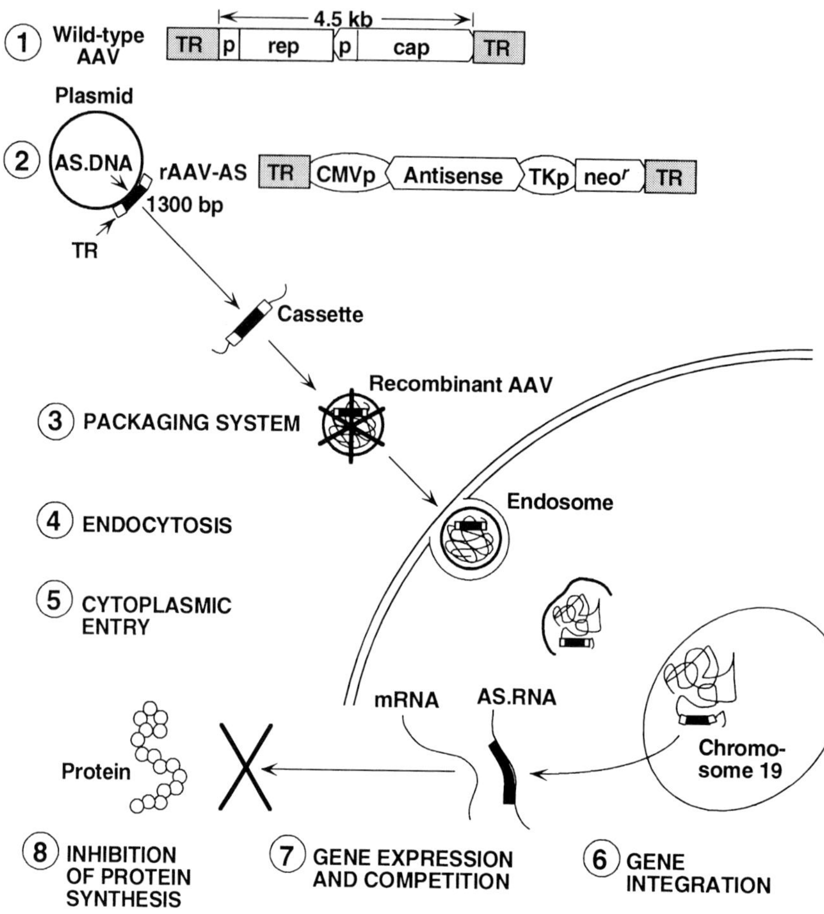

Fig. 1. Gene delivery of AS DNA with adeno-associated viral vector (AAV). 1. Wild-type AAV (simplified): terminal repeats (TR), native promoters (p), genes necessary for replication (rep) and capsid (cap). 2. AAV-based plasmid: The plasmid contains the terminal repeats (TR) characteristic of AAV. A cassette contains CMV promoter-driven antisense and TK promoter-driven neo^r gene for selection by G418 (geneticin). 3. Virus packaging: The packaging cell line is transfected with AAV-based plasmid and helper plasmid with *rep* and *cap* genes, and transduced with adenovirus as a helper virus. 4. Endocytosis: The viral vector fuses with the cell membrane by binding to adhesion molecules and becomes an endosome within the bilipid layer. 5. Cytoplasmic entry: The vesicle "opens" in the cytoplasm, releasing the vector that is transported to, and enters the nucleus. 6. Gene integration: AAV integrates with chromosome 19. It is not known if the addition of foreign DNA interferes with this integration. 7. Gene expression and competition: Genomic DNA in chromosome 19 produces an antisense RNA (AS-RNA). This competes with the natural mRNA and decreases its translation. 8. Inhibition of protein synthesis: A binding of antisense RNA to the mRNA prevents sliding through the ribosomal assembly to produce protein. The result of this gene delivery system should be reduction in the amount of protein, specifically targeted by the antisense.

the transgene for at least 8 weeks, had a decreased number of AT_1 receptors and reduced calcium response to angiotensin II stimulation.

Expression *in vivo* was tested first by direct injection into the brain. An AAV with the arginine vasopressin promoter (AVP) to drive a *lac*Z reporter gene was constructed. The vector expressed β-galactosidase in neurons of the paraventricular nucleus and supraoptic nucleus. The expression was in magnocellular cells which normally express AVP [59]. The expression was observed at 1 day, 1 week and after 1 month with no diminution of signal. This is an example of how AAV can be developed for specific tissue and/or cell gene expression and shows that AAV vectors can deliver foreign genes into adult brain for long periods of time.

Next, to test for effectiveness *in vivo*, rAAV-AT_1R-AS was microinfused into the lateral ventricles of adult male SHR. Control rats received AAV with *gfp* reporter gene but without the AS gene ('mock' vector) in vehicle which was artificial cerebrospinal fluid. Blood pressure was measured by tailcuff method. There was a significant decrease in systolic blood pressure (SBP) in one group of rats that received the rAAV-AS vector. No effect was observed in the controls. SBP decreased by 23 ± 2 mm Hg in the first week after administration. This drop in blood pressure was prolonged in 4 rats for 9 weeks, whereas controls had no reduction in blood pressure [63]. This was considerably longer than the longest effect observed with AS-ODN. Recombinant AAV-*gfp* expression in hypothalamus of the control rat group was de-

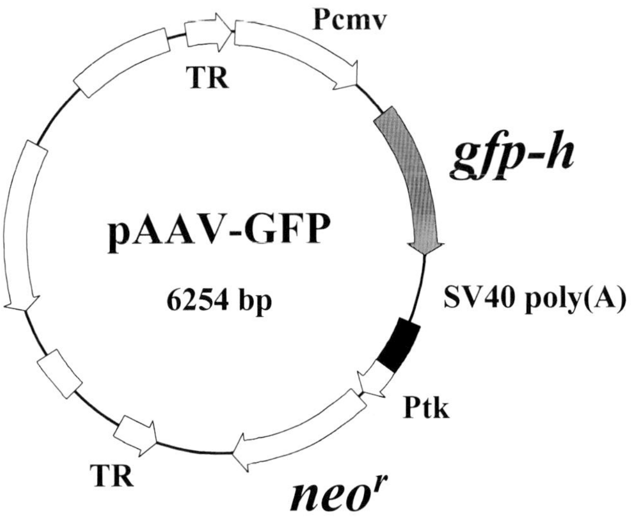

Fig. 2. Schematic diagram of the recombinant AAV vector containing the *gfp* gene. In the recombinant AAV-*gfp* vector almost all of the parental wild-type AAV genome (except the terminal repeats – TR) has been deleted and replaced with *gfp*, the gene encoding *Aequorea Victoria* green fluorescent protein gene, driven by a CMV promoter (Pcmv). The *neo*mycin resistance (*neo*^{r}) gene is under the control of thymidine kinase promoter (Ptk). The *gfp* serves as a reporter gene *in vitro* or *in vivo*, and the *neo*^{r} serves for selection *in vitro*.

tectable by RT-nested PCR 11 months after injection. Further, intra-cardiac injection of rAAV-AS in SHR, significantly reduced blood pressure and slowed the development of hypertension for several weeks [63, 64]. This result demonstrates that rAAV-AS in a single application, is effective in chronically inhibiting renin-angiotensin system. This encourages further research on gene regulation in cardiovascular diseases and to explore the most effective routes of delivery applicable to humans.

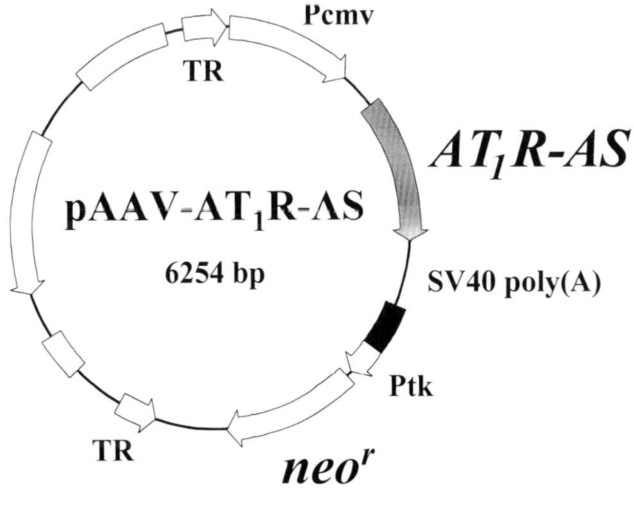

Fig. 3. Plasmid vector pAAV-AT$_1$R-AS contains a 750 bp fragment of the AT$_{1A}$ receptor cDNA in antisense orientation. Abbreviations: TR – AAV terminal repeat; Pcmv – human cytomegalovirus early promoter; SV40 poly (A) – polyadenylation signal from simian virus; Ptk – thymidine kinase promoter; *neo*^{r} – *neo*mycin phosphotransferase gene from Tn5. Other promoters have been substituted for CMV, including the arginine vasopressin (AVP), neuron specific enolase (NSE) and glial fibrillary acid protein (GFAP) promoters.

Acknowledgements

This work has been supported by NIH (MERIT) grant HL-23774 and grants from the American Heart Association, Florida Affiliate. Thanks to Nick Muzyczka, Ed Meyer, Birgitta Kimura, Leping Shen and Sara Galli for help in developing these methods and to Gayle Butters for word processing.

References

1. Wagner RW: Gene inhibition using antisense oligodeoxynucleotides. Nature 372: 333–335, 1994
2. Crooke ST: Therapeutic applications of oligonucleotides. Annu Rev Pharmacol Toxicol 32: 329–376, 1992
3. Stein CA, Cheng Y-C: Antisense oligonucleotides: Is the bullet really magical?. Science 261: 1004–1012, 1993
4. Chiasson BJ, Hooper ML, Murphy PR, Robertson HA: Antisense oligonucleotide eliminates *in vivo* expression of c-fos in mammalian brain. Eur J Pharmacol 277: 451–453, 1992
5. Wahlestedt C, Pich EM, Koob GF, Yee F, Heilig M: Modulation of anxiety and neuropeptide Y-Y1 receptors by antisense oligodeoxynucleotides. Science 259: 528–531, 1993
6. Wahlestedt C, Golanov E, Yamamoto S, Yee F, Ericson H, Yoo H, Inturrisi CE, Reis DJ: Antisense oligonucleotides to NMDA R1 receptor channel protect cortical neurons from excitotoxicity and reduce focal ischemic infarctions. Nature 363: 260–263, 1993
7. McCarthy MM, Masters DB, Rimvall K, Schwartz-Giblin S, Pfaff DW: Intracerebral administration of antisense oligodeoxynucleotides to GAD65 and GAD67 mRNAs modulates reproductive behavior in the female rat. Brain Res 636: 209–220, 1994
8. Phillips MI: Antisense inhibition and adeno-associated viral vector delivery for reducing hypertension. Hypertension 29: 177–187, 1997
9. Jeunemaitre X, Soubrier F, Kotelevtsev YV, Lifton RP, Williams CS, Charru A, Hunt SC, Hopkins PN, William RR, Lalouel JM, Corvol P: Molecular basis of human hypertension: Role of angiotensinogen. Cell 71: 169–180, 1992
10. Kim HS, Krege JH, Kluckman KD, Hagaman JR, Hodgin JB, Best CF, Jennette JC, Coffman TM, Maeda N, Smithies O: Genetic control of blood pressure and the angiotensinogen locus. Proc Natl Acad Sci 92: 2735–2739, 1995
11. Tanimoto K, Sugiyama F, Goto Y, Ishida J, Takimoto E, Yagami K, Fukamizu A, Murakami K: Angiotensinogen-deficient mice with hypotension. J Biol Chem 269: 31334–31337, 1994
12. Phillips MI, Mann JFE, Haebara H, Hoffman WE, Dietz R, Schelling P, Ganten D: Lowering of hypertension by central saralasin in the absence of plasma renin. Nature 270: 445–447, 1977
13. Nazarali AJ, Gutkind JS, Correa FM, Saavedra JM: Decreased angiotensin II receptors in subfornical organ of spontaneously hypertensive rats after chronic antihypertensive treatment with enalapril: Am J Hypertens 3: 59–61, 1990

14. Phillips MI, Kimura B: Brain angiotensin in the developing spontaneously hypertensive rat. J Hypertens 6: 607–612, 1988
15. Gyurko R, Wielbo D, Phillips MI: Antisense inhibition of AT_1 receptor mRNA and angiotensinogen mRNA in the brain of spontaneously hypertensive rats reduces hypertension of neurogenic origin. Reg Pep 49: 167–174, 1993
16. Simons RW: Naturally occurring antisense RNA control – a brief review. Gene 72: 35–44, 1988
17. Bennett CF, Condon TP, Grimm S, Chan H, Chiang MY: Inhibition of endothelial cell adhesion molecule expression with antisense oligonucleotides. J Immunol 152: 3530–3540, 1994
18. Bacon TA, Wickstrom E: Walking along human c-myc mRNA with antisense oligonucleotides: maximum efficacy at the 5′ *cap* region. Oncogene Res 6: 13–19, 1991
19. Cowsert LM, Fox MC, Zon G, Mirabelli CK: *In vitro* evaluation of phosphorothioate oligonucleotides targeted to the E2 mRNA of papillomavirus: Potential treatment for genital warts. Antimicrob Agents Chemother 37: 171–177, 1993
20. Lima WF, Monia BP, Ecker DJ, Freier SM: Implication of RNA structure on antisense oligonucleotide hybridization kinetics. Biochemistry 31: 12055–12061, 1992
21. Sczakiel G, Homann K, Rittner K: Computer-aided search for effective antisense RNA target sequences of the human immunodeficiency virus type 1. Antisense Res Dev 3: 45–52, 1993
22. Jaroszewski JW, Syi JL, Ghosh M, Ghosh K, Cohen JS: Targeting of antisense DNA: comparison of activity of anti-rabbit beta-globin oligodeoxyribonucleoside phosphorothioates with computer predictions of mRNA folding. Antisense Res Dev 3: 339–348, 1993
23. Campbell JM, Bacon TA, Wickstrom E: Oligodeoxynucleotide phosphorothioate stability in subcellular extracts, culture media, ser and cerebrospinal fluid. J Biochem Biophys Meth 20: 259–267, 1990
24. Li B, Hughes JA, Phillips MI: Uptake and efflux of intact antisense phosphorthioate deoxyoligonucleotide directed against angiotensin receptors in bovine adrenal cells. Neurochem Int 31: 393–403, 1997
25. Iversen PL, Zhu S, Meyer A, Zon G: Cellular uptake and subcellular distribution of phosphorothioate oligonucleotides into cultured cells. Antisense Res Dev 2: 211–222, 1992
26. Loke SL, Stein CA, Zhang XH, Mori K, Nakanishi M, Subasinghe C, Cohen JS, Neckers LM: Characterization of oligonucleotide transport into living cells. PNAS 86: 3474–3478, 1989
27. Wagner RW, Matteucci MD, Lewis JG, Gutierrez AJ, Moulds C, Froehler BC: Antisense gene inhibition by oligonucleotides containing C-5 propyne pyrimidines. Science 260: 1510–1513, 1993
28. Iversen PL, Mata J, Tracewell WG, Zon G: Pharmacokinetics of an antisense phosphorothioate oligodeoxynucleotide against rev from human immunodeficiency virus type 1 in the adult male rat following single injections and continuous infusion. Antisense Res Dev 4: 43–52, 1994
29. Phillips MI, Wielbo D, Gyurko R: Antisense inhibition of hypertension: A new strategy for renin-angiotensin candidate genes. Kidney Int 46: 1554–1556, 1994
30. Ambuehl P, Gyurko R, Phillips MI: A decrease of angiotensin receptor number in rat brain nuclei by antisense oligonucleotides against the angiotensin AT_{1A} receptor. Reg Pep 59:171–182, 1995
31. Wielbo D, Sernia C, Gyurko R, Phillips MI: Antisense inhibition of hypertension in the spontaneously hypertensive rat. Hypertension 25: 314–319, 1995
32. Morishita R, Gibbons GH, Kaneda Y, Ogihara T, Dzau VJ: Pharmacokinetics of antisense oligodeoxyribonucleotides (cyclin B1 and CDC 2 kinase) in the vessel wall *in vivo*: Enhanced therapeutic utility for restenosis by HVJ-liposome delivery. Gene 149: 13–19, 1994
33. Colige A, Sokolov BP, Nugent P, Baserge R, Prockop DJ: Use of an antisense oligonucleotide to inhibit expression of a mutated human procollagen gene (COL1A1) in transfected mouse 3T3 cells. Biochemistry 32: 7–11, 1993
34. Tomita N, Morishita R, Higaki J, Kaneda Y, Mikami H, Ogihara T: *In vivo* transfer of antisense oligonucleotide against rat angiotensinogen with HVJ-liposome delivery resulted in reduction of blood pressure in SHR. Hypertension 24: 397–402, 1994
35. Wielbo D, Simon A, Phillips MI, Toffolo S: Inhibition of hypertension by peripheral administration of antisense oligodeoxynucleotides. Hypertension 28: 147–151, 1996
36. Zelles T, Mohuczy D, Phillips MI: Inhibition of angiotensin receptor (AT_1) expression in NG108-15 cells transfected with AT_1 receptor antisense in an adeno-associated viral vector. Soc Neurosci 41.18: (abstr) 83, 1996
37. Mohuczy D, Gelband C, Phillips MI: Antisense inhibition of AT_1 receptor in vascular smooth muscle cells using adeno-associated virus-based vector. Hypertension 33(1 Pt 2): 354–359, 1999
38. Mohuczy D, Tang XP, Kimura B, Phillips MI: Adeno-associated virus-based vector with angiotensinogen cDNA is effective in rat hepatoma cells. FASEB J 13: A484, 398.18, 1999
39. Tang XP, Mohuczy D, Zhang Y, Kimura B, Galli SM, Phillips MI: Intravenous angiotensinogen antisense in AAV-based vector decreases hypertension. Am J Physiol 277: H2392–H2399, 1999
40. Mulligan RC: The basic science of gene therapy. Science 260: 926–932, 1993
41. Iyer SN, Lu D, Katovich MJ, Raizada MK: Chronic control of high blood pressure in spontaneously hypertensive rat by delivery of angiotensin type 1 receptor antisense. PNAS 93: 9960–9965, 1996
42. Brody SL, Jaffe HA, Eissa NT, Daniel C: Administration of an adenovirus containing the human CFTR cDNA to the respiratory tract of individuals with cystic fibrosis. Nat Genet 8: 42–51, 1994
43. Quantin B, Perricaudet LD, Tajbakhsh S, Mandel JL: Adenovirus as an expression vector in muscle cells *in vivo*. PNAS 89: 2581–2584, 1992
44. Le Gal La Salle G, Robert JJ, Berrard S, Ridoux V, Stratford-Perricaudet LD, Perricaudet M, Mallet J: An adenovirus vector for gene transfer into the neurons and glia in the brain. Science 259: 988–990, 1993
45. Lu D, Yang H, Raizada MK: Attenuation of Ang II actions by adenovirus delivery of AT_1 receptor antisense in neurons and SMC. Am J Physiol 274 (2 Pt 2): H719–H727, 1998
46. Burcin MM, Schiedner G, Kochanek S, Tsai SY, O'Malley BW: Adenovirus-mediated regulable target gene expression *in vivo*. PNAS 96: 355–360, 1999
47. Kochanek S, Clemens PR, Mitani K, Chen HH, Chan S, Caskey CT: A new adenoviral vector: Replacement of all viral coding sequences with 28 kb of DNA independently expressing both full-length dystrophin and beta-galactosidase. PNAS 93: 5731–5736, 1996
48. Muzyczka N, McLaughin S: Use of adeno-associated virus as a mammalian transduction vector. In: Y. Gluzman, S.H. Hughes (eds). Current Communications in Molecular Biology: Viral Vectors. Cold Spring Harbor Laboratory Press, NY, 1988, pp 39–44
49. Ponnazhagan S, Nallari ML, Srivastava A: Suppression of human alpha-globin gene expression mediated by the recombinant adeno-associated virus 2-based antisense vectors. J Exp Med 179: 733–738, 1994
50. Chatterjee S, Johnson PR, Wong KK: Dual-target inhibition of HIV-1 *in vitro* by means of an adeno-associated virus antisense vector. Science 258: 1485–1488, 1992
51. Muzyczka N: Use of adeno-associated virus as a general transduction vector for mammalian cells. In: Current Topics in Microbiology and Immunology, vol. 158. Springer-Verlag, Berlin, 1992, pp 97–129
52. Samulski RJ, Zhu X, Xiao X, Brook JD, Housman DE, Epstein N, Hunter LA: Targeted integration of adeno-associated virus (AAV) into human chromosome 19. EMBO J 10: 3941–3950, 1991
53. Linden RM, Winocour E, Berns KI: The recombination signals for adeno-associated virus site-specific integration. PNAS 93: 7966–7972, 1996

54. Lebkowski JS, McNally MM, Okarma TB, Lerch B: Adeno-associated virus: A vector system for efficient introduction and integration of DNA into a variety of mammalian cell types. Mol Cell Biol 8: 3988–3996, 1988
55. Flotte TR, Carter B, Conrad C, Guggino W, Reynolds T, Rosenstein B, Taylor G, Walden S, Wetzel R: A phase I study of an adeno-associated virus-CTFR gene vector in adult CF patients with mild lung disease. Human Gene Therapy. 7: 1145–1159, 1996
56. Gyurko R, Phillips MI: Antisense expression vector decreases angiotensin receptor binding in NG108-15 cells. Exp Biol: (abstr) 1915, 1995
57. Gyurko R, Wu P, Sernia C, Meyer E, Phillips MI: Antisense expression vector decreases angiotensinogen synthesis in H-4 hepatoma cells. American Heart Association 48th Annual Council for High Blood Pressure (abstr) 1994
58. Zolotukhin S, Potter M, Hauswirth WW, Guy J, Muzyczka N: A 'humanized' green fluorescent protein cDNA adapted for high-level expression in mammalian cells. J Virol 70: 4646–4654, 1996
59. Wu P, Du B, Phillips MI, Bui J, Terwilliger EF: Adeno-associated viral vector mediated transgene integration in non-dividing cells. J Virol 72: 5919–5926, 1998
60. Mohuczy D, Phillips MI: Adeno-associated virus vector as a highly efficient transporter of exogenous DNA into cells. FASEB J 10: A447, 1996
61. Xiao X, Li J, Samulski RJ: Production of high-titer recombinant adeno-associated virus vectors in the absence of helper adenovirus. J Virol 72: 2224–2232, 1998
62. Conway J, Rhys C, Zolotukhin I, Zolotukhin S, Muzyczka N, Hayward G, Byrne B: High-titer recombinant adeno-associated virus production utilizing a recombinant herpes simplex virus type 1 vector expressing AAV-2 rep and Cap. Gene Ther 6: 986–993, 1999
63. Phillips MI, Mohuczy-Dominiak D, Coffey M, Wu P, Galli SM, Zelles T: Prolonged reduction of high blood pressure with an *in vivo*, non-pathogenic, adeno-associated viral vector delivery of AT_1-R mRNA antisense. Hypertension 29: 374–380, 1997
64. Kimura B, Mohuczy D, Phillips MI: Injection of AT_1 receptor antisense in adeno-associated virus (AAV) attenuates hypertension in the spontaneously hypertensive rat (SHR). FASEB J 12: A90, 1998

Molecular and Cellular Biochemistry **212:** 155–169, 2000.

Angiotensin II induces gene transcription through cell-type-dependent effects on the nuclear factor-κB (NF-κB) transcription factor

Allan R. Brasier,[1] M. Jamaluddin,[1] Youqi Han,[1] Cam Patterson[2] and Marschall S. Runge[2]

Department of Internal Medicine, [1]Sealy Center for Molecular Science; [2]Sealy Center for Molecular Cardiology, The University of Texas Medical Branch Galveston, TX, USA

Abstract

The vasopressor octapeptide, angiotensin II (Ang II), exerts homeostatic responses in cardiovascular tissues, including the heart, blood vessel wall, adrenal cortex and liver (a major source of circulating plasma proteins). One of the effects of Ang II is to induce expression of regulatory, structural and cytokine genes that play important roles in long-term control of blood pressure, vascular remodeling, cardiac hypertrophy and inflammation. The identification of nuclear signaling pathways and target transcription factors has provide important insight into cellular responses and the spectrum of genes controlled by Ang II. Here we will review how Ang II activates the transcription factors, Activator Protein 1 (AP-1), Signal Transducer and Activator of Transcription (STATs), and Nuclear Factor-κB (NF-κB). NF-κB is of particular interest because it is an important mediator of resynthesis of the Ang II precursor, angiotensinogen AGT. Through this positive feedback loop, long-term changes in the activity of the renin angiotensin system occur. Although NF-κB is ubiquitously expressed, surprisingly the mechanism for Ang II-inducible NF-κB regulation differs between aortic smooth muscle cells (VSMCs) and hepatocytes. In VSMC, Ang II induces nuclear translocation of cytoplasmic transactivatory NF-κB proteins through proteolysis of its inhibitor, IκB. By contrast, in hepatocytes, Ang II induces large nuclear isoforms of NF-κB1 to bind DNA through a mechanism independent of changes in IκB turnover. NF-κB activation depends upon the activity of DAG-sensitive PKC isoforms and ROS signaling pathway. These observations indicate that significant differences exist in Ang II signaling depending upon cell-type involved and suggest the possibility that tissue-selective modulation of Ang II effects is possible in the cardiovascular system. (Mol Cell Biochem **212**. 155–169, 2000)

Key words: angiotensin II, NF-κB1, NF-κB, myocyte, hepatocyte, Renin Angiotensin System (RAS), transcription factor, STAT, AP-1

Abbreviations: AGT – angiotensinogen; Ang II – angiotensin II (amino acids 1–8 of angiotensinogen precursor); APRE – acute phase response element (of the rat angiotensinogen gene); AP-1 – activator protein-1; AT_1 – type 1 angiotensin II receptor; IL-6 – interleukin-6; iRAS – intravascular renin-angiotensin system; JNK – cJun N terminal kinase; lRAS - local renin-angiotensin system; NF-κB – nuclear factor κB; PKC – protein kinase C; STATs – signal transducers and activators of transcription; TNFα – tumor necrosis factor alpha

Address for offprints: A.R. Brasier, Division of Endocrinology, MRB 8.138, University of Texas Medical Branch, 301 University Boulevard, Galveston, TX 77555-1060, USA

Introduction

Angiotensin II (Ang II) is a potent vasopressor octapeptide formed from sequential proteolysis of its precursor, angiotensinogen (AGT), within the extracellular space. In the first rate-limiting step in Ang II formation, AGT circulating in plasma is proteolyzed into the inactive decapeptide, angiotensin 1 (Ang I [amino acids 1–10 of the AGT precursor]), by an aminopeptidase (Renin, E.C. 3.4.23.15) secreted from specialized juxtaglomerular cells of the kidney (reviewed in [1]). Angiotensin I is then rapidly and stoichiometrically converted into the potent vasopressor octapeptide, angiotensin II (Ang 1–8), the major effector peptide of the intravasular RAS (iRAS), by the effects of angiotensin converting enzyme (ACE, E.C. 3.4.15.1). Ang II binds with high affinity to the seven transmembrane-spanning type 1 angiotensin (AT_1) receptor through which it exerts its homeostatic effects on target tissues. Of particular importance, Ang II induces vasoconstriction and aldosterone secretion that results acutely in increased peripheral vascular resistance and blood volume, respectively, and chronically, in long-term changes in blood pressure [2]. Ang II stimulation also induces long-term changes in cardiovascular tissues including proliferation, hypertrophy or acquisition of a secretory phenotype. The cellular mechanisms by which these responses occur have been the subject of intense study; here we will review the genetic effects of Ang II in responsive tissues (vascular smooth muscle and liver cells).

Renin-Angiotensinsin System (RASs)

The classic view of the circulating renin-angiotensin system (termed here the intravascular RAS [iRAS]) has held that Ang II formation is controlled by regulated secretion of kidney-derived renin in response to acute hypotensive insults (reviewed in [3]). Although this is true for acute regulation of iRAS, a separate mechanism has been identified that influences long-term changes in RAS activity. Surprisingly, under normal conditions, AGT circulates at concentrations (0.6 μmol/L) lower than the Michaelis-Menten Equilibrium constant (K_m) of renin (1 μmol/L). This creates a situation where both AGT and renin concentrations are rate-limiting for maximal velocity of Ang I formation [1], so that changes in AGT concentration influence the long-term activity of the RAS. Various observations support this mechanism. Work in our lab and others has shown that AGT is encoded by a highly inducible gene regulated by myriad physiological systems than influence long-term changes in blood pressure. For example, in ambulatory patients, a positive correlation exists between AGT and blood pressure [4]. Steroid administration, a potent inducer of AGT gene expression and circulating AGT levels increased blood pressure in humans [5]. Mice with reduced copies of *AGT* show a linear, ~ 8 mm Hg, reduction in blood pressure for each copy of AGT deleted [6]. Transgenic mice overexpressing human *AGT* and *renin* genes are severely hypertensive [7]. Finally, gene linkage studies in humans have associated polymorphisms of AGT alleles with hypertensive disease states, including heritable forms of essential hypertension [8, 9], preeclampsia [10, 11] and an intermediate phenotype of hypertension characterized by a blunted vascular response to Ang II [12]. Taken together, these observations indicate that circulating AGT is an important determinant of RAS activity as it relates to long-term changes in blood pressure control. Of relevance here is that one of the important regulators of AGT expression is Ang II itself; this 'positive feedback loop' probably allows for resynthesis of depleted AGT as a result of its intravascular proteolysis (reviewed in [13]).

In addition to the classic iRAS, where Ang II production and effects are limited to the intravascular space, local RASs (lRASs) are likely to play an important role in cardiovascular physiology. The existence of lRAS is supported by an arterio-venous 'step-up' in AII concentration and the local expression of AGT and factors required for its conversion into Ang II (reviewed in [14]). lRASs are self-contained Ang II-generating systems that appear to be important in local organ function. In this system locally-generated (or systemically delivered) angiotensinogen is sequentially proteolyzed into Ang II. Although renin may be involved in the initial rate-limiting conversion of AGT into Ang I, its requirement may be replaced by cathepsin D, pepsin, or other renin-like aspartyl proteases (reviewed in [14]). lRASs have been described for adipose tissue [15–17], brain [18, 19], kidney [20, 21], adrenal [22, 23], arterial wall [10, 24, 25] and myocardium [17, 26, 27]. Regulation of lRAS activity differs from that of the iRAS, where AGT expression is under the control of tissue-specific regulatory influences. Evidence for such local, tissue-specific control includes: (i) the kidney, where glucocorticoids, potent inducers of hepatic AGT expression, do not stimulate its expression [28, 29]; and (ii) the arterial wall, where AGT is regulated by dietary sodium whereas its expression in surrounding periaortic fat is not sodium-sensitive [25]. The mechanisms controlling AGT expression in extrahepatic tissues is at present, descriptive and await definitive characterization. Moreover, the precise physiological role of lRAS has been difficult to separate from that of the iRASs, although there is reason to support the hypothesis that lRASs participate in the process of myocardial hypertrophy following pressure overload, vascular growth/reactivity, and control of kidney autoregulation.

Ang II Intracellular signaling

Effects of Ang II in responsive tissues is primarily mediated by the type 1 AT receptor (AT_1). AT_1 is a seven transmembrane-spanning receptor coupled to pertussis toxin-sensitive G proteins and binds specifically to Ang II with an equilibrium disassociation constant of ~ 2 nmol/L [30, 31]. A single class of receptor binding sites is expressed in vascular smooth muscle cells (USMCs), liver, brain myocardium and adrenal cortex [30, 32]. The presence of a single class of binding sites in normal tissues, sensitivity to AT_1 (but not AT_2)-specific receptor antagonists, and genetic studies have suggested that AT_1 is the primary receptor for Ang II. In particular, it was recently reported that mice with homozygous deletion of the duplicate AT_1 genes have marked hypotension and alterations in renal morphology [33]. These mice are phenotypic copies of those deficient in the angiotensinogen locus [6]. Together, these data indicate that AT_1 is the primary mediator of Ang II effects in normal tissues. In hepatocytes, Ang II stimulates the generation of *myo*-inositol 1,4,5 trisphosphate (IP_3) via phospholipase C resulting in increased intracellular calcium release. Similarly, Ang II inhibits the actions of glucagon and other hormones that activate Protein Kinase A, an effect that may be mediated by PLC-catalyzed hydrolysis of phosphatidylcholine [34]. Further downstream, Ang II activates translocation of typical and novel protein kinase C (PKC) isoforms (ε, δ and α [35]). The effects of Ang II on VSMCs have been reviewed [36], where it induces early activation of phospholipase isoforms. Of these, phospholipase D (PLD), responsible for conversion of phosphatidylcholine to choline and phosphatidic acid, exhibits a prolonged activation and therefore represents a major source of the diacylglycerol production required for activation of PKCs [36].

Ang II also stimulates phospholipase A2 (PLA_2), a family of enzymes that produces arachidonic acid from membrane phospholipids. Through PLA_2, arachidonic acid metabolites may assist in the assembly of a VSMC NAD(P)H oxidase system. Activation of this system produces a delayed and sustained production of reactive oxygen species in Ang II-stimulated VSMC, which serve as essential secondary messengers mediating the mitogenic effects of Ang II [37].

NAD(P)H oxidases generate O_2^- by an NADH- or NADPH-derived reduction of molecular oxygen. O_2^-. This system is best characterized in phagocytic cells, in which O_2^- is produced in large amounts by an oxidase which preferentially utilizes NADPH as its substrate. The components of the phagocytic oxidase include the membrane-bound heterodimeric flavocytochrome b558, consisting of the proteins gp91phox and p22phox, and the cytoplasmic components p47phox, p67phox, and Rac1 or Rac2. In contrast with the phagocytic oxidase, the VSMC oxidase utilizes NADH as its preferred substrate and produces O_2^- in relatively low but sustained levels. In VSMC, O_2^-, being highly unstable, rapidly dismutates to H_2O_2, which appears to be the primary mitogenic ROS in this cell type. p22phox, p47phox and Rac2 are believed to be components of the VSMC oxidase, but gp91phox and p67phox have not been detected [37, 38]. Recently, a VSMC-specific protein with similarity to gp91phox, called Mox1, was identified as a component of the VSMC O_2^--generating enzyme [38]. The presence of cell type-specific components of the NAD(P)H oxidase may account for the differences in substrate preferences and ROS generation seen in phagocytic and non-phagocytic cells. Presumably, the VSMC oxidase is optimized such that its stimulation by Ang II and thrombin elicits mitogenesis, whereas activation of the phagocytic oxidase has cytotoxic effects.

In VSMC and cardiac myocytes and fibroblasts, Ang II induces tyrosine phosphorylation of intracellular proteins. The exciting discovery that receptor-associated tyrosine kinases (JAK and TYK kinases) are associated with, and are regulated by, AT_1 [39–41] has shed new light on genetic effects of Ang II through the signal transducers and activators of transcription. AT_1 signaling is also coupled to activation of the mitogen-activated protein kinases (MAPK), ERK1 and ERK2, requiring the atypical protein kinase C, PKCz [42]. In adrenal glomerulosa cells phospholipase C is activated via a pertussis toxin-sensitive system and is responsible for the rapid formation of IP_3 and mobilization of intracellular calcium that initiates aldosterone secretion [43] and mitogenesis mediated by early gene synthesis [44].

General mechanisms for hormone-inducible gene expression

Expression of genes important in homeostatic growth responses is tightly regulated through multiple mechanisms. In many cases, the process of transcriptional regulation, where mRNA is synthesized from a DNA template, is the predominant mechanism for gene expression control (although in particular circumstances effects on mRNA stability in translation control are possible). *Cis* regulatory elements, control regions located in flanking sequences of non-coding DNA, are subject to assembly in chromatin (nucleosomes, linker histone H1 and higher order structures), and simultaneously are targets for binding by sequence-specific transcription factors. Studies of gene promoters maintained within their native chromatin contexts have shown that inducible transcription initiation is a coupled process involving transcription factor binding, nucleosomal reconfiguration, pre-initiation complex formation, promoter clearance and transcript elongation, resulting in pre-mRNA transcript formation (reviewed in [45]). The process of gene activation initiated through transcription factor binding activates genes via two mechanisms. In 'derepression', binding induces the nucleosomal rearrangement over DNA regions where DNA binding proteins are occluded from binding. In 'activation' binding of the transcription factor recruits interactions with additional factors and bridging 'co-activators' to couple with the preformed pre-initiation complex. These processes are schematically diagrammed in Fig. 1.

The derepression/activation mechanism is relevant to hormone-induced transcription where its presence has been documented for effecting glucocorticoid receptor-inducible changes in nucleosomal positioning for the retrovirus mouse mammary tumor virus (MMTV) promoter. In the absence of hormones, MMTV expression is repressed through the assembly of nucleosomes [46]. Under basal conditions, two nucleosomes constitutively bind over regions corresponding to the NF-1 and TATA boxes, occluding interaction with their cognate proteins. Following glucocorticoid stimulation, the glucocorticoid receptor binds to an upstream enhancer, repositioning the nucleosomes (derepression) allowing NF-1 and TATA box binding proteins to assemble (activation) and transcribe the MMTV provirus at a high rate. This phenomenon can be generalized to Ang II-inducible transcription factors. For example, we have recently shown that the Ang II-inducible transcription factor, NF-κB, induces a similar mechanism of derepression/activation on a target cytokine gene promoter called interleukin-8 (IL-8; ref. [47]). Subsequent sections of this review will show that Ang II is a potent activator of IL-8 transcription. Analysis of the proteins binding to the native IL-8 gene through the technique of ligation-mediated PCR, shows inducible TATA box binding following activation of the inducible transcription factor, NF-κB; see ref. [47]. Thus the derepression/activation mechanism may be relevant to at least a subset of Ang II-inducible target genes. Subsequent sections, will delineate the effects of Ang II on three transcription factor families that are linked to homeostatic vascular responses.

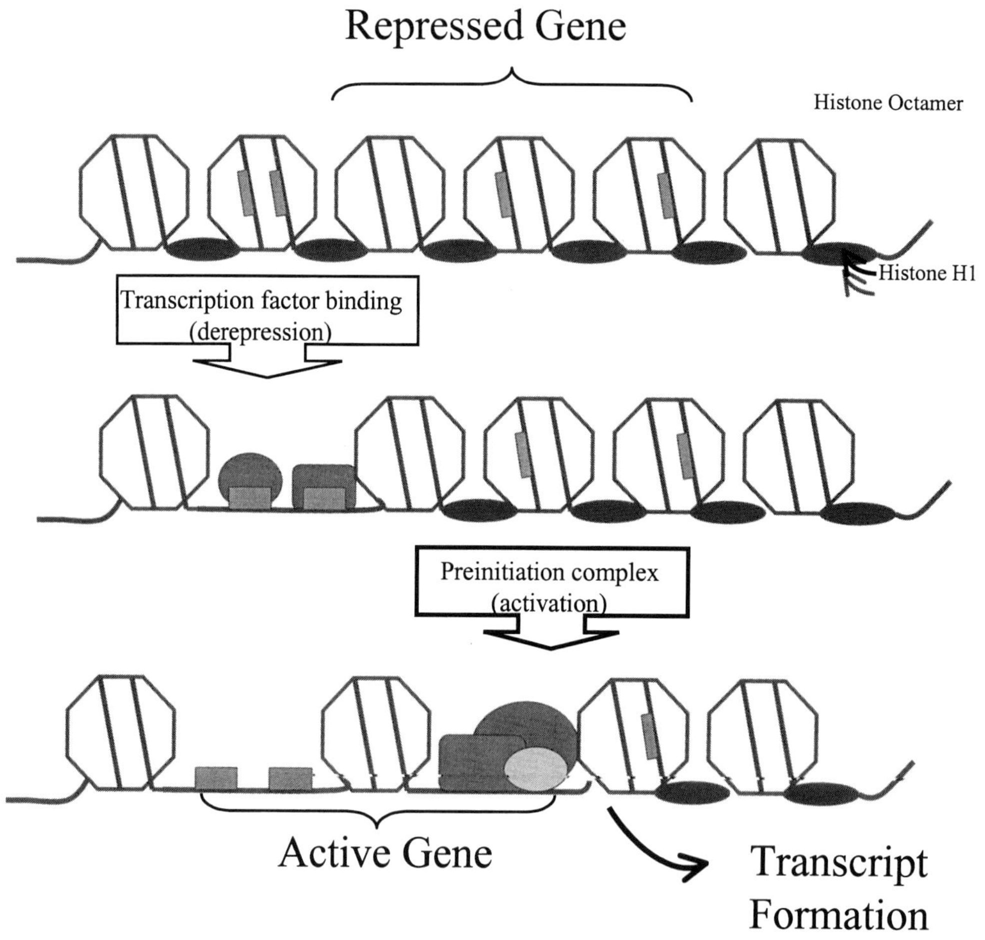

Fig. 1. Schematic view of coupled process of transcription activation. Transcription of genes within native chromatin context is a multistep process coordinated by protein, nucleosomal and DNA interactions. At the top is a schematic view of a repressed gene promoter bound by core nucleosomes and linker H1 histones. Following inducible transcription factor binding, energy-dependent modifications of nucleosomal positioning occur ('derepression'). This allows recruitment of additional transcription factors and binding of the preinitiation complex over the TATA box ('activation'). Finally, RNA polymerase with associated factors is recruited, melts the duplex at the start of transcription, clears the transcription start site and elongates the pre-mRNA transcript (reviewed in [45]).

agonist was unable to induce JNK activity [57]. These pathways are schematically shown in Fig. 2.

Ang II-inducible transcription factors: STATs

The signal transducers and activators of transcription (STATs) are a family of cytoplasmic transcription factors originally discovered to be the targets of cytokine/growth factor receptors that lack intrinsic kinase activity (interleukin-6 (IL-6), interferon (IFN), and growth hormone (GH)). At least six distinct protein members have been identified that are each encoded by separate genes and physically associate with the cytoplasmic domains of cytokine/growth factor receptors (reviewed in [60, 61]). STAT activation occurs following ligand-induced activation of specific receptor-associated tyrosine kinases (JAK and TYK kinases). JAK and TYK phosphorylate receptor-associated STATs on tyrosine residues in a manner which produces the simultaneous activities of DNA-binding and nuclear translocation. Although Ang II was known to activate tyrosine phosphorylation of cellular proteins shown to be important for its intracellular signaling, such as PLC-γ [36], it was a surprising observation that the G-protein-coupled AT_1 also directly activated JAK2/TYK2 in VSMC [41]. JAK2/TYK2 phosphorylation was rapidly detectable within 5–30 min and was coincident with STAT1 phosphorylation (with kinetics indistinguishable from that induced by the prototypic activator, IFNγ) and STAT 2 phosphorylation (also indistinguishable from its prototypic inducer, IFNα). STAT3 phosphorylation, however, was delayed and required approximately 1 h for maximal effects. JAK2/STAT 1/STAT 3 are physically associated with the AT_1 receptor in VSMC [41] and in cardiac myocytes [39]. Baker's group has extended these observations to link STAT activation with early gene (c-Fos) expression, where Ang II-inducible STATs are a major component of the complex that binds to the *c-Fos*, *sis*-inducing regulatory element [62]. In addition to tyrosine kinase activity (an event necessary and sufficient for STAT activation), Ang II induces serine phosphorylation of the transactivation domain of STAT3 [63]. Serine kinase activity was sensitive to the MAP kinase inhibitor, PD98059. Although not directly demonstrated in this study, serine phosphorylation of STAT is a potent inducer of its transcriptional activation potential [64]. These mechanisms are schematically diagrammed in Fig. 3.

Together, these observations indicate that Ang II controls nuclear abundance of both transcription factors AP-1 (through increased synthesis) and STATs (through cytoplasmic to nuclear translocation) and, in addition, controls their transactivation potential through signals generated via the MAP kinase pathway. This coordinate signaling process may play role in hypertrophic responses to pressure overload in the heart [27, 65, 66] and cellular proliferation in the vessel wall in response to injury [50, 67].

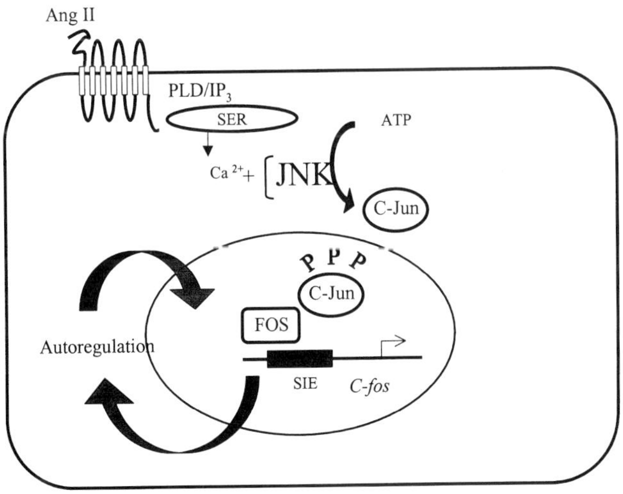

Fig. 2. Mechanism for Ang II effects on AP-1. Schematic view of relevant signaling pathways for AP-1 regulation. JNK activation is linked to intracellular calcium spike [57] which results in the N-terminal phosphorylation of C-Jun at serine residues. Phospho-C-Jun as a homodimer or heterodimer with C-Fos increases transcription at target sites, including the C-Fos and C-Jun promoters themselves, resulting in enhanced synthesis of members of the AP-1 transcription factor complex. PLD – phospholipase D – IP_3, *myo*-inositol trisphosphate; SER – smooth endoplasmic reticulum; SIE – sis inducing element (of C-Fos promoter).

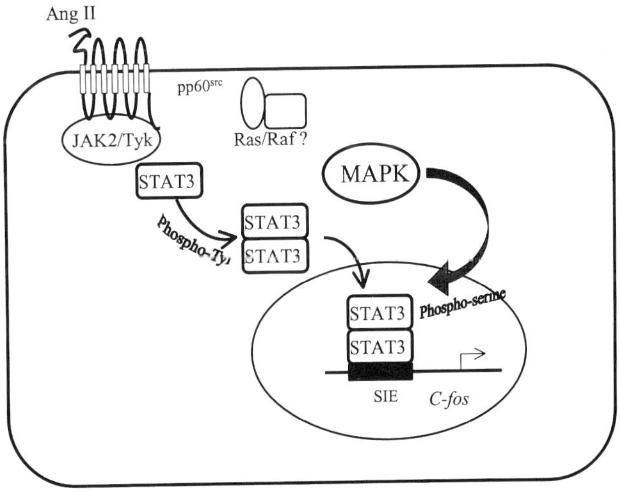

Fig. 3. Mechanism for Ang II effects on STAT in VSMC and cardiomyocytes. Schematic view of relevant signaling pathways for STAT regulation. The tyrosine kinases JAK2/Tyk and STAT3 are physically associated with the AT_1 receptor cytoplasmic domain. Following ligand activation, STAT3 is tyrosine phosphorylated, allowing the protein to dimerize and translocate into the nucleus. Mitogen activated protein kinase (MAPK), perhaps through $pp60^{src}$ /RAS/Raf activation serine phosphorylates the STAT3 activation domain, increasing its transcriptional activation potential.

tion to tyrosine kinase activity (an event necessary and sufficient for STAT activation), Ang II induces serine phosphorylation of the transactivation domain of STAT3 [63]. Serine kinase activity was sensitive to the MAP kinase inhibitor, PD98059. Although not directly demonstrated in this study, serine phosphorylation of STAT is a potent inducer of its transcriptional activation potential [64]. These mechanisms are schematically diagrammed in Fig. 3.

Together, these observations indicate that Ang II controls nuclear abundance of both transcription factors AP-1 (through increased synthesis) and STATs (through cytoplasmic to nuclear translocation) and, in addition, controls their transactivation potential through signals generated via the MAP kinase pathway. This coordinate signaling process may play role in hypertrophic responses to pressure overload in the heart [27, 65, 66] and cellular proliferation in the vessel wall in response to injury [50, 67].

Ang II-inducible transcription factor: NF-κB

Like STATs, NF-κB is a ubiquitiously-expressed cytoplasmic transcription factor whose primary mode of regulation is through cytoplasmic-to-nuclear translocation. NF-κB is composed of heterodimerising trans-activator subunits rel A (p65), c- rel and the DNA binding subunits NF-κB1 (p50) and NF-κB2 (p50B), (reviewed in [13, 68, 69]). In unstimulated cells NF-κB subunits are associated with the inhibitory proteins, IκB, resulting in inactivation of DNA binding activity [70]. After stimulation with the prototypical NF-κB activating agents (cytokines TNFα/IL-1), the multicomponent IκB kinase (IKK) complex phosphorylates IκB at serine residues 32 and 36 in its interval regulatory domain [71]. Phospho-IκB is thereby selectively targeted for degradation through dual proteasome [72, 73] and calpain-mediated pathways [74]. IκB proteolysis liberates NF-κB for rapid translocation to the nucleus where it binds to and activates target genes. Target genes for NF-κB transactivation include intercellular adhesion protein and cytokines (IL-8 [47], IL-6 [75]), which may play important roles in inducing vascular injury and inflammation in atherosclerosis. In fact, activated nuclear NF-κB has been observed in activated VSMC [76] and atherosclerotic lesions [77]. In the liver, acute phase reactants, AGT itself being one, are also important NF-κB target genes. AGT has particular significance because its increased expression activates 'positive feedback loops' in both the iRAS and lRAS ([78, 79], reviewed in [13]). Our lab has studied the regulation of NF-κB in Ang II- responsive tissues, VSMC, and liver epithelial cells. Surprisingly, Ang II activates the NF-κB family in a highly cell type-dependent manner.

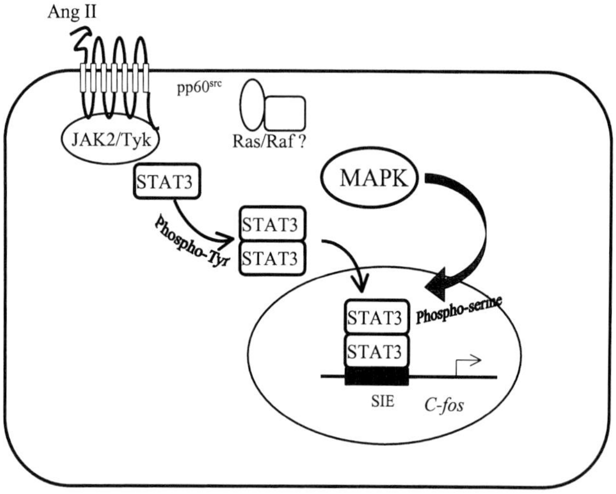

Fig. 3. Mechanism for Ang II effects on STAT in VSMC and cardiomyocytes. Schematic view of relevant signaling pathways for STAT regulation. The tyrosine kinases JAK2/Tyk and STAT3 are physically associated with the AT_1 receptor cytoplasmic domain. Following ligand activation, STAT3 is tyrosine phosphorylated, allowing the protein to dimerize and translocate into the nucleus. Mitogen activated protein kinase (MAPK), perhaps through $pp60^{src}$ /RAS/Raf activation serine phosphorylates the STAT3 activation domain, increasing its transcriptional activation potential.

Mechanisms for Ang II regulation of NF-κB in VSMC

IL-6 is a multifunctional cytokine that mediates B lymphocyte proliferation/induction of antibody synthesis, and is a mediator of the hepatic acute-phase response [80, 81]. In the vessel wall, locally secreted IL-6 also plays an additional important role as a VSMC growth factor through a paracrine mechanism involving production of platelet-derived growth factor [82, 83, 84]. IL-6 is encoded by a highly inducible promoter that is a target for tissue-specific and cytokine-inducible transcription factors [80, 85]. Both human and rat VSMC inducibly secrete IL-6 in response to inflammatory cytokines IL-1/TNF [81], bacterial cell wall-derived lipopolysaccharide [81], platelet-derived thrombin [86], and Ang II [82, 83]. Although Ang II was known to induce IL-6 expression by VSMC, the mechanism(s) responsible for its transcriptional activation were unknown until our recent report [75]; these essential findings will be briefly presented.

Ang II stimulates VSMC to express IL-6 in a rapid, but transient, manner. As shown in (Fig. 4), quiescent VSMC were stimulated for various times and changes in steady-state levels of IL-6 were measured using a Reverse Transcriptase-PCR (RT-PCR) technique. Ang II increased expression of IL-6 as early as 15 min, peaking at 1 h and falling thereafter. To show that the changes in IL-6 mRNA corresponded to

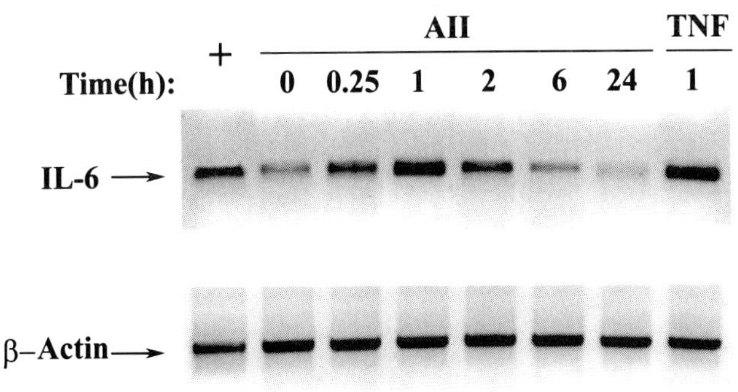

Fig. 4. Kinetics of IL-6 gene expression by Ang II in VSMC. Quiescent VSMC were stimulated with Sar1 Ang II for indicated times and changes in IL-6 mRNA abundance were quantified by RT-PCR [75]. Shown is an ethidium bromide stained agarose gel of the reaction products. Each sample was separately amplified for the internal control, β-actin, shown at bottom. Lane 1: +, indicates PCR reaction using plasmid template as a positive control. Lane 8: VSMC were stimulated with TNFα for 1 h. Ang II induces rapid but transient increase in IL-6 mRNA with a peak in expression at 1 h.

changes in IL-6 protein secretion, cell culture supernatants from Ang II-stimulated VSMCs (for 3 h) were assayed by enzyme-linked immunosorbent assay. Untreated control cultures produced 100 ± 80 ng/ml/3 h IL-6, whereas Ang II-stimulated (100 nmol/L) produced 850 ± 120 ng/ml/3 h, and those stimulated with 1 mmol/L accumulated 970 ± 120 ng/ml/3 h (× ± S.D., n = 4 experiments, $p < 0.05$ for each treatment compared to control, Student's *t*-test). Because the induction of IL-6 transcripts was sensitive to the effects of the transcription inhibitor Actinomycin D, we used transient transfection assays of IL-6 promoter/luciferase reporter plasmids to map the *cis* regulatory elements required for the Ang II induction. Ang II treatment induced IL-6 transcription in a manner completely dependent on the NF-κB motif. To determine whether Ang II was capable of activating other NF-κB-driven cytokine genes, we transfected wild-type and mutant IL-8 promoter/luciferase reporter constructs into VSMCs (Fig. 5). The IL-8 gene promoter contains an AP-1 binding site between –127 and –119 nucleotides and an NF-κB binding site between –80 and –70 nucleotides [87]. Ang II induced –162 IL-8/LUC promoter activity rapidly (reporter gene measurements were made after 6 h stimulation). Deletion of the AP-1 binding site reduced promoter activity, but did not completely abolish it. Further deletion of the NF-κB binding site completely abolished IL-8 transcription. These results indicate that IL-8 is driven by combined activities of AP-1 and NF-κB transcription factors in response to Ang II.

To determine the effects of Ang II on NF-κB binding, gel mobility shift assays were preformed on nuclear extracts from VSMC stimulated for various times. As shown in Fig. 6, Ang II induced increased NF-κB binding rapidly. Although not shown well here, in a previous study, we demonstrated that the induction of NF-κB binding was biphasic, with an early rapid phase detectable at 15 min, followed by a later peak at

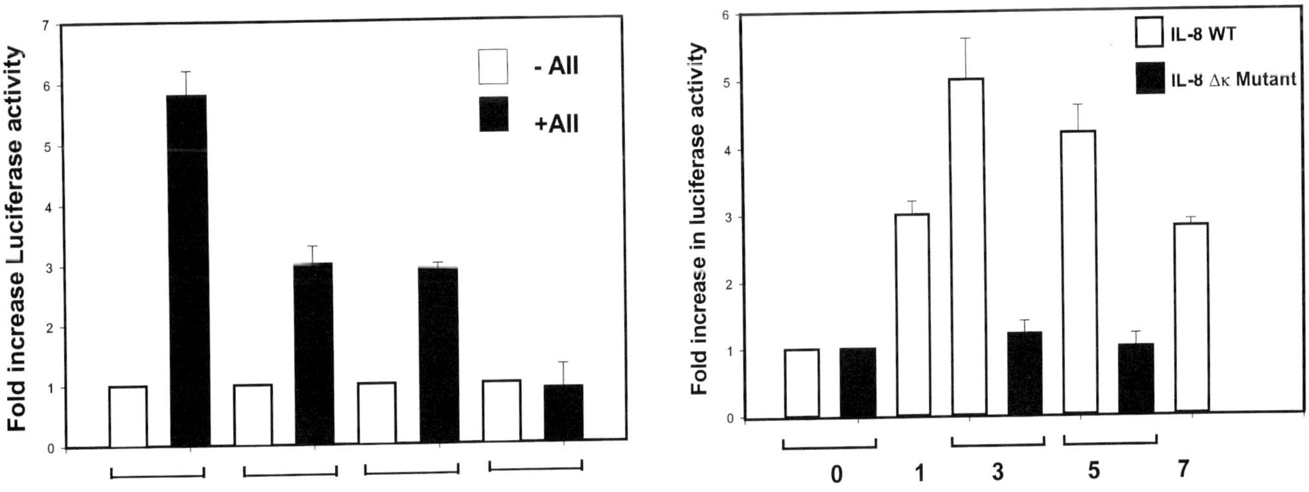

Fig. 5. Induction of IL-8 gene expression in VSMC. Left panel: Reporter genes containing the inducible control region of the IL-8 gene (nucleotides –162 to +22), driving expression of the luciferase gene and the constitutively expressed internal control alkaline phosphatase gene were transiently transfected into VSMC and serum starved. Cells were either left untreated (Ang II [–AII]) or stimulated with Ang II (+AII) for 5 h. Reporter gene activity was extracted and measured. Shown is the mean ± S.D. of the fold increase in normalized (luciferase/alkaline phosphatase) reporter activity from independently transfected in triplicate plates from a representative experiment. Although deletion of the AP-1 site reduces IL-8 induction, deletion of the NF-κB site completely abolishes inducible activation. Right panel: Site-directed mutagenesis of the NF-κB site was performed (IL-8Δκ /LUC contains a site mutation of the NF-κB binding site). Wild type (WT) and mutant promoters were stimulated with Ang II for the indicated times prior to measurement of luciferase reporter activity. Lacking the NF-κB site, IL-8 is not inducible by Ang II. Ang II induces IL-8 luciferase reporter activity in a manner that requires the NF-κB binding site.

24 h. The early phase of NF-κB translocation was dependent on weak simultaneous proteolysis of the IκBα and β inhibitors. During this early period nuclear Rel A rapidly increased within 0.25 h; compared to control levels, Rel A abundance increased 3.0 ± 1 fold (× ± S.D., n = 3, $p < 0.05$). To demonstrate that Rel A associates with IκBα and that this association was regulated by Ang II, we performed non-denaturing coimmunoprecipitation assays where Rel A was precipitated from cytoplasmic extracts, followed by fractionation of the immunoprecipitates on SDS PAGE and detection of associated IκBα by Western immunoblot. A time course of IκBα association with Rel A is shown in Fig. 7. IκBα is associated with Rel A in the cytoplasm of unstimulated cells. The abundance of Rel A-associated IκBα is reduced to 25 of 50% within 15 min of treatment. Moreover, in Western blots of nuclear extracts, we observed that the relative steady-state changes in Rel A were increased in the nuclear compartment following Ang II stimulation and exactly paralleled the changes seen in NF-κB binding [75]. These data indicate that the mechanism for initial Rel A nuclear translocation depends on IκBα proteolysis.

By contrast, we identified a later phase of NF-κB binding that was associated with enhanced processing of the p105 precursor into the mature 50 kDa NF-κB1 form. In Western immunoblots of cytoplasmic extracts of control cells, the majority of NF-κB1 was cytoplasmic, including discrete 105, 84 and the mature 50 kDa forms, with little of the 50 kDa form being nuclear. Following Ang II treatment, however, the abundance of 50 kDa NF-κB1 increased monotonically over 24 h. These data indicate Ang II perhaps has a biphasic effect on NF-κB1: initially by promoting its translocation, and secondarily by increasing its processing from the p105 precursor.

To determine the requirement for NF-κB in IL-6 stimulation, we previously reported an experiment where cells were treated with the IκB kinase inhibitor, Nα-Tosyl-Phe Chloromethyl Ketone (TPCK, [73]). Gel mobility shift assay of control and Ang II-stimulated VSMC in the absence or presence of 50 μmol/L of protease inhibitor TPCK. Pretreatment with TPCK, but not the unrelated serine protease inhibitor, PMSF, completely blocked the appearance of inducible NF-κB binding. Parallel-treated plates of VSMC were subjected to RT-PCR, to detect changes in endogenous IL-6 gene transcription. IL-6 mRNA was induced in the Ang II-treated cells, but not in Ang II-treated cells pretreated with TPCK. These data indicate that Rel•ANF-κB1 translocation is required for endogenous IL-6 gene expression following Ang II stimulation. This mechanism is schematically presented in Fig. 8.

These data are the first to link Ang II stimulation with cytokine expression through pleiotropic effects on the NF-κB transcription factor. Independently, others have shown that Ang II is a potent inducer of IL-6 gene expression and NF-κB activation in human saphenous vein endothelial cells [88], although the effect was not mapped to regions on the IL-6 promoter. At the cellular level, injured and stimulated VSMC express growth factor, adhesion proteins and remodeling genes, which are potential targets for NF-κB activation. This process may be important in the early stages of atherogenesis.

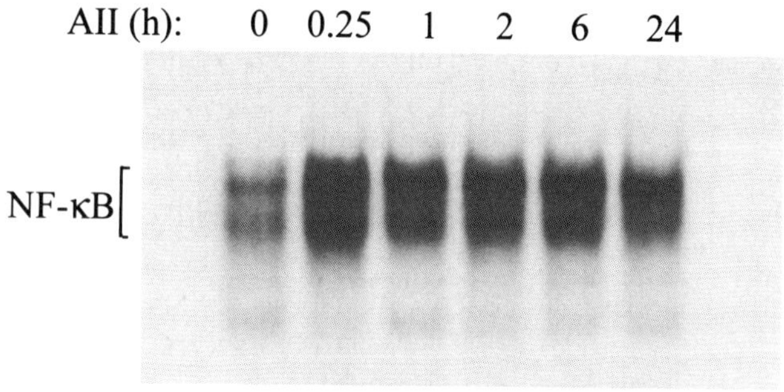

Fig. 6. Induction of NF-κB binding in VSMC. Gel mobility shift assay of rat VSMC stimulated for indicated times (at top) with Ang II. Nuclear extracts were prepared at the indicated times and used to bind the radiolabeled NF-κB binding site from the rat AGT promoter [104]. Shown is an autoradiogram of the bound complexes only. Ang II induces the rapid appearance of sequence-specific NF-κB binding containing the potent Rel A transactivator, that persists to 24 h. These observations have been reproduced in rat [75] and human [66] VSMC.

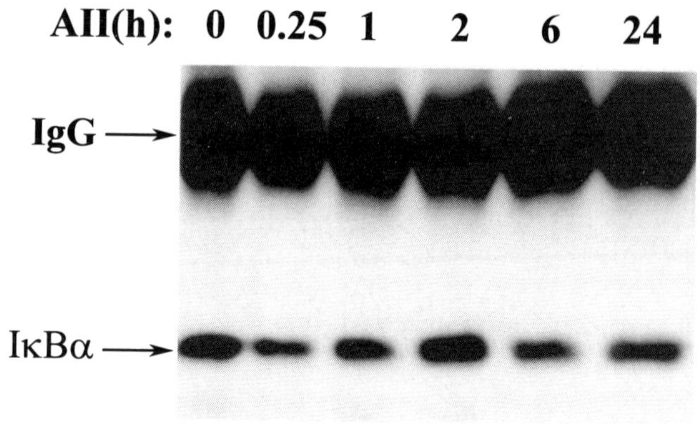

Fig. 7. Ang II effects on Rel A-IκBα association. A non-denaturing co-immunoprecipitation assay was used to immunoprecipitate Rel A-associated proteins. Cytoplasmic extracts taken from VSMC stimulated for the indicated times (at top) were precipitated with Rel A antibody. Immunoprecipitates were then fractionated on SDS-PAGE, and IκBα association was detected through Western immunoblot. The location of IκBα is shown at left. IκBα is associated with Rel A in unstimulated cells, and its association is decreased by 50% 15 min following Ang II stimulation. At later times, IκBα is resynthesized.

Ang II regulation of NF-κB in liver epithelial cells

Activation of the intravascular RAS, initiates a biological positive feedback loop in the hepatocyte, stimulating the

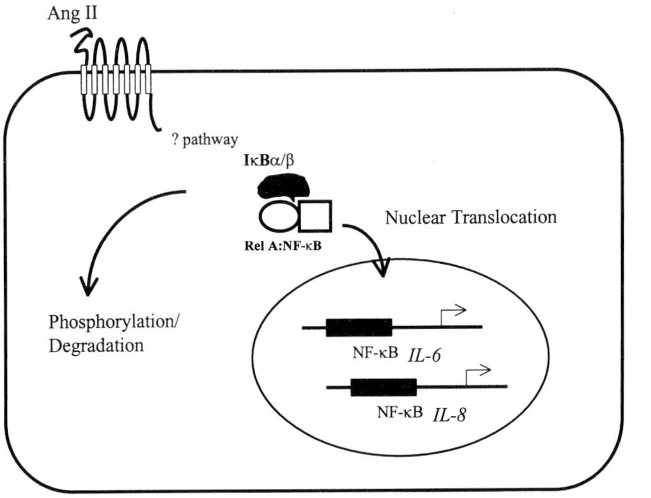

Fig. 8. Schematic of mechanism for NF-κB activation in VSMC. Schematic view of experimentally defined pathway for NF-κB regulation [75]. The IκB inhibitors, IκBα and IκBβ, tether and inactivate NF-κB (Rel A•NF-κB1 heterodimers) in the cytoplasm of cells. Following Ang II-stimulation, IκBs are degraded, allowing NF-κB to enter the nucleus and stimulate transcription of target genes, including the cytokines IL-6 and IL-8.

synthesis of its own precursor, AGT, through a mechanism involving enhanced mRNA production [89–91]. Nuclear run-on assays [92] and promoter mapping studies [79] have demonstrated that Ang II activates AGT transcription over physiologically-relevant concentrations of hormone. Previously, we reported the surprising findings that the Ang II-inducible *cis* element mediating AGT transcription was a well-characterized cytokine-inducible enhancer responsible for mediating the transcriptional induction of AGT during the acute phase response (this element is termed the acute phase response element (APRE) ([79], reviewed in [13]).

Ang II activates transcription of native rat AGT gene promoter over a physiologically-relevant range of 10–100 nMol/L [79]. Using transient transfections with site-directed mutations in selective *cis* control elements, we mapped the Ang II effect to a multihormone-inducible enhancer in the rat AGT gene located –615 to –470 nucleotides 5′ to the transcription start site [79]. Within the multihormone-inducible enhancer, site-directed mutations of the APRE that interfere with NF-κB transcription factor binding also abolish Ang II responsiveness. When tested as an isolated enhancer, we observed that the APRE functions as a rapidly-inducible Ang II-inducible enhancer in a sequence-specific fashion. For example, Ang II rapidly increases wild-type APRE sequences-driving reporter activity, but not the NF-κB binding site mutations (APRE M2 or APRE M6) with an apparent peak at 6 h and falling thereafter. These data indicate that Ang II increases hepatic AGT gene expression in an NF-κB-dependent manner.

That AT_1 mediates this effect is suggested by experiments in AT_1-deficient cells where stimulation occurred only when the type 1 Ang II receptor is expressed [79]; in a separate study, Ang II effects were blocked by the inclusion of 10-fold molar excess of the specific (competitive) AT_1 receptor antagonist, Dup 753 [93]. Together, these data indicate that Ang II activates NF-κB through an AT_1-dependent pathway.

Previously we have reported the characterization of the NF-κB complexes induced by Ang II. In gel mobility shift assays using highly purified nuclear extracts, Ang II induces sequence-specific NF-κB binding to radiolabeled APRE in a sequence-specific fashion [79, 93]. Mutations that interfere with transcriptional activation by Ang II also disrupt NF-κB binding, indicating that NF-κB binding mediates the Ang II effect. To demonstrate that this effect was not due to an artifactual stimulation produced by a transformed liver cell, Ang II infusions of freshly-isolated primary rat hepatocyte cultures also produced a rapid, fourfold increase in sequence-specific NF-κB binding to the APRE [79, 93]. Antibodies against the transcriptional activator subunit, Rel A, quantitatively 'supershift' the nucleoprotein complex, indicating that Rel A participates in this complex.

In side-by-side comparisons between Ang II and TNF-α inducible binding we observed differences in NF-κB complex mobility (shown in Fig. 9). Whereas Ang II treatment

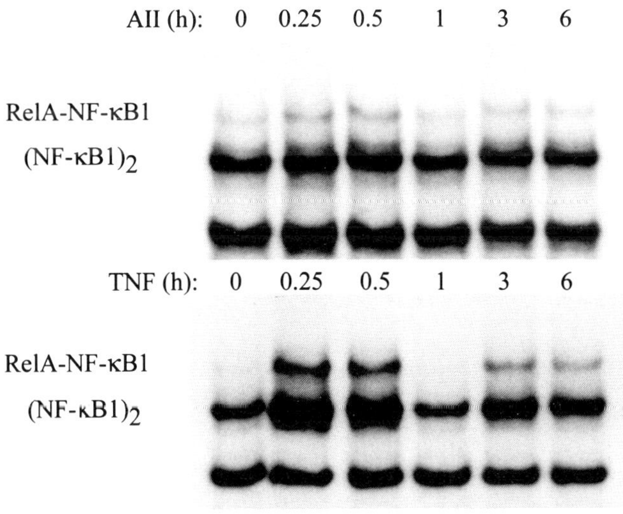

Fig. 9. Comparison of NF-κB activation in hepatocytes by Ang II and TNFα. Top: Ang II induces binding of NF-κB homodimeric complexes. Gel mobility shift assay of nuclear protein binding the APRE WT from cells stimulated for the indicated times (in h, top) with AII (100 nMol/L). Shown is an autoradiogram of the bound complexes only. Binding of the Rel A•NF-κB1 heterodimer is faintly detectable in control cells, whereas NF-κB1 homodimeric binding is strong. After treatment, NF-κB1 binding increases weakly in parallel with changes in transcriptional activity (for comparison see ref. [93]). n.s. – a nonspecific DNA-binding complex. Bottom: TNFα stimulation induces the Rel A•NF-κB1 heterodimer binding. Autoradiogram of representative time course. TNFα induces Rel A•NF-κB1 binding rapidly at 15 and 30 min, followed by a nadir at 1 h (due to the NF-κB-IκB autoregulatory loop [105]). The Rel A•NF-κB1 heterodimer then reaccumulates in the nucleus at 3 and 6 h.

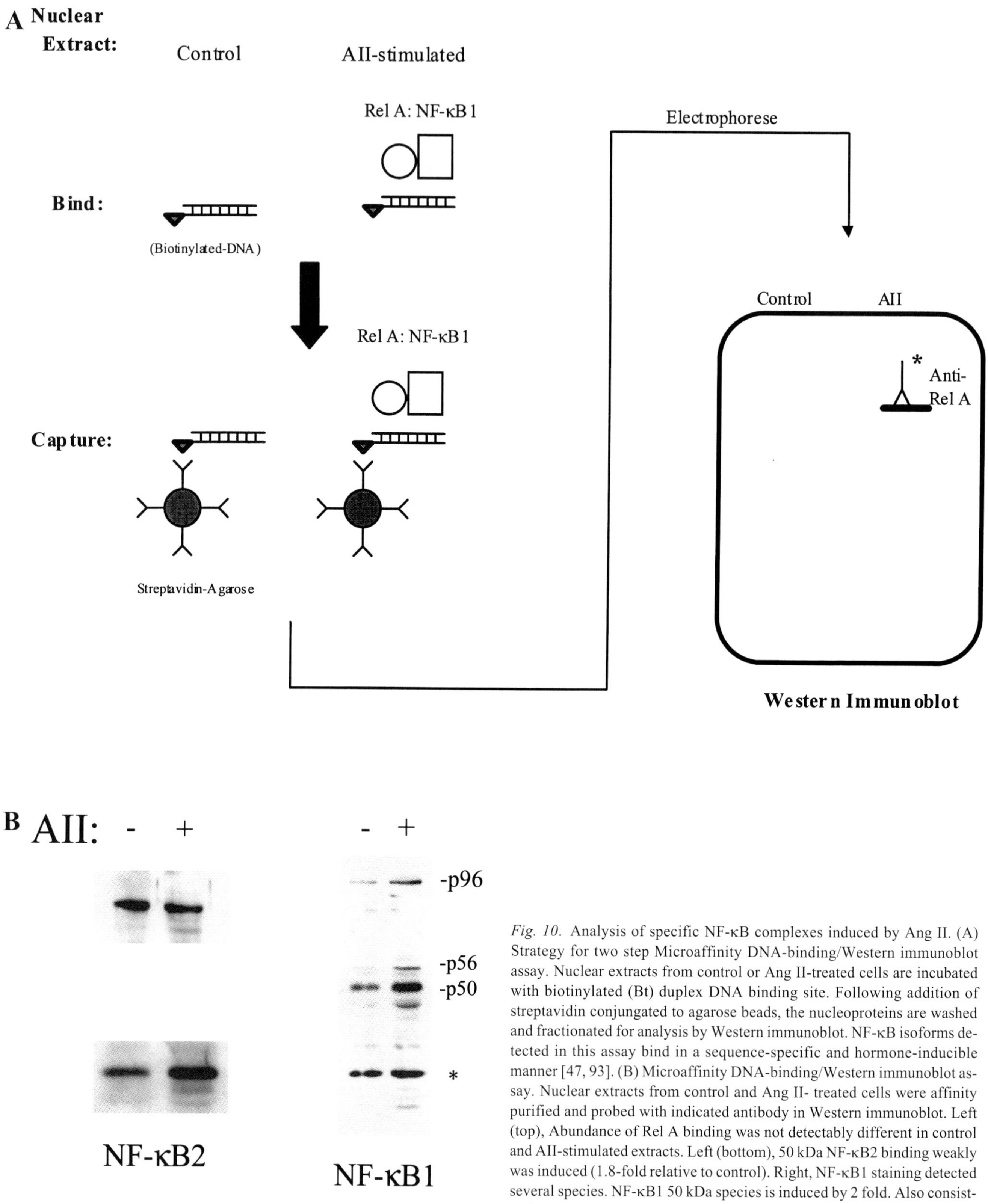

Fig. 10. Analysis of specific NF-κB complexes induced by Ang II. (A) Strategy for two step Microaffinity DNA-binding/Western immunoblot assay. Nuclear extracts from control or Ang II-treated cells are incubated with biotinylated (Bt) duplex DNA binding site. Following addition of streptavidin conjungated to agarose beads, the nucleoproteins are washed and fractionated for analysis by Western immunoblot. NF-κB isoforms detected in this assay bind in a sequence-specific and hormone-inducible manner [47, 93]. (B) Microaffinity DNA-binding/Western immunoblot assay. Nuclear extracts from control and Ang II- treated cells were affinity purified and probed with indicated antibody in Western immunoblot. Left (top), Abundance of Rel A binding was not detectably different in control and AII-stimulated extracts. Left (bottom), 50 kDa NF-κB2 binding weakly was induced (1.8-fold relative to control). Right, NF-κB1 staining detected several species. NF-κB1 50 kDa species is induced by 2 fold. Also consistently seen were two larger isoforms at 56 and 96 kDa (latter increasing 4.8 fold, p56 could not be accurately quantified due to its low signal in control lanes). Asterisks are specific NF-κB1 breakdown products. (Reprinted with permission from [93]).

weakly increased binding of the NF-κB1 complex (2–4 fold), upon TNF-α stimulation, Rel A: NF-κB1 complexes are rapidly induced at 15 and 30 min, followed by a nadir at 1 h (the nadir in binding is due to overshoot resynthesis of the IκBα inhibitor [70]). These data indicate that the two agonists induce qualitatively distinct NF-κB binding profiles.

In Western immunoblots of cytoplasmic extracts cells stimulated by Ang II, there were no detectable changes in either IκBα or IκBβ steady-state levels in the cytoplasm [93], indicating that Ang II activates NF-κB1 binding independently of detectable changes in cytoplasmic IκB abundance. This finding is in contrast to the effects of Ang II on IκBα levels in VSMC (Fig. 7; ref. [75]). Moreover, in Western immunoblots of nuclear extracts from treated cells, inducible NF-κB1 binding also occurs independently of changes in NF-κB1 processing or nuclear abundance [75].

To rigorously determine the composition of the NF-κB subunits regulated by hormone, we applied a technique for microaffinity capture of DNA binding proteins followed by their detection using Western blotting (schematically shown in Fig. 10A). Use of this specific technique avoids inherent problems associated with epitope masking in native complexes and can detect NF-κB isoforms without the bias introduced by UV crosslinking. Nuclear extracts from the Ang II-treated cells produce a distinct pattern of inducible NF-κB1 binding (Fig. 10B). We observed the presence of several additional, larger Ang II-inducible complexes specifically detected by the NF-κB1 antibody (of 56 and 96 kDa). Together, these data indicated that Ang II induces NF-κB1 isoform binding to the APRE (without changes in Rel A binding).

To assess the role of DAG-sensitive PKC isoforms in Ang II stimulation, PKC isoforms were depleted by chronic agonist exposure [93]. In this condition, PKC downregulation completely abolished Ang II- induced APRE transcription. Nuclear extracts from PKC-downregulated cells were then assayed for changes in APRE binding in the microaffinity isolation assay. PKC hydrolysis blocked inducible NF-κB1 p96, p56 and p50 binding, indicating an absolute requirement for the DAG-sensitive PKC isoforms in Ang II-induced NF-κB1 binding.

These observations suggest that inducible NF-κB1 binding mediates the effects of Ang II on AGT transcription in hepatocytes. Although the 50-kDa form of NF-κB1 is thought to be transcriptionally silent, in separate studies, others have demonstrated the presence of a widely-expressed, alternatively-spliced murine NF-κB1 98 kDa isoform (muNF-κB1 p98), an isoform disrupted in its ability to be retained in the cytoplasm [94]. Expression of muNF-κB1 transactivates NF-κB-driven reporter genes. Together, these findings indicate that NF-κB1 isoforms containing the COOH terminus also contain latent transcriptional activity (and may indicate a mechanism by which NF-κB1 p96 could activate the APRE). The mechanism by which the COOH terminal transactivation domain of NF-κB1 activates transcription has not been fully investigated. These forms may also directly associate with the CBP/p300 coactivators (important coactivators for the Rel A transcription factor subunit [95, 96]). Of interest is that others have shown a downstream STAT binding site (recognizing STAT3 and 6 isoforms) in the rAGT gene that mediates Ang II induction in neonatal rat cardiac myocytes [66]. These discrepant observations may indicate that mechanisms exist for cell type-specific regulation of AGT in liver and vascular tissues.

Ang II effects of NF-κB transcription is antioxidant-sensitive

Reactive oxygen species (ROS) are ubiquitous, highly diffusable and reactive molecules produced as a result of reduction of molecular oxygen, and include species such as hydrogen peroxide, superoxide anion, and hydroxyl radical [97]. Recently, a role of ROS in cellular responses to growth factor signaling has been described for platelet derived growth factor [98], basic fibroblast derived growth factor [99], thrombin [67], and Ang II [37]. ROS are important signals for inducing VSMC proliferation and chemotaxis [98, 100] and likely play an important role in the vascular pathology seen in atherosclerosis. Because Ang II is a potent inducer of ROS formation through its effects on the NADH/NADPH oxidase system and ROS are known to be important regulators of NF-κB [87, 101], we sought to determine the contribution of ROS in NF-κB activation. Addition of the free-radical scavenger, dimethylsulfoxide (DMSO) at concentrations known to inhibit hormone dependent intracellular oxidation (2% v/v, [87]), blocked the transcriptional effects of Ang II (Fig. 11). These data indicate a requirement for ROS in Ang II-inducible NF-κB transcriptional activity in hepatocytes. Further studies will be required to determine whether the effect of ROS is exerted through influences on nuclear translocation, DNA binding or transcriptional activation.

Coordination of intracellular signaling: Potential targets and physiological consequences of Ang II-inducible transcription factors

Here we review evidence indicating that Ang II controls the expression of transcription factors AP-1, STATs, and NF-κB through multiple mechanisms. Each transcription factor [67] activates gene clusters that account for early and late cellular responses to Ang II. There are surprisingly few well-characterized Ang II-responsive genes; some of the reported

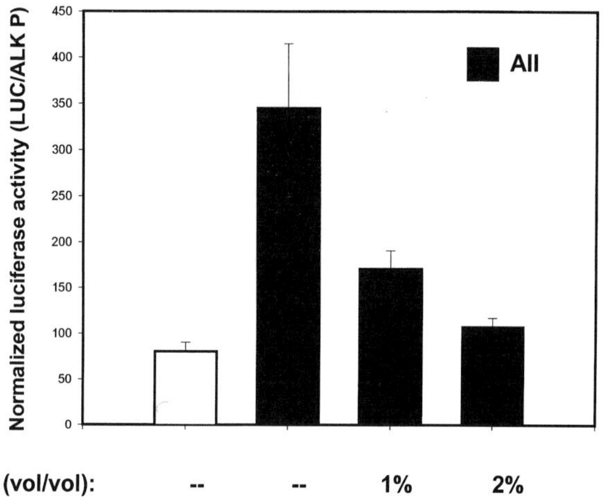

Fig. 11. Requirement for ROS in Ang II-induced transcription. Transient transfection assay using the rat AGT NF-κB binding site driving expression of the luciferase reporter gene. Cells were stimulated in the absence or presence of Ang II (Sar[1] AII, 100 nmol/L for 6 h) in the presence of increasing concentrations (indicated in vol %) of the free radical scavenger, dimethylsulfoxide (DMSO). Ang II-inducible transcription was blocked by 2% (vol/vol) DMSO, concentrations sufficient for inhibition of receptor activated ROS [87].

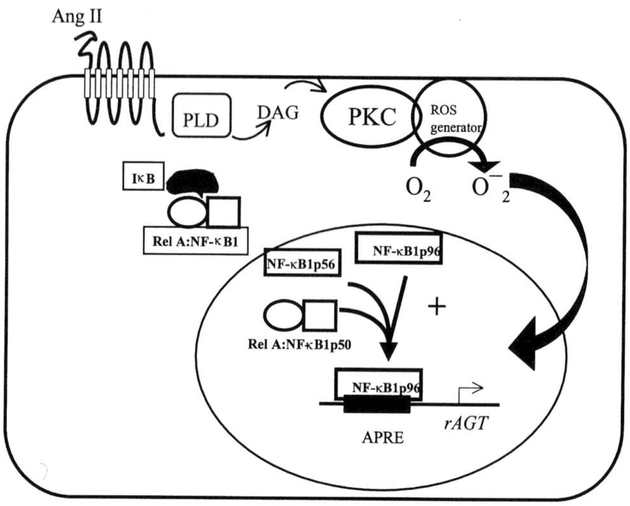

Fig. 12. Schematic of mechanism for NF-κB activation in liver cells. In liver cells, Ang II activates a diacyglycerol-sensitive PKC signal transduction cascade that targets nuclear NF-κB1 isoforms (96, 56, and 50 kDa) through an ROS-generating system of an unknown type. This signal activates NF-κB1 DNA-binding activity present in the nucleus and is associated with transcriptional activation. In this cell type, Ang II stimulated NF-κB1 recruitment independently of changes in Rel A nuclear abundance or IκB proteolysis [93].

targets for Ang II are tabulated in Table 1. Genetic effects of Ang II are likely to play important roles in long-term control of blood pressure (through enhanced AGT synthesis), hypertrophic responses to pressure overload in the heart [27, 65, 66] and cellular proliferation in the vessel wall [50]. One of the surprising biologic effects of Ang II is that its properties closely overlap those of inflammatory cytokines. The prototypical Ang II-inducible transcription factors AP-1, STATs and NF-κB are well-known, cytokine-inducible transcription factors. Ang II activates cytokine production (IL-6 and IL-8) in VSMC, acute-phase reactant synthesis in the liver, and ROS formation in stimulated cells, all characteristics of proinflammatory cytokines (such as TNFα). These findings probably indicate that one of the physiologic effects of Ang II is to induce inflammation, a characteristic feature of atherosclerotic lesions [102, 103].

We find it interesting that expression of particular genes are controlled by several Ang II-inducible transcription factors. One example would be the *c-Fos* promoter, where both AP-1 and STATs control its expression [62]. The cytokine IL-8 is controlled through combinatorial effects of AP-1 and NF-κB that bind to adjacent regions in its proximal promoter [87]. Likewise, AGT is controlled by NF-κB [13] and STATs [66]. Interactions between inducible transcription factors may allow for differences in magnitude or duration of gene expres-

Table 1. Target genes and tissues for AP-1, STATs and Nf-κB in Ang II-responsive tissues

		Transcriptional activator		
		AP-1	STAT	NF-κB
Cell cycle regulation genes				
	AP-1	VSMC [50, 51], CM [27, 49], AG [44], Hep [52], PTC [53]	CF [62], CM [39]	
	c-Myc	VSMC [106]		
Inflammation genes				
Adhesion molecules	*(I,V-CAM)*			EC [107], VSMC [108]
Cytokines	*(IL-8, IL-6)*			VSMC [75, 76, 88]
Acute phase reactants	*(AGT, a2-MG, SAA)*		Hep [109–111]	Hep [13, 78, 79]

Ang II-responsive genes are classified according to transcription factor primarily responsible for its activation. For each entry, the cell type in which the phenomenon has been observed is indicated. Abbreviations for genes: I,V-CAM are intercellular and vascular cellular adhesion molecules; α2-MG, alpha 2 macroglobulin, SAA, serum amyloid A precursor. Abbreviations for cell type are: AG – adrenal glomerulosa cells; CF – cardiac fibroblasts; CM – cardiac myocytes; EC – endothelial cells; Hep – Hepatocytes; PTC – proximal tubular cells; VSMC – vascular smooth muscle cells.

sion following Ang II-exposure in certain cell types. Further identification of Ang II inducible transcription factors, their targets and their interactions are needed.

Acknowledgements

This work was supported in part by National Heart, Lung, and Blood Institute Grant R01 55630-03 (to A.R.B.) and National Institutes of Environmental Health P30 ES06676 (R.S. Lloyd, UTMB). A.R.B. is an Established Investigator of the American Heart Association.

References

1. Reid I A, Morris BJ, Ganong WG: The renin-angiotensin system. Annu Rev Physiol 40: 377–410, 1978
2. Corvol P, Jeunemaitre X: Molecular genetics of human hypertension: Role of angiotensinogen. Endo Rev 18: 662–677, 1997
3. Peach MJ: Renin-angiotensin system: Biochemistry and mechanisms of action. Physiol Rev 57: 313–370, 1977
4. Walker WG, Whelton PK, Saito H, Russel RP, Hermann J: Relation between blood pressure and renin, renin substrate, AII, aldosterone and urinary sodium and potassium in 574 ambulatory subjects. Hypertension 1: 287–291, 1979
5. Krakoff LR: Measurement of plasma-renin substrate by radioimmunoassay of angiotensin. I. Concentration in syndromes associated with steroid excess. J Clin Endocrinol Metab 37: 110–117, 1973
6. Kim HS, Krege JH, Kluckman KD, Hagaman JR, Hodgin JB, Best CF, Jennette JC, Coffman TM, Maeda N, Smithies O: Genetic control of blood pressure and the angiotensinogen locus. Proc Natl Acad Sci USA 92: 2735–2739, 1995
7. Fukamizu A, Sugimura K, Takimoto E, Sugiyama F, Seo MS, Takahashi S, Hatae T, Kajiwara N, Yagami K, Murakami K: Chimeric renin-angiotensin system demonstrates sustain increase in blood pressure of transgenic mice carrying both human renin and human angiotensinogen genes. J Biol Chem 268: 11617–11621, 1993
8. Jeunemaitre X, Soubrier F, Kotelevtsev YV, Lifton RP, Williams CS, Charru A, Hunt SC, Hopkins PN, Williams RR, Lalouel JM: Molecular basis of human hypertension: Role of angiotensinogen. Cell 71: 169–180, 1992
9. Caulfield M, Lavender P, Farrall M, Munroe P, Lawson M, Turner P, Clark AJ: Linkage of the angiotensinogen gene to essential hypertension. New Engl J Med 330: 1629–1633, 1994
10. Morgan T, Craven C, Nelson L, Lalouel JM, Ward K: Angiotensinogen T235 expression is elevated in decidual spiral arteries. J Clin Invest 100: 1406–1415, 1997
11. Ward K, Hata A, Jeunemaitre X, Helin L, Nelson C, Namikawa C, Farrington PF, Ogasawara M, Suzumori K, Tomoda S: A molecular variant of angiotensinogen associated with preeclampsia. Nature Genet 4: 59–61, 1998
12. Hopkins PN, Lifton RP, Hollenberg SM, Jeunemaitre X, Hallouin M-C, Skuppin J, Williams CS, Dluhy RG, Lalouel JM, Williams RR, Williams GH: Blunted renal vascular response to angiotensin II is associated with a common variant of the angiotensinogen gene and obesity. J Hypertens 14: 199–207, 1996
13. Brasier AR, Li J: Mechanisms for inducible control of angiotensinogen gene transcription. J Hypertens 27: 465–475, 1996
14. Campbell DJ: Circulating and tissue angiotensin systems. J Clin Invest 79: 1–6, 1987
15. McGehee RE Jr, Ron D, Brasier AR, Habener JF: Differentiation-specific element: A cis-acting developmental switch required for the sustained transcriptional expression of the angiotensinogen gene during hormonal-induced differentiation of 3T3-L1 fibroblasts to adipocytes. Mol Endocrinol 7: 551–560, 1993
16. Cassis LA, Saye J, Peach MJ: Localization and regulation of rat angiotensinogen messenger RNA. Hypertension 11: 591–596, 1988
17. Campbell DJ, Habener JF: Angiotensinogen gene is expressed and differentially regulated in multiple tissues of the rat. J Clin Invest 78: 31–39, 1986
18. Ganten D, Hermann K, Bayer C, Unger T, Lang RE: Angiotensin synthesis in the brain and increased turnover in hypertensive rats. Science 221: 869–871, 1983
19. Sernia C: Location and secretion of brain angiotensinogen. Reg Pep 57: 1–18, 1995
20. Celio MR, Inagami T: Angiotensin II immunoreactivity coexists with renin in the juxtaglomerular granular cells of the kidney. Proc Natl Acad Sci USA 78: 3897–3900, 1981
21. Ingelfinger JR, Zuo W, Fon EA, Ellison KE, Dzau V: *In situ* hybridization evidence for angiotensinogen messenger RNA in the rat proximal tubule. An hypothesis for the intrarenal renin angiotensin system. J Clin Invest 85: 417–423, 1990
22. Aguilera G, Schirar A, Baukal A, Catt KJ: Circulating angiotensin II and adrenal receptors after nephrectomy. Nature (London) 289: 507–509, 1981
23. Racz K, Pinet F, Gasc J-M, Guyene T-T, Corvol P: Coexpression of renin, angiotensiongen and their messenger ribonuclei acids in adrenal tissues. J Clin Endocrinol Metab 75: 730–737, 1992
24. Swales JD, Abramovici D, Beck F, Bing RF, Loudon M, Thurston H: Arterial wall renin. J Hypertens 1(suppl 1): 17–22, 1983
25. Naftilan A, Zuo W, Ingelfinger JR, Ryan TJ, Pratt RE, Dzau V: Localization and differential regulation of angiotensinogen mRNA expression in the vascular wall. J Clin Invest 87: 1300–1311, 1991
26. Sadoshima J, Xu Y, Slayter HS, Izumo S: Autocrine release of angiotensin II mediates stretch-induced hypertrophy of cardiac myocytes *in vitro*. Cell 75: 977–984, 1993
27. Sadoshima J, Izumo S: Molecular characterization of angiotensin II-induced hypertrophy of cardiac myocytes and hyperplasia of cardiac fibroblasts. Critical role of the AT1 receptor subtype. Circ Res 73: 413–423, 1993
28. Ellison KE, Ingelfinger JR, Pivor M, Dzau V: Androgen regulation of rat renal angiotensinogen messenger RNA expression. J Clin Invest 83: 1941–1945, 1989
29. Kalinyak JE, Perlman AJ: Tissue-specific regulation of angiotensinogen mRNA accumulation by dexamethasone. J Biol Chem 262: 460–464, 1987
30. Gunther S, Alexander RW, Atkinson WJ, Gimbrone M Jr: Functional angiotensin II receptors in cultured vascular smooth muscle. J Cell Biol 92: 289–298, 1982
31. Murphy TJ, Alexander RW, Griendling KK, Runge MS, Bernstein KE: Isolation of a cDNA encoding the vascular type-1 angiotensin II receptor. Nature (Lond) 351: 233–236, 1991
32. Bouscarel B, Blackmore PF, Exton JH: Characterization of the angiotensin II receptor in primary cultures of rat hepatocytes. J Biol Chem 263: 14913–14919, 1988
33. Tsuchida S, Matsusaka T, Chen X, Okubo S, Niimura F, Nishimura H, Fogo A, Utsunomiya H, Inagami T, Ichikawa I: Murine double nullizygotes of the angiotensin type 1a and 1b receptor genes duplicate severe abnormal phenotypes of angiotensinogen nullizygotes. J Clin Invest 101: 755–760, 1998
34. Diaz-Laviada I, Larrodera P, Nieto JL, Cornet ME, Diaz-Meco MT, Sanchez MJ, Guddal PH, Johansen T, Haro A, Moscat J: Mechanism of inhibition of adenylate cyclase by phospholipase c-catalyzed hydrolysis of phosphatidylcholine. J Biol Chem 266: 1170–1176, 1991

35. Maloney JA, Tsygankova O, Szot A, Yang L, Li Q, Williamson JR: Differential translocation of protein kinase C isozymes by phorbol esters, EGF, and ANG II in rat liver WB cells. Am J Physiol 274: C974–C982, 1998
36. Griendling KK, Ushio-Fukai M, Lassegue B, Alexander RW: Angiotensin II signaling in vascular smooth muscle. Hypertension 29: 366–373, 1997
37. Griendling KK, Minieri CA, Ollerenshaw JD, Alexander RW: Angiotensin II stimulates NADH and NADPH oxidase activity in cultured vascular smooth muscle cells. Circ Res 74: 1141–1148, 1994
38. Suh Y-A, Arnold RS: Cell transformation by the superoxide-generating oxidase Mox1. Nature (London) 401: 79–82, 1999
39. McWhinney CD, Hunt RA, Conrad KM, Dostal DE, Baker KM: The type 1 angiotensin II receptor couples to STAT1 and STAT3 activation through Jak2 kinase in neonatal rat cardiac myocytes. J Mol Cell Cardiol 29: 2513–2524, 1997
40. Venema RC, Benema VJ, Eaton DC, Marrero MB: Angiotensin II-induced tyrosin phosphorylation of signal transducers and activators of transcription 1 is regulated by Janus-activated kinase 2 and Fynk kinases and mitogen activated protein kinase phosphatase 1. J Biol Chem 273: 30795–30800, 1998
41. Marrero MB, Schieffer B, Paxton WG, Heerdt L, Berk BC, Delafontaine P, Bernstein KE: Direct stimulation of Jak/STAT pathway by the angiotensin II AT1 receptor. Nature (London) 375: 247–250, 1995
42. Liao D-F, Monia B, Dean N, Berk BC: Protein kinase C-ζ mediates angiotensin II activation of ERK1/2 in vascular smooth muscle cells. J Biol Chem 272: 6146–6150, 1997
43. Kojima I, Kojima K, Kreutter D, Rasmussen H: The temporal integration of the aldosterone secretory response to angiotensin occurs via two intracellular pathways. J Biol Chem 259: 14448–14457, 1984
44. Clark AJ, Balla T, Jones MR, Catt KJ: Stimulation of early gene expression by angiotensin II in bovine adrenal glomerulosa cells: Roles of calcium and protein kinase C. Mol Endocrinol 6: 1889–1898, 1992
45. Kingston RE, Bunker CA, Imbalzano AN: Repression and activation by multiprotein complexes that alter chromatin structure. Genes Dev 10: 905–920, 1996
46. Beato M: Chromatin structure and the regulation of gene expression: Remodeling at the MMTV promoter. J Mol Med 74: 711–724, 1996
47. Brasier AR, Jamaluddin M, Casola A, Duan W, Shen Q, Garofalo R: A promoter recruitment mechanism for TNFα-induced IL-8 transcription in type II pulmonary epithelial cells: Dependence on nuclear abundance of Rel A, NF-κB1 and c-Rel transcription factors. J Biol Chem 273: 3551–3561, 1998
48. Angel P, Karin M: The role of Jun, Fos and the AP-1 complex in cell proliferation and transformation. Biochim Biophys Acta 1072: 129–157, 1991
49. Neyses L, Nouskas J, Luyken J, Fronhoffs S, Oberdorf S, Pfeifer U, Williams RS, Sukhatme VP, Vetter H: Induction of immediate-early genes by angiotensin II and endothelin-1 in adult rat cardiomyocytes. J Hypertens 11: 927–934, 1993
50. Taubman MB, Berk BC, Izumo S, Tsuda T, Alexander RW, Nadal Ginard B: Angiotensin II induces c-fos mRNA in aortic smooth muscle. Role of Ca^{2+} mobilization and protein kinase C activation. J Biol Chem 264: 526–530, 1989
51. Naftilan A, Pratt RE, Eldridge CS, Lin HL, Dzau V: Angiotensin II induces c-fos expression in smooth muscle via transcriptional control. Hypertension 13: 706–711, 1989
52. Gonzalez-Espinosa C, Garcia-Sainz JA: Angiotensin II and active phorbol esters induce proto-oncogene expression in isolated rat hepatocytes. Biochim Biophys Acta 1136: 309–314, 1992
53. Wolf G, Neilson EG: Angiotensin II induces cellular hypertrophy in cutlured murine proximal tubular cells. Am J Physiol 269: F768–F777, 1999
54. Ruiz-Ortega M, Egido J: Angiotensin II modulates cell growth-related events and synthesis of matrix proteins in rela interstitial fibroblasts. Kidney Int 52: 1497–1510, 1997
55. Blume A, Herdegen T, Unger T: Angiotensin peptides and inducible transcription factors. J Mol Med 77: 339–357, 1999
56. Angel P, Hattori K, Smeal T, Karin M: The jun proto-oncogen is positively autoregulated by its product. Cell 55: 875–885, 1988
57. Zohn IE, Yu H, Cox AD, Earp HS: Angiotensin II stimulates calcium-dependent activation of c-Jun N-terminal kinase. Mol Cell Biol 15: 6160–6168, 1995
58. Karin M: The regulation of AP-1 activity by mitogen-activated protein kinase. J Biol Chem 270: 16483–16486, 1995
59. Kudoh S, Komuro I, Mizuno T, Yamazaki T, Zou Y, Shiojima I, Takekoshi N, Yazaki Y: Angiotensin II stimulates c-Jun NH2-terminal kinase in cultured cardiac myocytes of neonatal rats. Circ Res 80: 139–146, 1997
60. Darnell JE Jr, Kerr , Startk GR: Jak-STAT pathways and transcriptional activation in response to IFNs and other extracellular signaling proteins. Science 264: 1415–1421, 1994
61. Darnell JE Jr: STATS and gene regulation. Science 277: 1630–1635, 1997
62. Bhat GJ, Thekkumkara TJ, Thomas WG, Conrad KM, Baker KM: Angiotensin II stimulates sis-inducing factor-like DNA binding activity: Evidence that the AT1a receptor activates transcription factor Stat 91 and/or a related protein. J Biol Chem 269: 31443–31449, 1994
63. Bhat GJ, Baker KM: Angiotensin II stimulates rapid serine phosphorylation of transcription factor STAT3. Mol Cell Biochem 170: 171–176, 1997
64. Wen Z, Zhong A, Darnell JE Jr: Maximal activation of transcription by Stat1 and Stat3 requires both tyrosine and serine phosphorylation. Cell 82: 241–250, 1995
65. Pan J, Fukuda K, Kodama H, Makino S, Takashi T, Sano M, Hori S, Ogawa S: Role of angiotensin II in activation of the JAK/STAT pathway induced by acute pressure overload in the rat heart. Circ Res 81: 611–617, 1997
66. Mascareno E, Dhar M, Siddiqui MAQ: Signal transduction and activator of transcription (STAT) protein-dependent activation of angiotensinogen promoter: A cellular signal for hypertrophy in cardiac muscle. Proc Natl Acad Sci USA 95: 5590–5594, 1998
67. Patterson C, Ruef J, Madamanchi NR, Barry-Lane P, Hu Z, Horaist C, Ballinger CA, Brasier AR, Bode C, Runge MS: Stimulation of a vascular smooth muscle cell NAD(P)H oxidase by thrombin. Evidence that p47(phox) may participate in forming this oxidase *in vitro* and *in vivo*. J Biol Chem 274: 19814–19822, 1999
68. Brasier AR, Han Y, Sherman CT: Transcriptional regulation of angiotensinogen gene expression. Vit Horm 57: 217–247, 1999
69. Siebenlist U, Franzoso G, Brown K: Structure, regulation and function of NF-κB. Annu Rev Cell Biol 10: 405–455, 1994
70. Han Y, Brasier AR: Mechanism for biphasic Rel A: NF-κB1 nuclear translocation in tumor necrosis factor α-stimulated hepatocytes. J Biol Chem 272: 9823–9830, 1997
71. Maniatis T: Catalysis by a multiprotein IκB kinase complex. Science 278: 818–819, 1997
72. Chen JZ, Parent L, Maniatis T: Site-specific phosphorylation of IκBα by a novel ubiquitination-dependent protein kinase activity. Cell 84: 853–862, 1996
73. Palombella VJ, Rando OJ, Goldberg AL, Maniatis T: The ubiquitin-proteasome pathway is required for processing the NF-κB1 precursor protein and the activation of NF-κB. Cell 78: 773–785, 1994
74. Han Y, Weinman SA, Boldogh S, Brasier AR: TNFα-inducible IκBα proteolysis and NF-κB activation mediated by cytosolic m-Calpain. Biol Chem 274: 787–794, 1999
75. Han Y, Brasier AR: Angiotensin II induces IL-6 transcription in vascular smooth muscle cells through pleiotropic activation of NF-κB transcription factors. Circ Res 84: 695–703, 1999

76. Bourcier T, Sukhova G, Libby P: The nuclear factor-κB signaling pathway participates in dysregulation of vascular smooth muscle cells *in vitro* and in human atherosclerosis. J Biol Chem 272: 15817–15824, 1997
77. Brand K, Page S, Rogler G, Bartsch A, Brandl R, Kneuchel R, Page M, Kaltschmidt C, Baeuerle PA, Neumeier D: Activated transcription factor nuclear factor-κB is present in the atherosclerotic lesion. J Clin Invest 97: 1715–1722, 1996
78. Brasier AR, Li J, Wimbish KA: Tumor necrosis factor activates angiotensinogen gene expression by the Rel A transactivator. J Hypertens 27: 1009–1017, 1996
79. Li J, Brasier AR: Angiotensinogen gene activation by AII is mediated by the Rel A (NF-κB p65) transcription factor: One mechanism for the renin angiotensin system (RAS) positive feedback loop in hepatocytes. Mol Endocrinol 10: 252–264, 1996
80. Akira S, Kishimoto T: IL-6 and NF-IL6 in acute-phase response and viral infection. Immunol Rev 127: 25–50, 1992
81. Loppnow H, Libby P: Proliferating or interleukin-1 activated human vascular smooth muscle cells secrete copious IL-6. J Clin Invest 85: 731–738, 1990
82. Gaumond F, Fortin D, Stankova J, Rola-Pleszczynski M: Differential signaling pathways in platelet-activating factor-inducted proliferation and interleukin-6 production by rat vascular smooth muscle cells. J Cardiovasc Pharmacol 30: 169–175, 1997
83. Beasley D: Phorbol ester and interleukin-1 induce interleukin-6 gene expression in vascular smooth muscle cells via independent pathways. J Cardiovasc Pharmacol 29: 323–330, 1997
84. Ikeda U, Ikeda M, Oohara T: Interleukin-6 stimulates c-myc expression and proliferation of cultured vascular smooth muscle cells. Am J Physiol 260: H1713–H1717, 1991
85. Yasukawa K, Hirano T, Watanabe Y: Structure and expression of human B-cell stimulatory factor 2 (BSF-2) gene. EMBO J 6: 2939, 1987
86. Loppnow H, Bil R, Hirt S, Schonbeck W, Herzberg M, Werdan K, Rietschel ET, Brandt E, Flad HD: Platelet derived IL-1 induces cytokine production, but not proliferation of human vascular smooth muscle cells. Blood 91: 134–141, 1998
87. Vlahopoulos S, Boldogh I, Brasier AR: NF-κB dependent induction of interleukin-8 gene expression by tumor necrosis factor α: Evidence for an antioxidant sensitive activating pathway distinct from nuclear translocation. Blood 94: 1878–1889, 1999
88. Kranzhofer R, Schmidt J, Pfeiffer CAH, Hagl S, Libby P, Kubler W: Angiotensin induces inflammatory activation of human vascular smooth muscle cells. Arteriosc Thromb Vasc Biol 19: 1623–1629, 1999
89. Hilgenfeldt U, Schwind S: Angiotensin II is the mediator of the increase in hepatic angiotensinogen synthesis after bilateral nephrectomy. Am J Physiol 265: E414–E418, 1993
90. Nakamura A, Iwao H, Fukui K, Kimura S, Tamaki T, Nakanishi S, Abe Y: Regulation of liver angiotensinogen and kidney renin mRNA levels by angiotensin II. Am J Physiol 258: E1–E6, 1990
91. Kohara K, Brosnihan KB, Ferrario CM, Milsted A: Peripheral and central angiotensin II regulates expression of genes of the renin-angiotensin system. Am J Physiol 262: E651–E657, 1992
92. Eggena P, Zhu JH, Clegg K, Barrett JD: Nuclear angiotensin receptors induce transcription of renin and angiotensinogen mRNA. Hypertension 22: 496–501, 1993
93. Jamaluddin M, Meng T, Sun J, Boldogh S, Han Y, Brasier AR: Angiotensin II induces constitutive Nuclear factor-κB1 (NF-κB1) isoforms to bind the angiotensinogen gene acute phase response element: A stimulus-specific pathway for NF-κB activation in hepatocytes. Mol Endocrinol 14: 99–113, 2000
94. Grumont RJ, Fecondo J, Gerondakis S: Alternate RNA splicing of murine *nfkb1* generates a nuclear isoform of the p50 precursor NF-κB1 that can function as a transactivator of NF-κB-regulated transcription. Mol Cell Biol 14: 8460–8470, 1994
95. Wadgaonkar R, Phelps KM, Haque Z, Williams AJ, Silverman ES, Collins T: CREB binding protein is a nuclear integrator of nuclear factor-κB and p53 signaling. J Biol Chem 274: 1879–1882, 1999
96. Perkins ND, Felzien LZ, Betts JC, Leung K, Beach DH, Nabel GJ: Regulation of NF-κB by cyclin-dependent kinases associated with the p300 coactivator. Science 275: 523–527, 1997
97. Khan AU, Wilson T: Reactive oxygen species as cellular messengers. Chem Biol 2: 437, 1995
98. Sundaresan M, Yu Z-X, Ferrans V, Irani K, Finkel T: Requirement for generation of H_20_2 for platelet-derived growth factor signal transduction. Science 270: 296–299, 1995
99. Lo YY, Cruz TF: Involvement of reactive oxygen species in cytokine and growth factor induction of c-fos expression in chrondrocytes. J Biol Chem 270: 11727–11730, 1995
100. Rao GN, Berk BC: Circ Res 70: 593–599, 1992
101. Ginn-Pease ME, Whisler RL: Redox signals and NF-κB activation in T cells. Free Rad Biol Med 25: 346–1998
102. Alexander RW: Hypertension and the pathogenesis of atherosclerosis. Oxidative stress and the mediation of arterial inflammatory response: A new perspective. J Hypertens 25: 155–161, 1995
103. Ross R: The pathogenesis of atherosclerosis: A perspective for the 1990s. Nature (London) 362: 801–809, 1993
104. Kume N, Cybulsky MI, Gimbrone MA Jr: Lysophosphatidylcholine, a component of atherogenic lipoproteins, induces mononuclear leukocyte adhesion molecules in cultured human and rabbit arterial endothelial cells. J Clin Invest 90: 1138–1144, 1992
105. Han Y, Meng T, Murray NR, Fields AP, Brasier AR: IL-1 Induced NF-κB-IκBα autoregulatory feedback loop in hepatocytes: A role for PKCα in post-transcriptional regulation of IκBα resynthesis. J Biol Chem 274: 939–947, 1999
106. Naftilan AJ, Pratt RE, Dzau VJ: Induction of platelet-derived growth factor A-chain and c-myc gene expressions by angiotensin II in cultured rat vascular smooth muscle cells. J Clin Invest 83: 1419–1424, 1989
107. Pastore L, Tessitore A, Martinotti S, Toniato E, Alesse E, Bravi MC, Ferri C, Desideri G, Gulino A, Santucci A: Angiotensin II stimulates intercellular adhesion molecule-1 (ICAM-1) expression by human vascular endothelial cells and increases soluble ICAM-1 release *in vivo*. Circulation 100: 1646–1652, 1999
108. Tummala PE, Chen XL, Sundell CL, Laursen JB, Hammes CP, Alexander RW, Harrison DG, Medford RM: Angiotensin II induces vascular cell adhesion molecule-1 expression in rat vasculature: A potential link between the renin-angiotensin system and atherosclerosis. Circulation 100: 1223–1229, 1999
109. Wegenka UM, Buschmann J, Lutticken C, Heinrich PC, Horn F: Acute-phase response factor, a nuclear factor binding to acute-phase response elements, is rapidly activated by interleukin-6 at the postranslational level. Mol Cell Biol 13: 276–288, 1993
110. Hocke GM, Barry D, Fey GH: Synergistic action of interleukin-6 and glucocorticoids is mediated by the interleukin-6 response element of the rat a2-macroglobulin gene. Mol Cell Biol 12: 2282–2294, 1992
111. Akira S, Nishio Y, Inoue M, Wang X-J, Wei S, Matsusaka T, Yoshida K, Sudo T, Naruto M, Kishimoto T: Molecular cloning of APRF, a novel IFN-stimulated gene factor 3 p91-related transcription factor involved in the gp130-mediated signaling pathway. Cell 63–71, 1994

Molecular and Cellular Biochemistry **212**: 171–175, 2000.

The role of Jak/STAT signaling in heart tissue renin-angiotensin system

Eduardo Mascareno and M.A.Q. Siddiqui

Center for Cardiovascular and Muscle Research, Department of Anatomy and Cell Biology, State University of New York Downstate Medical Center, Brooklyn, NY, USA

Abstract

The involvement of the Renin Angiotensin System (RAS) and the role of its primary effector, angiotensin II (Ang II), in etiology of myocardial hypertrophy and ischemia is well documented. In several animal models, the RAS is activated in cardiac cell types that express the receptor AT_1, and/or AT_2, through which the Ang II mediated effects are promoted. In this article, we briefly review recent experimental evidence on the critical role of a prominent signaling pathway, the Jak/STAT pathway in activation and maintenance of the local RAS in cardiac hypertrophy and ischemia. Recent studies in our laboratory document that the promoter of the prohormone angiotensinogen (Ang) gene serves as the target site for STAT proteins, thereby linking the Jak/STAT pathway to activation of heart tissue autocrine Ang II loop. STAT5A and STAT6, are selectively activated when the heart is subjected to ischemic injury, whereas activation of STAT3 and STAT5A is involved in myocardial hypertrophy. Blockage of RAS activation by treatment with specific inhibitor promotes a remarkable recovery in functional hemodynamics of the myocardium. Thus, activation of selective sets of STAT proteins constitutes the primary signaling event in the pathogenesis of myocardial hypertrophy and ischemia. (Mol Cell Biochem **212**: 171–175, 2000)

Key words: renin angiotensin system, mycocardial hypertrophy, ischemic heart, Jak/STAT

Introduction

The renin-angiotensin (RAS) system is responsible for maintaining normal homeostatic levels of a number of hemodynamic properties, such as, blood pressure, blood volume, and vascular tone. Perturbation of the RAS activity under pathophysiological conditions can be directly correlated with the development of systemic hypertension, cardiac hypertrophy, and cardiac failure [1–4]. At the molecular level, the octapeptide Angiotensin II (Ang II) is the one component of the RAS that is prominently involved in virtually all aspects of RAS activity and, as such, plays an important role in the etiology of the cardiovascular diseases above [5–8]. The binding of Ang II to the AT_1 receptor, a transmembrane domain G-protein coupled receptor, can activate several different intracellular signal transduction pathways that use the well-known signal transducers protein kinase A, protein kinase C, MAP kinase, and src [9–13].

Jak/STAT pathway

In addition to the signal transduction pathways above, Ang II also activates the Jak/STAT pathway. The components of the Jak/STAT pathway are present in a latent state in the cytoplasm of unstimulated cells. Binding of ligand leads to activation of the Janus kinase (Jak), a tyrosine kinase that phosphorylates the STAT proteins and allows them to translocate to the nucleus. Within the nucleus, the phosphorylated STATs function as transcription factors [14] that recognize and bind, in a sequence-specific fashion, to cis-regulatory elements in the promoter of target genes. One well-studied example of such sequence element is the gamma interferon-activated sequence (GAS) [15]. Initially, the Jak/STAT pathway was believed to be important for its ability to regulate the immune response. Studies of mice lacking STAT1 [16, 17], STAT4 [18], and STAT6 [19] genes showed impaired function of both innate and acquired immune responses [20].

Address for offprints: M.A.Q. Siddiqui, Center for Cardiovascular and Muscle Research, Department of Anatomy and Cell Biology, State University of New York, Downstate Medical Center, Brookly, NY 11203, USA

More recent studies have shown the Jak/STAT pathway to be a more widely used signal transduction pathway than what was initially suggested. In addition to the lymphokines and cytokines used to stimulate the Jak/STAT pathway in early studies, other ligands, such as, growth hormones [21, 22], growth factors PDGF [23] and EGF [24], prolactin [25], leptin [26, 27], and Ang II [28–30] were reported to activate the Jak/STAT pathway.

As mentioned above, STAT proteins are also transcription factors. This raises the question as to the identity of the nuclear genes that activate STAT specific targets. Studies to-date show that the specificity of STAT mediated signaling is regulated at different levels. For instance, the specificity of STAT proteins for target DNA sequences is regulated through their amino termini [31], the DNA binding domain [32], the Sh2 domain [33], the status of their phosphorylation [34–37], or the sequence of target cis-regulatory elements [38]. Of these regulatory elements, the interferon-stimulated response element (ISRE) was initially identified in the promoter of genes induced following treatment with interferon [39], as was the γ-interferon-inducible element (GAS), in the guanylate-binding protein gene promoter [40]. Subsequently, these sequences were found to be conserved in a host of genes which respond to STAT dependent activation of transcription. The diverse ways in which the specificity of STAT action can be regulated appears to parallel the diversity of genes identified as targets of the activated Jak/STAT pathway. These target genes encode proteins of with widely different functions and whose expression is common to many different cell types. Among these genes are the early activated gene (EAG), c-fos [41], Jun-B [42], C-reactive protein [43], human heme oxygenase [44], metalloproteinases-1 (TIMP-1) [45], protease serine inhibitor gene [46], angiotensinogen (Ang) [30], and probably many others yet to be identified.

Activation of Jak/STAT pathway by Ang II

Activation of the Jak/STAT pathway by Ang II has been observed in vascular smooth muscle cells [28], cardiac fibroblasts [29], and cardiac muscle cells [30]. Among the components of the Jak/STAT pathway that are activated by Ang II are Tyk2, Jak1 and Jak2 kinases, STAT1, STAT2, STAT3, STAT5A, and STAT6 proteins [30, 47, 48]. The understanding of the mechanisms underlying activation of the multiple signal transducers by a single ligand, namely Ang II, in the heart is complicated by the fact that the Jak/STAT signaling can also be activated by other ligand-activatable receptor systems, such as, the cytokine receptor family [49–51], an important member of which is the gp130 receptor [52]. Diverse extracellular ligands, such as IL-6 [52], cardiotrophin [53], leukemia inhibitory factor (LIF) [54], and Ang II, all of which potentiate cellular growth, transduce their signal predominantly through a common mediator, STAT3 protein. This would suggest that STAT3 is a key mediator of the 'cell growth' signal and its target genes may have a pivotal role in control of cellular growth. STAT3 may also serve as an important mediator of growth signaling in the cardiovascular system. Transgenic mice overexpressing the IL-6 ligand and a subunit of the gp130 receptor exhibited a phenotype characterized by cardiac hypertrophy [52]. The role of STAT3 in gp130-mediated signaling was also evident in the cre-lox mice in which only cells of the ventricular heart chamber null for gp130 cytokine receptor expression exhibited cardiac hypertrophy induced by trans-aortic constriction. No phosphorylation (activation) of STAT3 was observed in the ventricle chamber-restricted gp130 knockout mice (GP CKO) [55], suggesting that STAT3 may be involved in gp130-dependent signal transduction pathway mediating myocardial cell survival under hypertrophic conditions. However, using a similar approach to induce pressure overload hypertrophy, Pan *et al.* [56] found that STAT3 is phosphorylated within 1 h of abdominal aortic constriction, but the control mice used in these studies also showed activation of STAT3 after 2 days. While these contradictions could be interpreted as due to different animal model systems employed in these studies or the varying degrees of aortic constrictions used. A more accurate picture of STAT3 function in gp130-dependent growth signaling would emerge from studies on cre-lox null mice models for STAT3 expression in heart ventricle cells.

STAT-mediated regulation of the autocrine RAS loop and myocardial hypertrophy

We have shown that activation of the Jak/STAT pathway by Ang II is linked to the maintenance of the autocrine loop of the heart tissue RAS during cardiac hypertrophy [30]. Studies in rat primary cardiac cultures and spontaneously hypertensive rats (SHR) showed that three members of the STAT family, STAT3, STAT5A and STAT6, are directly involved in transcriptional activation via binding to the consensus St-Domain, a cis-element that resembles the GAS domain in the promoter of the human, rat, and mouse Ang genes. Activation of STAT6 via Ang II in the myocardium was further established in our studies on STAT6–/– null mice by examining the STAT-St-domain interaction after acute treatment (1 h) of Ang II (10^{-7} M). Figure 1 shows that nuclear extract from the myocardium of STAT6–/– null mice exhibited a marked reduction in the formation of DNA-protein complexes as compared with wild type mice treated under similar conditions suggesting that STAT6 is predominantly involved in regulation of Ang II expression. The residual binding activity is due to STAT3 and/or STAT5A. Whether STAT6–/– mice develop myocardial hypertrophy following chronic administration of Ang II is yet to be determined. Recent experiments (unpublished) in our laboratory also indicate that Jak2 phosphorylation is a prerequisite for upregulation of the Ang promoter activity. We have observed both STAT3 and STAT5A are activated in human left ventricular hypertrophy (LVH). Taken together, the Ang II mediated activation of the Jak/STAT pathway in cardiomyocytes appears to play a critical role in maintenance of the autocrine loop of Ang gene in the heart, and thereby, link the local RAS and the Jak/STAT pathway with these pathological processes i.e. myocardial hypertrophy, and ischemia.

Jak/STAT pathway in ischemic heart

The molecular mechanisms that underlie cellular dysfunction in the myocardium subjected to ischemia and reperfusion are not known. Several studies using isolated working heart as an in vivo model have confirmed the involvement of AT_1 receptor in pathogenesis of ischemia/reperfusion injury and in the resulting myocardial dysfunction. Administration of the AT_1 receptor antagonist, losartan, causes a functional recovery of the heart, suggesting that the signal transduction pathways associated with the AT_1 receptor promote cardiovascular damage in ischemia. We have identified components of the Jak/STAT signaling pathway activated during ischemia/reperfusion (Fig. 2). Analysis of the whole cell extract from rat myocardium revealed that STAT1, STAT3, STAT5A, and STAT6 are activated at higher levels during ischemia/reperfusion relative to that in ischemia alone. However, when we examined the nuclear localization of these STAT proteins, we observed that only STAT5A and STAT6 interact with the St-domain, suggesting that activated STAT3 and STAT1 target other genes. Consistent with this notion, other studies have suggested that STAT3 has a cardioprotective role [55, 57].

In the ischemic rat heart, phosphorylated STAT5A and STAT6 promote activation of the tissue RAS via their role in transcriptional up-regulation of the Ang gene. These same STAT proteins were also shown to be activated in human heart tissues with left ventricular ischemic injury. The underlying mechanisms are thus evolutionarily conserved. Our studies have also documented that blockage of the overactive RAS by treatment with losartan, an AT_1 blocker, tyrphostin AG490, an inhibitor of Jak2 phosphorylation, and by preconditioning resulted in loss of STAT/Ang promoter interaction. Hearts subjected to ischemia/reperfusion in the presence of tyrphostin AG490 indeed retain better functional (Mascareno *et al.*, submitted for publication). Thus, these findings not only established that activation of the Jak/STAT signaling pathway significantly contributes to the etiology of myocardial ischemia, but also suggest that interference with the STAT/Ang promoter interaction is likely to potentiate recovery in cardiac function.

Conclusions

The RAS, which plays a pivotal role in physiological functions of the cardiovascular system, has a complex role in eti-

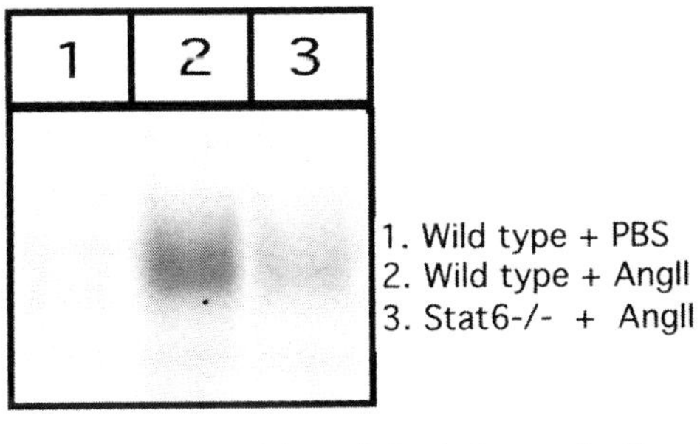

Fig. 1. Ischemia and ischemia/reperfusion induces activation of STATs proteins: In order to establish the role of myocardial ischemia as inducer of Jak/STAT activation, whole cell extract from normal heart (BL), heart treated with 30 min ischemia (I-30), plus 60 min reperfusion (I-30 R60), or 120 min reperfusion (I-30 R120), or 60 min ischemia (I-60) only, were incubated with monoclonal anti-phosphotryosine (4G10). After immunoprecipitation the proteins were isolated, and run in a SDS acrylamide gel followed by transfer to nitrocellulose membrane. The membrane was then tested against polyclonal antibody for STAT1, STAT3, STAT5A and STAT6 (Santa Cruz).

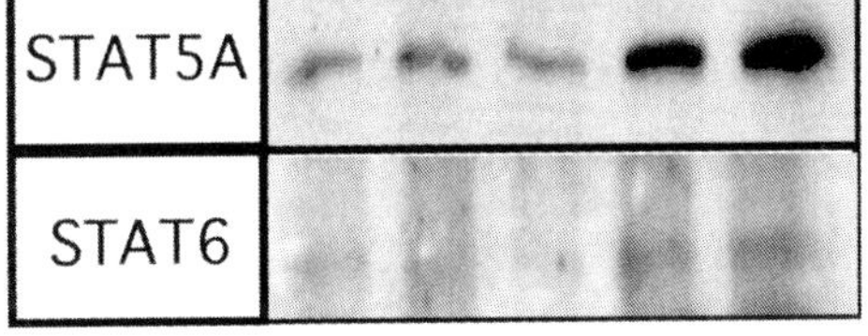

Fig. 2. STAT6 is required for Ang II mediated signaling in rat heart: The Ang II dependent activation of STAT proteins was tested by acutely injecting Ang II peptide (10^{-7} M) in the superior *vena cava* to wild type mice or STAT6–/– null mice. After 60 min hearts from wild type and STAT6–/– null mice were excised to make nuclear extracts. These nuclear extracts were tested for their ability to bind the conserved GAS domain present in the Ang promoter in gel mobility shift assay (GMSA). Lane 1 wild type mice injected with buffer phosphate saline, lane 2 wild type mice injected with Ang II peptide, and lane 3 STAT6–/– injected with Ang II peptide.

ology of diseases, such as hypertrophy and ischemic injury. Activation of the RAS is coupled to a prominent signaling pathway, the Jak/STAT pathway. The focus of this article is to enumerate the components of the Jak/STAT pathway in maintaining the autocrine loop of the heart tissue RAS. Our studies suggest that activation of a selective seta of STAT proteins is the highlight of the signaling event in the onset of rat cardiac hypertrophy and ischemia/reperfusion and the underlying mechanism is conserved in humans. Inhibition of Jak2 by tyrphostin AG490, afforded significant cardioprotection to the heart subjected to ischemia/reperfusion. This led us to believe that Jak2 mediated phosphorylation of specific STAT proteins plays a prominent role in the overall signaling cascade and thereby offers the potential for therapeutic approach for recovery of the function of the disease heart.

References

1. Jeunemaitre X, Soubrier F, Kotelevtsev YV, Lifton RP, Williams CS, Charru A, Hunt SC, Hopkins PN, Williams RR, Lalouel JM, Corvol P: Molecular basis of human hypertension: role of angiotensinogen. Cell 71: 169–180, 1992
2. Kimura S, Mullins JJ, Bunnemann B, Metzger R, Hilgenfeldt U, Zimmermann F, Jacob H, Fuxe K, Ganten D, Kaling M: High blood pressure in transgenic mice carrying the rat angiotensinogen gene. EMBO J 11: 821–827, 1992
3. Pfeffer JM, Pfeffer MA, Mirsky I, Braunwald E: Regression of left ventricular hypertrophy and prevention of left ventricular dysfunction by captopril in the spontaneously hypertensive rat. Proc Natl Acad Sci USA 79: 3310–3314, 1982
4. DiBona GF, Jones SY, Brooks VL: Ang II receptor blockade and artrial baroreflex regulation of renal nerve activity in cardiac failure. Am J Phys 269: R1189–R1196, 1995
5. Reid JL, MacFadyen RJ, Squire IB, Lees KR: Angiotensin-converting enzyme inhibitors in heart failure: Blood pressure changes after the first dose. Am Heart J 126: 794–797, 1993
6. Waeber B, Brunner HR: Cardiovascular hypertrophy: Role of angiotensin II and bradykinin. J Cardiol Pharm 27(suppl 2): S36–S40, 1996
7. Bulpitt CJ: Quality of life with ACE inhibitors in chronic heart failure. J Cardiol Pharm 27(suppl 2): S31–S35, 1996
8. Eberhardt RT, Kevak RM, Kang PM, Frishman WH: Angiotensin II receptor blockade: an innovative approach to cardiovascular pharmacotherapy. J Clin Pharm 33: 1023–1038, 1993
9. Sadoshima J, Izumo S: Molecular characterization of angiotensin II induced hypertrophy of cardiac myocytes and hyperplasia of cardiac fibroblasts. Critical role of the AT1 receptor subtype. Circ Res 73: 413–423, 1993
10. Duff JL, Marrero M, Paxton W, Schieffer B, Bernstein KE, Berk BC: Angiotensin II signal transduction and the mitogen activated protein kinase pathway. Cardiovasc Res 30: 511–517, 1995
11. Booz GW, Baker KM: Molecular signalling mechanisms controlling growth and function of the cardiac fibroblasts. Cardiovasc Res 30: 537–543, 1995
12. Schieffer B, Paxton W, Marrero M, Bernstein K: Importance of tyrosine phosphorylation in Angiotensin II type 1 receptor signaling. Hypertension 27: 476–480, 1996
13. Bernstein K, Marrero M: The importance of tyrosine phosphorylation in angiotensin II signaling. Trends Cardiovasc Med 6: 179–187, 1996
14. Ihle JN: Stats: Signal transducers and activators of transcription. Cell 84: 331–334, 1996
15. Zhong Z, Wen Z, Darnell JE Jr: STAT3: A STAT family member activated by tyrosine phosphorylation in response to epidermal growth factor and interleukin-6. Science 264: 95–98, 1994
16. Durbin JE, Hackenmiller R, Simon MC, Levy DE: Targeted disruption of the mouse STAT1 gene results compromised innate immunity to viral diseases. Cell 84: 443–450, 1996
17. Meraz MA, White JM, Sheeba KC, Bach EA, Rodig SJ, Dighe AS, Kaplan DH, Schreiber RD: Targeted disruption of the STAT1 gene in mice reveals unexpected physiologic specificity in the Jak/STAT signaling pathway. Cell 84: 431–442, 1996
18. Kaplan MH, Sun YL, Hoey T, Grusby MJ: Impaired IL-12 responses and enhanced development of Th2 cells in STAT4-deficient mice. Nature 382: 174–177, 1996
19. Kaplan MH, Schindler U, Smiley ST, Grusby MJ: STAT6 is required for mediating responses to IL-4 and for development of Th2 cells. Immunity 4: 313–319, 1996
20. Liu X, Robinson GW, Wagner KU, Garret L, Wynshaw-Boris A, Hennighausen L: STAT5A is mandatory for adult mammary gland development and lactogenesis. Genes Dev 11: 179–186, 1997
21. Wood TJ, Sliva D, Lobie PE, Pircher TJ, Gouilleux F, Wakao H, Gustafsson JA, Groner B, Norstedt G, Haldosen LA: Mediation of growth hormone-dependent transcriptional activation by mammary gland factor/STAT 5. J Biol Chem 270: 9448–9453, 1995
22. Gouilleux F, Pallard C, Dusanter-Fourt I, Wakao H, Haldosen LA, Norstedt G, Levy D, Groner B: Prolactin, growth hormone, erythropoietin and granulocyte-macrophage colony stimulating factor induce MGF-STAT5 DNA binding activity. EMBO J 14: 2005–2013, 1995
23. Novak U, Mui A, Miyajima A, Paradiso L: Formation of STAT5-containing DNA binding complexes in response to colony-stimulating factor-1 and platelet-derived growth factor. J Biol Chem 271: 18350–18354, 1996
24. Leaman DW, Pisharody S, Flickinger TW, Commane MA, Schlessinger J, Kerr IM, Levy DE, Stark GR: Roles of JAKs in activation of STATs and stimulation of c-fos gene expression by epidermal growth factor. Mol Cell Biol 16: 369–375, 1996
25. Gouilleux F, Wakao H, Mundt M, Groner B: Prolactin induces phosphorylation of Tyr694 of STAT5 (MGF), a prerequisite for DNA binding and induction of transcription. EMBO J 13: 4361–4369, 1994
26. Ghilardi N, Ziegler S, Wiestner A, Stoffel R, Heim MH, Skoda RC: Defective STAT signaling by the leptin receptor in diabetic mice. Proc Natl Acad Sci USA 93: 6231–6235, 1996
27. Baumann H, Morella KK, White DW, Dembski M, Bailon PS, Kim H, Lai CF, Tartaglia LA: The full-length leptin receptor has signaling capabilities of interleukin 6-type cytokine receptors. Proc Natl Acad Sci USA 93: 8374–8378, 1996
28. Marrero MB, Schieffer B, Paxton WG, Heerdt L, Berk BC, Delafontaine P, Bernstein KE: Direct stimulation of Jak/STAT pathway by the angiotensin II AT1 receptor. Nature 375: 247–250, 1995
29. Bhat GJ, Thekkumkara TJ, Thomas WG, Conrad KM, Baker KM: Angiotensin II stimulates sis-inducing factor-like DNA binding activity. Evidence that the AT1A receptor activates transcription factor-STAT91 and/or a related protein. J Biol Chem 269: 31443–31449, 1994
30. Mascareno E, Dhar M, Siddiqui MAQ: Signal Transduction and Activator of Transcription (STAT) protein-dependent activation of angiotensinogen promoter: A cellular signal for hypertrophy in cardiac muscle. Proc Natl Acad Sci USA 95: 5590–5594, 1998
31. Xu X, Sun YL, Hoey T: Cooperative DNA binding and sequence-selective recognition conferred by the STAT amino-terminal domain. Science 273: 794–797, 1996
32. Horvath CM, Wen Z, Darnell JE Jr: A STAT protein domain that determines DNA sequence recognition suggests a novel DNA-binding domain Genes Dev 9: 984–994, 1995

33. Hemmann U, Gerhartz C, Heesel B, Sasse J, Kurapkat G, Grotzinger J, Wollmer A, Zhong Z, Darnell JE Jr, Graeve L, Heinrich PC, Horn F: Differential activation of acute phase response factor/STAT3 and STAT1 via the cytoplasmic domain of the interleukin 6 signal transducer gp130. II. Src homology SH2 domains define the specificity of STAT factor activation. J Biol Chem 271: 12999–13007, 1996
34. Karin M: Signal transduction from the cell surface to the nucleus through the phosphorylation of transcription factors. Curr Opin Cell Biol 6: 415–424, 1994
35. Gotoh A, Takahira H, Mantel C, Litz-Jackson S, Boswell HS, Broxmeyer HE: Steel factor induces serine phosphorylation of STAT3 in human growth factor-dependent myeloid cell lines. Blood 88: 138–145, 1996
36. David M, Petricoin E III, Benjamin C, Pine R, Weber MJ, Larner AC: Requirement for MAP kinase (ERK2) activity in interferon alpha- and interferon beta-stimulated gene expression through STAT proteins. Science 269: 1721–1723, 1995
37. Zhong Z, Wen Z, Darnell JE Jr: STAT3: A STAT family member activated by tyrosine phosphorylation in response to epidermal growth factor and interleukin-6. Science 264: 95–98, 1994
38. Seidel HM, Milocco LH, Lamb P, Darnell JE, Stein RB, Rosen J: Spacing of palindromic half sites as a determinant of selective STAT (signal transducers and activators of transcription) DNA binding and transcriptional activity. Proc Natl Acad Sci USA 92: 3041–3045, 1995
39. Darnell JE Jr, Kerr IM, Stark GM: Jak-STAT pathways and transcriptional activation in response to IFNs and other extracellular signaling proteins. Science 264: 1415–1421, 1994
40. Lew DJ, Decker T, Strehlow I, Darnell JE Jr.: Overlapping elements in the guanylate-binding protein gene promoter mediate transcriptional induction by alpha and gamma interferons. Mol Cell Biol 11: 182–191, 1991
41. Hill CS, Treisman R: Differential activation of c-fos promoter elements by serum, lysophosphatidic acid, G proteins and polypeptide growth factors. EMBO J 14: 5037–5047, 1995
42. Kojima H, Nakajima K, Hirano T: IL-6-inducible complexes on an IL-6 response element of the junB promoter contain STAT3 and 36 kDa CRE-like site binding protein(s). Oncogene 12: 547–554, 1996
43. Zhang D, Sun M, Samols D, Kushner I: STAT3 participates in transcriptional activation of the C-reactive protein gene by interleukin-6. J Biol Chem 271: 9503–9509, 1996
44. Lavrovsky Y, Drummond GS, Abraham NG: Downregulation of the human heme oxygenase gene by glucocorticoids and identification of 56b regulatory elements. Biochem Biophys Res Commun 218: 759–765, 1996
45. Bugno M, Graeve L, Gatsios P, Koj A, Heinrich PC, Travis J, Kordula T: Identification of the interleukin-6/oncostatin M response element in the rat tissue inhibitor of metalloproteinases-1 (TIMP-1) promoter. Nucl Acid Res 23: 5041–5047, 1995
46. Kordula T, Travis J: The role of STAT and C/EBP transcription factors in the synergistic activation of rat serine protease inhibitor-3 gene by interleukin-6 and dexamethasone. Biochem J 313: 1019–1027, 1996
47. McWhinney CD, Dostal DE, Baker KM: Angiotensin II activates STAT5 through Jak2 kinase in cardiomyocytes. J Mol Cell Cardiol 30: 751–761, 1998
48. Pan J, Fukuda K, Saito M, Matsuzaki J, Kodama H, Sano M, Hori S, Ogawa: Biphasic activation of the Jak/STAT pathway by angiotensin II in rat cardiomyocytes. Circ Res 84: 1127–1136, 1998
49. Ihle JN, Kerr IM: Jaks and STATs in signaling by the cytokine receptor superfamily. Trends Genet 11: 69–74, 1995
50. Heim MH: The Jak-STAT pathway: Specific signal transduction from the cell membrane to the nucleus. Eur J Clin Invest 26: 1–12, 1996
51. Schindler C, Darnell JE Jr. : Transcriptional responses to polypeptide ligands: The Jak-STAT pathway. Annu Rev Biochem 64: 621–651, 1995
52. Hirota H, Yoshida K, Kishimoto T, Taga T: Continuous activation of gp130, a signal transducing receptor component for interleukin 6 related cytokines, causes myocardial hypertrophy in mice. Proc Natl Acad Sci USA 92: 4862–4866, 1995
53. Pennica D, King KL, Shaw KJ, Luis E, Rullamas J, Luoh SM, Darbonne WC, Knutzon DS, Yen R, Chien KR, Baker JB, Wood W: Expression cloning of cardiotrophin 1, a cytokine that induces cardiac myocyte hypertrophy. Proc Natl Acad Sci USA 92: 1142–1146, 1995
54. Kodama H, Fukuda K, Pan J, Makino S, Baba A, Hori S, Ogawa S: Leukemia inhibitory factor, a potent cardiac hypertrophic cytokine, activates the Jak/STAT pathway in rat cardiomyocytes. Circ Res 81: 656–663, 1997
55. Hirota H, Chen J, Betz U, Rajewsky K, Gu Y, Ross J, Muller W, Chien KR: Loss of a gp130 cardiac muscle cell survival pathway is a critical event in the onset of heart failure during biomechanical stress. Cell 97: 189–198, 1999
56. Pan J, Fukuda K, Kodama H, Makino H, Makino S, Takahashi T, Sano M, Hori S, Ogawa S: Role of angiotensin II in activation of the Jak/STAT pathway induced acute pressure overload in the rat heart. Circ Res 81: 611–617, 1997
57. Takeda K, Kaisho T, Yoshida N, Takeda J, Kishimoto T, Akira S: STAT3 activation is responsible for IL-6 dependent T cell proliferation through preventing apoptosis: Generation and characterization of T cell specific STAT3 deficient mice. J Immunol 161: 4652–4660, 1998

Molecular and Cellular Biochemistry **212:** 177–182, 2000.

Rho plays an important role in angiotensin II-induced hypertrophic responses in cardiac myocytes

Ryuichi Aikawa, Issei Komuro, Ryozo Nagai and Yoshio Yazaki
Department of Cardiovascular Medicine, University of Tokyo Graduate School of Medicine, Tokyo, Japan

Abstract

Angiotensin II (Ang II) evokes a variety of hypertrophic responses such as activation of protein kinases, reprogramming of gene expressions and an increase in protein synthesis in cardiac myocytes. In this study, we examined the role of Rho family small GTP binding proteins (G proteins) in Ang II-induced cardiac hypertrophy. Ang II strongly activated extracellular signal-regulated protein kinases (ERKs) in cardiac myocytes of neonatal rats. Although Ang II-induced activation of ERKs was completely suppressed by an Ang II type 1 receptor antagonist, CV-11974, this activation was not inhibited by the pretreatment with C3 exoenzyme, which abrogates Rho functions. Overexpression of Rho GDP dissociation inhibitor (Rho-GDI), dominant negative mutants of Rac1 (D.N.Rac1), or D.N.Cdc42 had no effects on Ang II-induced activation of transfected ERK2. The promoter activity of skeletal α-actin and c-*fos* genes was increased by Ang II, and the increase was partly inhibited by overexpression of Rho-GDI and the pretreatment with C3 exoenzyme. Ang II increased phenylalanine incorporation into cardiac myocytes by approximately 1.4 fold as compared with control, and this increase was also significantly suppressed by the pretreatment with C3 exoenzyme. These results suggest that the Rho family small G proteins play important roles in Ang II-induced hypertrophic responses in cardiac myocytes. (Mol Cell Biochem **212**: 177–182, 2000)

Key words: angiotensin II, Rho family proteins, ERK, c-*fos*, skeletal α-actin, protein synthesis

Abbreviations: Ang II – angiotensin II; DMEM – Dulbecco's modified Eagle's medium; D.N.RhoA – dominant negative mutant of RhoA; D.N.Rac1 – dominant negative mutant of Rac1; D.N.Cdc42 – dominant negative mutant of Cdc42; ERK – extracellular signal-regulated kinase; FBS – fetal bovine serum; G protein – guanine nucleotide-binding protein; HA-ERK2 – hemagglutinin (HA)-tagged ERK2; JNK – c-Jun NH2-terminal kinase; MAPK – mitogen-activated protein kinase; MEK – mitogen-activated protein kinase/ERK kinase; MBP – myelin basic protein; PHE – phenylephrine; PKC – protein kinase C; Rho-GDI – Rho GDP dissociation inhibitor; SDS-PAGE – sodium dodecyl sulfate-polyacrylamide gel electrophoresis

Introduction

Clinical studies have demonstrated that cardiac hypertrophy is not only an adaptational state before cardiac failure, but also an independent risk factor of morbidity and mortality due to heart diseases [1]. Thus, it has become even more important to understand the underlying mechanisms of the development of cardiac hypertrophy. Cardiac hypertrophy is thought to be induced by mechanical load and humoral factors such as angiotensin II (Ang II) [2], endothelin-1(ET-1) [3], and phenylephrine [4]. We and others have reported that Ang II induces a variety of hypertrophic responses such as activation of protein kinases [5–8], reprogramming of gene expressions [6, 7, 9], and an increase in protein synthesis [6, 9] in cultured cardiac myocytes. It remains unclear, however, how Ang II induces such intracellular events.

Address for offprints: I. Komuro, Department of Cardiovascular Medicine, University of Tokyo Graduate School of Medicine, Tokyo, Japan

The Rho family small GTP binding proteins consisting of Rho, Rac and Cdc42 subfamilies, regulate many aspects of cytoskeletal functions [10]. Rho has been shown to participate in the formation of focal adhesions and actin stress fibers, as well as in mediating redistribution of cytoskeletal components [10, 11]. There is also strong evidence indicating that Rho plays a pivotal role in integrin-mediated signaling events. First, activated Rho stimulates stress fiber formation [12], and plating of cells onto fibronectin-coated dishes in serum-free medium results in the rapid formation of stress fibers [13]. Second, both Rho and adhesion to fibronectin activate phosphatidylinositol-4-phosphate 5-kinase [13]. Third, specific inhibition of Rho using botulinum C3 exoenzyme blocks certain cellular responses, which is similar to the loss of integrin-mediated cell adhesion [13]. Last, overexpression of constitutively active Rho restores the ability of suspended cells to respond to growth factors [13]. Recently, integrin ligation has been demonstrated to induce activation and translocation to the nucleus of extracellular-regulated kinases (ERKs) [14]. In addition, Rho proteins are required for the activation of ERKs in NIH 3T3 cells plated on fibronectin [15]. Although the molecular mechanism by which Rho family proteins regulate integrins still remains to be determined, all these observations suggest that Rho is critically involved in integrin-mediated intracellular signalings.

In this study, we examined the role of Rho family small G proteins in the development of Ang II-induced hypertrophic responses. We demonstrate that the Rho family proteins play important roles in Ang II-induced expression of skeletal a-actin and c-fos genes, and an increase in protein synthesis in cardiac myocytes.

Materials and methods

[γ-^{32}P] ATP was purchased from Du Pont-New England Nuclear Co. (Boston, MA, USA). Dulbecco's modified Eagle's medium (DMEM), fetal bovine serum (FBS), and genistein were from GIBCO BRL Co. (Gaithersburg, MD, USA). An anti-ERKs (αY91) antibody was provided by K. Tobe [16]. C3 exoenzyme was a kind gift from S. Narumiya [17]. An anti-hemagglutinin (HA) polyclonal antibody was from MBL (Japan). Ang II, insulin, myelin basic protein (MBP) and other reagents were purchased from Sigma Chemical Co. (St. Louis, MO, USA)

cDNA plasmids

HA-tagged ERK2 (HA-ERK2) was kind gifts from M. Karin [18]. Rho-GDI, various mutants of RhoA, Rac1 and Cdc42 were from J.S. Gutkind [19]. All plasmid DNA was prepared by using QIAGEN plasmid DNA preparation kits (Germany).

Cell culture

Primary cultures of cardiac myocytes were prepared from ventricles of 1-day old Wistar rats as described previously [20] according to the method of Simpson [4]. In brief, cardiomyocytes were plated at a field density of 1×10^5 cells/cm^2 on silicone dishes with 2 ml of culture medium (DMEM with 10% FBS) [20]. Twenty four h after seeding, the culture medium was changed to a solution of serum-free DMEM for 48 h before treatment.

Transfection

Twenty four h after plating the cells on culture dishes, we transfected DNA by using the calcium phosphate method as described previously [20]. For each dish, 2 μg of HA-ERK2 plasmid DNA was transfected with 6 μg of the relevant other plasmids such as D.N.RhoA, Rho-GDI, or empty vector. After 15 h of transfection, the culture medium was removed and cells were washed with phosphate-buffered saline (PBS) for twice and were maintained in serum-free DMEM for 48 h before stretching or treatment with chemical reagents. In case of luciferase assay, 2 μg of c-*fos* or skeletal α-actin promoter luciferase plasmid was transfected into cardiomyocytes with Rho-GDI or empty vector by using the calcium phosphate method. After 6 or 12 h of transfection, the culture medium was removed, and cells were washed with PBS for twice and were maintained in serum-free DMEM for 48 h before treatment with chemical reagents.

Kinase assay of endogenous ERKs

Endogenous ERK activity was measured using MBP as a substrate [9, 16]. In brief, cardiomyocytes were lysed with buffer A (25 mM Tris-HCl pH 7.4, 25 mM NaCl, 1 mM sodium orthovanadate, 10 mM sodium pyrophosphate, 10 nM okadaic acid, 0.5 mM EGTA and 1 mM phenylmethylsulfonyl fluoride), and the lysates were incubated with an anti-ERKs polyclonal antibody for 1 h at 4°C. After incubation, the immune complex was precipitated using protein A sepharose beads, washed, resuspended and incubated with 25 μg MBP at 25°C for 10 min in 25 μl of kinase buffer (25 mM Tris-HCl pH 7.4, 10 mM $MgCl_2$, 1 mM DTT, 40 μM ATP, 2 μCi [γ-^{32}P]ATP, 2 μM PKI, and 0.5 mM EGTA). After incubation, the reaction was terminated by adding Laemmili sample buffer (0.002% bromophenol blue, 0.01 M sodium phosphate buffer pH 7.0, 10% glycerol, 0.4% SDS, 1% 2-mercaptoethanol) to the samples, and the samples were boiled for 5 min. The supernatants were subjected to SDS-PAGE, and the gel was washed with 7% acetic acid for 30 min and 3% glycerol for 30 min, dried and then subjected to autoradiography.

Kinase assay of transfected HA-tagged ERK2

After stimulation, cardiomyocytes into which HA-ERK2 was transfected were lysed with the lysis buffer A and the lysates were incubated with anti-HA polyclonal antibody for 1 h at 4°C. After incubation, the immunecomplex was precipitated using protein A sepharose, resuspended in 25 μl of the kinase buffer and incubated with 25 μg MBP as a substrate at 25°C for 10 min. After incubation, the reaction was terminated by adding the Laemmli sample buffer to the samples. The supernatants were subjected to SDS-PAGE, and the gel was dried and subjected to autoradiography.

c-fos *and skeletal α-actin promoter analysis (Luciferase assay)*

After transfection of 1 μg c-*fos* or skeletal α-actin-luciferase reporter gene, cardiac myocytes were washed with PBS twice and maintained in serum-free media for two days. Transfected cardiomyocytes were lysed with 100 μl of the lysis buffer of the luciferase assay kit (Toyo Ink Inc. Tokyo, Japan). An aliquot of supernatant was added to a buffer containing luciferin according to the luciferase assay kit. Luciferase activity was measured by a Berthold Lumat LB9501 luminometer.

Amino acid uptake into cardiac myocytes

After starvation for 48 h, cardiac myocytes were incubated with 10^{-6} Ang II for 24 h. The relative amount of protein synthesis was determined by assessing the incorporation of the radioactivity into a trichloroacetic acid-insoluble fraction as previously reported [5, 6].

Results

Rho is not involved in Ang II-induced activation of ERKs in cardiac myocytes

Rho family members such as RhoA, Rac1 and Cdc42 have been reported to regulate the activity of mitogen-activated protein kinase (MAPK) subfamilies such as c-Jun NH2-terminal kinase (JNK) and p38MAPK in Hela and COS-7 cells [19, 21]. It has also been reported that Rho proteins are involved in the integrin-dependent activation of ERKs [15], and that Rho proteins activate ERKs through activation of p21-activated kinase [22]. To elucidate the role of Rho protein in the development of Ang II-induced hypertrophic responses in cardiac myocytes, we first examined the effect of C3 exoenzyme on Ang II-induced ERK activation in cardiac myocytes. C3 exoenzyme selectively ADP-ribosylates small G proteins of the Rho subfamily at asparagine-41 thereby blocking their action [23]. After incubated with 10 μg/ml C3 exoenzyme for 36 h, cultured cardiomyocytes of neonatal rats were treated with Ang II (10^{-6} M) for 10 min. A phase-contrast microscopic analysis revealed that the C3 exoenzyme treatment did not change the morphology of cardiac myocytes (data not shown). Incubation of Ang II activated markedly (4–5 fold) ERKs as described previously [5, 9]. Although pretreatment with an angiotensin II type 1 receptor specific antagonist, CV-11974 (10^{-6} M, for 60 min) completely inhibited Ang II-induced activation of ERKs, C3 exoenzyme had only marginal effects on the increase in the ERK activity by Ang II (Fig. 1A). These results suggest that Rho is not involved in Ang II-indused activation of ERKs in cardiac myocytes.

We next examined the role of other Rho family proteins in Ang II-induced ERK activation using different methods. We transfected HA-ERK2 into cardiomyocytes with Rho GDP dissociation inhibitor (Rho-GDI), dominant negative mutant of Rac1 (D.N.Rac1), or D.N.Cdc42, and examined the activity of transfected ERK2 after exposure to Ang II. Ang II increased the activity of the transfected ERK2 (Fig. 1B) as well as of the endogenous ERKs (Fig. 1A) in cardiac myocytes. Ang II-induced activation of ERK2 was not affected by overexpressing Rho-GDI, D.N.Rac1 or D.N.Cdc42 (Fig. 1B).

Rho is important for Ang II-induced expression of c-fos *and skeletal α-actin genes in cardiac myocytes*

Expression of immediate early response genes and fetal genes is a genetic response of cardiac myocytes to Ang II [6, 7, 9, 24]. We examined the role of Rho proteins in Ang II-induced gene expression in cardiac myocytes. A reporter gene containing c-*fos* gene promoter was transfected into cardiac myocytes. The luciferase activity of the c-*fos* reporter gene was increased by Ang II, and its increase was partly inhibited by the pretreatment with C3 exoenzyme (Fig. 2). When Rho-GDI was transfected with the c-*fos* reporter gene into cardiac myocytes, Ang II mildly increased the reporter luciferase activity (Fig. 2). Next, we examined the role of Rho proteins in Ang II-induced fetal gene expression in cardiac myocytes. Ang II-induced activation of skeletal α-actin promoter was also significantly suppressed by the pretreatment with C3 exoenzyme or overexpression of Rho-GDI (Fig. 2). These results suggest that Rho proteins play a critical role in the Ang II-induced reprogramming of gene expressions.

Rho is involved in Ang II-induced increase in phenylalanine incorporation into cardiomyocytes

To determine whether Rho proteins are involved in Ang II-induced cardiac hypertrophy, we examined the relative pro-

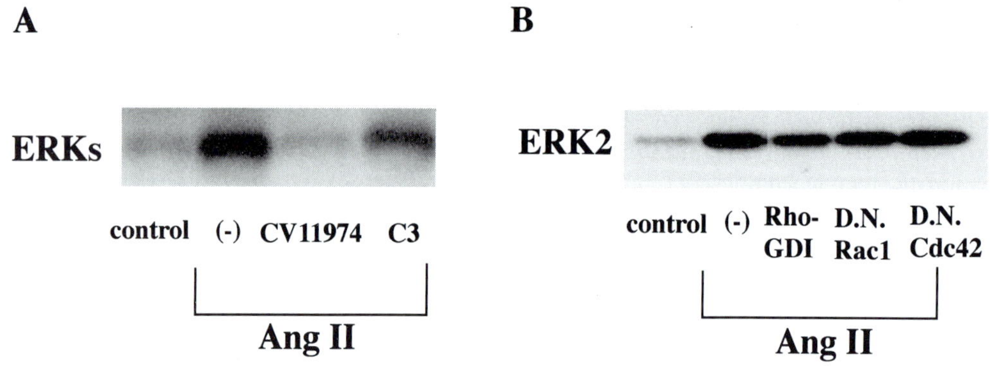

Fig. 1. Rho is not involved in Ang II-induced activation of ERKs in cardiomyocytes. (A) Cultured cardiac myocytes were incubated with Ang II (10^{-6} M, for 10 min). Some cardiomyocytes were pretreated with 10 μg/ml C3 exoenzyme for 36 h or CV-11974 (10^{-6} M, for 60 min). ERKs were immunoprecipitated from cell lysates using a polyclonal antibody against ERKs (αY91), and the kinase activity was assayed using MBP as a substrate. After electrophoresis, the gel was dried and subjected to autoradiography. A representative autoradiogram from three independent experiments was shown. (B) Rac1 and Cdc42 are not involved in Ang II-induced activation of transfected ERK2 in cardiomyocytes. After transfection of HA-ERK2 with either Rho-GDI, D.N.RhoA, D.N.Rac1 or D.N.Cdc42, cardiac myocytes were incubated with Ang II (10^{-6} M) for 10 min. HA-ERK2 was immunoprecipitated from cell lysates using a polyclonal antibody against HA, and the kinase activity was measured using MBP as a substrate. After electrophoresis, the gel was dried and subjected to autoradiography. The data represent the average percentage of the control (100%) from 4 independent experiments.

tein synthesis after stimulation of Ang II in the presence or absence of C3 exoenzyme. As described in the previous paper [5], 10^{-6} Ang II for 24 h increased phenylalanine incorporation into cardiac myocytes by approximately 1.4 fold as compared with control (Fig. 3). This increase was significantly (60%) suppressed by the pretreatment with 10 μg/ml C3 exoenzyme for 36 h, suggesting that Rho proteins play a critical role in the Ang II-induced cardiac hypertrophy.

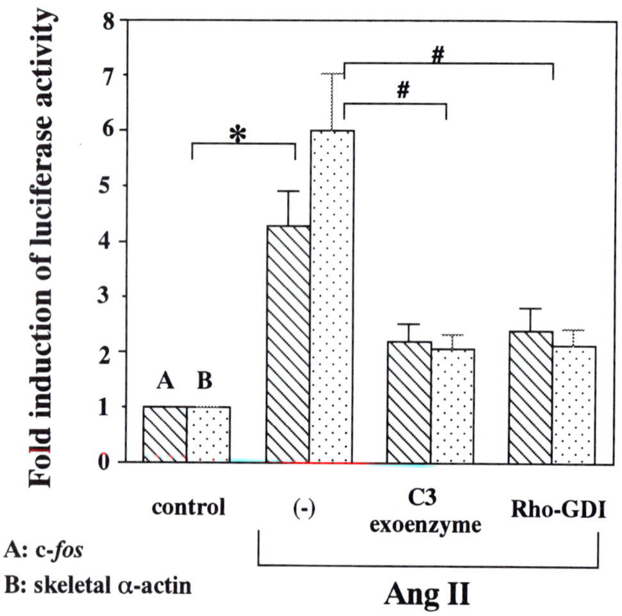

Fig. 2. Rho proteins are important for Ang II-induced expression of c-*fos* and skeletal α-actin genes. c-*fos* (A) or skeletal α-actin (B)-luciferase reporter gene was transfected into cultured cardiac myocytes by calcium phosphate method. Functions of Rho proteins were inhibited by either pretreatment by 10 μg/ml C3 exoenzyme for 36 h or overexpression of Rho-GDI. Cardiac myocytes were incubated by 10^{-6} M Ang II for 48 h and lysed with 100 μl of lysis buffer. Luciferase activity was measured by a Berthold Lumat LB9501 luminometer. The data represent the average percentage of the controls (100%) from four independent experiments (mean ± S.E.). *$p < 0.05$ vs. control; $^{\#}p < 0.05$ vs. Ang II without any inhibition of Rho function.

Discussion

We and others have reported that Ang II activates Raf-1 kinase-MEK-ERKs cascade [5, 7, 8, 25]. Moreover, JNK, a subfamily of MAPK, is also activated by mechanical stress and Ang II in cardiomyocytes [26, 27]. Although the activation of protein kinase cascade of phosphorylation has been reported to be important for the development of hypertrophic responses [5, 7, 25, 26], it is not completely clear how Ang II is converted into these biochemical signals.

A growing body of evidence has suggested that Rho family proteins are critically involved in regulation of cytoskeleton such as formation of actin microfilaments and focal adhesions [12, 28]. It has recently been reported that Rho proteins play a critical role in mechanical stress-induced hypertrophic responses [29] and that Ang II partly mediates mechanical stress-induced cardiac hypertrophy [5]. Therefore, we examined whether Rho family small G proteins are involved in Ang II-induced ERK activation in cardiomyocytes. In the present study, Ang II-induced activation of ERKs was not suppressed by overexpression of Rho-GDI or pretreatment with C3 exoenzyme in cardiac myocytes. These results suggest that Rho proteins are not involved in ERK activation induced by Ang II.

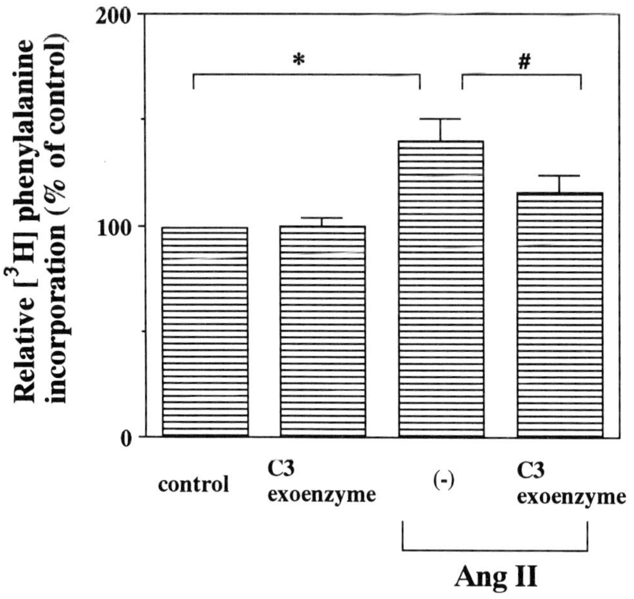

Fig. 3. Rho is required for Ang II-induced increase in protein synthesis. After pretreatment with 10 μg/ml C3 exoenzyme for 36 h, cardiac myocytes were incubated by 10^{-6} M Ang II for 24 h and [^{3}H] phenylalanine (1 μCi/ml) was added 2 h before harvest. The total radioactivity of incorporated [^{3}H] phenylalanine into proteins was determined by liquid scintillation counting. The data represent the average percentage of the control (100%) from 3 independent experiments (mean ± S.E.). *$p < 0.05$ vs. control; $^{\#}p < 0.05$ vs. Ang II without any inhibition of Rho function.

Hypertrophy of neonatal ventricular myocytes is associated with a number of phenotypic changes including reprogramming of gene expressions, changes in cell shape, and organization of contractile proteins into sarcomeric units [24, 30, 31]. It has recently been reported that Rho is required for gene expression which is induced by activation of Gαq and α1-adrenergic receptor, and that Rho-dependent pathways are involved in phenylephrine (PHE)-induced atrial natriuretic factor (ANF) gene expression by interacting with Ras cascades in cardiac myocytes [32]. In addition, the activation of ERKs through Ras is required for PHE-induced gene expression, and Ras but not Rho plays a critical role in actin organization in cardiac myocytes [33, 34]. We showed that Rho proteins are important for Ang II-induced expressions of c-*fos* and skeletal α-actin genes (Figs 2A and 2B), indicating that Rho plays an important role in the Ang II-induced reprogramming of gene expressions as well as PHE-induced gene expressions. Although inhibition of Rho protein functions did not affect Ang II-induced activation of ERKs, Ang II-induced gene expression was partly suppressed by the pretreatment with C3 exoenzyme and overexpression of Rho-GDI. These results suggest that the role of Rho proteins in gene expression may be different from their role in ERK activation. Recent studies have demonstrated that Rho proteins are involved in transcriptional activation of c-*fos* gene by regulating serum response factor (SRF) in several cell types including cardiac myocytes, and activation of SRF-linked signaling pathway does not correlate with activation of the ERKs [35, 36]. Taken together, Rho proteins may regulate these gene expressions without the involvement of activation of ERKs in cardiomyocytes.

The Rho family of small G proteins play a critical role in Ang II-induced hypertrophic responses such as expression of specific genes and an increase in protein synthesis. It needs further investigation to elucidate how Rho proteins are involved in the regulation of Ang II-induced cardiac hypertrophy.

Acknowledgements

We wish to thank Dr. K. Tobe for advice and F. Masuo, M. Nakajima and C. Kuwabara for the excellent technical assistance. This work was supported by a Grant-in-Aid for Scientific Research, Developmental Scientific Research and Scientific Research on Priority Areas from the Ministry of Education, Science, Sports and Culture of Japan (to I.K.).

References

1. Levy D, Garrison RJ, Savage DD, Kannel WB, Castelli WP: Prognostic implications of echocardiographically determined left ventricular mass in the Framingham heart study. New Engl J Med 322: 1561–1566, 1990
2. Baker KM, Booz GW, Dostal DE: Cardiac actions of angiotensin II: Role of an intracardiac renin-angiotensin system. Annu Rev Physiol 54: 227–241, 1992
3. Ito H, Hirata Y, Hiroe M, Tsujino M, Adachi S, Takamoto T, Nitta M: Endothelin-1 induces hypertrophy with enhanced expression of muscle-specific genes in cultured neonatal rat cardiomyocytes. Circ Res 69: 209–215, 1991
4. Simpson P: Norepinephrine-stimulated hypertrophy of cultured rat myocardial cells is an alpha 1 adrenergic response. J Clin Invest 72: 732–738, 1983
5. Yamazaki T, Komuro I, Kudoh S, Zou Y, Shiojima I, Mizuno T, Takano H: Mechanical stress activates protein kinase cascade of phosphorylation in neonatal rat cardiac myocytes. J Clin Invest 96: 438–446, 1995
6. Takano H, Komuro I, Zou Y, Kudoh S, Yamazaki T, Yazaki Y: Activation of p70 S6 protein kinase is necessary for angiotensin II-induced hypertrophy in neonatal rat cardiac myocytes. FEBS Lett 379: 255–259, 1996
7. Sadoshima J, Xu Y, Slayter HS, Izumo S: Autocrine release of angiotensin II mediates stretch-induced hypertrophy of cardiac myocytes *in vitro*. Cell 75: 977–984, 1993
8. Zou Y, Komuro I, Yamazaki T, Aikawa R, Kudoh S, Shiojima I, Hiroi Y: Protein kinase C, but not tyrosine kinases or Ras, plays a critical role in angiotensin II-induced activation of Raf-1 kinase and extracellular signal-regulated protein kinases in cardiac myocytes. J Biol Chem 271: 33592–33597, 1996
9. Yamazaki T, Tobe K, Hoh E, Maemura K, Kaida T, Komuro I, Tamemoto H: Mechanical loading activates mitogen-activated protein kinase and S6 peptide kinase in cultured rat cardiac myocytes. J Biol Chem 268: 12069–12076, 1993

10. Hall A: Small GTP-binding proteins and the regulation of the actin cytoskeleton. Annu Rev Cell Biol 10: 31–54, 1994
11. Ridley AJ, Hall A: The small GTP-binding protein rho regulates the assembly of focal adhesions and actin stress fibers in response to growth factors. Cell 70: 389–399, 1992
12. Nobes CD, Hall A: Rho, rac, and cdc42 GTPases regulate the assembly of multimolecular focal complexes associated with actin stress fibers, lamellipodia, and filopodia. Cell 81: 53–62, 1995
13. Chong LD, Traynor-Kaplan A, Bokoch GM, Schwartz MA: The small GTP-binding protein Rho regulates a phosphatidylinositol 4-phosphate 5-kinase in mammalian cells. Cell 79: 507–513, 1994
14. Chen Q, Kinch MS, Lin TH, Burridge K, Juliano RL: Integrin-mediated cell adhesion activates mitogen-activated protein kinases. J Biol Chem 269: 26602–26605, 1994
15. Renshaw MW, Toksoz D, Schwartz MA: Involvement of the small GTPase rho in integrin-mediated activation of mitogen-activated protein kinase. J Biol Chem 271: 21691–21694, 1996
16. Tobe K, Kadowaki T, Hara K, Gotoh Y, Kosako H, Matsuda S, Tamemoto H: Sequential activation of MAP kinase activator, MAP kinases, and S6 peptide kinase in intact rat liver following insulin injection. J Biol Chem 267: 21089–21097, 1992
17. Morii N, Narumiya S: Preparation of native and recombinant Clostridium botulinum C3 ADP-ribosyltransferase and identification of Rho proteins by ADP-ribosylation. Meth Enzymol 256: 196–206, 1995
18. Minden A, Lin A, Smeal T, Derijard B, Cobb M, Davis R, Karin M: c-Jun N-terminal phosphorylation correlates with activation of the JNK subgroup but not the ERK subgroup of mitogen-activated protein kinases. Mol Cell Biol 14: 6683–6688, 1994
19. Coso OA, Chiariello M, Yu JC, Teramoto H, Crespo P, Xu N, Miki T: The small GTP-binding proteins Rac1 and Cdc42 regulate the activity of the JNK/SAPK signaling pathway. Cell 81: 1137–1146, 1995
20. Komuro I, Kaida T, Shibazaki Y, Kurabayashi M, Takaku F, Yazaki Y: Stretching cardiac myocytes stimulates proto-oncogene expression. J Biol Chem 265: 3595–3598, 1990
21. Minden A, Lin A, Claret FX, Abo A, Karin M: Selective activation of the JNK signaling cascade and c-Jun transcriptional activity by the small GTPases Rac and Cdc42Hs. Cell 81: 1147–1157, 1995
22. Frost JA, Xu S, Hutchison MR, Marcus S, Cobb MH: Actions of Rho family small G proteins and p21-activated protein kinases on mitogen-activated protein kinase family members. Mol Cell Biol 16: 3707–3713, 1996
23. Valencia A, Chardin P, Wittinghofer A, Sander C: The ras protein family: Evolutionary tree and role of conserved amino acids. Biochemistry 30: 4637–4648 (review) 1991
24. Karns LR, Kariya K, Simpson PC: M-CAT, CArG, and Sp1 elements are required for alpha 1-adrenergic induction of the skeletal alpha-actin promoter during cardiac myocyte hypertrophy. J Biol Chem 270: 410–417, 1995
25. Yamazaki T, Komuro I, Kudoh S, Zou Y, Shiojima I, Hiroi Y, Mizuno T: Endothelin-1 is involved in mechanical stress-induced cardiomyocyte hypertrophy. J Biol Chem 271: 3221–3228, 1996
26. Komuro I, Kudo S, Yamazaki T, Zou Y, Shiojima I, Yazaki Y: Mechanical stretch activates the stress-activated protein kinases in cardiac myocytes. FASEB J 10: 631–636, 1996
27. Kudoh S, Komuro I, Mizuno T, Yamazaki T, Zou Y, I. S, Takekoshi N: Angiotensin II stimulates c-Jun NH2-terminal kinase in cultured cardiac myocytes of neonatal rats. Circ Res 80: 139–146, 1997
28. Narumiya S: The small GTPase Rho: Cellular functions and signal transduction. J Biochem 120: 215–228, 1996
29. Aikawa R, Komuro I, Yamazaki T, Zou Y, Kudoh S, Zhu W, Kadowaki T: Rho family small G proteins play critical roles in mechanical stress-induced hypertrophic responses in cardiac myocytes. Circ Res 84: 458–466, 1999
30. Komuro I, Katoh Y, Kaida T, Shibazaki Y, Kurabayashi M, Takaku F, Yazaki Y: Mechanical loading stimulates cell hypertrophy and specific gene expression in cultured rat cardiac myocytes. J Biol Chem 266: 1265–1268, 1991
31. Chien KR, Knowlton KU, Zhu H, Chien S: Regulation of cardiac gene expression during myocardial growth and hypertrophy: Molecular studies of an adaptive physiologic response. FASEB J 5: 3037–3046, 1991
32. Sah VP, Hoshijima M, Chien KR, Brown JH: Rho is required for Galphaq and alpha1-adrenergic receptor signaling in cardiomyocytes. J Biol Chem 271: 31185–31190, 1996
33. Thorburn J, Frost JA, Thorburn A: Mitogen-activated protein kinases mediate changes in gene expression, but not cytoskeletal organization associated with cardiac muscle cell hypertrophy. J Cell Biol 126: 1565–1572, 1994
34. Thorburn J, Xu S, Thorburn A: MAP kinase- and Rho-dependent signals interact to regulate gene expression but not actin morphology in cardiac muscle cells. EMBO J 16: 1888–1900, 1997
35. Hill CS, Wynne J, Treisman R: The Rho family GTPases RhoA, Rac1, and CDC42Hs regulate transcriptional activation by SRF. Cell 81: 1159–1170, 1995
36. Ueyama T, Sakoda T, Kawashima S, Hiraoka E, Hirata K, Akita H, Yokoyama M: Activated RhoA stimulates c-fos gene expression in myocardial cells. Circ Res 81: 672–678, 1997

Molecular and Cellular Biochemistry **212**: 183–186, 2000.

Mechanisms of angiotensin II-induced platelet-derived growth factor gene expression

Levon M. Khachigian,[1] Yoh Takuwa[2] and Tucker Collins[3]

[1]*Centre for Thrombosis and Vascular Research, The University of New South Wales, and Department of Haematology, Prince of Wales Hospital, Sydney, Australia;* [2]*Department of Physiology, Kanazawa University School of Medicine, Kanazawa, Japan;* [3]*Vascular Research Division, Department of Pathology, Brigham and Women's Hospital & Harvard Medical School, Boston, MA, USA*

Abstract

Angiotensin II induces the expression of platelet-derived growth factor A-chain and B-chain in cultured vascular smooth muscle cells at the level of transcription. The renin-angiotensin system has also been implicated in the increased expression of platelet-derived growth factor in the mechanically-injured artery wall. This review is concerned with recent developments in our understanding of the signaling and transcriptional mechanisms mediating the inducible expression of one vasoconstrictor by another. (Mol Cell Biochem **212**: 183–186, 2000)

Key words: angiotensin, platelet-derived growth factor, transcriptional regulation, Egr-1

Angiotensin II (ATII) is an octapeptide produced from angiotensinogen by the action of renin and angiotensin-converting enzyme (ACE) [1]. It stimulates a wide variety of physiologic systems that increase blood pressure, including vascular smooth muscle constriction, adrenal production of aldosterone and increased salt and water reabsorption by the kidney. ATII has long been implicated in a number of cardiovascular disease states associated with vascular smooth muscle cell growth, including atherosclerosis, post-angioplasty restenosis, and vascular hypertrophy in primary hypertension. ATII interacts with and signals through two known high-affinity receptor subtypes, designated AT1 and AT2. The availability of selective inhibitors for each of these receptor subtypes has demonstrated that most of the hemodynamic manifestations of angiotensin II are mediated by the AT1 receptor [2]. Signaling through the AT2 receptor is presently not well understood but evidence suggests that this receptor is involved in growth inhibition [3] and apoptosis [4, 5]. The abundant expression of the AT2 receptor in fetal tissues and immature brain suggest its function in development.

The AT1 receptor consists of seven-transmembrane, G protein-coupled receptor whose helices interact to form a ligand-binding pocket [2]. The AT1 receptor comprises two isoforms, AT1a and AT1b, which are strikingly similar in amino acid structure, pharmacological specificity and signal transduction mechanism. Upon binding ATII, the AT1 receptor triggers a series of signal transduction pathways that initiate a flow of information into the cell. Activation of the AT1 receptor involves heterotrimeric G-proteins and the dissociation of these G-proteins into the $G\alpha$ and $G\beta\gamma$ subcomplexes. These components act as intracellular signals by inducing effector molecules, such as adenylate cyclase, although the precise role of G proteins in ATII signaling is uncertain.

The AT1 receptor also activates signal transduction pathways that are engaged by growth factor and cytokine receptors. ATII stimulates the rapid tyrosine phosphorylation of phospholipase C. These events result in the accumulation of inositol 1,4,5-trisphosphate and diacylglycerol (DAG). This leads to the mobilization of calcium from intracellular stores and the stimulation of protein kinase C. ATII signal transduction also involves the mitogen-activated protein kinase pathway. Like other seven transmembrane receptors, the AT1 receptor lacks intrinsic kinase activity. When ATII binds to the AT1 receptor, a soluble tyrosine kinase, possibly pp60c-

Address for offprints: L.M. Khachigian, Centre for Thrombosis and Vascular Research, School of Pathology, The University of New South Wales, Sydney, NSW 2052, Australia src, is recruited to the cell surface [6]. This kinase activates a guanine nucleotide exchange factor (GEF) that activates p21ras by stimulating the exchange of bound GDP for GTP on p21ras. Activated p21ras then recruits Raf-1 kinase into an active plasma membrane bound complex. Activated Raf-1 kinase, in turn, phosphorylates and activates MEK that then activates MAP kinase. MAP kinase then activates nuclear transcription factors that induce the expression of early growth response genes. Finally, ATII stimulates the recruitment and tyrosine phosphorylation of the JAK family of intracellular kinases. STATs (signal transducers and activators of transcription) are substrates of JAK and become phosphorylated on tyrosine in response to ATII. STAT proteins form complexes with associated components that then move to the nucleus and induce gene transcription [7]. Collectively, these diverse signaling events serve as links between the cell surface receptor and the nuclear transcriptional events leading to alterations in smooth muscle cell growth.

A second growth-regulatory molecule implicated in inappropriate smooth muscle growth contributing to vascular pathologic settings is platelet-derived growth factor (PDGF) [8]. PDGF consists of an A-chain and/or a B-chain held together in homodimeric or heterodimeric configuration by conserved disulfide bonds. The PDGF A-chain and B-chain genes reside on separate chromosomes and are independently regulated. The receptors for PDGF consist of α- and β-isoforms. The α-receptor binds either the PDGF A-chain or B-chain, whereas the β-receptor binds only to the B-chain. When engaged by PDGF, its receptors play different biological roles in vascular smooth muscle cells. For example, both PDGFR-α and PDGFR-β stimulate proliferation, whereas only the PDGFR-β receptor appears to mediate migration [9]. This suggests that PDGF B-chain may have greater biologically potency in vascular smooth muscle cells than PDGF-A. In these cells, the PDGF A-chain and B-chain are inducibly expressed by exposure to ATII at the level of transcription. This review will focus on the signaling and transcriptional mechanisms controlling PDGF-A and PDGF–B expression by ATII.

Insight into the molecular mechanisms mediating the expression of PDGF-A by ATII was obtained by transient transfection analysis in smooth muscle cells isolated from the arteries of adult rats. ATII increased reporter gene expression driven by segments of the PDGF-A promoter bearing G+C-rich elements for the zinc finger transcription factor, early growth response factor-1 (Egr-1) [10]. This immediate-early gene product is rapidly induced by ATII at the transcriptional level [11]. Electrophoretic mobility shift assays revealed that Egr-1 protein accumulates in the nuclei of cells exposed to ATII and binds to this distinct site in the PDGF-A promoter. This interaction is sequence-specific, time-dependent and comparable to cells incubated with phorbol ester [11], a known activator of Egr-1 [12]. Inhibition of ATII-induced Egr-1 and PDGF-A expression by the specific MEK1/2 inhibitor, PD98059 [11], indicates that MEK/ERK is a critical component of the ATII signal transduction pathway that leads to increased PDGF-A expression. The mechanisms upstream of MEK regulating ATII-induction of Egr-1 and PDGF-A, however, presently remain unresolved.

ATII receptor antagonists have widely been used by investigators to shed light on the biological effects of ATII *in vivo*. Losartan, an AT1 receptor antagonist, helped demonstrate that this specific receptor subtype mediates ATII activation of ERK and the expression of Egr-1 and PDGF-A [11]. The AT2 receptor antagonist PD123319 failed to influence ATII-induction of PDGF-A. Preclinical studies using losartan have strongly implicated a role for ATII in the processes of vasoconstriction, aldosterone release, inhibition of renin release, vasopressin release and noradrenaline release [13]. ATII-induced vascular wall hypertrophy in rats is blocked by both losartan and PD123319 [14]. However, whereas losartan restored normal arterial pressure, PD123319 had no effect [14]. That the AT1 receptor mediates ATII activation of the PDGF-A promoter via the MEK/ERK/Egr-1 pathway is consistent with previous findings indicating that Egr-1 induction and neointima formation after arterial injury can be inhibited by TCV-116, an antagonist of AT1 receptor function [15].

The vasodilator nitric oxide (NO) is an inhibitor of smooth muscle cell replication and function in the vessel wall [16–18]. Produced from L-arginine via the metabolic activity of NO synthase (NOS), NO also inhibits adhesion molecule expression, leukocyte adhesion and platelet aggregation [19]. The antiproliferative effects of NO may, at least in part, be due to its capacity to down-regulate Cdk2 and cyclin A gene transcription [20]. Since Egr-1 binds and activates the promoters of many genes whose products influence smooth muscle cell migration and proliferation [21–23], antagonism of Egr-1 function by NO would provide an important mechanism controlling Egr-1-dependent gene expression in the vessel wall. ATII induction of Egr-1 and PDGF-A is blocked by the NO donor, 2-morpholinosydnonimine (SIN-1). ATII signaling is also blocked by overexpression of eNOS. Conversely, it is augmented by the L-arginine analogue and NOS inhibitor, N-nitro-L-arginine methyl ester (L-NAME) [11]. The capacity of NO to inhibit AT1-dependent PDGF-A expression is consistent with NO inhibition of smooth muscle cell migration through the AT1 receptor [24] given the chemotactic activity of PDGF-A for this cell type [25]. The mechanism with which NO modulates Egr-1 and PDGF-A expression is presently not known but may involve Ras-dependent signaling. NO donors block SMC proliferation by preventing Ras-dependent activation of Raf-1 [26].

PDGF B-chain expression is stimulated via the AT1 receptor in cultured newborn rat medial smooth muscle cells and neointimal smooth muscle cells exposed to ATII [27]. In contrast, ATII is unable to modulate PDGF-B expression in smooth muscle cells isolated from adult rats. The heterogenous nature of smooth muscle cells in general is likely to account for differences in the capacity of ATII to induce PDGF-B expression in adult, newborn and neointimal cells. For example, neointimal smooth muscle cells share many phenotypic similarities with newborn rat medial SMCs [28–31]. Newborn rat SMCs and neointimal SMCs both express PDGF-B gene, as well as PDGF-A gene, and secrete increased amounts of PDGF activity compared with adult rat SMCs [29–31]. Responsiveness of both neointimal and newborn rat smooth muscle cells to ATII with induction of PDGF-B gene expression provides further evidence for the phenotype similarity of these smooth muscle subtypes.

A number of studies have shown that renin, angiotensinogen, ACE and AT1 mRNA are upregulated in rat arteries following balloon injury [32, 33]. Since inducible PDGF-B and PDGF-A mRNA expression in balloon-injured rat carotid arteries can be attenuated by TCV-116 [34], ATII may play a direct role in the activation of both chains of PDGF in the response of the vessel wall to mechanical injury. Indeed, Wong *et al.* have shown that PDGF-A expression after balloon injury is blocked by perindopril inhibition of ACE [35].

ATII activates several members of the MAP kinase superfamily (ERK, JNK and p38) in newborn and adult smooth muscle cells [27]. However, ERK and JNK, but not p38, are involved in ATII-induced PDGF-B mRNA expression [27]. ATII-induced PDGF-B mRNA expression is abrogated by PD98059, but not by the p38 inhibitor SB202580. Similarly, transient transfection analysis has revealed that ATII-activation of the PDGF-B promoter is abolished by the coexpression of a dominant negative forms of either ERK, JNK but not p38 [27]. Additionally, dominant negative Ras abolishes the stimulatory effects of ATII on ERK activity and PDGF-B expression. The specific transcription factor(s) involved in ATII induction of PDGF-B have not yet been determined. It is possible that this process involves Egr-1, whose activation is ERK-, JNK- and Ras-dependent. Egr-1 has previously been found to interact with an atypical site in the PDGF-B promoter which overlaps with an element for a second zinc finger protein, Sp1 [21]. Inducible expression of both PDGF genes may involve the displacement of Sp1 by Egr-1. Future studies should ascertain the role of Egr-1 in ATII-induced PDGF-B expression and establish whether additional transcriptional regulators or inhibitors of signaling such as NO are also involved.

In summary, recent investigations have revealed a direct link between ATII and PDGF in vascular smooth muscle cells and the involvement of more than one MAP kinase signaling pathway. ATII may stimulate the local production of the PDGF at sites of vascular injury. This process of growth factor induction and vasoconstrictor production may be antagonized by NO, an inhibitor of smooth muscle replication and vasodilator. This balance may be part of a homeostatic control mechanism that maintains quiescence in the vessel wall.

References

1. Morgan L, Pipkin FB, Kalsheker N: Angiotensinogen: Molecular biology, biochemistry and physiology. Int J Biochem Cell Biol 28: 1211–1222, 1996
2. Inagami T: Recent progress in molecular and cell biological studies of angiotensin receptors. Curr Opin Nephrol Hypertens 4: 47–51, 1995
3. Goto M, Mukoyama M, Suga S, Matsumoto T, Nakagawa M, Ishibashi R, Kasahara M, Sugawara A, Tanaka I, Nakao K: Growth-dependent induction of angiotensin II type 2 receptor in rat mesangial cells. Hypertension 30: 358–362, 1997
4. Ma J, Nishimura H, Fogo A, Kon V, Inagami T, Ichikawa I: Accelerated fibrosis and collagen deposition develop in the renal interstitium of angiotensin type 2 receptor null mutant mice during ureteral obstruction. Kidney Int 53: 937–944, 1998
5. Yamada T, Horiuchi M, Dzau VJ: Angiotensin II type 2 receptor mediates programmed cell death. Proc Natl Acad Sci USA 93: 156–160, 1996
6. Schieffer B, Drexler H, Ling BN, Marrero MB: G protein coupled receptors control vascular smooth muscle cell proliferation via pp60c-src and p21ras. Am J Physiol 272: C2019–C2030, 1997
7. Ihle JN, Thierfelder W, Teglund S, Stravapodis D, Wang D, Feng J, Parganas E: Signaling through the cytokine receptor superfamily. Ann NY Acad Sci 865: 1–9, 1998
8. Ross R: The pathogenesis of atherosclerosis: A perspective for the 1990s. Nature 362: 801–809, 1993
9. Koyama N, Hart CE, Clowes AW: Different functions of the platelet-derived growth factor-alpha and -beta receptors for the migration and proliferation of cultured baboon smooth muscle cells. Circ Res 75: 682–691, 1994
10. Gashler A, Sukhatme V: Early growth response protein 1 (Egr-1): Prototype of a zinc-finger family of transcription factors. Prog Nucl Acid Res 50: 191–224, 1995
11. Day FL, Rafty LA, Chesterman CN, Khachigian LM: Angiotensin II (ATII)-inducible platelet-derived growth factor A-chain gene expression is p42/44 extracellular signal-regulated kinase-1/2 and Egr-1 dependent and modulated via the ATII type 1 but not type 2 receptor – induction by ATII antagonized by nitric oxide. J Biol Chem 274: 23726–23733, 1999
12. Khachigian LM, Williams AJ, and Collins T: Interplay of Sp1 and Egr-1 in the proximal PDGF-A promoter in cultured vascular endothelial cells. J Biol Chem 270: 27679–27686, 1995
13. Dickstein K, Timmermans P, Segal R: Losartan: A selective antagonist II type 1 (AT1) receptor antagonist for the treatment of heart failure. Exp Opin Invest Drugs 7: 1897–1914, 1998
14. Sabri A, Levy BI, Poitevan P, Caputo L, Faggin E, Marotte F, Rappaport L, Samuel JL: Differential roles of AT1 and AT2 receptor subtypes in vascular trophic and phenotypic changes in response to stimulation with angiotensin II. Arterioscler Thromb Vasc Biol 17: 257–264, 1997
15. Kim S, Kawamura M, Wanibuchi H, Ohta K, Hamaguchi A, Omura T, Yulimura T, Miura K, Iwao H: Angiotensin II type 1 receptor blockade inhibits the expression of immediate-early genes and fibronectin in rat injured artery. Circulation 92: 88–95, 1995
16. Chen L, Daum G, Forough R, Clowes M, Walter U, Clowes AW: Overexpression of human endothelial nitric oxide synthase in rat vascular smooth muscle cells and in balloon-injured carotid artery. Circ Res 82: 862–870, 1998
17. Seki J, Nishio M, Kato Y, Motoyama Y, Yoshida K: FK409, a new nitric oxide donor, suppresses smooth muscle proliferation in the rat model of balloon angioplasty. Atherosclerosis 117: 97–106, 1995
18. Janssens S, Flaherty D, Nong Z, Varenne O, van Pelt N, Haustermans C, Zoldelyi P, Gerard R, Collen D: Human endothelial nitric oxide gene transfer inhibits vascular smooth muscle cell proliferation and neointima formation after balloon injury in rats. Circulation 97: 1274–1281, 1998

19. Cooke JP, Dzau VJ: Nitric oxide synthase: Role in the genesis of vascular disease. Ann Rev Med 48: 489–509, 1997
20. Guo K, Andres V, Walsh K: Nitric oxide-induced downregulation of Cdk2 activity and cyclin A gene transcription in vascular smooth muscle cells. Circulation 97: 2066–2072, 1998
21. Khachigian LM, Lindner V, Williams AJ, Collins T: Egr-1-induced endothelial gene expression: A common theme in vascular injury. Science 271: 1427–1431, 1996
22. Khachigian LM, Collins T: Inducible expression of Egr-1-dependent genes: a paradigm of transcriptional activation in vascular endothelium. Circ Res 81: 457–461, 1997
23. Khachigian LM, Collins T: Early growth response factor 1: A pleiotropic mediator of inducible gene expression. J Mol Med 76: 613–616, 1998
24. Dubey RK, Jackson EK, Luscher TF: Nitric oxide inhibits angiotensin II-induced migration of rat aortic smooth muscle cell: Role of cyclic nucleotides and angiotensin-1 receptors. J Clin Invest 96: 141–149, 1995
25. Ferns GAA, Sprugel KH, Seifert RA, Bowen-Pope DF, Kelly JD, Murray M, Raines EW, Ross R: Relative platelet-derived growth factor receptor subunit expression determines cell migration to different dimeric forms of PDGF. Growth Factors 3: 315–324, 1990
26. Yu SM, Hung LM, Lin CC: cGMP-elevating agents suppress proliferation of vascular smooth muscle cells by inhibiting the activation of epidermal growth factor signaling pathway. Circulation 95: 1269–1277, 1997
27. Deguchi J-O, Masatoshi M, Nakaoka T, Collins T, Takuwa Y: Angiotensin II stimulates platelet-derived growth factor-B chain expression in newborn rat vascular smooth muscle cells through Ras, extracellular signal-regulated protein kinase, and c-Jun N-terminal protein kinase mechanisms. Circ Res 85: 565–574, 1999
28. Gordon D, Mohai LG, Schwartz SM: Induction of polyploidy in cultures of neonatal rat aortic smooth muscle cells. Circ Res 59: 2268–2274, 1989
29. Majesky MW, Benditt EP, Schwartz SM: Expression and developmental control of platelet-derived growth factor A-chain and B-chain/Sis genes in rat aortic smooth muscle cells. Proc Natl Acad Sci USA 85: 1524–1528, 1988
30. Lemire JM, Covin CW, White S, Giachelli CM, Schwartz SM: Characterization of cloned aortic smooth muscle cells from young rats. Am J Pathol 144: 1068–1081, 1994
31. Majesky MW, Giachelli CM, Reidy MA, Schwartz SM: Rat carotid neointimal smooth muscle cells reexpress a developmentally regulated mRNA phenotype during repair of arterial injury. Circ Res 71: 759–768, 1992
32. Rakugi H, Jacob HJ, Krieger JE, Ingelfinger JR, Pratt RE: Vascular injury induces angiotensinogen gene expression in the media and neointima. Circulation 93: 283–287, 1993
33. Viswanathan M, Strimberg C, Seltzer A, Saavedra JM: Balloon angioplasty enhances the expression of angiotensin II AT1 receptors in the neointima of rat aorta. J Clin Invest 90: 1707–1712, 1992
34. Abe J, Deguchi J, Mastsumoto T, Takuwa N, Noda M, Ohno M, Makuchi M, Kurokawa K, Takuwa Y: Stimulated activation of platelet-derived growth factor receptor in vivo in balloon-injured arteries: A link between angiotensin II and intimal thickening. Circulation 96: 1906–1913, 1997
35. Wong J, Rauhoft C, Dilley RJ, Agrotis A, Jennings GL, A B: Angiotensin-converting enzyme inhibition abolishes medial smooth muscle PDGF-AB biosynthesis and attenuates cell proliferation in injured carotid arteries: relationships to neointima formation. Circulation 96: 1631–1640, 1997

Molecular and Cellular Biochemistry **212**: 187–201, 2000.

Transactivation of EGF receptor induced by angiotensin II regulates fibronectin and TGF-β gene expression via transcriptional and post-transcriptional mechanisms

Hiroaki Matsubara, Yasutaka Moriguchi, Yasukiyo Mori, Hiroya Masaki, Yoshiaki Tsutsumi, Yasunobu Shibasaki, Yoko Uchiyama-Tanaka, Schoichiro Fujiyama, Yoko Koyama, Atsuko Nose-Fujiyama, Satoshi Iba, Eriko Tateishi and Toshiji Iwasaka

Department of Medicine II, Kansai Medical University, Moriguchi, Osaka, Japan

Abstract

The signaling cascade elicited by angiotensin II (Ang II) resembles that characteristic of growth factor, and recent evidence indicates transactivation of epidermal growth factor receptor (EGF-R) by G protein-coupled receptors. Here, we report the involvement of EGF-R in Ang II-induced synthesis of fibronectin and TGF-β in cardiac fibroblasts. Ang II stimulated fibronectin mRNA levels dose-dependently with a maximal increase (~ 5-fold) observed after 12 h of incubation. Ang II-, or calcium ionophore-induced fibronectin synthesis was completely abolished by tyrosine kinase inhibitors and intracellular Ca^{2+} chelating agents. Ang II-induced fibronectin mRNA was not affected by PKC inhibitors or PKC depletion, whereas specific inhibition of EGF-R function by a dominant negative EGF-R mutant and tyrphostin AG1478 abolished induction of fibronectin mRNA. We isolated the rat fibronectin gene including the 5′-flanking region and found that the AP-1 binding site present in the promoter region was responsible for the Ang II responsiveness of this gene. Gel retardation assay revealed the binding of nuclear protein to the AP-1 site, which was supershifted with anti-*c-fos* and anti-*c-jun* but not anti-ATF-2 antibodies. Conditioned medium from Ang II-treated cells contained TGF-β bioactivity and addition of neutralizing TGF-β antibody modestly (46%) inhibited induction of fibronectin. Ang II-induced synthesis of TGF-β was also abolished by inhibition of EGF-R function. The effect of TGF-β was exerted by stabilizing fibronectin mRNA without affecting the promoter activity and required *de novo* protein synthesis. We concluded that Ang II-induced expression of fibronectin and TGF-β is mediated by downstream signaling of EGF-R transactivated by Ca^{2+}-dependent tyrosine kinase, and that Ang II-induced fibronectin mRNA expression is regulated by two different mechanisms ; transcriptional control by binding of *c-fos/c-jun* complex to the AP-1 site, and post-transcriptional control by mRNA stabilization due to autocrine and/or paracrine effects of TGF-β. Thus, this study suggested that the action of Ang II on extracellular matrix formation should be interpreted in association with the EGF-R signaling cascade. (Mol Cell Biochem **212**: 187–201, 2000)

Key words: angiotensin II receptors, angiotensin II type 1 receptor, angiotensin II type 2 receptor, angiotensin II, AT1 receptor, epidermal growth factor receptor

Abbreviations: Ang II – angiotensin II; AT1-R – angiotensin type 1 receptor; ERK – extracellular signal-regulated kinase; PLC – phospholipase C; PKC – protein kinase C; VSMC – vascular smooth muscle cells; LPA – lysophosphatidic acid; EGF-R – epidermal growth factor receptor; VSMC – vascular smooth muscle cells; MEK – ERK kinase

Address for offprints: H. Matsubara, Department of Medicine II, Kansai Medical University, Fumizonocho 10-15, Moriguchi, Osaka 570-8507, Japan

Introduction

Cardiac fibroblasts isolated from neonatal rat hearts have abundant high-affinity, angiotensin II (Ang II) receptors, which are classified pharmacologically as belonging to the Ang II type 1 receptor (AT1-R) subtype [1, 2]. These cells have been used for the study to examine AT1-R-mediated signaling [2–4] and extracellular matrix remodeling in heart failure [5]. AT1-R stimulation was found to stimulate DNA synthesis and cell proliferation [1] and to also increase the synthesis of extracellular matrix proteins [4, 6], suggesting that cardiac fibroblasts contribute to remodeling of the cardiac interstitium under a variety of physiological and pathological conditions. Ang II, acting via AT1-R, initiates early biochemical events, including rapid production of diacylglycerol and inositol 1,4,5-triphosphate by phospholipase C (PLC)-mediated hydrolysis of inositol phospholipids and activation of protein kinase C (PKC) [7–10]. Ang II also induces a rapid increase in expression of the growth-associated nuclear proto-oncogenes similarly to cellular events mediated by peptide growth factors, and stimulates tyrosine phosphorylation of multiple substrates, including p44 and p42 mitogen-activated protein/extracellular signal-regulated kinases (ERKs) [11–16].

Although it was reported that PKC played a dominant role in the Ang II-induced activation of ERK in vascular smooth muscle cells (VSMC) [17, 18] or cardiac myocytes [13], other studies indicated that calcium signaling rather than PKC plays a critical role in ERK activation in these cells [14–16]. Certain aspects of signal transduction characteristic of Ang II stimulation resemble those evoked by growth factors. Activation of phospholipase C-γ, tyrosine kinases, ERK and expression of nuclear proto-oncogenes exemplify phenomena common to Ang II and growth factors signaling [9, 12]. Recently, cross-talk between G protein-coupled receptors and growth factor receptors with intrinsic tyrosine kinase activity was reported. In vascular smooth muscle cells (VSMC), stimulation of the AT1-R causes transactivation of platelet-derived growth factor β receptor (β PDGF-R) [19] or epidermal growth factor receptor (EGF-R) [20], resulting in transmission of AT1-R-mediated ERK signaling in association with formation of adapter protein complex containing the SH2 domain, such as Grb2 or Shc. Stimulation of Rat-1 cells with endothelin-1, lysophosphatidic acid (LPA) or thrombin induced a rapid increase in tyrosine phosphorylation of EGF-R and $p185^{neu}$ leading to activation of the ERK [21]. Stimulation of Cos-7 cells with Gi- or Gq-coupled receptors caused phosphorylation of EGF-R associated with assembly of Shc and Grb2 [22, 23], and calcium-dependent EGF-R activation by stimulation of voltage-sensitive calcium channels was also found in PC12 cells [24, 25]. Ang II activated tyrosine phosphorylation of the insulin-like growth factor-1 receptor and insulin receptor substrate-1 in VSMC [26]. In addition, m1 muscarinic acetylcholine receptor was also shown to transactivate EGF-R and its downstream signaling, resulting in modulation of potassium channel function in human embryonic kidney 293 cells [27]. Thus, additional tyrosine kinases that phosphorylate receptor tyrosine kinases ligand-independently appear to contribute in a general or cell type-specific way to mitogenic signaling mediated through G protein-coupled receptors. We [28] and Eguchi *et al.* [20] recently showed that Ang II-induced ERK activation and its mitogenic signals are dominantly mediated by downstream signaling of transactivated EGF-R.

Two thirds of the myocardial cell population is composed of non-myocyte cells, the majority of which are fibroblasts [5]. Cardiac fibroblasts are responsible for the production and deposition of extracellular matrix proteins such as fibronectin and collagen types 1 and III [5, 6]. Fibronectin is an extracellualr matrix and plasma protein with various functions that plays important roles in cardiac remodeling as well as cell adhesion and migration [29]. Although fibronectin synthesis in cardiac fibroblasts was reported to be induced in response to Ang II [2, 6], the signal transduction mechanism responsible for this induction remains to be determined. In the present study, we found that Ang II-induced synthesis of fibronectin and TGF-β is regulated through downstream signaling of EGF-R transactivated by Ca^{2+}/calmodulin-dependent tyrosine kinases. Induction of fibronectin mRNA by Ang II was regulated by two different mechanisms ; transcriptional control by binding of *c-fos*/*c-jun* complex to the AP-1 site, and post-transcriptional control by autocrine and/or paracrine effects of TGF-β, which is exerted by increasing the mRNA stability via *de novo* protein synthesis and up-regulates fibronectin mRNA levels by about 46%.

Materials and methods

Materials

GF109203X, BAPTA-AM, genistein, ST638, and W7 were purchased from Calbiochem. TMB-8, nifedipine, calmidazolium chloride were purchased from Calbiochem. A23187, phorbol 12-myristate 13-acetate (PMA) and EGTA were purchased from Sigma. Losartan was kindly provided by DuPont Merck Pharmaceutical (Wilmington, DE, USA). PD123319 was kindly provided by Parke-Davis and Warner-Lambert Co. (Ann Arbor, MI, USA). Fibronectin antibody was purchased from Telios Pharmaceuticals, San Diego, CA, USA.

Cell culture

Cardiac fibroblasts were prepared from ventricles of 1–2-day-old Wistar rats and grown as previously described [30, 31].

Subcultured fibroblasts from passages 4 and 5, used in this experiment, were > 99% positive for immunostaining with vimentin antibody and were negative for desmin (for myocytes), smooth muscle α-actin (for VSMC) and a polyclonal antibody against von Willebrand factor (for endothelial cells) (Sigma). Subconfluent cells were serum-starved for 24 h and used for the experiments.

Northern blotting

Total RNA was extracted by guanidium isothiocyanate-cesium chloride centrifugation, fractionated on 1% agarose/formaldehyde gels, and transferred to nylon membranes as previously reported [30–33]. Blots were then hybridized with a random-primed ^{32}P-labeled cDNA probes consisting of fibronectin, TGF-β and glyceraldehyde-3-phosphate dehydrogenase (GAPDH) as an internal control. Hybridized signals were measured by scanning densitometry and fibronectin mRNA levels were arbitrarily normalized relative to the GAPDH mRNA levels.

Stable transfection and CAT assay

Cardiac fibroblasts were cotransfected with a 5:1 molar ratio of the fibronectin promoter-chloramphenicol acetyltransferase (CAT) chimeric constructs and pSV2neo DNAs (~ 15 μg/10 cm dish) using LIPOFECTAMINE PLUS™ reagent according to the manufacture's instruction (GIBCO BRL) followed by incubation with fresh medium for 48 h as previously reported [34]. Selection of G418 (500 μg/ml) resistance was started 48 h later. After 8–12 days, resistance foci appeared and pooled foci of stable transformants were subjected to the standard differential protocol [28]. The stable transformants were serum-deprived for 24 h and then exposed with Ang II (100 nmol/L) for further 12 h. In each transfection, 5 μg of Rous sarcoma virus βGal was also cotransfected. CAT activity was normalized for transfection efficiency by β-galactosidase activity and for cell density by protein concentration. CAT activities were determined by a dual diffusion assay and normalized for cell density by protein concentration as previously reported [35, 36]. CAT activity was expressed relative to the activity of a promoterless CAT construct.

[^{35}S] methionine labeling of cells and immunoprecipitation

Quantification of fibronectin secreted into medium was performed as previously reported [6]. Briefly, fibroblasts were made quiescent by serum depletion for 12 h. After 12 h of incubation with or without Ang II (0.1 μmol/L), 100 μCi/ml of [^{35}S] methionine was added for another 12 h. For immunoprecipitation, aliquots (100 μl) of the medium were diluted with 900 μl of RIPA buffer (50 mmol/L Tris, 150 mmol/L NaCl, 1% Nondiet P-40, 0.5% deoxycholate, and 0.1% SDS) and treated with normal rabbit serum (10 μl) to remove non-specific binding, followed by incubation with rabbit antibody to fibronectin for 16 h at 40°C. Immune complexes were removed by protein A-Sepharose beads (Pharmacia LKB, Uppsala, Sweden) and analyzed by SDS-PAGE [6].

Creation of mutations in the AP-1 site

Mutation in the AP-1 site at –453 present in the 5′ flanking region was created by PCR overlap extension mutagenesis as previously reported [36]. Briefly, two DNA fragments having overlapping ends were first amplified with two sets of primers (A and B, C and D) by using p564-CAT as a template for making the mutations. Primers A and D were T3 and T7 primers for the pBS vector, respectively. Primers B and C had the following DNA sequence: B, 5′-GTGACCGTGGTGGTGTTTGAGAACA TTTATCCA-3′ (antisense, nt –453 to –421); C, 5′-TCAAACACCACCACGGTCAC-3′ (sense, nt –439 to –421). Primers B contained a mutation in the AP-1 site at the 3′ end, and primers B and C were designed to overlap at the 5′ end. The two respective PCR products were mixed and amplified again with primers A and D. After confirming the DNA sequence, the PCR product was subcloned into the 5′ end of the CAT gene.

Gel retardation assay

The gel retardation assay and preparation of nuclear extract from cells exposed to Ang II (100 nmol/L) for 45 min was performed as previously reported [36]. The final protein concentration was 1–0.5 mg/ml. The oligonucleotides corresponding to DNA fragment containing AP-1 and mutated AP-1 sequence (between nt –438 and –473) or ATF/CRE motif (nt –143 to –178) were synthesized, labeled with [γ-^{32}P]ATP using T4 kinase (Takara Shuzo, Kyoto, Japan), and annealed to make double-strand DNA. Nuclear extracts were incubated with the probe for 30 min at room temperature in the binding buffer, consisting of 0.5 mmol/L Hepes, 60 mmol/L KCl, 0.1 mmol/L EDTA, 0.5 mmol/L DTT, 0.5 mmol/L PMSF, 12% glycerol, and 2 μg of poly (dI-dC). The mixture was loaded on a 6% polyacrylamide gel in, and electrophoresed as previously reported [36]. Supershift assays were performed with rabbit polyclonal anti-*c-fos* IgG raised against the amino acids 128–152 portion of *c-fos* and rabbit anti-*c-jun* IgG raised against the amino acids 247–263 portion of *c-jun* (Santa Cruz Inc.). Each antibody (1 μg each) was added to the samples after the initial binding reaction and then incubated for 1 h at room temperature.

Quantification of latent and active TGF-β

The bio-assay for TGF-β released into medium was performed by measuring proliferation inhibition of mink lung epithelial cell line ATCC CCL-64 as reported elsewhere [37]. Briefly, CCL-64 cells were seeded in 24-well plates in DMEM containing 10% FCS. After incubation for 24 h, the medium was discarded and changed to DMEM containing 1% FCS. After additional incubation for 24 h, control or test samples were added, and cells were incubated for another 24 h. Subsequently, 1 μCi/mL of [^{3}H]thymidine (NEN) was pulsed for 6 h and the amount of [^{3}H]thymidine incorporated into TCA-precipitable materials was measured by a liquid scintillation counter. To determine whether cells are producing latent TGF-β, latent TGF-β was converted to the active form by treating culture supernatant with 10 N HCl (final pH = 2.0) for 30 min at room temperature, followed by neutralization with equimolar NaOH. TGF-β activities were determined on basis of the standard curve (TGF-β concentration vs. percent of inhibition) which was drawn by using TGF-β1 (R&D System Inc., Minneapolis, MN, USA). Culture supernatants were serially diluted to fall within the range of the standard curve. Specificity of the observed effects was proven by competition of the TGF-β activity with a purified rabbit anti-TGF-β IgG and a control IgG (R&D System Inc.).

Membrane preparation and binding assay

The binding assay was performed as previously reported [6, 31]. Briefly, membrane fractions were incubated with ^{125}I-[Sar^1, Ile^8] Ang II in the assay volume for 3 h at 22°C. B_{max} and Kd values, AT1-R and AT2-R densities were calculated by Scatchard analyses and non-linear least-squares regression analysis on basis of inhibition by CGP42112A.

Measurement of dehydrogenase activities (LDH)

A spectrophotometric enzyme assay (DRI-CHEM slide LDH-P, Fuji Film) was performed to measure LDH release in the medium [28]. In brief, 10 μl of triplicate medium and reagent were mixed for 2 min at room temperature. Absorbance at 540 nm was determined and LDH activity (U/L) was automatically calculated by FD3030 analyzer. One unit of LDH activity is defined as that amount of enzyme that will catalyze the formation of 1 μmole of NADH per min [28].

Statistical analysis

The results are expressed as means ± S.E. Statistical analyses were performed with an one-way ANOVA followed by pairwise contrasts (control vs. conditions) using Dunnett's multiple comparison test. Data were considered statistically significant when p was < 0.05.

Results

Ang II-induced fibronectin mRNA level is mediated by AT1-R stimulation

Ang II (100 nmol/L) significantly stimulated fibronectin mRNA accumulation after 6 h of incubation, reaching a maximal increase after 12 h and thereafter maintained a similar level until 24 h (Fig. 1). Ang II-induced fibronectin mRNA levels were increased dose-dependently with a maximal peak at 100 nmol/L (data not shown). Therefore, subsequent experiments were performed with 100 nmol/L Ang II stimulation for 12 h. To determine which Ang II receptor subtypes mediates fibronectin mRNA accumulation, Ang II receptor densities were quantified in cardiac fibroblasts and the effects of the AT1-R antagonist losartan or the AT2-R antagonist PD123319 were examined. Cardiac fibroblasts (passage 4) expressed only AT1-R (Kd: 0.21 ± 0.01 nmol/L, B_{max}: 178 ± 5 fmol/mg protein) and AT2-R expression was not detectable in these cells. Induction of fibronectin mRNA by Ang II was completely blocked by losartan but not by PD123319 (Fig. 1), demonstrating that Ang II-induced fibronectin mRNA accumulation is mainly mediated by AT1-R, and AT2-R is not involved in this induction mechanism by Ang II. We previously reported that AT2-R is substantially expressed in primary culture of cardiac fibroblasts from cardiomyopathic hamsters during heart failure, whereas AT2-R expression is abruptly down-regulated with repeated passages of cells [6]. Since the primary culture of neonatal rat cardiac fibroblasts also express detectable amounts of AT2-R, it is likely that down-regulation of AT2-R in the cells (passage 4–5) used in this study is due to repeated passages of cells.

Phorbol ester-sensitive protein kinase C is not involved in Ang II-induced fibronectin mRNA accumulation

Stimulation of AT1-R results in generation of inositol triphosphate (IP_3) and diacylglycerol which induce the release of Ca^{2+} from intracellular stores and PKC activation, respectively [9]. We next examined the effects of the PKC inhibitors GF109203X and calphostin C on Ang II-induced fibronectin mRNA levels. Pretreatment with GF109203X (1 μmol/L) or

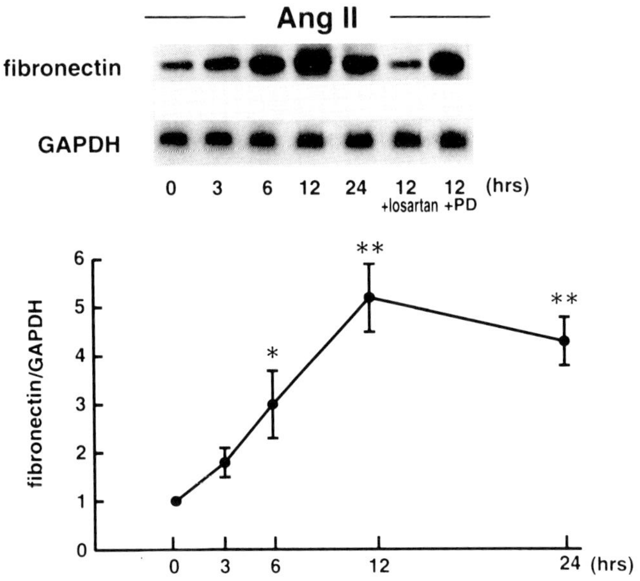

Fig. 1. Time-dependent effects of Ang II on fibronectin mRNA accumulation in cardiac fibroblasts. Serum-deprived fibroblasts were stimulated with Ang II (100 nmol/L) for the indicated periods. Total RNA (20 μg) was separated on agarose formaldehyde gels and transferred onto nylon membranes for hybridization with ^{32}P-labeled fibronectin or GAPDH cDNA probes. Exposure time was 48 h for fibronectin mRNA and 12 h for GAPDH mRNA. To test the effects of losartan (5 μmol/L) or PD123319 (5 μmol/L), the cells were pretreated with these inhibitors for 30 min and then exposed to Ang II (100 nmol/L) for 12 h. Densities of fibronectin mRNA signals were measured by densitometry and normalized relative to those of GAPDH mRNA signals. The results are arbitrarily indicated as values relative to fibronectin mRNA levels in the unstimulated cells (control). Results shown are means ± S.E. of 4 separate experiments and representative data are shown. $^{*}p < 0.05$; $^{**}p < 0.01$ vs. the control.

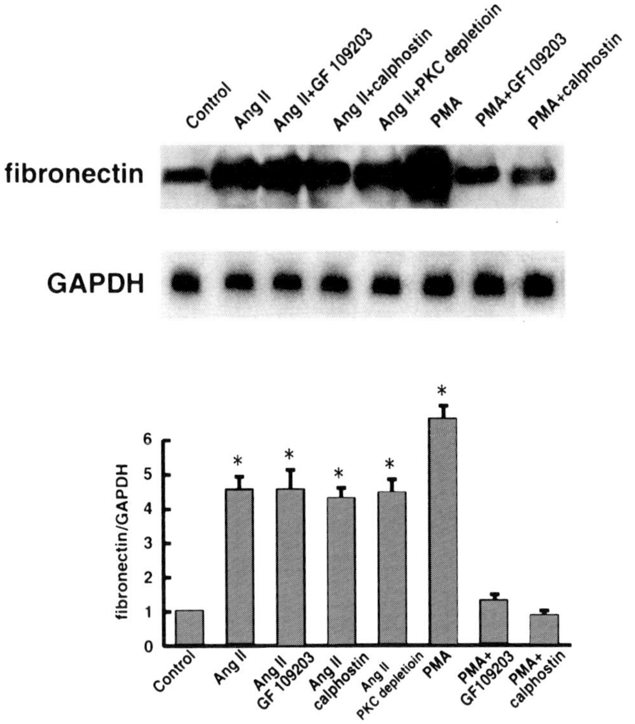

Fig. 2. Effects of PKC on Ang II-induced fibronectin mRNA levels. Serum-deprived fibroblasts were pretreated with or without the PKC inhibitor GF109203X (1 μmol/L) for 30 min or calphostin C (100 nmol/L) for 60 min and then stimulated with either Ang II (100 nmol/L) or PMA (100 nmol/L) for 12 h. PKC depletion experiments were performed by incubating cells with PMA (10 μmol/L) for 24 h. Quantification of mRNA levels was performed as described in Fig. 1. Results shown are means ± S.E. of 4 separate experiments and representative data are shown. $^{*}p < 0.01$ vs. the control.

calphostin C (50 nmol/L) completely inhibited fibronectin mRNA accumulation in response to 100 nmol/L PMA, whereas no significant inhibition of Ang II-induced fibronectin mRNA level was observed (Fig. 2). Depletion of phorbol ester-sensitive PKC by 24-h incubation with 10 μmol/L of PMA also did not inhibit Ang II-induced accumulation of fibronectin mRNA (Fig. 2). These results suggest that PKC-mediated pathways do not play a dominant role in Ang II-induced fibronectin mRNA accumulation, and that GF109203X-, calphostin C- and phorbol ester-sensitive PKC are not involved in this mechanism.

Ang II-induced fibronectin mRNA accumulation is stimulated via calcium and calmodulin-dependent pathway

We have previously shown that exposure of fibroblasts to Ang II markedly increases intracellular Ca^{2+} levels, and this increase was blocked by the intracellular Ca^{2+} chelator BAPTA-AM (10 μmol/L) but not by extracellular Ca^{2+} chelation by EGTA. These observations indicated that Ca^{2+} mobilization after AT1-R stimulation is mainly caused by release of Ca^{2+} from intracellular stores [38]. As shown in Fig. 3, Ang II-induced fibronectin mRNA levels were completely inhibited by pretreatment with BAPTA-AM (10 μmol/L) and TMB8 (100 μmol/L) commonly used as intracellular Ca^{2+} chelators, but not by pretreatment with EGTA. These Ca^{2+} chelators did not affect the basal expression level of fibronectin mRNA (data not shown). In agreement with these results, addition of the Ca^{2+} ionophore A23187 also stimulated fibronectin mRNA accumulation (Fig. 3). Elevation of cytosolic Ca^{2+} activates a variety of enzymes through interaction with calmodulin. To examine whether calmodulin mediates fibronectin mRNA levels in response to Ang II, fibroblasts were preincubated with the calmodulin inhibitors W-7 (100 μmol/L) and calmidazolium (10 μmol/L). These drugs completely blocked Ang II-induced fibronectin mRNA accumulation (Fig. 3). These calmodulin inhibitors at the concentrations used in the present study did not affect the basal level of fibronectin mRNA and did not have any effect on the PMA-induced increase in fibronectin mRNA level (data not shown), suggesting that Ang II stimulates fibronectin mRNA accumulation through a Ca^{2+}/calmodulin-dependent mechanism.

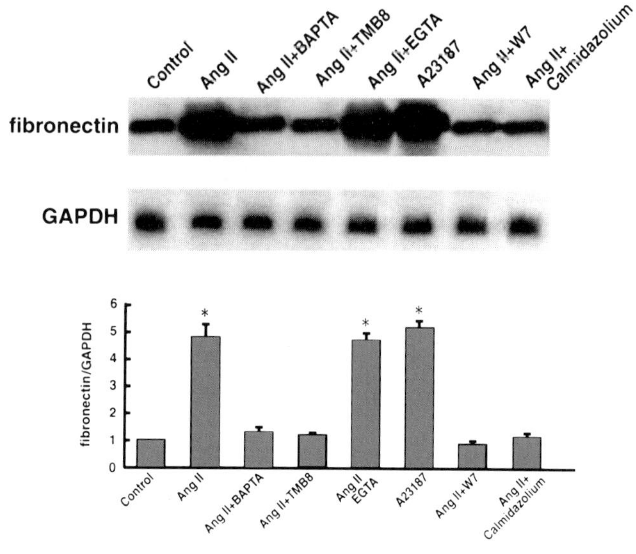

Fig. 3. Effects of Ca^{2+} signal and Ca^{2+}/calmodulin inhibitors on Ang II- or A23187-induced ERK activation. Serum-deprived fibroblasts were pretreated with or without intracellular Ca^{2+} chelator BAPTA-AM (10 μmol/L) for 30 min, another intracellular Ca^{2+} chelator TMB-8 (100 μmol/L) for 30 min, the extracellular Ca^{2+} chelator EGTA (5 mmol/L) for 3 min, the calmodulin inhibitors W7 (10 μmol/L) and calmidazolium (10 μmol/L) for 30 min and then stimulated with Ang II (100 nmol/L) or A23187 (10 μmol/L) for 12 h. Quantification of mRNA levels was performed as described in Fig. 1. Results shown are means ± S.E. of 4 separate experiments and representative data are shown. $^{*}p < 0.01$ vs. the control.

Ang II-induced fibronectin mRNA accumulation is activated by Ca^{2+}/calmodulin-dependent protein tyrosine kinases

To determine whether tyrosine kinase activity is required for Ca^{2+}-dependent fibronectin mRNA accumulation, cells were pretreated with genistein (100 μmol/L) and ST638 (100 μmol/L), protein kinase inhibitors with a strong preference for tyrosine-specific kinases [39, 40], and then stimulated by either Ang II or A23187. These inhibitors completely abolished both Ang II- and A23187-induced fibronectin mRNA accumulation with no effects on basal fibronectin mRNA levels (Fig. 4). On the other hand, these tyrosine kinase inhibitors did not have any effect on PMA-induced increase in fibronectin mRNA level (data not shown). These findings suggest that protein tyrosine kinases activated downstream of the Ca^{2+}/calmodulin pathway are closely involved in Ang II-induced fibronectin mRNA accumulation.

We also examined the effects of these kinase inhibitors on cell toxicity at the concentrations used in the above experiments. Although we also measured lactic dehydrogenase activities (LDH) in culture medium from cells treated with these drugs, no LDH activities were detected. These results suggested that complete inhibition of Ang II- or A23187-induced fibronectin mRNA accumulation by these drugs was not due to nonspecific cytotoxicity.

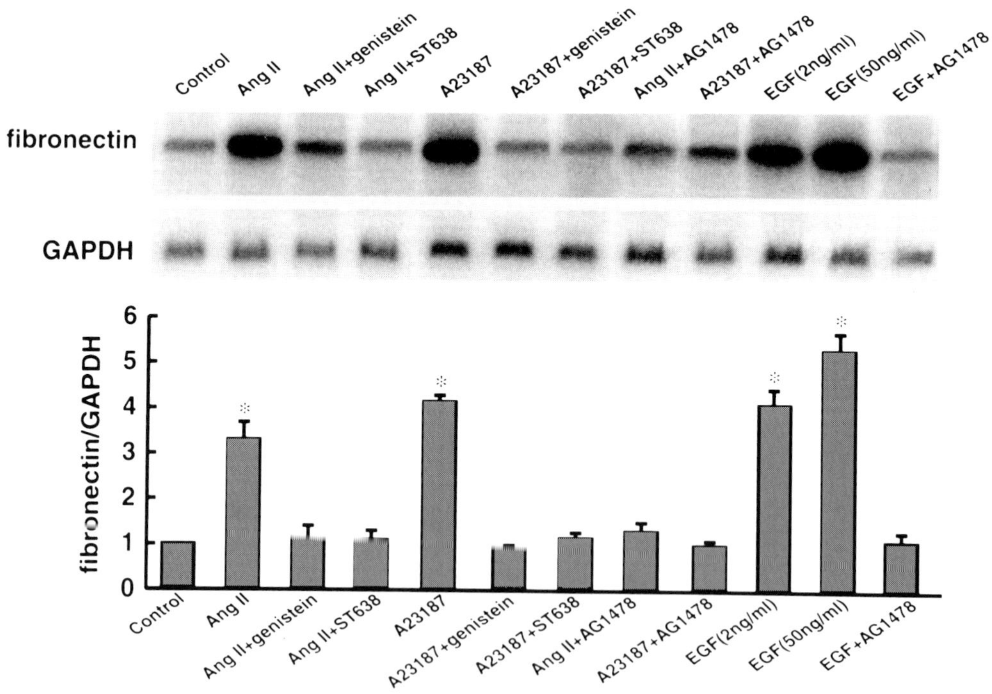

Fig. 4. Effects of tyrosine kinase inhibitors and EGF-R antagonist on Ang II- and A23187-induced ERK activation. Serum-deprived fibroblasts were pretreated with or without the tyrosine kinase inhibitors genistein (100 μmol/L) or ST638 (100 μmol/L) for 45 min, AG1478 (250 nmol/L) for 20 min and then challenged with 100 nmol/L Ang II or EGF for 12 h. Quantification of mRNA levels was performed as described in Fig. 1. Results shown are means ± S.E. of 4 separate experiments and representative data are shown. $^{*}p < 0.01$ vs. the control.

Ang II-induced transactivation of EGF-R is involved in induction of fibronectin mRNA

We recently found that in cardiac fibroblasts Ca^{2+}/calmodulin-dependent tyrosine kinases activated by AT1-R stimulation cause tyrosine-phosphorylation of EGF-R, but not PDGF-R, which mediates AT1-R/Ras/ERK signaling [28]. Therefore, we next examined whether this EGF-R transactivation mechanism is involved in AT1-R-mediated fibronectin mRNA accumulation. Specific inhibition of EGF-R function by the specific EGF-R antagonist tyrphostin AG1478 [20, 28] markedly inhibited both Ang II- and A23187-induced fibronectin mRNA accumulation (Fig. 4). EGF also dose-dependently stimulated fibronectin mRNA levels, and this increase was blocked by AG1478 (Fig. 4). However, AG1478 did not affect the basal expression level of fibronectin mRNA (data not shown).

The role of EGF-R in AT1-R-mediated fibronectin mRNA accumulation was further analyzed by specific inhibition of the EGF-R signal. A dominant-negative EGF-R mutant lacking the cytoplasmic domain of human EGF-R (HEGFR 533del) was constructed and stably transfected into cardiac fibroblasts. This mutant inhibited the downstream signaling of rat EGF-R by formation of signaling-defective heterodimers with the wild-type receptor, and we have shown that cardiac fibroblasts stably expressing HEGFR 533del abolished Ang II- or EGF-induced EGF-R phosphorylation [28]. Using these cloned cells, we examined Ang II- or EGF-induced effects on fibronectin mRNA accumulation. In HEGFR 533del cells, fibronectin mRNA accumulations at low and higher ligand concentrations (2 and 50 ng/ml EGF) was completely and partially attenuated, respectively, compared with that in the control transfected cells, while Ang II-induced expression of fibronectin mRNA was completely blocked (Figs 5A and 5B). We next examined the involvement of ERK activated downstream of EGF-R on Ang II-induced fibronectin mRNA levels. Inhibition of ERK activity by MEK inhibitor PD98059 greatly (74%, $p < 0.05$) inhibited Ang II-induced fibronectin mRNA accumulation (Fig. 5A), suggesting an important role of ERK activation in induction of fibronectin mRNA following Ang II stimulation. To test whether PD98059 causes cell injury at the concentrations used in this study, we measured the LDH levels in culture medium. We did not detect the LDH activity in culture medium from Ang II-stimulated cells pretreated with PD98059 (10 μmol/L), indicating the

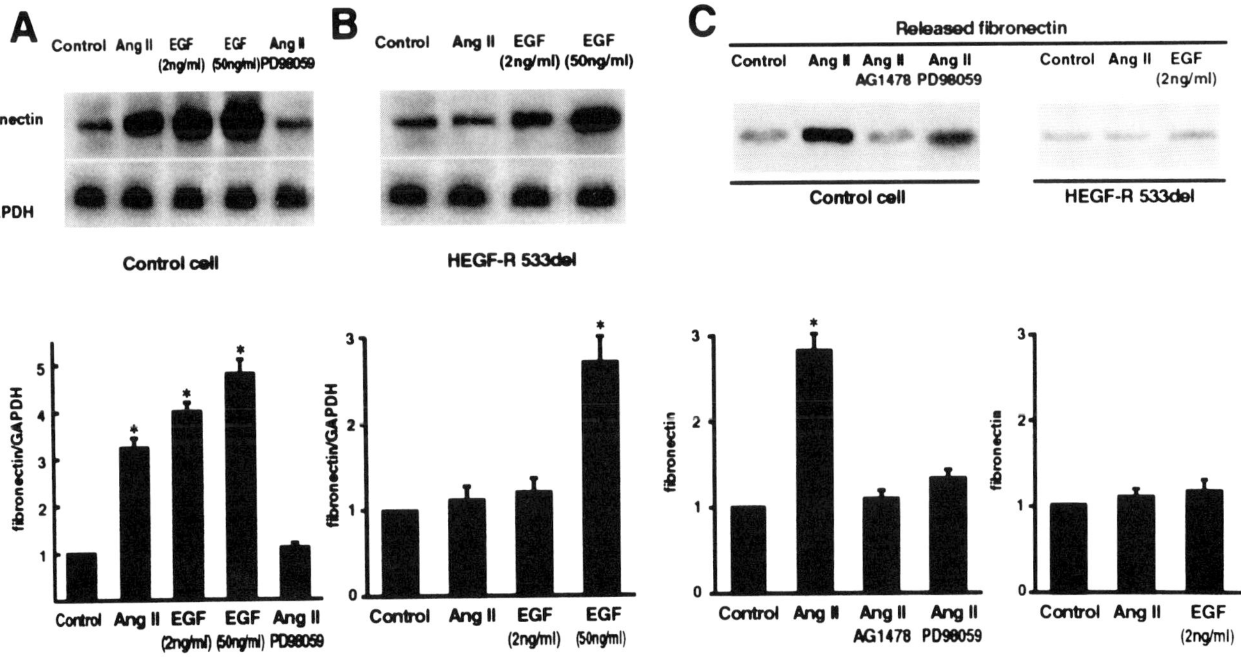

Fig. 5. Specific inhibition of Ang II-induced fibronectin mRNA induced by dominant negative EGF-R mutant and MEK inhibitor, and quantification of fibronectin protein released into culture medium. *Panels A and B*: Cardiac fibroblasts stably transfected with pcDNA expression vector alone (control) (panel A) or pcDNA containing HEGFR 533del (panel B) were starved of serum for 12 h and then stimulated with Ang II (100 nmol/L) or EGF (2 and 50 ng/ml) for 12 h. The effect of MEK inhibitor PD98059 (10 μmol/L) was examined by pretreatment with PD98059 for 30 min. Quantification of mRNA levels was performed as described in Fig. 1. Results shown are means ± S.E. of 4 separate experiments and representative data are shown. $^*p < 0.01$ vs. the control. *Panel C*: Cardiac fibroblasts (control cells) or HEGFR 533del cells incubated in the presence or absence of Ang II (100 nmol/L) for 12 h were subsequently incubated with [^{35}S]-methionine for further 12 h. AG1478 (250 nmol/L) and PD98059 (10 μmol/L) were added 20 min prior to exposure to Ang II. Culture supernatants were used for immunoprecipitation of fibronectin as described in Materials and methods. Signals of 220 KDa representing fibronectin on SDS-PAGE were measured by densitometry and the value in the control was arbitrarily expressed as 1 unit. Values represent means ± S.E. of 6 separate experiments. $^*p < 0.01$ vs. controls (unstimulated cells).

effect by this drug was not due to induction of apoptosis or nonspecific cytotoxicity.

Ang II-induced fibronectin mRNA expression results in an increase in fibronectin protein release

We also examined whether fibronectin mRNA accumulation by Ang II results in the induction at fibronectin protein level. Fibronectin protein secreted into the medium was analyzed by immunoprecipitation using anti-fibronectin antibody as previously described [6]. Immunoprecipitated fibronectin secreted into medium was increased by 2.8-fold over the control by 0.1 μmol/L of Ang II. Pretreatment with AG1478 completely abolished Ang II-induced fibronectin secretion, while PD98059 modestly (67%, $p < 0.05$) inhibited it (Fig. 5C). To further confirm the role of EGF-R transactivation, we tested the effects of Ang II and EGF on the fibronectin release from cells expressing HEGF-R 533del. As shown in Fig. 5C, fibronectin release induced by Ang II (100 nmol/L) and EGF (2 ng/ml) was completely abolished in HEGF-R 533del cells. Although we tested the effect of PD123319 on Ang II-stimulated release of fibronectin, no significant effect was observed in fibronectin release from cells pretreated with PD123319 (data not shown).

AP-1 binding sequence in the fibronectin gene promoter confers responsiveness to Ang II

To investigate the effects of Ang II on fibronectin gene transcription, a rat genomic EMBL3 library was screened for the 5′-flanking region of the fibronectin gene with a ^{32}P-end-labeled DNA fragment synthesized by PCR on the basis of the sequence data reported by Patel *et al.* [41]. We isolated the rat fibronectin gene containing 1908 bp of 5′-flanking region from the rat genomic library. The sequence downstream of position –1080 relative to the transcriptional start site (+1) was identical to that reported by Patel [41]. Computer analysis of the sequence revealed the presence of binding sites for several transcription factors such as AP-1 (TGACGCA at –453), ATF/CRE (TGACGTCA at –160), AP-2 (CCCCAGGC at –364) and PEA2 (GACCGCA at –257). To examine the promoter activity of the rat fibronectin gene in cardiac fibroblasts, we constructed 5′ sequential deletion mutants of the fibronectin promoter using restriction enzyme sites, and fused these mutants to the CAT gene as a reporter (Fig. 6A). We established stable transformants of these chimeric CAT constructs because of the low efficiency of transient transfection rate into cardiac fibroblasts, and tested Ang II responsiveness of each construct. The longest chimeric construct (p1908-CAT) had sufficient promoter activity, resulting in 2.8-fold higher CAT activation than the basal activity of the promoter-less construct. When cells transfected with p1908-CAT were exposed to Ang II, the CAT activity showed an increase of 3.1-fold relative to that in unstimulated cells, indicating that the Ang II-responsive element is present in the 1908 bp of 5′-flanking region. Using stable transformants of 5′ sequential deletion mutants, we found a significant decrease in the basal CAT activity with deletions between –1908 and –1079, –564 and –414, –202 and –123, –123 and +0, suggesting the presence of several positive cis-regulatory elements (Fig. 6A).

Exposure to Ang II of cells stably transfected with p1079-CAT, p882-CAT, p746-CAT or p564-CAT induced significant increases (about 3-fold) in CAT activity to an extent similar to that in p1908-CAT, whereas in cells transfected with p414-CAT, p202-CAT or p123-CAT the Ang II-induced increases were abolished, suggesting the presence of an Ang II-responsive element in the promoter region between nt –564 to –414. Since the AP-1 site (TGACGCA) was located at nt –453 of the fibronectin gene and *c-fos* and *c-jun* mRNA expression were up-regulated in response to Ang II in cardiac fibroblasts [28], we tested the involvement of the AP-1 site by site-directed mutation (TG<u>ACGC</u>A TG<u>GATA</u>A) in this motif of p564-CAT (termed p564m-CAT) followed by stable transfection into the cells. Interestingly, we found that the responsiveness to Ang II was completely inhibited in cells transfected with p564m-CAT (Fig. 6A), indicating a critical role of the AP-1 site in the Ang II-induced increase of fibronectin gene transcription. To further study the role of EGF-R transactivation in Ang II-stimulated fibronectin promoter activity, we tested the effects of EGF-R antagonist AG1478 or MEK inhibitor PD98059. As shown in Fig. 5A, treatment of p1908-CAT with AG1478 completely abolished Ang II-stimulated fibronectin promoter activity, whereas its activity was partially inhibited by PD98059 (64% inhibition in promoter activity).

c-fos *and* c-jun *heterodimer induced by Ang II binds to the AP-1 site of the fibronectin gene promoter*

We have previously shown that *c-fos* and *c-jun* mRNA expression are up-regulated in response to Ang II in cardiac fibroblasts [28]. To test whether Ang II-induced *c-fos* and *c-jun* complex binds the AP-1 binding site of the fibronectin gene promoter to enhance gene transcription, we performed gel retardation assay using an oligonucleotide (nt –438 to –463) including the AP-1 binding site as a probe. The nuclear extract from unstimulated control cells contained no protein capable of binding to the probe, whereas we found a retarded band when the probe was incubated with nuclear extract from cells stimulated with Ang II (Fig. 6B). An oligonucleotide corresponding to the AP-1 (TGACGCA) sequence inhibited the binding of nuclear extract, while a

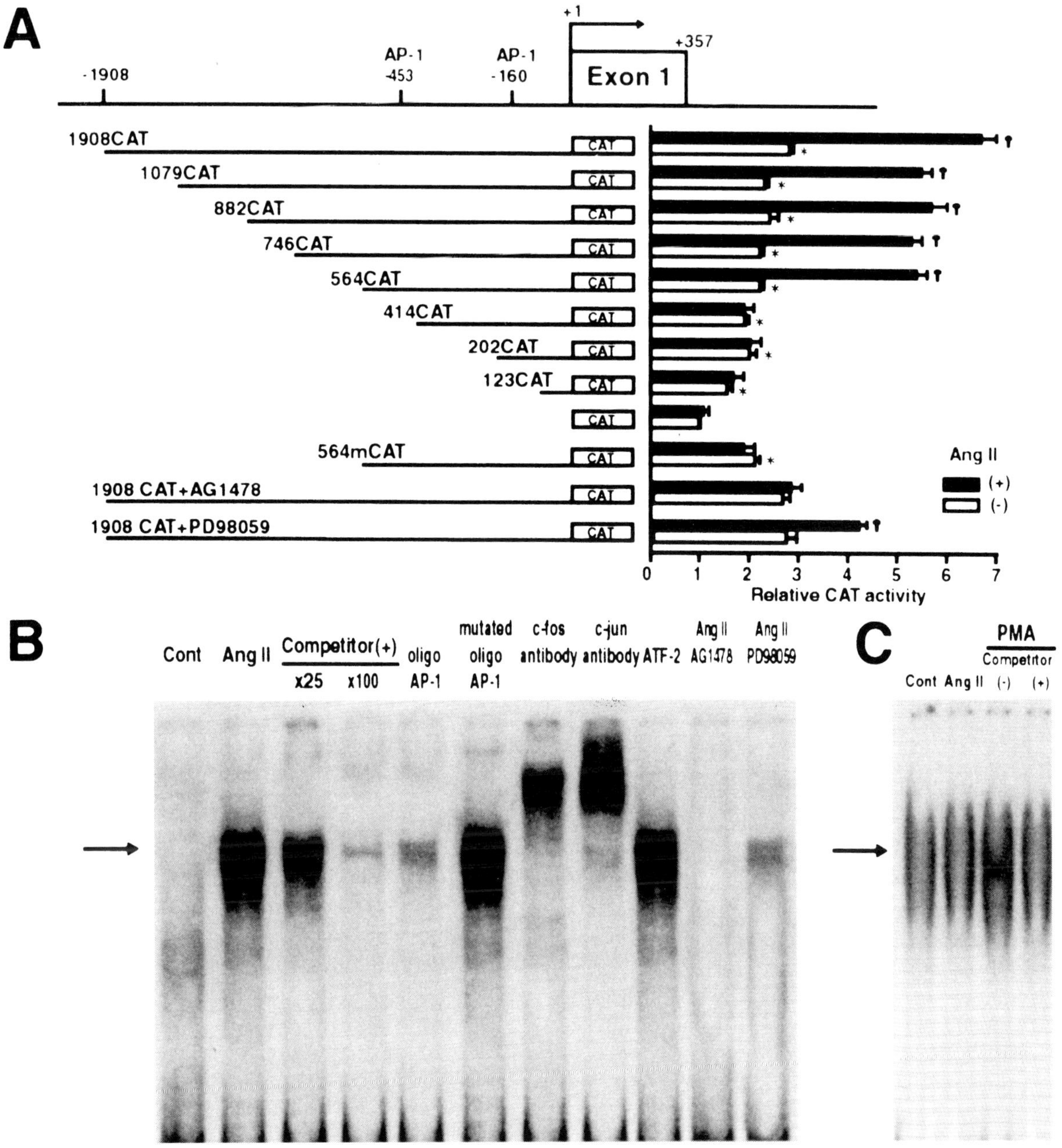

Fig. 6. Ang II responsiveness of fibronectin promoter regions and the binding of AP-1 complex analyzed by the gel mobility shift assay. *Panel A*: The CAT constructs containing 1908 bp of the 5′-flanking region and progressively shorter promoter sequences were stably transfected into cardiac fibroblasts. Transfected cells were stimulated by Ang II (100 nmol/L) for 12 h after 24 h serum depletion and relative CAT activities were determined as described in the Materials and methods section. AG1478 (250 nmol/L) and PD98059 (10 μmol/L) were added 30 min prior to exposure to Ang II. Values represent means ± S.E. of 6 separate experiments. 564 mCAT contained a mutation in the AP-1 site at nt –453. *$p < 0.05$ vs. the CAT activity of cells stably transfected with promoterless CAT construct; $^{\dagger}p < 0.05$ vs. the CAT activity of unstimulated cells in the absence of Ang II. *Panel B*: Double stranded oligonucleotide containing an AP-1 site (nt –438 to –473) was used as a probe in the gel shift assay. Nuclear extract was obtained from cells exposed to Ang II (100 nmol/L) for 45 min. AG1478 (250 nmol/L) and PD98059 (10 μmol/L) were added 20 min prior to exposure to Ang II. An unlabeled oligo corresponding to nt –438 to –466 was used as the competitor at 25 × and 100 × molar excess. AP-1 and mutated AP-1 oligos were used at 200 × molar excess. Oligo AP-1 indicates AP-1 sequence present at –453 (TGACGCA) and mutated oligo AP-1 indicates the mutated sequence TGGATAA. Supershift assays were performed with rabbit polyclonal anti-*c-fos* IgG raised against amino acids 128–152 of *c-fos* and rabbit anti-*c-jun* IgG raised against amino acids 247–263 of *c-jun*. Each antibody was added to the samples after the initial binding reaction and then incubated for 1 h at room temperature. *Panel C*: The oligonucleotides (nt –143 to –178) including the ATF/CRE motif was used as a probe in the gel shift assay. Nuclear extract was obtained from cells exposed to Ang II (100 nmol/L) or PMA (1 μmol/L) for 45 min. The unlabeled ATF/CRE oligo corresponding to –143 to –178 was used as a competitor at 200 × molar excess.

mutated oligo (TGGATGAA) did not affect binding. Furthermore, protein which bound to the AP1 oligonucleotide was supershifted when preincubated with anti-*c-fos* and anti-*c-jun* antibodies, but not anti-ATF2 antibody (Fig. 6B). To further study the role of EGF-R transactivation in Ang II-stimulated gel mobility of AP-1 complex, we tested the effects of EGF-R antagonist AG1478 or MEK inhibitor PD98059. As shown in Fig. 6B, treatment with AG1478 abolished Ang II-stimulated binding of AP-1 complex to the AP-1 sequence, whereas its binding was not completely inhibited by PD98059.

ATF/CRE motif (TGACGTCA) present at nt –160 and this motif is known to stimulate transcription of several genes [42]. Promoter analyses using CAT reporter constructs (Fig. 6A) revealed that the ATF/CRE motif present in the promoter region of the fibronectin gene did not confer Ang II responsiveness. To test whether nuclear extract from Ang II-treated cells binds the ATF/CRE motif, we performed gel retardation assay using an oligonucleotide (nt –143 to –178) including the ATF/CRE motif. As shown in Fig. 6C, the binding activity of nuclear extract to the ATF/CRE motif was not enhanced by Ang II treatment, whereas PMA induces the binding protein specific to the ATF/CRE motif.

TGF-β secreted into medium following Ang II treatment induces fibronectin expression

Since Ang II is known to induce the synthesis and secretion of TGF-β, resulting in stimulation of fibronectin expression in mesangial cells [43], we tested whether this autocrine and/or paracrine effect of TGF-β also occurs in cardiac fibroblasts. TGF-β activity was measured using the standard CCL-64 cell bioassay, where TGF-β secreted by cells into culture medium was measured by its ability to inhibit cell growth. As shown in Fig. 7A, the mean amount of active TGF-β present in six control cell cultures was 0.18 ng/ml (open bar). When stimulated with 100 nmol/L Ang II, a 6.7-fold increase to 1.2 ng/ml was seen. This increased activity was completely abrogated by neutralizing anti-TGF-β antibody, but not by non-immune control rabbit IgG. Acid treatment of conditioned medium resulted in significant increases in active TGF-β in both control and Ang II-treated cultures. However, acid treatment of control conditioned medium resulted in 0.8 ng/ml active TGF-β, while active TGF-β in Ang II-treated culture media increased to 2.9 ng/ml (hatched bars). This result indicated that Ang II increased not only the production of both latent and active TGF-β but also stimulated the conversion of latent to active TGF-β.

To study the role of EGF-R transactivation and ERK activation in Ang II-stimulated release of TGF-β, we determined Ang II-stimulated TGF-β activity using HEGF-R 533del cells or the cells pretreated with MEK inhibitor PD98059. Ang II-stimulated TGF-β activities in the incubation medium were completely inhibited in HEGF-R 533del cells and by treatment with PD98059 (Fig. 7A), indicating that ERK activated downstream of EGF-R is a major mediator to regulate the release of TGF-β by Ang II.

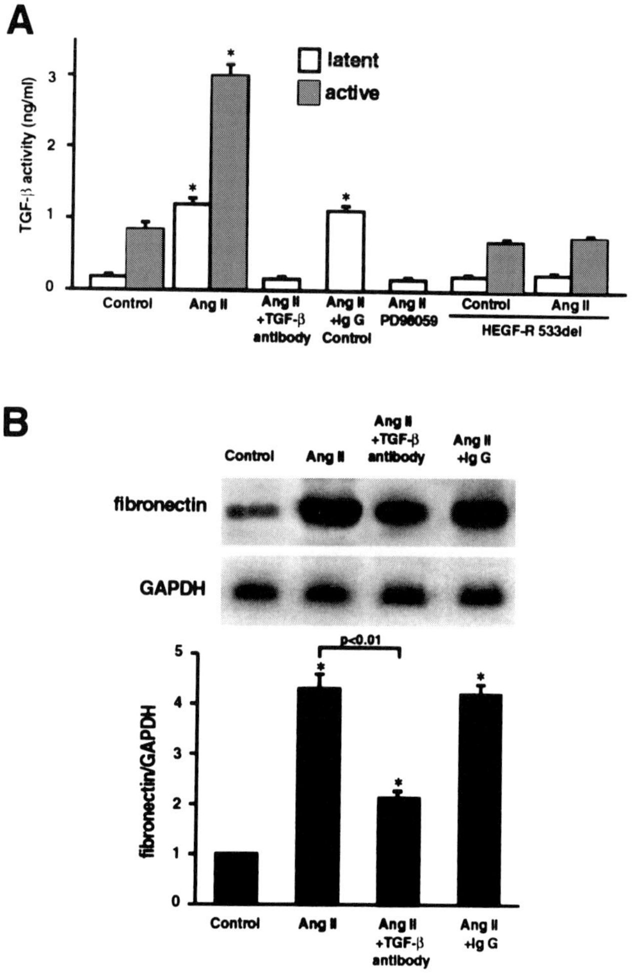

Fig. 7. Effects of Ang II on TGF-β bioactivity before and after acidification of culture medium measured by standard CCL-64 cell bioassay, and the effects of anti-TGF-β neutralizing antibody on Ang II-induced fibronectin mRNA. *Panel A*: Cardiac fibroblasts or HEGFR 533del cells incubated with serum-free medium for 12 h were stimulated with Ang II (100 nmol/L) for 12 h and then TGF-β activity in culture supernatant was measured by bioassay as described in the Materials and methods section. PD98059 (10 μmol/L) was added 20 min prior to exposure to Ang II. Open bars represent TGF-β activity in culture supernatant from six fibroblast cultures. Hatched bars represent TGF-β activity of the same culture supernatants after transient acidification to activate latent TGF-β. The black bar represents the TGF-β activity in the presence of both Ang II and neutralizing antibody to TGF-β. *$p < 0.01$ vs. latent TGF-β activity in the control. *Panel B*: Serum-deprived cells were preincubated with anti-TGF-β neutralizing antibody (100 μg/ml) or control IgG (100 μg/ml) and then exposed to Ang II (100 nmol/L) for 12 h. Fibronectin mRNA levels were quantified as described in Fig. 1. Values represent means ± S.E. of 4 separate experiments. *$p < 0.01$ vs. controls (unstimulated cells).

To assess whether secreted TGF-β mediated the effect of Ang II on fibronectin mRNA accumulation, Ang II-treated cells were coincubated with neutralizing anti-TGF-β antibody (100 μg/ml) or control IgG (100 μg/ml) (Fig. 7B). Interestingly, treatment of cultures with neutralizing antibody significantly (46%, $p < 0.01$) abolished the induction of fibronectin mRNA expression. Control IgG had no effect. Addition of a sufficient amount of anti-TGF-β neutralizing antibody was confirmed by the finding that the conditioned medium from cells incubated with Ang II and anti-TGF-β neutralizing antibody for 24 h contained control levels of TGF-β activity (0.15 ng/ml) as measured by bioassay. These results suggested that Ang II-induced fibronectin mRNA synthesis is mediated by autocrine and/or paracrine effect of TGF-β, and that this action of TGF-β is not sufficient to cause the full induction of fibronectin mRNA.

Ang II stimulates TGF-β mRNA expression through EGF-R transactivation, and TGF-β regulates fibronectin expression by stabilizing its mRNA with no effect on transcriptional activity

We next examined the effects of Ang II-induced EGF-R transactivation on TGF-β mRNA expression and the mechanism by which TGF-β regulates fibronectin synthesis. Addition of Ang II (100 nmol/L) stimulated TGF-β mRNA accumulation with a time course similar to that of fibronectin mRNA induction (significant increase at 6 h and peak level at 12 h with a 3.3-fold increase). Pretreatment with the EGF-R inhibitor AG1478 completely abolished Ang II-induced TGF-β mRNA accumulation (Fig. 8A) without affecting the basal level of TGF-β mRNA expression (data not shown). To further study the role of EGF-R transactivation in Ang II-stimulated TGF-β mRNA expression, we determined TGF-β mRNA levels in HEGF-R 533del cells or the cells pretreated with MEK inhibitor PD98059. Ang II-stimulated TGF-β mRNA accumulations were completely inhibited in HEGF-R 533del cells and by treatment with PD98059 (Fig. 8A), indicating that ERK activated downstream of EGF-R is a major mediator to regulate TGF-β mRNA accumulation by Ang II.

We also examined the role of TGF-β in regulating fibronectin gene expression. TGF-β (1, 10 and 100 ng/ml) did not significantly increase the CAT activity of the longest chimeric construct (p1908-CAT) (data not shown). Therefore, we next tested the effect of TGF-β on fibronectin mRNA stability by inhibiting gene transcription with actinomycin D. In unstimulated control cells, the half life of fibronectin mRNA was 14 ± 0.1 h. When cells were stimulated with Ang II or TGF-β, the half life was significantly ($p < 0.01$) increased to 23.6 ± 0.1 h and 23.2 ± 0.1 h, respectively, compared with that in controls (Fig. 8B). Moreover, we tested the involvement of *de novo* protein synthesis in Ang II- or TGF-β-induced stabilization of fibronectin mRNA. Pretreatment of cells with cycloheximide completely reversed Ang II- or TGF-β-induced increases in mRNA stability to the control level (Fig. 8B). These observations suggested that TGF-β secreted into the medium by Ang II contributes to the accumulation of fibronectin mRNA mainly by increasing its mRNA stability rather than affecting its gene transcription, and that this post-transcriptional effect is exerted via *de novo* protein synthesis.

Discussion

The major new findings of this study were that Ang II-induced synthesis and secretion of fibronectin and TGF-β are mediated by downstream signaling of EGF-R transactivated by Ca^{2+}/calmodulin-dependent tyrosine kinase, in which ERK-mediated pathway plays an important role, and that Ang II-induced fibronectin mRNA expression is regulated by two different mechanisms ; transcriptional control by binding of *c-fos/c-jun* complex to the AP-1 site, and post-transcriptional control by autocrine and/or paracrine effects of TGF-β, which is exerted by increasing the mRNA stability and requires *de novo* protein synthesis.

Fibronectin is important for cell adhesion and cell migration, events which occur in wound healing, organogenesis and cardiac remodeling, and it is possible that this protein molecule plays an important role in the remodeling of cardiac interstitium secondary to myocardial hypertrophy [44, 45]. A direct role for Ang II in remodeling is supported by the observation that cultured rat cardiac fibroblasts demonstrate increased fibronectin, collagen types 1 and 3 transcript expression, as well as secretion of new collagen following stimulation with Ang II [2, 4]. However, little is known about the mechanism by which Ang II stimulates collagenous protein synthesis in cardiac fibroblasts. Recently, we and Eguchi *et al.* clearly showed that in cardiac fibroblasts [28] or VSMC [16, 20], Ang II-induced ERK activation and *c-fos* expression is mainly mediated through downstream signaling of EGF-R transactivated by Ang II in a Ca^{2+}/calmodulin-dependent, PKC-independent manner, and that Ang II as well as the Ca^{2+} ionophore A23187 induced tyrosine phosphorylation of the EGF-R, which was sufficient to recruit the adaptor proteins that are involved in Ras activation. In this study, we further extended these previous studies and found for the first time that such a Ca^{2+}/calmodulin-dependent transactivation mechanism of EGF-R is operating on both Ang II-induced fibronectin and TGF-β gene expression.

We also showed that the AP-1 sequence present in the promoter region plays an important role in Ang II responsiveness of the fibronectin gene, and that AP-1 complex (containing *c-fos/c-jun*) is associated with this cognate recognition sequence. A serum response element (SRE) mediates *c-fos* induction by growth factors, cytokines and other stimuli that

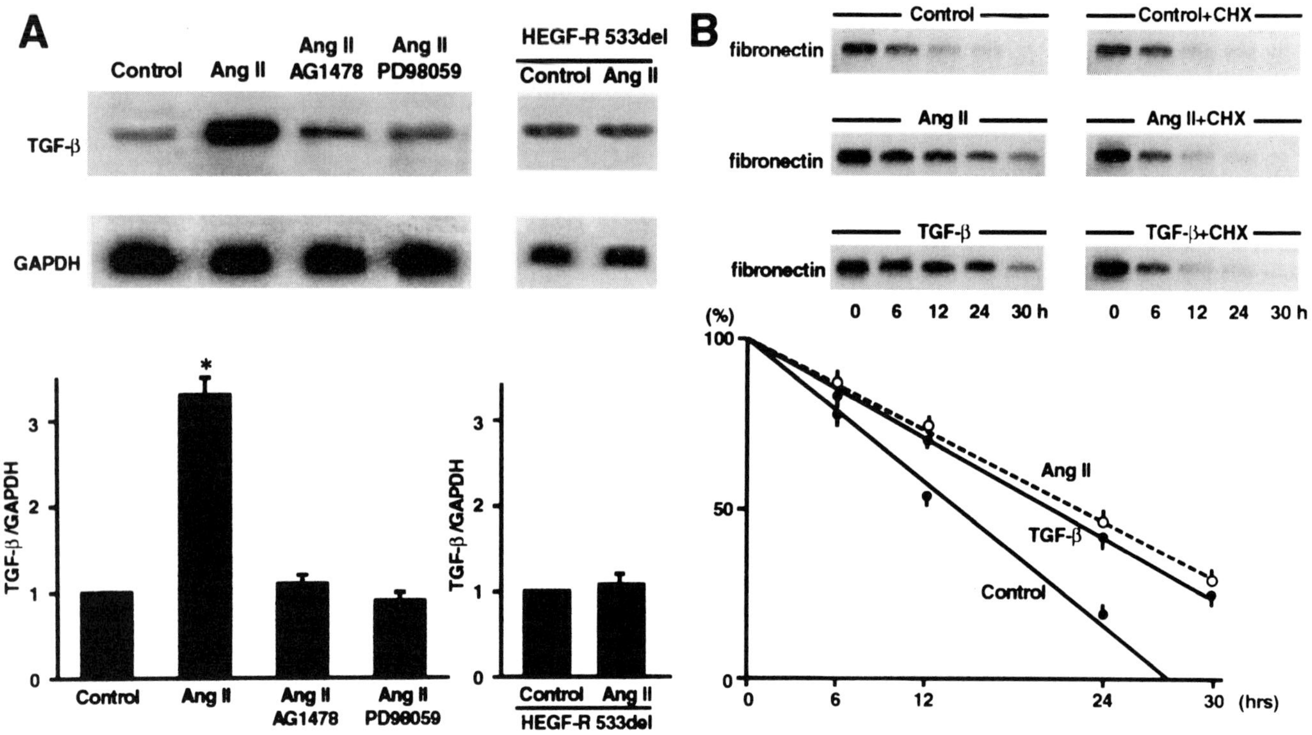

Fig. 8. Effects of Ang II-induced EGF-R transactivation and MEK inhibitor on TGF-β mRNA accumulation and mRNA stability. *Panel A*: Serum-deprived cardiac fibroblasts or HEGFR 533del cells were pretreated with or without AG1478 (250 nmol/L) and PD98059 (10 μmol/L) for 20 min and then challenged with 100 nmol/L Ang II for 12 h. TGF-β mRNA was quantified as described in Fig. 1. *p < 0.01 vs. the controls. *Panel B*: Cells stimulated for 12 h with serum-free medium containing Ang II (100 nmol/L) or TGF-β (10 ng/ml) were treated with actinomycin D (5 μg/ml), and then the time-dependent changes in fibronectin mRNA level were analyzed by Northern blotting as described in Fig. 1. To examine the effects of cycloheximide (CHX), cells were pretreated for 4 h with cycloheximide (5 μg/ml) and then exposed to Ang II (100 nmol/L) or TGF-β (10 ng/ml). Values represent means ± S.E. of 6 separate experiments, and a representative Northern blot is shown.

activate ERK [10, 46]. The *c-fos* promoter contains SRE, and induction of *c-fos* expression occurs upon formation of a ternary complex factor composed of p62TCF and the serum response factor at the SRE [47]. ERK was shown to phosphorylate p62TCF (also known as Elk-1 or SAP-1), resulting in enhanced ternary complex formation [48]. Recent studies have shown that Jak-STAT [3, 49], Rho-dependent pathways [50] or c-Jun N-terminal kinase (JNK) [51] also induce *c-fos* gene expression. As Ang II-induced *c-fos* expression in cardiac fibroblasts [28] or VSMC [20] is completely inhibited by specific inhibition of EGF-R function, it is likely that Ang II-stimulated *c-fos* expression is mainly regulated by phosphorylated p62TCF after EGF-R-mediated ERK activation. *c-Jun* is one of the major components of the transcriptional factor AP-1, which regulates expression of many genes with an AP-1 binding site in their promoter regions [52]. The transcriptional activating activity of *c-jun* is regulated at the post-translational level by phosphorylation of *c-jun*. *c-Jun* is phosphorylated at two serine residues within N-terminal transactivation domain by JNK [42]. JNK was reported to be activated by downstream signaling of EGF via phosphatidylinositol 3-kinase- dependent mechanism in Hela cells [53, 54]. We also found that stimulation of EGF-R is able to stimulate JNK activity in cardiac fibroblasts (S. Murasawa, unpublished observation). In this study, we showed that Ang II-induced transcriptional activity of fibronectin gene and binding of *c-fos*/*c-jun* complex to the AP-1 site are completely abolished by specific inhibition of EGF-R function. This finding suggests that in cardiac fibroblasts the transcriptional activating activity of *c-jun* is enhanced by downstream signaling of EGF-R. Considering that inhibition of ERK activity by MEK inhibitor partially inhibited the Ang II-stimulated transcriptional activity of fibronectin gene and binding of *c-fos*/*c-jun* complex, it is likely that JNK pathway activated downstream of EGF-R is also involved in the transcriptional regulation of fibronectin gene.

ATF-2, a member of the family of CRE binding proteins, is not activated by agents that increase cAMP. ATF-2 can form a complex not only with *c-jun* but also with itself and some other members of the ATF family, and the complex binds the ATF/CRE motif to enhance gene transcription [42]. Although the fibronectin gene has an ATF/CRE motif in the 5′-flanking region, the results from the promoter analyses (Fig. 6A) and gel retardation assays (Fig. 6B) indicated that the ATF/

CRE motif is not involved in Ang II responsiveness of this gene. Expression of ANF [55], endothelin-1 [56] and collagenase genes containing AP-1 sites are regulated by AP-1 activity, while the *c-jun* promoter and adhesion molecule E-selectin are up-regulated by binding of ATF-2/*c-jun* complex to the ATF/CRE motif [42]. This study demonstrated for the first time that fibronectin is included in the gene family regulated by AP-1 complex. Since collagenase activity is also stimulated by Ang II [42], it is interesting that both collagenase and fibronectin genes are transcriptionally regulated by AP-1 activity. Further studies are needed to define the interaction between Ang II-induced collagenous protein synthesis and degradation.

An important function of Ang II in the heart may be stimulation of cardiac fibroblasts to release growth factors. Ang II can directly modulate vascular smooth muscle growth by stimulating release of platelet-derived growth factors, TGF-β and basic fibroblast growth factors [57]. Primary cultures of neonatal cardiac fibroblasts secrete significant amounts of TGF-β when stimulated with Ang II [58], which was shown in turn to cause fibrillar collagen synthesis [59]. Although similar paracrine and/or autocrine action of TGF-β to induce synthesis of fibronectin and collagen proteins was reported in glomerular mesangial cells [43], it remains unclear how TGF-β regulates the expression of fibronectin. In the present study, we clearly demonstrated that Ang II increases the level of TGF-β mRNA and production of both latent and active TGF-β into medium, and that this induction is mediated via ERK activated downstream of EGF-R transactivated by Ang II. We also found that TGF-β did not affect the promoter activity of the fibronectin gene but stimulated fibronectin mRNA accumulation by stabilizing its mRNA metabolism, and that this action requires *de novo* protein synthesis. TGF-β has been shown to induce promoter activity of the collagen type 1 gene through a nuclear factor 1 (NF1) binding site in mouse NIH 3T3 and rat osteosarcoma cells [60], whereas post-transcriptional regulation of collagen type 1 gene has also been reported in lung fibroblasts [61] As there is no NF1 site in the 1.9 kb of promoter region of the fibronectin gene, it is plausible that the effect of TGF-β is exerted at the post-transcriptional level by stabilization of fibronectin mRNA. Eghbali *et al.* [59] reported that in cardiac fibroblasts the effect of TGF-β on collagen type 1 synthesis required *de novo* synthesis of proteins by examining the effect of cycloheximide, and that TGF-β did not stimulate the induction of proto-oncogenes *c-fos*, *c-jun* or *Egr-1*. We also confirmed these findings in our cells (Y. Moriguchi, unpublished observation). Thus, this study demonstrated that induction of fibronectin by Ang II is regulated by both transcriptional and post-transcriptional mechanisms and the latter is due to autocrine and/or paracrine effect of TGF-β. Although our study also indicated that the effect of TGF-β on fibronectin mRNA stabilization required *de novo* protein synthesis, identification of the proteins produced and the mechanism of stabilization of fibronectin mRNA remain to be determined.

In summary, we demonstrated that the expression of fibronectin as well as TGF-β, a strong mediator of collagenous protein synthesis, is increased through an AT1-R-mediated transactivation mechanism of EGF-R induced by a Ca^{2+}-dependent tyrosine kinase. We also showed that ERK activation via EGF-R transactivation is a major mediator to regulate the synthesis and release of fibronectin and TGF-β, whereas pathway(s) (probably JNK) other than ERK activated downstream of EGF-R is also partially involved in the transcriptional control of fibronectin gene. Recently, we have shown that a Ca^{2+}-dependent tyrosine kinase, Pyk2/CAKβ/RAFTK, is partially involved in this transactivation mechanism by examining the effects of dominant negative Pyk2 [38]. However, the Pyk2-mediated action alone cannot sufficiently account for transactivation of EGF-R and an other mechanism is likely to be responsible for tyrosine phosphorylation of EGF-R. The direct involvement of the EGF-R in Ang II-induced fibronectin and TGF-β expression presents a novel paradigm for cross-talk between AT1-R and growth factor receptor signaling pathways. Therefore it is important for us to interpret the effects of Ang II on extracellular matrix formation in association with the signaling cascade regulating cellular proliferation and/or differentiation by growth factors.

Acknowledgements

This study was supported in part by research grants from the Ministry of Education, Science and Culture, Japan, Study Group of Molecular Cardiology, Naito Foundation, Clinical Pharmacology Foundation and Japan Medical Association and Japan Smoking Foundation, Japan Heart Foundation. Dr. Satoshi Murasawa is Research Fellow of the Japan Society for the Promotion of Science.

References

1. Schorb W, Booz GW, Dostal DE, Conrad KM, Chang KC, Baker KM: Angiotensin II is mitogenic in neonatal rat cardiac fibroblasts. Circ Res 72: 1245–1254, 1993
2. Villarreal FJ, Kim NN, Ungab GD, Printz MP, Dillmann W: Identification of functional angiotensin II receptors on rat cardiac fibroblasts. Circulation 88: 2849–2861, 1993
3. Bhat GJ, Thekkumkara TJ, Thomas WG, Conrad KM, Baker KM: Angiotensin II stimulates sis-inducing factor-like DNA binding activity. J Biol Chem 269: 31443–31449, 1994
4. Crabos M, Roth M, Hahn AWA, Erne P: Characterization of angiotensin II receptors in cultured adult rat cardiac fibroblasts. J Clin Invest 93: 2372–2378, 1994

5. Weber KT, Brilla CG: Pathological hypertrophy and cardiac interstitium. Fibrosis and renin-angiotensin-aldosterone system. Circulation 83: 1849–1865, 1991
6. Ohkubo N, Matsubara H, Nozawa Y, Mori Y, Satoshi Murasawa S, Kijima K, Maruyama K, Masaki H, Tsutsumi Y, Shibazaki Y, Iwasaka T, Inada M: Angiotensin type 2 receptors are re-expressed by cardiac fibroblasts from failing myopathic hamster hearts and inhibit cell growth and fibrillar collagen metabolism. Circulation 96: 3954–3962, 1997
7. Matsubara H, Inada M: Molecular insights into angiotensin II type 1 and type 2 receptors : Expression, signaling and physiological function and clinical application of its antagonists. Endo J 45: 137–150, 1998
8. Inagami T, Kitami Y: Angiotensin II receptor: Molecular cloning, functions and regulation. Hypertens Res 17: 87–97
9. Berk BC, Corson MA: Angiotensin II signal transduction in vascular smooth muscle, role of tyrosine kinases. Circ Res 80: 607–616, 1997
10. Booz GW, Dostal DE, Singer HA, Baker KM: Involvement of protein kinase C and Ca^{2+} in angiotensin II-induced mitogenesis of cardiac fibroblasts. Am J Physiol 267: C1308–C1318, 1994
11. Schorb W, Peeler TC, Madigan NN, Conrad KM, Baker KM: Angiotensin II -induced protein tyrosine phosphorylation in neonatal rat cardiac fibroblasts. J Biol Chem 269: 19626–19632, 1994
12. Bernstein KE, Marrero MB: The importance of tyrosine phosphorylation in angiotensin II signaling. Trend Cardiovasc Med 6: 179–187, 1996
13. Zou Y, Komuro I, Yamazaki T, Aikawa R, Kudoh S, Shiojima I, Hiroi Y, Mizuno T, Yazaki Y: Protein kinase C, but not tyrosine kinases or Ras, plays a critical role in angiotensin II -induced activation of Raf-1 kinase and extracellular signal-regulated protein kinases in cardiac myocytes. J Biol Chem 27: 33592–33597, 1996
14. Sadoshima J, Izumo S: The heterotrimetric Gq protein-coupled angiotensin II receptor activates p21ras via the tyrosine kinase-Shc-Grb2-Sos pathway in cardiac myocytes. EMBO J 15: 775–787, 1996
15. Sadoshima J, Qiu Z, Morgan JP, Izumo S: Angiotensin II and other hypertrophic stimuli mediated by G protein-coupled receptors activate tyrosine kinase, mitogen-activated protein kinases, and 90-kD S6 kinase in cardiac myocytes. The critical role of Ca^{2+}-dependent signaling. Circ Res 76: 1–15, 1995
16. Eguchi S, Matsumoto T, Motley ED, Utsunomiya H, Inagami T: Identification of an essential signaling cascade for mitogen-activated protein kinase activation by angiotensin II in cultured rat vascular smooth muscle cells. J Biol Chem 271: 14169–14175, 1996
17. Molloy CJ, Taylor DS, Weber H: Angiotensin II stimulation of rapid protein tyrosine phosphorylation and protein kinase activation in rat aortic smooth muscle cells. J Biol Chem 268: 7338–7345, 1993
18. Tsuda T, Kawahara Y, Ishida Y, Yokoyama M: Angiotensin II stimulates two myelin basic protein/microtubule-associated protein 2 kinases in cultured rat vascular smooth muscle cells. Circ Res 71: 620–630, 1992
19. Linseman DA, Benjamin CW, Jones DA: Convergence of angiotensin II and platelet-derived growth factor receptor signaling cascades in vascular smooth muscle cells. J Biol Chem 270: 12563–12568, 1995
20. Eguchi S, Numaguchi K, Iwasaki H, Matsumoto T, Yamakawa T, Utsunomiya H, Motley ED, Kawakatsu H, Owada KM, Hirata Y, Marumo F, Inagami T: Calcium-dependent epidermal growth factor receptor transactivation mediates the angiotensin II-induced mitogen-activated protein kinase activation in vascular smooth muscle cells. J Biol Chem 273: 8890–8896, 1998
21. Daub H, Weiss FU, Wallasch C, Ullrich A: Role of transactivation of the EGF receptor in signaling by G-protein-coupled receptors. Nature 379: 557–560, 1996
22. Luttrell LM, Della Rocca J, van Biesen T, Luttrell DK, Lefkowitz RJ: Gβγ subunits mediate Src-dependent phosphorylation of the epidermal growth factor receptor. J Biol Chem 272: 4637–4644, 1997
23. Daub H, Wallasch C, Lankenau A, Herrlich A, Ullrich A: Signal characteristics of G protein-transactivated EGF receptor. EMBO J 16: 7032–7044, 1997
24. Zwick E, Daub H, Aoki N, Yamaguchi-Aoki Y, Tinhofer I, Maly K, Ullrich A: Critical role of calcium-dependent epidermal growth factor receptor transactivation in PC12 cell membrane depolarization and bradykinin signaling. J Biol Chem 272: 24767–24770, 1997
25. Rosen LB, Greenberg ME: Stimulation of growth factor receptor signal transduction by activation of voltage-sensitive calcium channels. Proc Natl Acad Sci USA 93: 1113–1118, 1996
26. Showkat Ali M, Schieffer B, Delafontaine P, Bernstein KE, Ling BN, Marrero MB: Angiotensin II stimulates tyrosine phosphorylation and activation of insulin receptor substrate 1 and protein-tyrosine phosphatase 1D in vascular smooth muscle cells. J Biol Chem 272: 12373–12379, 1997
27. Tsai W, Moreielli AD, Peralta EG: The m1 muscarinic acetylcholine receptor transactivates the EGF receptor to modulate ion channel activity. EMBO J 16: 4597–4605, 1997
28. Murasawa S, Mori Y, Nozawa Y, Gotoh N, Shibuya M, Masaki H, Maruyama K, Yoshiaki Tsutsumi Y, Moriguchi Y, Iwasaka T, Inada M, Matsubara H: Angiotensin II AT1 receptor-induced extracellular signal-regulated protein kinase activation is mediated by Ca^{2+}/calmodulin-dependent transactivation of epidermal growth factor receptor. Circ Res 82: 1338–1348, 1998
29. Hynes RO: In: Fibronectin. Academic Press, San Diego, CA, 1989 pp 181–246
30. Matsubara H, Kanasaki M, Murasawa S, Tsukaguchi Y, Nio Y, Inada M: Differential gene expression and regulation of angiotensin II receptor subtypes in rat cardiac fibroblasts and cardiomyocytes in culture. J Clin Invest 93: 1592–1601, 1994
31. Kijima K, Matsubara H, Komuro I, Yazaki Y, Inada M: Mechanical stretch induces enhanced expression of angiotensin II receptors in neonatal rat cardiac myocytes. Circ Res 79: 887–897, 1994
32. Suzuki J, Matsubara H, Inada M: Rat angiotensin II (type 1A) receptor mRNA regulation and subtype expression in myocardial growth and hypertrophy. Circ Res 73: 439–447, 1993
33. Nio Y, Matsubara H, Murasawa S, Kanasaki M, Inada M: Regulation of gene transcription of angiotensin II receptor subtypes in myocardial infarction. J Clin Invest 95: 46–54, 1995
34. Murasawa S, Matsubara H, Kijima K, Maruyama K, Ohkubo N, Mori Y, Iwasaka Y, Iwasaka T, Inada M: Down- regulation by cAMP of angiotensin II type 2 receptor gene expression in PC12 cells. Hypertens Res 19: 271–279, 1996
35. Murasawa S, Matsubara H, Urakami M, Inada M: Regulatory elements that mediate expression of the gene for thr angiotensin II type 1a receptor for the rat. J Biol Chem 268: 26996–27003, 1993
36. Murasawa S, Matsubara H, Mori Y, Kijima K, Maruyama K, Inada M: Identification of a negative cis-regulatory element and trans-acting protein that inhibit transcription of the angiotensin II type 1a receptor gene. J Biol Chem 270: 24282–24286, 1995
37. Cone JL, Brown DR, DeLarco JE: Improved method of purification of transforming growth factor, type β from platelets. Anal Biochem 168: 71–74, 1998
38. Murasawa S, Mori Y, Nozawa Y, Masaki H, Maruyama K, Matsubara H: Role of calcium-sensitive tyrosine kinase PYK2/CAKβ/RAFTK in angiotensin II-induced Ras/ERK signaling. Hypertension 32: 668–675, 1998
39. Levitzki A, Gazit A: Tyrosine kinase inhibition: An approach to drug development. Science 267: 1782–1787, 1995
40. Shiraishi T, Owada MK, Tatsuja M, Yamashita T, Watanabe K, Kakunaga T: Specific inhibitors of tyrosine-specific protein kinases: Properties

of 4-hydroxycinnamamide derivatives *in vitro*. Cancer Res 49: 2374–2378, 1989
41. Patel RS, Odermatt E, Schwarzbauer JE, Hynes RO: Organization of the fibronectin gene provides evidence for exon shuffling during evolution. EMBO J 6: 2565–2572, 1987
42. Force T, Pombo CM, Avruch JA, Bonventre JV, Kyriakis JM: Stress-activated protein kinases in cardiovascular disease. Circ Res 78: 947–953, 1996
43. Kagami S, Border WA, Miller DE, Noble NA: Angiotensin II stimulates extracellular matrix protein synthesis through induction of transforming growth factor-β expression in rat glomerular mesangial cells. J Clin Invest 93: 2431–2437, 1994
44. Dostal DE, Booz GW, Baker KM: Angiotensin II signaling pathways in cardiac fibroblasts: Conventional versus novel mechanisms in mediating cardiac growth and function. Mol Cell Biochem 157: 15–21, 1996
45. Mamuya WS, Brecher P: Fibronectin expression in the normal and hypertrophic rat hearts. J Clin Invest 89: 392–401, 1992
46. Karin M: The regulation of AP-1 activity by mitogen-activated protein kinases. J Biol Chem 270: 16483–16486, 1995
47. Treisman R: Ternary complex factors: Growth factor regulated transcriptional activators. Curr Opin Genet Dev 4: 96–101, 1994
48. Gille H, Kortenjann M, Thomae O, Moomaw C, Slaughter C, Cob MH, Shaw PE: ERK phosphorylation potentiates ELK-1-mediated ternary complex formation and transactivation. EMBO J 14: 951–962, 1995
49. Marrero MB, Schieffer B, Paxton WG, Heerdt L, Berk BC, Delafontaine P, Bernstein KE: Direct stimulation of Jak/STAT pathway by the angiotensin II AT1 receptor. Nature 375: 247–250, 1995
50. Hill CS, Wynne J, Treisman R: The Rho family GTPases RhoA, Rac1, and CDC42 Hs regulates transcriptional activation by SRF. Cell 81: 1159–1170, 1995
51. Cavigelli M, Dolfi F, Claret FX, Karin M: Induction of *c-fos* expression through JNK-mediated TCF/ELK-1 phosphorylation. EMBO J 14: 5957–5964, 1995
52. Angel P, Imagawa M, Chiu R, Stein B, Imbra RJ, Rahmsdorf HJ, Jonat C, Herrlich P, Karin M: Phorbol ester-inducible genes contain a common cic element recognized by a TPA-modulated trans-acting factor. Cell 49: 729–739, 1987
53. Logan S, Falasca M, Hu P, Schlessinger J: Phosphatidylinositol 3-kinase mediates epidermal growth factor-induced activation of the c-Jun N-terminal kinase signaling pathway. Mol Cell Biol 17: 5784–5790, 1997
54. Clarke N, Arenzana N, Hai T, Minden A, Prywes R: Epidermal growth factor induction of the c-jun promoter by a Rac pathway. Mol Cell Biol 18: 1065–1073, 1998
55. Kovacic-Milivojevic B, Gardner DG: Divergent regulation of the human atrial natriuretic peptide gene by c-fos. Mol Cell Biol 12: 292–301, 1992
56. Kudoh S, Komuro I, Mizuno T, Yamazaki T, Zou Y, Shiojima I, Takekoshi N, Yazaki Y: Angiotensin II stimulates c-Jun NH2-terminal kinases in cultured cardiac myocytes of neonatal rats. Circ Res 80: 139–146, 1997
57. Gibbons GH, Pratt RE, Dzau VJ: Vascular smooth muscle cell hypertrophy vs hyperplasia. Autocrine transforming growth factor-β1 expression determines growth response to angiotensin II. J Clin Invest 90: 456–461, 1992
58. Fisher SA, Absher M: Norepinephrine and Ang II stimulates secretion of TGF-β by neonatal rat cardiac fibroblasts *in vitro*. Am J Physiol 268: C910–C917, 1995
59. Eghbali M, Tomek R, Sukhatme VP, Woods C, Bhambi B: Differential effects of transforming growth factor-β1 and phorbol myristate acetate on cardiac fibroblasts. Regulation of fibrillar collagen mRNAs and expression of early transcriptional factors. Circ Res 69: 483–490, 1991
60. Rossi P, Karsenty G, Roberts AB, Roche NS, Sporn MB, Crombrugghe B: A nuclear factor binding site mediates the transcriptional activation of a type 1 collagen. Cell 52: 405–414, 1988
61. Keski-Oja J, Raghow R, Sawdey M, Loskutoff DJ, Postlethwaite AE, Kang AH, Moses HL: Regulation of mRNAs for type-1 plasminogen activator inhibitor, fibronectin and type I procollagen by transforming growth factor-β. J Biol Chem 263: 3111–3115, 1988

Molecular and Cellular Biochemistry **212**: 203–209, 2000.

Expression of renin-angiotensin system and extracellular matrix genes in cardiovascular cells and its regulation through AT1 receptor

Kouichi Tamura,[1,2] Yuqing E. Chen,[2] Qin Chen,[2] Nobuo Nyui,[1] Masatsugu Horiuchi,[2] Izumi Takasaki,[1] Nobuko Tamura,[3] Richard E. Pratt,[2] Victor J. Dzau[2] and Satoshi Umemura[1]

[1]*Department of Internal Medicine II;* [3]*Department of Dermatology, Yokohama City University School of Medicine, Yokohama, Japan;* [2]*Cardiovascular Research, Department of Medicine, Brigham & Women's Hospital, Harvard Medical School, Boston, MA, USA*

Abstract

Angiotensinogen (AGT) is a unique substrate of the renin-angiotensin system and fibronectin (FN) is an important component of the extracellular matrix. These play critical roles in the pathophysiological changes including cardiovascular remodeling and hypertrophy in response to hypertension. This study was performed to examine the regulation of AGT and FN gene in cardiac myocytes (CMs) and vascular smooth muscle cells (VSMCs) in response to mechanical stretch. Mechanical stretch significantly increased the AGT mRNA expression in CMs, while these stimuli did not affect FN mRNA levels. On the other hand, mechanical stretch upregulated FN mRNA levels in VSMCs, whereas no increase in AGT mRNA levels was observed in response to stretch stimuli. An angiotensin II type 1 (AT1) receptor antagonist (CV11974) significantly decreased these stretch-mediated increases in mRNA level and promoter activity of the AGT and FN gene, whereas angiotensin II type 2 (AT2) receptor antagonist (PD123319) did not affect the induction. These results indicate that mechanical stretch activates transcription of the AGT and FN gene mainly via AT1 receptor-pathway in CMs and VSMCs. Furthermore, mechanisms regulating AGT and FN gene seem to be different between CMs and VSMCs. (Mol Cell Biochem **212**: 203–209, 2000)

Key words: angiotensinogen, fibronectin, gene expression, transcriptional regulation, cardiomyocytes, vascular smooth muscle cells

Introduction

The renin-angiotensin system exerts a major influence on blood pressure as well as sodium and extracellular fluid balance through generation of angiotensin II (Ang II), which has a variety of actions such as vasoconstrictor activity and stimulation of the production and release of aldosterone. The renin-angiotensin system has been strongly implicated in the development of several cardiovascular diseases and accumulating evidence from biochemical and molecular studies of angiotensin suggests that distinct local renin-angiotensin system with different regulatory mechanisms from the classical plasma renin-angiotensin system may exist [1, 2]. A local renin-angiotensin system may exist and function in the brain, heart, adrenal gland, kidney, blood vessel wall, and adipose tissue. Whether all components of local renin-angiotensin system are physiologically relevant is controversial, and the exact role of these systems remains elusive, but it is interesting to speculate that a local renin-angiotensin system may augment the effects of Ang II on a particular tissue in specific physiological and pathophysiological processes. A number of studies indicate that the renin-angiotensin system plays a

Address for offprints: K. Tamura, Department of Internal Medicine II, Yokahama City University School of Medicine, 3–9, Fukuura, Kanazawa-ku, Yokohama 236-0004, Japan pivotal role in the development of cardiovascular remodeling and hypertrophy [1, 2].

Among the components of the renin-angiotensin system, angiotensinogen is a unique substrate of renin *in vivo*. Expression of the angiotensinogen gene is regulated in cell type-specific and differentiation-linked manners in cell culture systems [3–9]. Molecular variants of the human angiotensinogen gene are associated with essential hypertension [10], and recent studies have shown that several mutations of the proximal promoter region of the human angiotensinogen gene that affect the binding activity of transcription factors are associated with essential hypertension [11–13]. These results suggest that transcriptional regulation of the angiotensinogen gene is intimately involved in the pathogenesis of cardiovascular diseases.

Hemodynamic and endocrine factors are among the most important factors implicated in the physiology and pathophysiology of the cardiovascular tissues. Arterial hypertension evokes structural and functional changes of the vascular wall. Modifications of the extracellular matrix, including fibronectin (FN) and collagen, have been previously reported in vessel walls of hypertensive animals [14]. Activation and qualitative changes in the extracellular matrix participate in vascular wall remodeling and in the pathogenesis of atherosclerosis. Vascular remodeling in hypertension may be an adaptive response to increased transmural pressure [15, 16]. Mechanical stress seems to play a direct role in vascular remodeling, since mechanical stretch is able to increase protein synthesis by vascular smooth muscle cells (VSMCs) [17]. However, neuronal and humoral factors may be critical in hypertension-induced remodeling of vascular wall. Especially, several *in vivo* studies have reported that hypertension activates the vascular renin-angiotensin system including angiotensin-converting enzyme (ACE) [18], and infusion of pressor and subpressor doses of Ang II increases aortic FN mRNA in both hypertensive and normotensive animals [19, 20]. The aim of the present study was to investigate the regulation of AGT and FN gene expression in rat CMs and VSMCs subjected to mechanical stretch.

Materials and methods

Materials

AT1 receptor-specific antagonist CV11974 and AT2 receptor-specific antagonist PD123319 were supplied from Takeda Chemical and Parke Davis, respectively.

Cell culture

Primary cultures of CMs were prepared from the ventricles of 1–3-day-old Sprague-Dawley rats, and VSMCs were aseptically isolated from thoracic aortic explants of 5-week-old Sprague-Dawley rats (Oriental Kobo, Tokyo, Japan), as described previously [21, 22]. For cell-stretching experiments, we developed elastic culture dishes (15 × 30 × 10 mm, Ikemoto Chemical, Co, Tokyo, Japan) which were vulcanized with liquid silicone rubber consisting of methylvinyl polysiloxane and dimethylhydrogen silicone resin using platinum as a catalyst. The bottoms of these dishes are 0.5-mm-thick, and cultured cells are easily observed by microscopic examination. Cells were plated at a field density of 1×10^5 cells/cm^2 on plastic dishes or on laminine-coated (20 μg/ml) silicone dishes as described previously [21, 23, 24]. Uniaxial strain (20%) was applied by stretching the silicone dishes by 20%.

RNA isolation and Northern blot hybridization

Northern blot analysis was performed essentially as described previously [21, 22]. Total cellular RNA was isolated from VSMCs, using the single-step method by acid guanidinium thiocyanate-phenol-chloroform extraction. Twenty micrograms of each total RNA sample were denatured with 1 M glyoxal and 50% dimethyl sulfoxide, electrophoresed on a 1.2% agarose gel, and transferred to GeneScreen Plus (Nylon) membrane (Dupont). Filters were prehybridized for 1 h at 65°C in a solution consisting of 1% sodium dodecyl sulfate (SDS), 1 M NaCl, and 10% dextran sulfate. Hybridization proceeded for 24 h at 65°C in the same solution containing 200 μg/ml denatured salmon sperm DNA and 3×10^5 cpm/ml of the ^{32}P-labeled probes. Filters were washed twice with 2 × SSC (1 × SSC = 0.15 M NaCl, 0.015 M sodium citrate) for 5 min at room temperature, twice with 2 × SSC and 1% SDS for 30 min at 60°C, and twice with 0.1 × SSC for 30 min at room temperature. Dried filters were exposed to X-ray film or to the imaging plate of FUJIX BIO-Imaging Analyzer BAS2000 (Fuji Photo Film).

Plasmid construction

Plasmid pAGT-CAT was constructed by inserting a mouse AGT promoter fragment (–501 to +22 relative to the major transcription start site) into the *Bgl*II/*Hin*dIII sites upstream of the CAT coding sequences of pUCSV0CAT [4, 5], and plasmid pFN-CAT was constructed by inserting a 2044-bp *Pst*I/*Pst*I rat FN promoter fragment (–1908 to +136 relative to the major transcription start site) excised from pF1900CAT [25] into the *Bgl*II/*Hin*dIII sites upstream of the CAT coding sequences of pUCSV0CAT.

DNA transfection and CAT assay

The AGT or FN promoter-CAT chimeric construct (5 μg) and a β-galactosidase expression plasmid pCH110 (1 μg), which is used to normalize transfection efficiency, were transiently cotransfected into cells as described previously [21, 22]. Cells were incubated for 12 h, washed, and incubated for 12 h in fresh serum-free medium. Cells were then treated with Ang II or subjected to stretch for 12 h, and collected; cell extracts were prepared by freezing and thawing. The protein concentration was determined with bovine serum albumin as a standard. The reaction mixture containing 40 μg of the cell extracts, was incubated, and the labeled chloramphenicol and acetylated derivatives were separated by ascending thin layer chromatography as described previously [26]. The chromatograms were exposed to the imaging plate of a FUJIX BIO-Imaging Analyzer BAS2000. The conversion ratios of [^{14}C]chloramphenicol were measured with the computer analyzer of the BAS2000. DNA transfection was performed 4 times for each construct. To correct for transfection efficiency, β-galactosidase activity of cell extracts was measured. CAT activity was also normalized to the protein content of each extract.

Results

Effects of mechanical stretch on AGT and FN mRNA in CMs and VSMCs

We first examined whether mechanical stretch modulates AGT and FN gene expression in CMs and VSMCs. We found that stretching myocytes increased AGT mRNA accumulation but not FN mRNA level (Fig. 1A). On the other hand, expression of FN mRNA was increased in response to stretch in VSMCs, whereas AGT mRNA levels were not affected by stretch (Fig. 1B). These results indicate that mechanical stretch modulates the expression of AGT and FN genes differently in CMs and VSMCs.

Effects of Ang II receptor antagonists, transcriptional inhibitor, and protein synthesis inhibitor on stretch-mediated increases in AGT and FN mRNA

To examine a possible role of endogenous Ang II in stretch-mediated regulation of AGT and FN mRNA, we treated CMs and VSMCs with the Ang II receptor antagonists and analyzed the AGT and FN mRNA expression in response to stretch. Cells were preincubated for 30 min with CV11974, an AT1 receptor-specific antagonist, and then were subjected to mechanical stretch for 12 h. CV11974 significantly but not completely inhibited the stretch-mediated activation of AGT and FN mRNA (Fig. 2). In contrast, preincubation of cells with PD123319, an AT2 receptor-specific antagonist, did not affect the response to mechanical stretch at all. None of these Ang II receptor antagonists alone had any influence on the expression of AGT and FN mRNA. These results indicate that

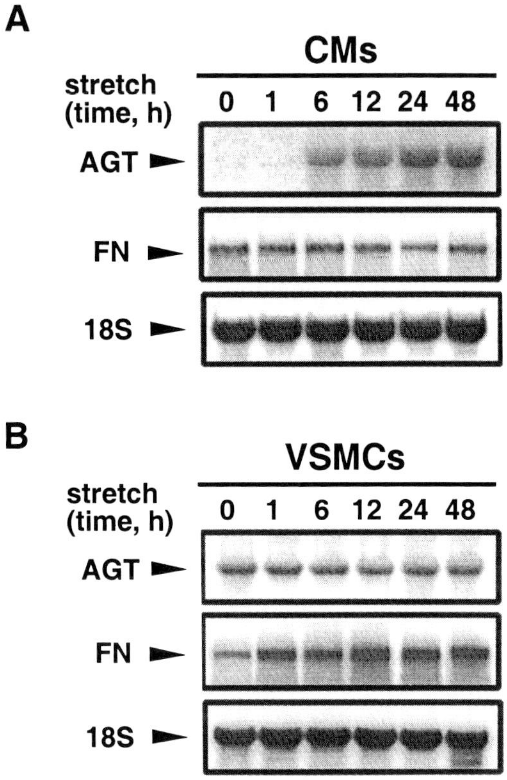

Fig. 1. Effects of stretching on angiotensinogen (AGT) and fibronectin (FN) mRNA expression in cardiac myocytes (CMs, A) and vascular smooth muscle cells (VSMCs, B). Cells were stimulated with mechanical stretch (20%) for indicated durations. Northern blot hybridization was performed as described in Materials and methods.

mechanical stretch activates AGT and FN mRNA expression, at least partly, through an AT1 receptor-dependent pathway in CMs and VSMCs.

To determine whether *de novo* RNA or protein synthesis was required for the stretch-induced increase in AGT and FN mRNA, cells were preincubated with actinomycin D or cycloheximide for 30 min and were subjected to stretch for 6 h. The RNA synthesis inhibitor actinomycin D almost abolished the stretch-mediated increase in AGT and FN mRNA, whereas the induction of AGT and FN mRNA by stretch was not significantly altered by the protein synthesis inhibitor cycloheximide (Fig. 3). These results indicate that *de novo* mRNA transcription is required for the induction of AGT and FN mRNA expression by stretch, but that the effects of stretch do not require *de novo* protein synthesis to stimulate the expression of AGT and FN mRNA.

Effects of mechanical stretch on transcription of AGT and FN genes

Since previous studies showed that expression of the AGT and FN gene is regulated at least partly at the level of tran-

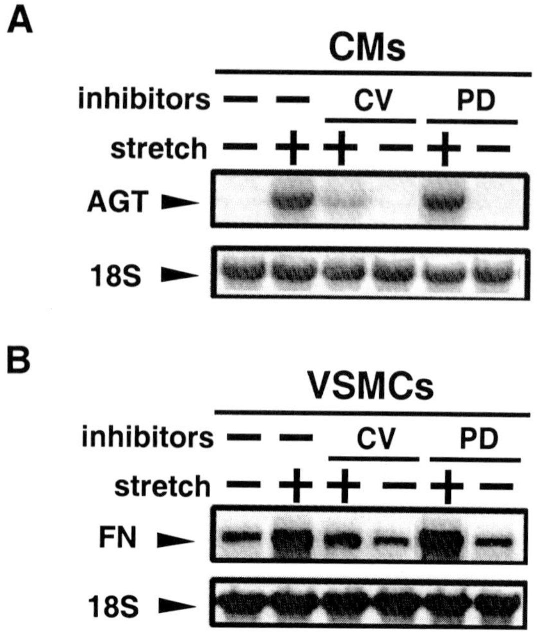

Fig. 2. Effects of Ang II receptor antagonists on stretch-induced angiotensinogen (AGT) and fibronectin (FN) mRNA expression in cardiac myocytes (CMs, A) and vascular smooth muscle cells (VSMCs, B). Cells were preincubated for 30 min with CV11974 (CV, 10 μM) or PD123319 (PD, 10 μM), followed by stretch (20%) for 24 h (CMs) or 12 h (VSMCs). Northern blot hybridization was performed as described in Materials and methods.

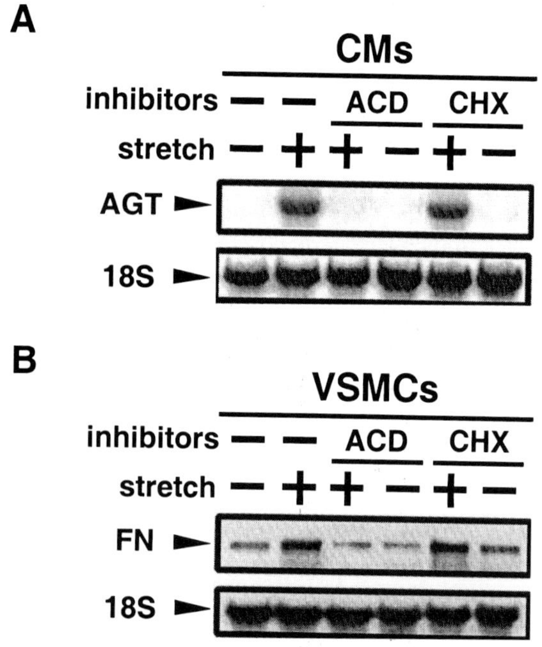

Fig. 3. Effects of protein synthesis inhibitor and transcriptional inhibitor on stretch-induced angiotensinogen (AGT) and fibronectin (FN) mRNA expression in cardiac myocytes (CMs, A) and vascular smooth muscle cells (VSMCs, B). Cells were preincubated for 30 min with actinomycin D (ACD, 1 μg/ml) or cycloheximide (CHX, 10 μg/ml), followed by stretch (20%) for 24 h (CMs) or 12 h (VSMCs). Northern blot hybridization was performed as described in Materials and methods.

scription [25, 27], we examined the effects of stretch on the transcriptional activity of the AGT and FN genes by transient transfection assay. As shown in Fig. 4, stretching CMs significantly increased CAT expression directed by the promoter region from –501 to +22 of the AGT gene (pAGT-CAT). Similarly, the FN promoter (–1908 to +136 relative to the major transcription start site) was able to confer the stretch-mediated increase in CAT expression in VSMCs.

Since the above results suggest that mechanical stretch stimulates the AGT and FN gene expression through activation of transcription, we finally examined which subtype of the Ang II receptor was involved in the activation of AGT and FN promoter in response to mechanical stretch. Cells were incubated for 30 min with Ang II receptor antagonists and were subjected to stretch for 6 h. Similar to the effect of CV11974 on the stretch-mediated induction of AGT and FN mRNA, both CV11974 significantly but not completely inhibited the stretch-mediated increase in AGT and FN promoter-CAT expression; however, PD123319 had no significant effect on the stretch-mediated activation of AGT and FN promoter.

Discussion

Mechanical stress exemplified by stretching cardiovascular cells stimulates the secretion of Ang II from these cells, and the secreted Ang II activates protein kinase and produces cardiovascular cell hypertrophy through an AT1 receptor-pathway. In the present study, we focused on the regulation of AGT and FN gene expression in response to stretching of cardiovascular cells, and found that mechanical stress specifically increased expression of the cardiac AGT and vascular FN genes.

Accumulating evidence suggests that humoral or neural factors, or both, and hemodynamic overload contribute to the pathogenesis of cardiovascular hypertrophy, and a number of studies suggest that the circulating and cardiovascular renin-angiotensin system plays an important role in the development of cardiovascular hypertrophy [1, 2]. All components of the renin-angiotensin system (e.g. renin, angiotensinogen, ACE, and Ang II receptor) have been identified in the heart and vascular walls at both the mRNA and protein levels [1, 2]. Our findings and others suggest that cardiac AGT and vascular FN are activated in response to hypertrophic stimulation *in vivo* and plays an important role in the development of cardiovascular hypertrophy.

By binding to the AT1 receptor, Ang II increases the activities of signal transduction pathways, the expression of immediate-early response genes, and the synthesis of proteins in cardiovascular cells [28, 29]. Mechanical stretch of cardiovascular cells also induces the secretion of Ang II and evokes these hypertrophic responses [30, 31]. After induc-

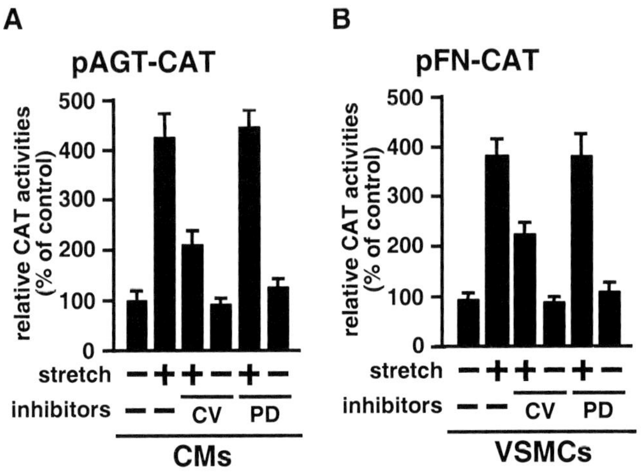

Fig. 4. Effects of Ang II receptor antagonists on stretch-induced stimulation of promoter activity. (A) Cardiac myocytes (CMs) were transfected with a plasmid pAGT-CAT (5 μg) and a β-galactosidase expression plasmid pCH110 (1 μg). Twenty-four h after the transfection, CMs were pretreated for 30 min with CV11974 (CV, 10 μM) or PD123319 (PD, 10 μM), and were subjected to stretch for an additional 12 h. (B) Vascular smooth muscle cells (VSMCs) were transfected with a plasmid pFN-CAT (5 μg) and a β-galactosidase expression plasmid pCH110 (1 μg). Forty-eight h after the transfection, VSMCs were pretreated for 30 min with CV11974 (CV, 10 μM) or PD123319 (PD, 10 μM), and were subjected to stretch for an additional 6 h. CAT assay was performed as described in Materials and methods. CAT activities were measured with an Imaging Analyzer BAS2000 and expressed relative to those achieved with cell extracts from control cells (the mean CAT activity of control cells is expressed as 100%). Bars represent means ± S.E. of 4 independent transfection experiments.

tion of immediate-early response genes, the expression of several genes changes in response to pressure overload. Since stretch-induced up-regulation of AGT mRNA was evident after 6 h of stimulation and since increased expression of AGT is maintained during the chronic phase of hypertrophy *in vivo* [32–34], the cardiac AGT gene may be classified as a stable late marker of cardiac hypertrophy similarly to the cardiac AT1 receptor [35, 36]. On the other hand, the FN mRNA levels started to increase 1 h after cyclic stretch and remained higher than control at 48 h, suggesting that the vascular FN induction is closely involved in the process of vascular hypertrophic response [37, 38].

In the present study, actinomycin D completely inhibited the stretch-mediated increases in AGT and FN mRNA. On the other hand, suppression by CV11974 of the stretch-mediated increases in AGT and FN mRNA expression and transcriptional activity were significant but not complete in cardiovascular cells. These results indicate that stretch activated AGT and FN gene transcription mainly via AT1 receptor-pathway. However, this incomplete inhibition of the stretch-mediated increase in mRNA expression and promoter activity of the AGT and FN genes suggests two possibilities. One possibility is that factors other than Ang II may also be involved in stretch-mediated transcription of the AGT and FN genes in cardiovascular cells, as we and others have reported with respect to the stretch-mediated activation of ERK and other protein kinases in CMs [23, 24, 39]. The other possibility is that Ang II is produced intracellularly and activates intracellular (e.g. nuclear) Ang II receptors [40]. A recent study have shown that the intracellular Ang II effect is totally inhibited by the concomitant injection of CV11974 but extracellular CV11974 does not influence the intracellular effects [41]. Therefore, it would be possible that mechanical stretch of cells produces Ang II intracellularly to the concentration enough to activate nuclear Ang II receptors which are insensitive to CV11974 in the medium.

The results in this study also suggest a possible existence of the cell type-specific regulatory mechanism for the AGT and FN genes in cardiovascular cells in response to mechanical stretch. We also recently found that the expression of AGT and FN genes is regulated differently in the cardiovascular tissues of hypertensive rats [42]. With respect to transcriptional regulation of AGT gene, we and others previously identified several novel proximal promoter regions which are important for cell type-specific activation of AGT gene in cultured cells and in hypertensive rats [8, 43]. On the other hands, our previous study showed that an AP-1-like element of the FN promoter plays a role in the Ang II-mediated activation of vascular FN gene [22]. Therefore, these different transcriptional mechanisms of the AGT and FN genes may explain the cell type-specific activation of these genes in cardiovascular cells. Experiments now in progress will define these promoter elements responsible for the mechanical stretch-induced upregulation of AGT and FN genes and will provide a molecular basis for the transcriptional machinery of these genes in cardiovascular cells in response to hypertrophic stimuli.

Acknowledgements

This work was supported by grants from the Ministry of Education, Science, and Culture of Japan, Uehara Memorial Foundation, Ichiro Kanehara Foundation, and Yokohama Foundation for Advancement of Medical Science. KT is supported by a Grant for Research Abroad of the Japan Society for the Promotion of Science.

References

1. Griendling KK, Murphy TJ, Alexander RW: Molecular biology of the renin-angiotensin system. Circulation 87: 1816–1828, 1993
2. Dzau VJ, Re R: Tissue angiotensin system in cardiovascular medicine: A paradigm shift? Circulation 89: 493–498, 1994

3. Tamura K, Umemura S, Fukamizu A, Ishii M, Murakami K: Recent advances in the study of renin and angiotensinogen genes: From molecules to the whole body. Hypertens Res 18: 7–18, 1995
4. Fukamizu A, Takahashi S, Seo MS, Tada M, Tanimoto K, Uehara S, Murakami K: Structure and expression of the human angiotensinogen gene: Identification of a unique and highly active promoter. J Biol Chem 265: 7576–7582, 1990
5. Tamura K, Tanimoto K, Takahashi S, Sagara M, Fukamizu A, Murakami K. Structure and expression of the mouse angiotensinogen gene. Jpn Heart J 33: 113–124, 1992
6. Tamura K, Tanimoto K, Ishii M, Murakami K, Fukamizu A: Proximal and core DNA elements are required for efficient angiotensinogen promoter activation during adipogenic differentiation. J Biol Chem 268: 15024–15032, 1993
7. McGehee RE Jr, Ron D, Brasier AR, Habener JF: Differentiation-specific element: A cis-acting developmental switch required for the sustained transcriptional expression of the angiotensinogen gene during hormonal-induced differentiation of 3T3-L1 fibroblasts to adipocytes. Mol Endocrinol 7: 551–560, 1993
8. Tamura K, Umemura S, Ishii M, Tanimoto K, Murakami K, Fukamizu A: Molecular mechanism of transcriptional activation of angiotensinogen gene by proximal promoter. J Clin Invest 93: 1370–1379, 1994
9. Tamura K, Umemura S, Iwamoto T, Yamaguchi S, Kobayashi S, Takeda K, Tokita Y, Takagi N, Murakami K, Fukamizu A, Ishii M: Molecular mechanism of adipogenic activation of the angiotensinogen gene. Hypertension 23: 364–368, 1994
10. Jeunemaitre X, Soubrier F, Kotelevtev YV, Lifton RP, Williams CS, Charru A, Hunt SC, Hopkins PN, Williams RR, Lalouel J-M, Corvol P: Molecular basis of human hypertension: Role of angiotensinogen. Cell 71: 169–180, 1992
11. Inoue I, Nakajima T, Williams CS, Quackenbush J, Puryear R, Powers M, Cheng T, Ludwig EH, Sharma AM, Hata A, Jeunemaitre X, Lalouel J-M: A nucleotide substitution in the promoter of human angiotensinogen is associated with essential hypertension and affects basal transcription. J Clin Invest 99: 1786–1797, 1997
12. Ishigami T, Umemura S, Tamura K, Hibi K, Nyui N, Kihara M, Yabana M, Watanabe Y, Sumida Y, Nagahara T, Ochiai H, Ishii M: Essential hypertension and 5β-upstream core promoter region of human angiotensinogen gene. Hypertension 30: 1325–1330, 1997
13. Ishigami T, Tamura K, Fujita T, Kobayashi I, Hibi K, Kihara M, Toya Y, Ochiai H, Umemura S: Angiotensinogen gene polymorphism near transcription start site and blood pressure: Role of a T-to-C transition at intron I. Hypertension 34: 430–434, 1999
14. Takasaki I, Chobanian AV, Sarzani R, Brecher P: Effect of hypertension on fibronectin expression in the rat aorta. J Biol Chem 35: 21935–21939, 1990
15. Dzau VJ, Gibbons GH: Vascular remodeling: mechanisms and implication. J Cardiovasc Pharmacol 21(suppl 1): S1–S5, 1993
16. Glagov S: Intimal hyperplasia, vascular modeling, and the restenosis problem. Circulation 89: 2888–2891, 1994
17. Grande JP, Glagov S, Bates SR, Horwitz AL: Effect of normolipemic and hyperlipemic serum on biosynthetic response to cyclic stretching of aortic smooth muscle cells. Arteriosclerosis 9: 446–452, 1989
18. Shiota N, Miyazaki M, Okunishi H: Increase of angiotensin converting enzyme gene expression in the hypertensive aorta. Hypertension 20: 168–174, 1992
19. Himeno H, Crawford DC, Hosoi M, Chovanian AV, Brecher P: Angiotensin II alters aortic fibronectin independently of hypertension. Hypertension 23: 823–826, 1994
20. Kim S, Ohta K, Hamaguchi A, Omura T, Tominaga K, Yukimura T, Miura K, Tanaka M, Iwao H: AT1 receptor-mediated stimulation by angiotensin II of rat aortic fibronectin gene expression *in vivo*. Br J Pharmacol 113: 662–663, 1994
21. Tamura K, Umemura S, Nyui N, Hibi K, Ishigami T, Kihara M, Toya Y, Ishii M: Activation of angiotensinogen gene in cardiac myocytes by angiotensin II and mechanical stretch. Am J Physiol 275: R1–R9, 1998
22. Tamura K, Nyui N, Tamura N, Fujita T, Kihara M, Toya Y, Takasaki I, Takagi N, Ishii M, Oda K, Horiuchi M, Umemura S: Mechanism of angiotensin II-mediated regulation of fibronectin gene in rat vascular smooth muscle cells. J Biol Chem 273: 26487–26496, 1998
23. Nyui N, Tamura K, Mizuno K, Ishigami T, Hibi K, Yabana M, Kihara M, Fukamizu A, Ochiai H, Umemura S, Murakami K, Ohno S, Ishii M: Stretch-induced MAP kinase activation in cardiomyocytes of angiotensinogen deficient mice. Biochem Biophys Res Commun 235: 36–41, 1997
24. Nyui N, Tamura K, Mizuno K, Ishigami T, Kihara M, Ochiai H, Kimura K, Umemura S, Ohno S, Taga T, Ishii M: gp130 is involved in stretch-induced MAP kinase activation in cardiac myocytes. Biochem Biophys Res Commun 245: 928–932, 1998
25. Nakajima T, Nakamura T, Tsunoda S, Nakada S, Oda K: E1A-responsive elements for repression of rat fibronectin gene transcription. Mol Cell Biol 12: 2837–2846, 1992
26. Tamura K, Tanimoto K, Murakami K, Fukamizu A: A combination of upstream and proximal elements is required for efficient expression of the mouse renin promoter in cultured cells. Nucl Acids Res 20: 3617–3623, 1992
27. Eggena P, Barrett JD: Regulation and functional consequences of angiotensinogen gene expression. J Hypertens 10: 1307–1311, 1992
28. Berk BC, Corson MA: Angiotensin II signal transduction in vascular smooth muscle: Role of tyrosine kinases. Circ Res 80: 607–616, 1997
29. Griendling KK, Ushio-Fukai M, Lassegue B, Alexander RW: Angiotensin II signaling in vascular smooth muscle. New concepts. Hypertension 29: 366–373, 1997
30. Li Q, Muragaki Y, Hatamura I, Ueno H, Ooshima A: Stretch-induced collagen sysnthesis in cultured smooth muscle cells from rabbit aortic media and a possible involvement of angiotensin II and transforming growth factor-β. J Vasc Res 35: 93–103, 1998
31. Sadoshima J, Xu Y, Slayter HS, Izumo S: Autocrine release of angiotensin II mediates stretch-induced hypertrophy of cardiac hypertrophy. Cell 75: 977–984, 1993
32. Baker KM, Chernin MI, Wixson SK, Aceto JF: Renin-angiotensin system involvement in pressure-overload cardiac hypertrophy in rats. Am J Physiol 259: H324–H332, 1990
33. Tamura K, Umemura S, Nyui N, Yamakawa T, Yamaguchi S, Ishigami T, Tanaka S, Tanimoto K, Takagi N, Sekihara H, Murakami K, Ishii M: Tissue specific regulation of angiotensinogen gene expression in spontaneously hypertensive rat. Hypertension 27: 1216–1223, 1996
34. Tamura K, Umemura S, Yamakawa T, Nyui N, Hibi K, Watanabe Y, Ishigami T, Yabana M, Tanaka S, Sekihara H, Murakami K, Ishii M: Modulation of tissue angiotensinogen gene expression in genetically obese hypertensive rats. Am J Physiol 272: R1704–R1711, 1997
35. Kijima K, Matsubara H, Murasawa S, Maruyama K, Mori Y, Ohkubo N, Komuro I, Yazaki Y, Iwasaka T, Inada M: Mechanical stretch induces enhanced expression of angiotensin II receptor subtypes in neonatal rat cardiac myocytes. Circ Res 79: 887–897, 1996
36. Suzuki J, Matsubara H, Urakami M, Inada M: Rat angiotensin II (type 1A) receptor mRNA regulation and subtype expression in myocardial growth and hypertrophy. Circ Res 73: 439–447, 1993
37. Kim S, Kawamura M, Wanibuchi H, Ohta K, Hamaguchi A, Omura T, Yukimura T, Miura K, Iwao H: Angiotensin II type 1 receptor blockade inhibits the expression of immediate-early genes and fibronectin in rat injured artery. Circulation 92: 88–95, 1995
38. Bardy N, Merval R, Benessiano J, Samuel J-L, Tedgui A: Pressure and angiotensin II synergistically induce aortic fibronectin expression in organ culture model of rabbit aorta: Evidence for a pressure-induced tissue renin-angiotensin system. Circ Res 79: 70–78, 1996

39. Yamazaki T, Komuro I, Kudoh S, Zou Y, Shiojima I, Mizuno T, Takano H, Hiroi Y, Ueki K, Tobe K, Kadowaki T, Nagai R, Yazaki Y: Angiotensin II partly mediates mechanical stress-induced cardiac hypertrophy. Circ Res 77: 258–265, 1995
40. Booz GW, Conrad KM, Hess AL, Singer HA, Baker KM: Angiotensin-II-binding sites on hepatocyte nuclei. Endocrinology 130: 3641–3649, 1992
41. Haller H, Lindschau C, Erdmann B, Quass P, Luft FC: Effects of intracellular angiotensin II in vascular smooth muscle cells. Circ Res 79: 765–772, 1996
42. Tamura K, Chiba E, Yokoyama N, Sumida Y, Yabana M, Tamura N, TakasaKi I, Takagi N, Ishii M, Horiuchi M, Umemura S: Renin-angiotensin system and fibronectin gene expression in Dahl Iwai salt-sensitive and salt-resistant rats. J Hypertens 17: 81–89, 1999
43. Morishita R, Higaki J, Tomita N, Aoki M, Moriguchi A, Tamura K, Murakami K, Kaneda Y, Ogihara T: Role of transcriptional cis-elements, angiotensinogen gene-activating elements, of angiotensinogen gene in blood pressure regulation. Hypertension 27: 502–507, 1996

Molecular and Cellular Biochemistry **212**: 211–217, 2000.

Regulation of angiotensin II receptors in the medullary thick ascending limb

Donna H. Wang and Jianping Li

Department of Medicine and Pharmacology and Toxicology, Michigan State University, East Lansing, MI, USA

Abstract

Angiotensin II (Ang II) is an important regulator of the function of medullary thick ascending limb of loop of Henle (MTAL). Recent studies showed that changes in Ang II receptor expression occur and underlie changes in the function of proximal tubules during altered sodium intake. The present experiment was designed to determine (1) whether expression of the type 1 Ang II (AT1) receptor in the MTAL is regulated by altered sodium intake, and (2) the specific pathway(s) mediating sodium-induced AT1 expression in the MTAL. Wistar rats were fed a normal sodium (0.5%, NS), low sodium (0.07%, LS), or high sodium (4%, HS) diet for 2 weeks. Northern blot analysis and radioligand binding showed that in rats fed a normal sodium diet the rank of order for both AT1 mRNA expression and receptor density was outer medulla > cortex > inner medulla. Sodium restriction significantly increased both AT1 mRNA expression and receptor density in the outer medulla. In contrast, neither AT1 mRNA expression nor receptor density in the outer medulla was altered by sodium loading. Losartan treatment (3 mg/kg/ per day by oral gavage for 2 weeks) prevented low sodium-induced upregulation of the AT1 receptor in the outer medulla, but it had no effect on AT1 expression in the outer medulla of rats fed a normal sodium diet. Highly purified suspensions of MTAL were isolated from rats fed a normal or low sodium diet. Low sodium intake significantly increased AT1 mRNA level by 184% and AT1 receptor density by 58% in MTALs. Primary cultures of MTAL cells were treated with PBS, Ang II (10^{-8} M), and Ang II + 17 octadecynoic (17 ODYA, 10 μM). Ang II caused about 2-fold increase in AT1 mRNA levels, and this increase was diminished by about 30% by the addition of 17 ODYA. We conclude that (1) sodium restriction but not sodium loading increases AT1 receptor expression in the MTAL, (2) low sodium-induced upregulation of the AT1 receptor in the MTAL is Ang II-dependent, and (3) Ang II-induced upregulation of the AT1 receptor in the MTAL is mediated, at least in part, by cytochrome P450 pathways. (Mol Cell Biochem **212**: 211–217, 2000)

Key words: angiotensin receptor, medullary thick ascending limb, sodium intake, primary cell culture, gene expression

Introduction

Angiotensin II (Ang II) has a number of actions on the kidney to regulate renal hemodynamics, electrolyte excretion, and cell growth [1, 2]. The intrarenal effects of Ang II are mediated to a substantial extent by the type 1 Ang II (AT1) receptor [3, 4]. Regulation of this receptor expression may therefore be a major factor in both normal and pathological renal control. It has been shown that sodium restriction increases Ang II receptor density and AT1 receptor mRNA levels in proximal tubules [5, 6]. Furthermore, Ang II-induced changes in ion transport correlate with changes in Ang II receptor density in proximal tubules [7], and this effect can be blocked by the AT1 receptor antagonist losartan [8]. These studies establish the importance of regulation of the AT1 receptor in the control of the function of proximal tubules. In contrast, the regulation and function of the AT1 receptor in the MTAL are largely unknown.

The MTAL plays a critical role in salt and water homeostasis by reabsorbing a sizeable fraction (20–25%) of filtered NaCl [9, 10]. Recently it has been shown that low concentrations (10^{-12} M–10^{-10} M) of Ang II inhibit whereas high concentrations (10^{-8} M–10^{-6} M) of Ang II stimulate apical membrane K^+ channels and Na^+-K^+-$2Cl^-$ cotransport activ-

Address for offprints: D.H. Wang, Department of Medicine, B316 Clinical Center, Michigan State University, East Lansing, MI 48824-1313, USA ity in the MTAL [11, 12], two transporters that are critical for transcellular NaCl reabsorption in this segment [10]. In addition, the MTAL reabsorbs most of the filtered $NaHCO_3$ that escapes the proximal tubule, a process important in the control of urinary net acid excretion [9]. Good *et al.* reported that Ang II (10^{-8} M in the bath) inhibits HCO_3^- absorption in the isolated, perfused MTAL of the rat [13]. The inhibition of HCO_3^- absorption by Ang II in the MTAL can be blocked by the AT1 receptor antagonist losartan [14]. These studies indicate that the AT1 receptor is critically involved in regulating sodium and acid-base balances in the MTAL under pathophysiological conditions. However, the regulation of the MTAL AT1 receptor, which may affect Ang II-mediated MTAL function, has not been studied. This study was designed to examine whether sodium restriction or loading alters AT1 receptor expression in the MTAL, and if so, what is the specific pathway(s) mediating sodium-induced AT1 expression in this segment.

Materials and methods

Animal groups

Seven-week-old male Wistar rats weighing between 150–200 g (Charles River Laboratories Inc., Wilmington, MA, USA) were randomly divided into five groups and pair-fed a normal sodium diet (0.5%, NS), low sodium diet (0.07%, LS), high sodium diet (4%, HS), normal sodium diet plus losartan (NS+LOS), and low sodium diet plus losartan (LS+LOS) for 2 weeks. Losartan (3 mg/kg per day) was dissolved in 0.5–1 ml water and given by oral gavage. This dose of losartan was chosen because it has previously been demonstrated to provide AT1 antagonism without lowering blood pressure [15]. An equal amount of water was given by oral gavage to non losartan-treated rats.

Suspension of MTAL

A modification of the method of Trinh-Trang-Tan *et al.* was used [16]. Briefly, the kidney was cut into thin slices along the cortico-medullary axis and placed in ice-cold Hanks solution (137 mM NaCl, 0.3 mM Na_2HPO_4, 4.2 mM $NaHCO_3$, 5.4 mM KCl, 0.44 mM KH_2PO_4, 0.4 mM $MgSO_4$, 0.5 mM $MgCl_2$, 1.2 mM $CaCl_2$, 5 mM glucose) gassed with 95% O_2-5% CO_2. All procedures were performed at 0–4°C. The inner stripe of the outer medulla was excised from each slice and the resulting pieces of inner stripe tissue were forced through stretched medical gauze to obtain small cubes of tissue. This tissue was enzymatically digested for 3 × 15 min in Hanks solution containing collagenase (0.025%, wt/vol), and gassed with the O_2-CO_2 mixture. The supernatant containing isolated cells and dissociated tubules was collected. The supernatant was poured over a 100 µm nylon sieve to collect the long tubule segments. These segments were washed with 20 ml gassed Hanks solution containing 1% bovine serum albumin (BSA) and 15 ml oxygenated Hanks-BSA solution. Tubule suspensions were centrifugated 5 min at low speed (80 × g). The supernatant was discarded and tubules were resuspended in 1.5–3.0 ml of Hanks-BSA solution. The purity of the preparation (> 98%) was confirmed visually under the light microscope and by immunocytochemical staining with anti-Tamm-Horsfall antibody [17].

Plasma renin activity

At the end of the 2-week diet and losartan treatment, the rats were decapitated and samples of plasma (3 ml) were collected and placed in chilled EDTA tubes. Plasma renin activity was determined by measuring production of Ang I using a commercially available radioimmunoassay kit (Incstar Co).

Northern blot analysis

AT1 cDNA probes (790-bp, from –180 to +610) were obtained by digesting a clone pUC19 containing 2.3 kb of the rat AT1A receptor with *Kpn* I and *Eco*RI [15]. The AT1 cDNA probes were labeled with [^{32}P]dCTP by use of a Multiprime DNA labeling system (Amersham, Arlington, IL, USA) to a specific activity of 3×10^8 cpm/µg. The labeled probes were separated from unincorporated nucleotides by use of a Mini-Spin G-50 DNA purification spin column (Amersham). Total RNA from the outer medulla or MTAL suspensions was extracted using the guanidine thiocyanate-phenol-chloroform extraction protocol [18]. Electrophoresis of 30 µg denatured RNA was carried out in a 1% agarose gel containing 2.2 mol/l formamide. RNA was transferred to a positively charged nylon membrane (Fisher, Houston, TX, USA), and the blot was prehybridized for 5 h at 42°C in hybridization buffer (50% deionized formamide, 5 × Denhardt's solution, 5 × saline sodium citrate (SSC), 0.5% sodium dodecyl sulfate, and 200 µg/ml denatured salmon sperm DNA) and then hybridized with the ^{32}P-labeled probes for 18–20 h at 42°C. At the end of this period, the blot was washed successively in 2 ×, 1 ×, and 0.5 × SSC containing 0.1% sodium dodecyl sulfate at 60°C. After exposure to X-ray film, autoradiographic signals were scanned with a XL laser densitometer (Ultrascan, Bromma, Sweden). To control for differences in RNA loading, blots were incubated at 90°C for 10 min in 20 mM tris (hydroxymethyl) aminomethane HCL (pH 8.0) to strip off cDNA probes and later rehybridized with ^{32}P-labeled 18S rRNA probe. The results are expressed as the ratio of AT1 mRNA to 18S rRNA.

Ligand binding assay

Tissue was centrifuged and the pellet was suspended in ice-cold assay buffer containing 100 mM NaCl, 5 mM EDTA, 10 mM Hepes (pH 7.5), 100 mM mannitol, 0.5% bovine serum albumin, 0.1 mM $MgSO_4$, 0.5% trypsin inhibitor, 0.005% aprotinin, 0.1 mM phenylmethylsulfonylfluoride for 10 min [19, 20]. An aliquot of the suspension was used for protein assay using modified Bradford dye-binding procedure [21]. To measure the AT1 receptor density, 30 µg protein was incubated with 125 pmol/L to 2 nmol/L ^{125}I-(Sar, Ile) Ang II (kindly provided by Dr. Robert C. Speth, Washington State University, USA) in a final volume of 200 µl assay buffer containing 0.1% bovine serum albumin in the presence or absence of PD123319 (10 µmol/L) [19]. Nonspecific binding was measured in the presence of 1 µmol/L unlabeled Ang II. Binding assay was performed for 120 min at room temperature and followed by immediate filtration through glass-fiber filters (Whatman GF/C, Hillsboro, OR, USA). The filter-bound radioactivity was counted in a gamma spectrometer (Beckman, LS 3801, Irvine, CA, USA). Receptor affinity and concentration were calculated by Scatchard analysis using the computerized curve-fitting program LIGAND [22].

Primary culture of MTAL cells

Kidneys were removed and transferred to a petri dish with Dulbecco's modified Eagles's medium (DMEM, 450 mg/dl glucose) containing 100 U/ml penicillin, 100 µg/ml streptomycin, and 2 mM supplemental glutamine [23]. The papilla and cortex were cut away and MTAL tubules were isolated as described above [16]. The tubule suspension was washed in PBS, and further purified by density-gradient centrifugation in a Percoll medium as previously described [24]. MTALs were harvested from the heaviest band (fraction 2), washed twice in PBS, and then incubated for 30 min in 0.5 mg/ml purified collagenase (Gibco) and 15 mg/ml deoxyribonuclease I at 37°C. After digestion, cells were washed in DMEM with 10% fetal calf serum (Gibco), and 10^4 cells were plated into each well of a 96-well plastic cell culture plate and incubated with 95% air/5% CO_2 atmosphere at 37°C. After reaching 70% confluence, the medium was replaced by serum-free DMEM, and the cells were cultured for 48 h to become quiescent [25]. The cells were stimulated with Ang II (10^{-8} M) or Ang II plus 17 octadecynoic (17 ODYA, 10 µM). The purity (> 98%) of the cell cultures was confirmed by immunocytochemistry using Tamm-Horsfall protein.

Statistical analysis

Results were expressed as means ± S.E. The data were analyzed by one-way ANOVA followed by the Tukey-Kramer multiple comparison test. Differences were considered statistically significant at the $p < 0.05$ level.

Results

Body weight and plasma renin activity

There was no significant difference in body weight among the five groups at the end of the experiment (NS, 285 ± 6 g; LS, 272 ± 6 g; HS, 282 ± 3 g; NS+LOS, 261 ± 10 g; LS+LOS, 275 ± 7 g). The ability of low or high sodium intake to modulate the renin-angiotensin system was confirmed by changes in plasma renin activity (PRA). PRA was increased ($p < 0.05$) in rats fed a low sodium diet and decreased ($p < 0.05$) in rats fed a high sodium diet compared with rats fed a normal sodium diet (NS, 32.7 ± 2.2 ng/ml/h; LS, 47.9 ± 3.8 ng/ml/h; HS, 2.1 ± 0.5 ng/ml/h). Losartan treatment also increased ($p < 0.05$) PRA in rats fed either a normal or low sodium diet (NS+LOS, 64.0 ± 6.2 ng/ml/h; LS+LOS, 69.8 ± 4.4 ng/ml/h).

AT1 expression in different renal regions of rats fed a normal sodium diet

AT1 mRNA content in the renal cortex, outer and inner medulla of the control rats was determined by Northern blot analysis (Fig. 1A). Densitometric analysis indicated that the ratio of AT1 mRNA-to-18S rRNA was significantly higher in the outer medulla, but lower in the inner medulla compared to the cortex. Consistent with the changes in AT1 mRNA levels, the AT1 receptor density was significantly higher in the outer medulla but lower in the inner medulla compared to the cortex (Fig. 1B). The calculated dissociation constants (Kd)

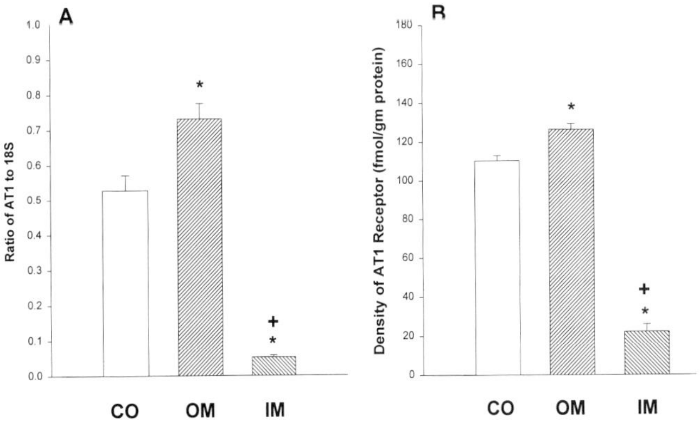

Fig. 1. (A) Northern blot analysis of AT1 mRNA content in the renal cortex (CO), outer medulla (OM), and inner medulla (IM) of rats fed a normal sodium diet. (B) Radioligand binding assay analysis of AT1 receptor density in CO, OM, and IM of rats fed a normal sodium diet. n = 5 rats per group. *$p < 0.05$ compared with CO; $^+p < 0.05$ compared with OM.

Table 1. B_{max} and Kd in 3 different renal regions of rats fed a normal sodium diet

	B_{max} (fmol/mg protein)	Kd (nmol/L)
Cortex	177 ± 4	0.93 ± 0.04
Outer medulla	209 ± 7^{+}	0.90 ± 0.04
Inner medulla	37 ± 4^{+}*	0.96 ± 0.02

N = 5 per group. $^{+}p < 0.05$ vs. cortex; $*p < 0.05$ vs. outer medulla.

Table 2. B_{max} and Kd in the outer medulla in response to altered sodium intake

	B_{max} (fmol/mg protein)	Kd (nmol/L)
NS	209 ± 7	0.90 ± 0.04
LS	246 ± 8^{+}	1.02 ± 0.09
HS	172 ± 24	0.76 ± 0.09

N = 5 per group. $^{+}p < 0.05$ vs. NS and HS. NS – normal sodium intake; LS – low sodium intake; HS – high sodium intake.

and maximal binding (B_{max}) of Ang II to its receptors in the cell membrane from the renal cortex, outer medulla, and inter medulla are summarized in Table 1. There was no significant difference in the binding constants among kidney preparations from the three regions (Table 1). However, total receptor density was significantly higher in the outer medulla but lower in the inner medulla compared to the cortex (Table 1).

AT1 expression in the outer medulla in response to altered sodium intake

Northern blot analysis showed that AT1 mRNA level in the outer medulla was significantly elevated by sodium restriction compared with control (Fig. 2A). However, AT1 mRNA content in this region was not altered by sodium loading (Fig. 2A). Likewise, radioligand binding showed that sodium restriction significantly increased the AT1 receptor density (Fig. 2B) as well as B_{max} (Table 2). Neither AT1 receptor density (Fig. 2B) nor B_{max} (Table 2) in the outer medulla was altered by sodium loading. Also, there was no significant difference in the Kd among medulla preparations in response to altered sodium intake (Table 2).

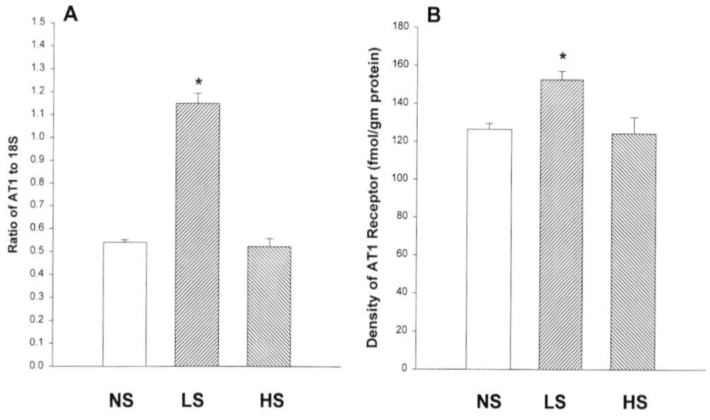

Fig. 2. (A) Northern blot analysis of AT1 mRNA content in the renal outer medulla of rats fed a normal (NS), low (LS), or high (HS) sodium diet. (B) Radioligand binding assay analysis of AT1 receptor density in the renal outer medulla of rats fed NS, LS, or HS sodium diet. n = 5 rats per group. $*p < 0.05$ compared with NS and HS.

AT1 expression in the outer medulla in response to losartan treatment

To determine whether activation of the renin angiotensin system by sodium restriction plays a role in low sodium-induced upregulation of AT1 expression in the outer medulla, losartan was given to rats fed either a normal or low sodium diet to block Ang II binding to AT1 receptors. Losartan treatment prevented low sodium-induced upregulation of the AT1 mRNA expression in the outer medulla, but had no effect on medullary AT1 expression in rats fed a normal sodium diet (Fig. 3).

AT1 expression in the MTAL in response to sodium restriction

Highly purified suspensions of MTAL were isolated from rats fed a normal or low sodium diet. Low sodium intake significantly increased both AT1 mRNA level and AT1 receptor density in the MTAL (Fig. 4), a result consistent with that found in the outer medulla. There was no significant dif-

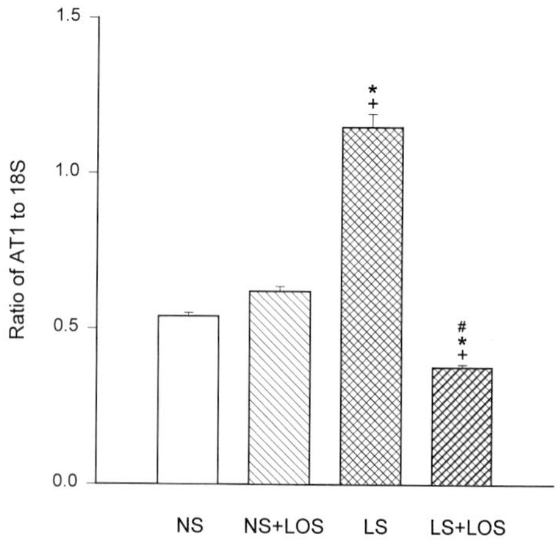

Fig. 3. Northern blot analysis of AT1 mRNA content in the renal outer medulla of rats fed a normal sodium diet (NS), NS plus losartan (NS+LOS), a low sodium diet (LS), and LS plus losartan (LS+LOS). n = 5 rats per group. $^{+}p < 0.05$ compared with NS; $*p < 0.05$ compared with NS+LOS; $^{\#}p < 0.05$ compared with LS.

ference in the binding constants of AT1 receptors between MTAL preparations from rats fed a low sodium (0.75 ± 0.12 nmol/L) and normal sodium (0.45 ± 0.01 nmol/L) diet.

AT1 expression in primary cultures of MTAL cells in response to Ang II and 17 ODYA

To determine whether Ang II has a direct effect on AT1 mRNA expression and the possible pathway mediating Ang II-induced AT1 expression in the MTAL, primary cultures of MTAL cells were treated with PBS (control), Ang II (10^{-8} M), and Ang II+17 ODYA (10 μM). Northern blot analysis of AT1 mRNA levels from 2 sets of experiments indicated that Ang II caused about 2 fold increases in AT1 mRNA levels, and Ang II-induced increase in AT1 mRNA levels is diminished by about 30% by the addition of 17 ODYA (Fig. 5).

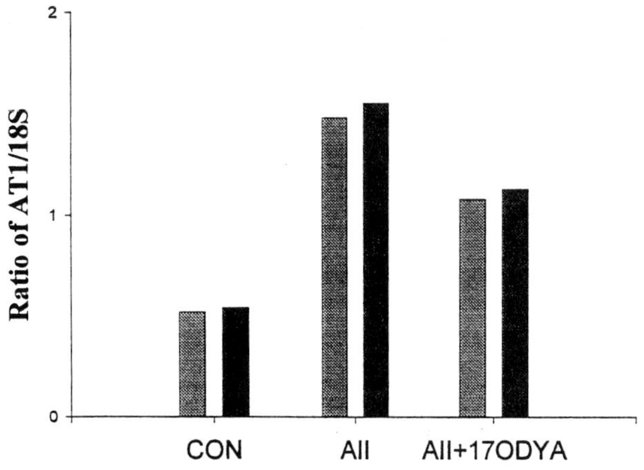

Fig. 5. Northern blot analysis of AT1 mRNA levels from two sets of experiments in which primary cultures of MTAL cells were treated with PBS (CON), Ang II (AII, 10^{-8} M), and AII + 17 octadecynoic (17 ODYA, 10 μM). Ang II causes about 2 fold increase in AT1 mRNA levels, and Ang II-induced increase in AT1 mRNA levels is diminished by about 30% by the addition of 17ODYA.

Discussion

Despite the fact that the MTAL contains functional AT1 receptors that may play a critical role in salt and water homeostasis, little information exists regarding how this receptor is regulated and what pathway(s) is involved in governing the expression of this receptor in the MTAL. This study was designed to examine directly the regulation of the AT1 receptor by altered sodium intake and the specific pathway(s) mediating sodium-induced AT1 expression in the MTAL. Our data demonstrate that (1) sodium restriction but not sodium loading increases AT1 receptor expression in the MTAL, (2) low sodium-induced upregulation of the AT1 receptor in the MTAL is Ang II-dependent, and (3) Ang II induced upregulation of the AT1 receptor in the MTAL is mediated, at least in part, by cytochrome P450 pathways. These findings may have several important pathophysiological implications.

Several lines of evidence suggest that altered sodium intake modulates Ang II receptor expression in the glomerulus and proximal tubule. With sodium restriction, glomerular Ang II receptor density and AT1 mRNA levels are decreased [6, 26–28]. In contrast, proximal tubule Ang II receptor density and AT1 mRNA levels are increased by sodium restriction [5, 6]. Since the outer medulla contains the highest AT1 receptor density compared with the cortex and inner medulla (Fig. 1), changes in AT1 receptor expression in the outer medulla may have critical impact on the renal function. Our data show that AT1 receptor mRNA levels and receptor density in the MTAL, a major component of the outer medulla, are increased in response to a low dietary sodium intake. The increase in AT1 receptor expression in the MTAL might underlie changes in the function of this segment during sodium restriction. This possibility is supported by the observations (1) that increased rates of hydrogen ion secretion induced by Ang II in early versus late proximal tubule segments correlate with higher Ang II receptor density in the early proximal tubule cells [7]; (2) that the magnitude of the transport inhibition induced by losartan correlates with Ang II receptor density in the early and late subsegments of the proximal convoluted tubule [8], i.e. losartan is a more effective transport inhibitor in the early subsegment, which has higher Ang II receptor density [8]; and (3) that increased bicarbonate reabsorption in the distal tubule of rats fed a low sodium diet is markedly attenuated by blockade of the AT1 receptor with losartan [29]. These studies support the concept that increased AT1 receptor expression in the MTAL contributes to low sodium-induced changes in ion transport in this segment.

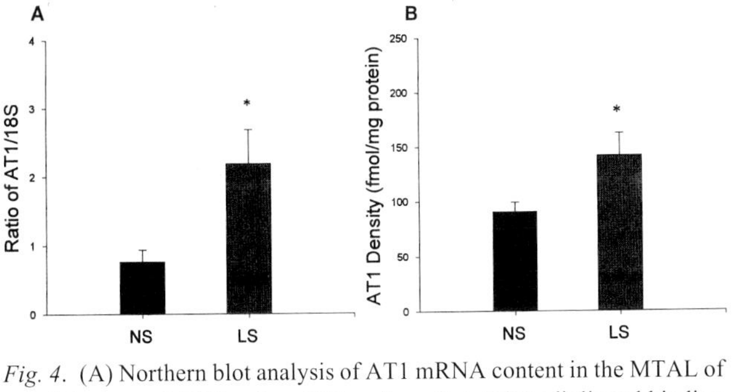

Fig. 4. (A) Northern blot analysis of AT1 mRNA content in the MTAL of rats fed a normal (NS) or low (LS) sodium diet. (B) Radioligand binding assay analysis of AT1 receptor density in the MTAL of rats fed a normal (NS) or low (LS) sodium diet. n = 5 rats per group. *$p < 0.05$ compared with NS.

In sodium depleted or replete states, altered neuro-endocrine responses, e.g. the renin angiotensin system, may affect AT1 receptor expression in the kidney. It has been shown that conditions that increase the activity of the renin-angiotensin system *in vivo* down-regulate glomerular Ang II receptors [28]. Conversely, conditions that decrease the activity of the renin-angiotensin system *in vivo* upregulate glomerular Ang II receptors [28]. These results have been interpreted as being a consequence of homologous regulation of Ang II receptors after prolonged agonist exposure or withdrawal. In contrast, we have now shown that Ang II receptors in the MTAL are regulated in the opposite direction compared with that in glomeruli, i.e. increased activity of the renin-angiotensin system upregulates AT1 receptor expression in the MTAL. In addition, the upregulation of the AT1 receptor during activation of the renin-angiotensin system by sodium restriction can be blocked by losartan, indicating that a positive feedback regulation of the AT1 receptor is present in the MTAL.

Although the renin-angiotensin system may play a role in AT1 receptor expression in the MTAL *in vivo*, simultaneous changes in other neuro-hormonal systems during sodium restriction may confound the result. The present study shows that Ang II has a direct effect on AT1 receptor expression in primary culture of MTAL cells, consistent with the hypothesis that Ang II mediates low sodium-induced upregulation of the AT1 receptor in this segment. The mechanisms responsible for this positive feedback regulation are unknown. However, it has been shown that in MTALs Ang II inhibits Cl^- and HCO_3^- transport through formation of P450 omega-hydroxylase metabolite, 20 hydroxyeicosatetraenoic acid (20HETE) [12, 13]. It is possible that cytochrome P450-dependent metabolites also play a role in Ang II-induced AT1 receptor regulation in the MTAL. Indeed, our data show that a selective inhibitor of cytochrome P450 enzymes, 17ODYA, partially blocks Ang II-induced increase in AT1 mRNA levels in cultured MTALs. These results demonstrate for the first time that Ang II-induced upregulation of the AT1 receptor in MTALs is mediated, at least in part, by cytochrome P450 pathways.

In conclusion, we have shown that in MTALs sodium restriction increases AT1 receptor expression via an Ang II-dependent mechanism. Since the AT1 receptor in MTALs is critically involved in regulating sodium and acid-base balances, increased AT1 receptor expression may serve as a compensatory mechanism to modulate MTAL function during sodium deprivation. In addition, the identification of the novel pathway responsible for Ang II-induced upregulation of the AT1 receptor in MTALs may provide basis for future therapeutic manipulation of AT1 receptor in the MTAL.

Acknowledgements

This study was supported in part by National Institutes of Health (grant HL-52279 and HL-57853). Dr. Wang is an established investigator of the American Heart Association. We are grateful to Dr. David Good for his input in the MTAL methods.

References

1. Guyton AC: Dominant role of the kidneys and accessory role of whole-body autoregulation in the pathogenesis of hypertension. Am J Hypertens 2: 575, 1989
2. Hall JE, Brands MW: The renin-angiotensin aldosterone systems. In: D.W. Seldin, F. Fiebisch (eds). The Kidney: Physiology and Pathophysiology, 2nd edn. Raven Press, New York, 1992
3. Chiu AT, Duncia JV, McCall DE, Wong PC, Price WA, Thoolen MJMC, Carini DJ, Johnson Al, Timmermans PBMWM: Nonpeptide angiotensin II receptor antagonists. III. Structure-function studies. J Pharmacol Exp Ther 250: 867–874, 1989
4. Wong PC, Price WA, Chiu AT, Duncia JV, Carini DJ, Wexler RR, Johnson AL, Timmermans PBMWM: Nonpeptide angiotensin II receptor antagonists. VIII. Characterization of functional antagonism displayed by DuP753, an oral active antihypertensive agent. J Pharmacol Exp Ther 252: 719–725, 1990
5. Douglas JG: Angiotensin receptor subtypes of the kidney cortex. Am J Physiol 253: F1–F7, 1987
6. Cheng HF, Becker BN, Burns KD, Harris RC: Angiotensin II upregulates type 1 angiotensin II receptors in renal proximal tubule. J Clin Invest 95: 2012–2019, 1995
7. Liu FY, Cogan G: Angiotensin II stimulation of hydrogen ion secretion in the rat early proximal tubule. Modes of action, mechanism, and kinetics. J Clin Invest 82: 601–607, 1988
8. Xie MH, Liu FY, Wong PC, Timmermans PB, Cogan MG: Proximal nephron and renal effects of DuP 753, a nonpeptide angiotensin II receptor antagonist. Kidney Intl 38:473–479, 1990
9. Good DW: The thick ascending limb as a site of renal bicarbonate reabsorption. Sem Nephrol 13: 225–235, 1993
10. Winters CJ, Reeves WB, Andreoli TE: A survey of transport properties of the thick ascending limb. Sem Nephrol 11: 236–247, 1991
11. Amlal H, Vernimmen C, LeGoff C, Paillard M, Bichara M: Control of Na^+-K^+(NH_4^+)-$2Cl^-$ cotransport by angiotensin II (AII)-activated transduction pathways in medullary thick ascending limb (MTAL) cells. J Am Soc Nephrol 4: 862, 1993
12. Lu M, Zhu Y, Balazy M, Reddy KM, Falck JR, Wang WH: Effect of angiotensin II on the apical K^+ channel in the thick ascending limb of the rat kidney. J Gen Physiol 108: 537–547, 1996
13. Good DW, George T, Wang DH: Angiotensin II inhibits HCO_3^- absorption via a cytochrome P-450-dependent pathway in MTAL. Am J Physiol 276: F726–F736, 1999
14. Wang DH, Li JP, Hu ZY, Qiu JX, Good DW: Function and regulation of angiotensin (Ang II) receptors in rat medullary thick ascending limb (MTAL). Hypertension 30: 505, 1997
15. Wang DH, Du Y: Distinct Mechanisms of upregulation of type 1 angiotensin II receptor gene expression in kidney and adrenal gland. Hypertension 26: 1134–1137, 1995
16. Trinh-Trang-Tan MM, Bouby N, Coutaud C, Bankir L: Quick isolation of rat medullary thick ascending limbs. Pflügers Arch 407: 228–234, 1986
17. Wang DH, Prewitt RL, Beebe SJ: Regulation of PDGF A: A possible mechanism for angiotensin II induced vascular growth. Am J Physiol 269: H356–H364, 1995
18. Chomczynski P, Sacchi N: Single-step method of RNA isolation by acid guanidinium thiocyanate-phenol-chloroform extraction. Anal Biochem 162: 156–159, 1987

19. Du Y, Guo DF, Inagami T, Speth RC, Wang DH: Regulation of angiotensin II receptor subtype and its gene expression in adrenal gland: Role of type 1 angiotensin II receptor. Am J Physiol 271: H440–H446, 1996
20. Wang DH, Du Y, Zhao HW, Granger JP, Speth R, DiPette DJ: Regulation of angiotensin receptor and its gene expression: Role in renal growth. J Am Soc Nephrol 8: 1–6, 1997
21. Bradford MM: A rapid and sensitive method for the quantitation of microgram quantities of protein utilizing the principal of protein-dye binding. Ann Biochem 72: 248–254, 1976
22. Munson PJ, Rodbard D: LIGAND: A versatile computerized approach to characterization of ligand binding system. Ann Biochem 107: 220–239, 1980
23. Wolf G, Ziyadeh FN, Helmchen U, Zahner G, Schroeder R, Stahl RAK: Ang II is a mitogen for a murine cell line isolated from medullary thick ascending limb of Henle's loop. Am J Physiol 268: F940–F947, 1995
24. Ziyadeh FN, Sharma K, Ericksen M, Wolf G: Stimulation of collagen gene expression and protein synthesis in murine mesangial cells by high glucose is mediated by autocrine activation of transforming growth factor-β. J Clin Invest 93: 536–542, 1994
25. Desgranges C, Campan M, Gadeau AP, Guerineau N, Mollard P, Razada G: Influence of 8-(N,N-diethyl-amino) octyl-3,4,5-trimethoxybenzoate (TMB-8) on cell cycle progression and proliferation of cultured arterial smooth muscle cells. Biochem Phamacol 41: 1045–1054, 1991
26. Gunter S, Gimbrone MA, Alexander RW: Regulation by angiotensin II of its receptors in resistance blood vessels. Nature 287: 230–232, 1980
27. Skorecki K, Ballermann BJ, Rennke HG, Brenner BM: Angiotensin II receptor regulation in isolated renal glomeruli. Fed Proc 42: 3064–3070, 1983
28. Bellucci A, Bilkes BM: Mechanism of sodium modulation of glomerular angiotensin receptors in the rat. J Clin Invest 74: 1593–1600, 1984
29. Levine DZ, Lacovitti M, Buckman S, Burns KD: Role of angiotensin II in dietary modulation of rat late distal tubule bicarbonate flux *in vivo*. J Clin Invest 97: 120–125, 1996

Molecular and Cellular Biochemistry **212**: 219–225, 2000.

Bradykinin (B_2) independent effect of captopril on the development of pressure overload cardiac hypertrophy

Marian Turcani[1] and Heinz Rupp[2]

[1]*Institute of Pathophysiology, Medical School, Comenius University, Bratislava, Slovak Republic;* [2]*Molecular Cardiology Laboratory, Department of Internal Medicine and Cardiology, Philipps University of Marburg, Marburg, Germany*

Abstract

Besides the reduction of angiotensin II formation, locally increased kinins may play a role in the cardiovascular action of angiotensin converting enzyme (ACE) inhibitors.

To characterize the contribution of bradykinin to the effects of ACE inhibition by captopril on the development of pressure overload hypertrophy, sham-operated rats and rats with ascending aortic constriction were treated with captopril (80 mg/kg/day) or captopril and B_2 kinin receptor antagonist HOE 140 (0.5 mg/kg/day) for 7 weeks. Left ventricular mass and geometry, hydroxyproline concentration and myosin isozymes (marker of a fetal phenotype) were assessed. Rats with aortic constriction exhibited a marked increase in left ventricular weight and diastolic pressure-volume relationship was shifted to smaller volumes. Signs of congestive heart failure were not apparent. The hydroxyproline concentration remained unaltered. However, the proportion of isomyosin V3 was increased ($p < 0.05$). Administration of captopril reduced ($p < 0.05$) systolic blood pressure, body and cardiac weight in all treated rats. The reduction of left ventricular weight was disproportionally higher in pressure overloaded rats, thus the relative left ventricular weight decreased by 15% ($p < 0.05$). Captopril augmented the isomyosin V1 expression ($p < 0.05$) in sham operated as well as pressure overloaded rats. The isomyosin V1 percentage was inversely related to the relative left ventricular weight. Two different ($p < 0.05$) correlation lines were detected for untreated and captopril treated rats. None of captopril associated effects were removed by simultaneously administered B_2 kinin receptor antagonist HOE 140.

Thus, stimulation of bradykinin B_2 receptor appears not to mediate the effects of captopril on cardiac growth and contractile proteins during the development of pressure overload hypertrophy. (Mol Cell Biochem **212**: 219–225, 2000)

Key words: heart hypertrophy, myosin isozymes, angiotensin converting enzyme inhibition, HOE 140, captopril, bradykinin

Introduction

Despite numerous studies and clear defined target structure, mechanism of action of angiotensin converting enzyme (ACE) inhibitors on cardiac hypertrophic growth is still controversial. Decreased stimulation of the angiotensin receptor seems not to be responsible for the reduction of hypertrophic growth of the overloaded ventricle [1–3] because angiotensin receptor type 1 blockade with losartan does not interfere with the adaptation of left ventricle to a persistent systolic pressure overload due to ascending aortic constriction [4]. An unaltered response of angiotensin II type 1 receptor knockout mice to aortic constriction support the notion that development of pressure overload hypertrophy does not depend on trophic effects of angiotensin II [5]. However, direct consequences of ACE inhibition involve not only a decreased formation of angiotensin II but also a reduction of kinin degradation with their consecutive accumulation in local tissue

Address for offprints: M. Turcani, Institute of Pathophysiology, Medical School, Comenius University, Sasinkova 4, 811 08 Bratislava, Slovak Republic and increased stimulation of B_2 kinin receptors, which mediate most of the effects of bradykinin [6].

To test the hypothesis that ACE inhibitors modulate hypertrophic cardiac growth through the augmented stimulation of bradykinin B_2 receptor we compared the development of pressure overload hypertrophy in rats treated with an ACE inhibitor captopril alone and in combination with HOE 140 (icatibant), the most potent, stable and long-lasting bradykinin B_2 receptor antagonist [7]. Pressure overload was induced by constriction of ascending aorta to minimize unloading of the left ventricle by peripheral effects of ACE inhibition [8]. The response of the heart to the increased systolic stress was evaluated after 7 weeks by assessing the cardiac gross morphometry. Fetal transformation of myosin molecules was followed by the analysis of myosin isozymes composition which appears to be influenced also by metabolic signals [9]. Furthermore, the hydroxyproline concentration was examined for monitoring interstitial or perivascular fibrosis.

Materials and methods

Male 3-week-old Wistar/WU rats were purchased from Charles-River (Kissleg, Germany) and were housed at 21–23°C on a 12:12-h light-dark cycle. The rats had free access to tap water and regular chow (Ssniff of Plange, Soest, Germany). Handling of animals and all experimental procedures were in accordance with the Institutional Guidelines of the University of Tübingen.

The ascending aorta was banded with a 3-0 silk suture ligature tied against a 0.8 mm blunt steel wire at a body weight of 90–100 g under Hypnorm® (fluanisone/fentanyl-dihydrogencitrate) anaesthesia (1 ml/kg, i.p.). The wire was removed, whereby the aorta was constricted to 60–70% of the original diameter [4]. Age-matched control animals underwent a right thoracotomy and the ascending aorta was isolated but not constricted. Six experimental groups with 10 animals each were used. One group of sham-operated animals as well as one group of rats with ascending aortic constriction remained without drug treatment. All other animals (4 groups) were treated with 80 mg/kg body wt/day captopril starting one day post surgery. Captopril (provided by Bristol-Myers Squibb, München, Germany) was given in the drinking water. The dose was maintained by monitoring daily water consumption and body weight. Two groups of captopril treated rats (sham-operated rats and rats with aortic constriction) had subcutaneously implanted osmotic minipumps with HOE 140 dissolved in 0.9% NaCl. The pumps released the B_2 kinin receptor antagonist in the dose 0.5 mg/kg body wt/day. This dose and mode of application effectively blocked the depressor response to exogenously applied bradykin [10]. The osmotic minipumps were implanted during the surgery for aortic constriction and were changed every 10 days. HOE 140 and osmotic minipumps were provided by Aventis (Frankfurt, Germany).

Systolic arterial blood pressure and heart rate were measured in conscious rats by the tail-cuff method.

For the assessment of the left ventricular passive pressure-volume relations, the atrial-ventricular groove was ligated with a silk string and the right ventricle was emptied by incision. The left ventricle was filled with a defined volume of saline and emptied in 0.05 ml steps while recording the passive ventricular pressure. Three reproducible pressure-volume curves were generated within 3–4 min after the ligation and within this time no effects of anoxia on the pressure-volume relation could be detected.

The heart and lungs were excised and weighed. The atria and major vessels were first trimmed and right ventricular free wall was dissected along the septum. Left ventricular free wall and septum were weighed together.

The proportion of myosin isozymes was determined by non-dissociating polyacrylamide gel electrophoresis in the presence of pyrophosphate [11]. Portions (30–50 mg) of the left ventricle frozen in liquid nitrogen were used. Myosin isozymes were stained with Coomassie brilliant blue R250 and the gels were scanned using a Quick Scan densitometer (Helena Laboratories, Beaumont, TX, USA). The isozymes V1, V2, V3 were quantitated by measuring peak heights.

Collagen content of ventricular tissue was assessed by determining the hydroxyproline concentration. A 70–100 mg portion of the left ventricle was freeze-dried and processed essentially as given by Stegemann [12].

Statistical analysis was performed with Statistica/w (Statsoft, Tulsa, OK, USA). Normality of distribution was checked by the Kolmogoroff-Smirnoff test and equality of variances according to Cochran. Multiple comparisons were made by two-way analysis of variance and Tukey HSD *post-hoc* test. Regression lines were compared by analysis of covariance. Statistical significance was assumed at $p < 0.05$.

Results

As expected, treatment with captopril decreased arterial systolic pressure in sham operated rats (Table 1). The hypotensive consequence of aortic constriction was, however, not further amplified by the ACE inhibition (Fig. 1A). Bradykinin B_2 receptor blockade by HOE140 had no effect on captopril induced hypotension. A positive chronotropic effect of captopril was detected only in sham operated animals (Table 1).

Captopril retarded body growth and the differences between rats treated with captopril and captopril plus HOE 140 were not statistically significant (Table 1). HOE 140 had also no effect on the captopril induced reduction of lung and heart weight. When the variance of lung weight was analyzed after correction for covariation in body weight, the effect of

Table 1. Body, heart and lung weight, heart rate and systolic arterial pressure

	Sham			Aortic constriction		
	Untreated	+Captopril	+Captopril +HOE 140	Untreated	+Captopril	+Captopril +HOE 140
	n = 10	n = 10	n = 10	n = 10	n = 10	n = 10
Body weight (g)	306 ± 11	258 ± 18*#	266 ± 19*#	291 ± 21	262 ± 12*#	261 ± 18*#
Lung weight (g)	1.47 ± 0.12	1.25 ± 0.21*	1.27 ± 0.18*	1.36 ± 0.16	1.23 ± 0.11*	1.23 ± 0.16*
Lung weight/body weight (mg/g)	4.79 ± 0.38	4.84 ± 0.64	4.78 ± 0.61	4.69 ± 0.39	4.68 ± 0.39	4.72 ± 0.53
Heart weight (mg)	917 ± 63	740 ± 64*#	742 ± 69*#	1253 ± 109*	989 ± 81#	1025 ± 115#
Heart weight/body weight (mg/g)	2.99 ± 0.17	2.87 ± 0.13#	2.79 ± 0.11#	4.31 ± 0.16*	3.76 ± 0.25*#	3.97 ± 0.46*#
Heart rate (beats/minute)	388 ± 16	418 ± 10*	405 ± 15*	386 ± 8	399 ± 16	393 ± 20
Arterial systolic pressure (mm Hg)	134 ± 6	120 ± 10*	119 ± 8*	122 ± 5*	121 ± 7*	117 ± 8*

Heart rate and systolic arterial pressure were measured using a tail-cuff method in unanesthetized rats. Values are means ± S.D. *$p < 0.05$ compared with sham-operated untreated rats; #$p < 0.05$ compared with untreated rats with aortic constriction.

treatment disappeared (ANCOVA: $F(2,53) = 0.1$; $p < 0.9$). In accordance, no significant differences were detected in the relative lung weight (Table 1).

By contrast, the effect of captopril on heart weight remained preserved after correction for covariation in body weight (ANCOVA: $F(2,53) = 5.84$; $p < 0.05$). Considering the unchanged relative right ventricular weight among the treatment groups, this might be caused mainly by changes in left ventricular weight (Table 2). Pressure overload and captopril treatment interacted in their effects on left ventricular growth (Fig. 1B). While growth inhibition by captopril was proportional to the reduced body mass in sham operated rats, this effect was significantly more pronounced in pressure overloaded ventricles. Aortic constriction reduced the size of left ventricular cavity volume (Table 2). Captopril did neither affect this variable in sham nor in pressure overloaded rats. However, captopril treatment was associated with an increased relative left ventricular chamber volume in sham op-

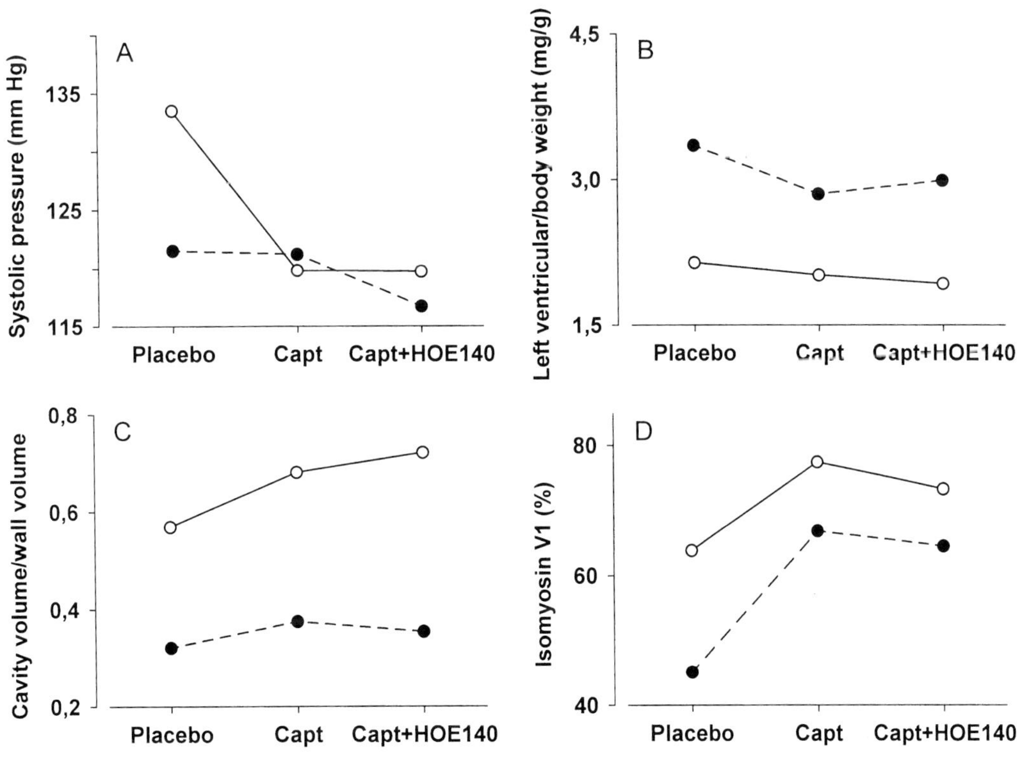

Fig. 1. Two-way interaction between captopril treatment and pressure overload for: (A) arterial systolic pressure (ANCOVA: $F(2,54) = 4.79$; $p < 0.05$); (B) left ventricular weight/body weight (ANCOVA: $F(2,54) = 3.39$; $p < 0.05$); (C) left ventricular cavity volume/wall volume (ANCOVA: $F(2,54) = 3.84$; $p < 0.05$); (D) isomyosin V1 (ANCOVA: $F(2,54) = 4.41$; $p < 0.05$). (O) Sham operated untreated or captopril and captopril+HOE 140 treated rats, (●) untreated or captopril and captopril+HOE 140 treated rats with ascending aortic constriction. Abbreviation: Capt – captopril.

Table 2. Cardiac morphometry

	Sham			Aortic constriction		
	Untreated	+Captopril	+Captopril +HOE 140	Untreated	+Captopril	+Captopril +HOE 140
	n = 10	n = 10	n = 10	n = 10	n = 10	n = 10
Left ventricular weight (mg)	655 ± 37	520 ± 42*#	514 ± 56*#	974 ± 92*	750 ± 83*#	778 ± 90*#
Left ventricular weight/body weight (mg/g)	2.14 ± 0.11	2.01 ± 0.07#	1.93 ± 0.09#	3.35 ± 0.15*	2.85 ± 0.26*#	2.98 ± 0.39*#
Right ventricular weight (mg)	184 ± 22	158 ± 24*	161 ± 19*	182 ± 17	172 ± 14	168 ± 18
Right ventricular weight/body weight (mg/g)	0.60 ± 0.07	0.61 ± 0.08	0.61 ± 0.07	0.63 ± 0.05	0.65 ± 0.06	0.63 ± 0.05
Atrial weight (mg)	78 ± 24	63 ± 17#	66 ± 15#	98 ± 14	68 ± 12#	70 ± 12#
Atrial weight/body weight (mg)	0.25 ± 0.07	0.24 ± 0.06#	0.25 ± 0.06#	0.34 ± 0.04*	0.26 ± 0.04#	0.27 ± 0.05#
Left ventricular cavity volume (mm^3)	372 ± 29	359 ± 25	369 ± 35#	311 ± 42*	279 ± 36*	271 ± 28*
Left ventricular cavity volume/wall volume	0.57 ± 0.06	0.68 ± 0.04*#	0.72 ± 0.1*#	0.32 ± 0.05*	0.37 ± 0.06*	0.35 ± 0.07*

Left ventricular cavity volume, and left ventricular cavity volume/ wall volume are given for the common intraventricular pressure 6 mm Hg. Values are means ± S.D. *$p < 0.05$ compared with sham operated untreated rats; #$p < 0.05$ compared with untreated rats with aortic constriction.

erated rats as assessed by the increase in the ratio of the cavity to wall volume. This captopril effect was modified by the pressure overload and no dilation was detected in rats with aortic constriction (Fig. 1C). Captopril treatment prevented the enlargement of cardiac atria in rats with ascending aortic constriction. Effects of captopril on mass and dimension of heart chambers were not affected by the HOE 140 treatment.

Although aortic constriction and captopril treatment had no effects on cardiac hydroxyproline concentration, myosin composition changed considerably (Table 3). Pressure overload diminished the percentage of isomyosin V1 and increased the percentage of isomyosin V3. Captopril enhanced the proportion of isomyosin V1 in sham operated and pressure overloaded rats as well. Yet the increment was more prominent in rats with aortic constriction (Fig. 1D). Isomyosin V1 percentage correlated ($p < 0.05$) inversely with the relative left ventricular weight. As documented by two different regression lines (Fig. 2; ANCOVA: $F(2,53) = 17.33$; $p < 0.05$) the changes in isomyosin V1 proportion could not be explained solely by the reduction of left ventricular weight associated wtih the captropil application. Treatment with HOE 140 did not eliminate the effects of ACE inhibition on isomyosin composition.

Discussion

Reduction of cardiac hypertrophic growth during treatment with ACE inhibitors was repeatedly described in rats with myocardial infarction [3, 13], pressure overloaded rats after constriction of the abdominal aorta above renal arteries [14, 15] as well as in spontaneously hypertensive rats [16, 17]. In all these models, ventricular unloading is a likely cause of decreased cardiac hypertrophic growth. However, antihypertrophic effect of ACE inhibition during a persistent pressure overload caused by ascending aortic constriction has not yet been described. By contrast, 2 weeks treatment with ramipril did not influence the development of left ventricular hypertrophy in adult rats with aortic arch constriction [18]. In the present study, treatment of 4 weeks old sham operated rats with captopril for 7 weeks reduced the overall body growth and proportionally the growth of the intrathoracic organs lungs and heart. Also the arterial systolic pressure was diminished. Disproportionally greater was, however, the antihypertrophic effect of captopril in pressure overloaded left ventricles. Because captopril had no effect on arterial systolic pressure in rats with ascending aortic constriction, a load independent effect of captopril on car-

Table 3. Myosin isozyme population and hydroxyproline concentration

	Sham			Aortic constriction		
	Untreated	+Captopril	+Captopril +HOE 140	Untreated	+Captopril	+Captopril +HOE 140
	n = 10	n = 10	n = 10	n = 10	n = 10	n = 10
Myosin V1 (%)	64 ± 7	77 ± 3*#	73 ± 4*#	45 ± 3*	67 ± 6#	65 ± 7#
Myosin V2 (%)	22 ± 3	15 ± 2*#	18 ± 2*#	28 ± 2*	20 ± 3#	21 ± 4#
Myosin V3 (%)	14 ± 4	8 ± 1*#	9 ± 2*#	27 ± 3*	13 ± 3#	14 ± 4#
Hydroxyproline (μg/mg dry weight)	2.57 ± 0.34	2.60 ± 0.24	2.71 ± 0.15	2.70 ± 0.33	2.63 ± 0.16	2.82 ± 0.52

Values are means ± S.D. *$p < 0.05$ compared with sham-operated untreated rats; #$p < 0.05$ compared with untreated rats with aortic constriction.

diomyocyte growth induced by the pressure overload appears likely.

While captopril did not change the concentric geometry of pressure overloaded ventricles, ACE inhibition was associated in sham operated animals with an eccentric remodeling of the left ventricle. A decline in the left ventricular cavity volume/wall volume ratio contributes to the compensation of pressure overload despite a lower degree of hypertrophy. In contrast, the increased left ventricular volume in sham operated animals might be a response of the left ventricle to the decreased pressure load during ACE inhibition. Captopril normalized the atrial weight in rats with aortic constriction. This effect could be a consequence of improved ventricular relaxation [19, 20] and reduced venous return [21] which accompanies the ACE inhibition.

A striking finding of this study was the profound effect of captopril on myosin isozyme composition. Captopril caused the left ventricular myosin to shift from the isomyosin V3 to the isomyosin V1 both in sham operated and pressure overloaded animals. This effect was more pronounced in rats with ascending aortic constriction and the fetal transformation of myosin molecules in the hypertrophied myocardium was prevented. The myosin isoform shift associated with the captopril treatment could partially be explained by the reduced left ventricular weight. However, the two different correlation lines between isomyosin V1 percentage and relative left ventricular weight (Fig. 2) suggest other mechanisms.

Ascending aortic constriction and unchanged arterial systolic pressure seem to exclude hemodynamic changes as a mechanism of antihypertrophic and myosin modulating effect of captopril. Decreased production of angiotensin II is not a probable mechanism either, since activation of angiotensin receptors type 1 appears not to play a major role in the development of pressure overload induced hypertrophy [4, 5, 22]. A third postulated mechanism of action of ACE inhibitors is the amplified stimulation of kinin receptors. The elevated local kinin concentration by ACE inhibition activates signal transduction pathways that generate second messengers such as cyclic guanosine monophosphate via an increase in nitric oxide or cyclic adenosine monophosphate via an increase in prostacyclin [6]. Nitric oxide [23, 24] and prostacyclin [24] are known to be antimitogenic and antiproliferative. They exert also intense metabolic effects [6]. Enhancement of glucose oxidation by bradykinin [26] and nitric oxide [27] might explain the myosin isoform shift caused by captopril.

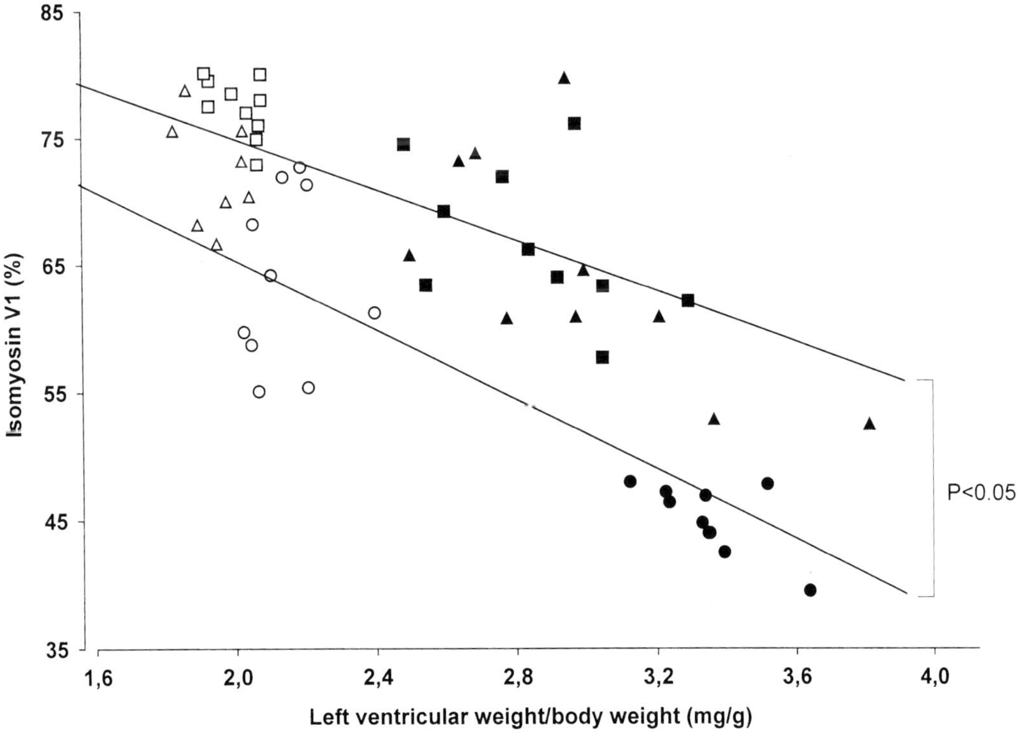

Fig. 2. Correlation between the left ventricular weight/body weight ratio and isomyosin V1 percentage. Untreated sham operated rats (O), sham operated rats treated with captopril (□), sham operated rats treated with captopril plus HOE 140 (Δ), untreated rats with ascending aortic constriction (●), rats with ascending aortic constriction treated with captopril (■), rats with ascending aortic constriction treated with captopril plus HOE 140 (▲). Correlation involving untreated sham operated rats and rats with aortic constriction: $y = 95.986–15.082\ x$, $r = –0.879$, $p < 0.0001$. Correlation involving treated sham operated rats and rats with aortic constriction: $y = 97.358–10.956\ x$, $r = –0.757$, $p < 0.0001$. Correlation lines are different by ANCOVA: $F(2,53) = 17.33$; $p < 0.05$.

However, none of captopril effects were removed by the bradykinin B_2 receptor blockade with HOE 140. Therefore, a bradykinin involvement in the captopril effects on pressure overload induced cardiac hypertrophy is unlikely. As an alternative explanation, a bradykinin independent stimulation of nitric oxide and prostacyclin activity by captopril could be postulated. A direct enhancement of nitric oxide and prostacyclin might be hypothesized. When angiotensin II is reduced by ACE inhibition, the actions of nitric oxide and prostacyclin generated by bradykinin independent mechanisms should be enhanced [28]. In addition, the free radical scavenging effect of sulphydryl group containing captopril might further increase nitric oxide [29, 30] and prostacyclin activity [31, 32].

In conclusion, captopril modifies the left ventricular response to the pressure overload caused by ascending aortic constriction. The degree of hypertrophy was reduced and the fetal transformation of myosin molecules was prevented. These captopril effects were probably not mediated by hemodynamic changes and increased stimulation of bradykinin B_2 receptors. As candidate mechanisms, bradykinin independent production of nitric oxide and/or prostacyclin should be considered.

Acknowledgements

The study was supported by the Alfred-Teufel-Stiftung, the German Research Foundation (Ru 245/7-1) and the Slovak Grant Agency (No. 1/7492/20). The experimental part of this study was performed at the Institute of Physiology II, University of Tübingen, Germany. The support of Professor R. Jacob is greatly appreciated.

References

1. Linz W, Schölkens BA, Ganten D: Converting enzyme inhibition specifically prevents the development and induces regression of cardiac hypertrophy in rats. Clin Exp Hypertens A11: 1325–1350, 1989
2. Peeler TC, Baker KM, Esmurdoc CF, Chernin MI: Angiotensin converting enzyme inhibition in Dahl salt-sensitive rats. Mol Cell Biochem 104: 45–50, 1991
3. Schieffer B, Wirger A, Meybrunn M, Seitz S, Holtz J, Riede UN, Drexler H. Comparative effects of chronic angiotensin-converting enzyme inhibition and angiotensin II type 1 receptor blockade on cardiac remodeling after myocardial infarction in the rat. Circulation 89: 2273–2282, 1994
4. Turcani M, Rupp H: Development of pressure overload induced cardiac hypertrophy is unaffected by long-term treatment with losartan. Mol Cell Biochem 188: 225–233, 1998
5. Harada K, Komuro I, Shiojima I, Hayashi D, Kudoh S, Mizuno T, Kijima K, Matsubara H, Sugaya T, Murakami K, Yazaki Y: Pressure overload induces cardiac hypertrophy in angiotensin II type 1A receptor knockout mice. Circulation 97: 1952–1959, 1998
6. Linz W, Wiemer G, Gohlke P, Unger T, Schölkens BA: Contribution of kinins to the cardiovascular actions of angioternsin-converting enzyme inhibitors. Pharmacol Rev 47: 25–49, 1995
7. Wirth K, Breipohl G, Stechl J, Knolle J, Henke S, Schölkens B: DesArg9-D-Arg[Hyp3,Thi5,D-Tic7,Oic8]bradykinin (desArg10-[Hoe140]) is a potent bradykinin B1 receptor antagonist. Eur J Pharmacol 205: 217–218, 1991
8. Weinberg EO, Schoen FJ, George D, Kagaya Y, Douglas PS, Litwin SE, Schunkert H, Benedict CR, Lorell BH: Angiotensin-converting enzyme inhibition prolongs survival and modifies the transition to heart failure in rats with pressure overload hypertrophy due to ascending aortic stenosis. Circulation 90: 1410–1422, 1994
9. Zarain-Herzberg A, Rupp H: Transcriptional modulators targeted at fuel metabolism of hypertrophied heart. Am J Cardiol 83: 31H–37H, 1999
10. Bao G, Gohlke P, Qadri F, Unger T: Chronic kinin receptor blockade attenuates the antihypertensive effect of ramipril. Hypertension 20: 74–79, 1992
11. Rupp H, Wahl R, Hansen M: Influence of diet and carnitine palmitoyltransferase 1 inhibition on myosin and sarcoplasmic reticulum. J Appl Physiol 72: 352–360, 1992
12. Stegemann H, Stalder K: Determination of hydroxyproline. Clin Chim Acta 18: 267–273, 1967
13. Pfeffer MA, Braunwald E: Ventricular remodeling after myocardial infarction: Experimental observations and clinical implications. Circulation 81: 1161–1172, 1990
14. Linz W, Schölkens BA, Ganten D: Converting enzyme inhibition specifically prevents the development and induces regression of cardiac hypertrophy in rats. Clin Exp Hypertens 11: 1325–1350, 1989
15. Baker KM, Chernin MI, Wixson SK, Aceto JF: Renin-angiotensin system involvement in pressure-overload cardiac hypertrophy in rats. Am J Physiol 259: H324–H332, 1990
16. Freslon JL, Giudicelli JF: Compared myocardial and vascular effects of captopril and dihydralazine during hypertension development in spontaneously hypertensive rats. Br J Pharmacol 80: 533–543, 1983
17. Linz W, Jessen T, Becker RHA, Schölkens BA, Wiemer G: Long-term ACE inhibition doubles lifespan of hypertensive rats. Circulation 96: 3164–3172, 1997
18. Zierhut W, Zimmer HG, Gerdes AM: Effect of angiotensin converting enzyme inhibition on pressure-induced left ventricular hypertrophy in rats. Circ Res 69: 609–617, 1991
19. Litwin SE, Katz SE, Weinberg EO, Lorell BH, Aurigemma GP, Douglas PS: Serial echocardiographic-Doppler assessment of left ventricular geometry and function in rats with pressure-overload hypertrophy. Chronic angiotensin-converting enzyme inhibition attenuates the transition to heart failure. Circulation 91: 2642–2654, 1995
20. Anning PB, Grocott-Mason RM, Lewis MJ, Shah AM: Enhancement of left ventricular relaxation in the isolated heart by an angiotensin-converting enzyme inhibitor. Circulation 92: 2660–2665, 1995
21. Raya TE, Gay G, Aguirre M, Goldman S: The importance of venodilatation in the prevention of left ventricular dilatation after chronic large myocardial infarction in rats: A comparison of captopril and hydralazin. Circ Res 64: 330–337, 1989
22. Mohabir R, Young SD, Strosberg AM: Role of angiotensin in pressure overload-induced hypertrophy in rats: Effects of angiotensin-converting enzyme inhibitors, an AT_1 receptor antagonist, and surgical reversal. J Cardiovasc Pharmacol 23: 291–299, 1994
23. Ishygai Y, Mori T, Ikeda T, Fukuzawa A, Shibano T: Role of bradykinin-NO pathway in prevention of cardiac hypertrophy by ACE inhibitor in rat cardiomyocytes. Am J Physiol 273: H2659–H2663, 1997
24. Ritchie RH, Marsh JD, Lancaster WD, Diglio CA, Schiebinger RJ: Bradykinin blocks angiotensin II-induced hypertrophy in the presence of endothelial cells. Hypertension 31(part 1): 39–44, 1998

25. Shirotani P, Yui Y, Hattori R, Kawai C: U-61,431F, a stable prostacyclin analogue, inhibits the proliferation of bovine vascular smooth muscle cells with little antiproliferative effect on endothelial cells. Prostaglandins 41: 97–110, 1991
26. Rösen P, Eckel J, Reinauer H: Influence of bradykinin on glucose uptake and metabolism studied in isolated cardiac myocytes and isolated perfused rat hearts. Hoppe Seyler Z Physiol Chem 364: 1431–1438, 1983
27. Henriksen EJ, Jacob S, Kinnick TR, Youngblood EB, Schmit MB, Dietze GJ: ACE inhibition and glucose transport in insulin-resistant muscle: Roles of bradykinin and nitric oxide. Am J Physiol 277: R332–R36, 1999
28. Cachofeiro V, Maeso R, Rodrigo E, Navarro J, Ruilope LM, Lahera V: Nitric oxide and prostaglandins in the prolonged effects of losartan and ramipril in hypertension. Hypertension 26: 236–243, 1995
29. Ota Y, Kugiyama K, Sugiyama S, Ohgushi M, Matsumura T, Doi H, Ogata N, Oka H, Yasue H: Impairment of endothelium-dependent relaxation of rabbit aortas by cigarette smoke extract – role of free radicals and attenuation by captopril. Atherosclerosis 131: 195–202, 1997
30. Ruiz FJ, Salom MG, Ingles AC, Quesada T, Vicente E, Carbonell LF: N-acetyl-L-cysteine potentiates depressor response to captopril and enalaprilat in SHRs. Am J Physiol 267: R767–R772, 1994
31. Van Gilst VM, Wijingarden JV, Scholtens E, deGraeff PA, deLangen CDJ, Wesseling H: Captopril induced increase in coronary flow: An SH-dependent effect on arachidonic metabolism. J Cardiovasc Pharmacol 9(suppl 2): S31–S36, 1987
32. Moroi M, Akatsuka N, Fukazawa M, Hara K, Ishikawa M, Aikawa J, Namiki A, Yamaguchi T: Endothelium-dependent relaxation by angiotensin-converting enzyme inhibitors in canine femoral arteries. Am J Physiol 266: H583–H589, 1994

Molecular and Cellular Biochemistry **212:** 227–235, 2000.

Role of cardiac renin-angiotensin system in sarcoplasmic reticulum function and gene expression in the ischemic-reperfused heart

Satoshi Takeo,[1] Yoshihisa Nasa,[1] Kouichi Tanonaka,[1] Fuminari Yamaguchi,[1] Ken-ichi Yabe,[1] Hideki Hayashi[1] and Naranjan S. Dhalla[2]

[1]*Department of Pharmacology, Tokyo University of Pharmacy and Life Science, Tokyo, Japan;* [2]*Institute of Cardiovascular Sciences, St. Boniface General Hospital Research Centre, University of Manitoba, Faculty of Medicine, Winnipeg, Manitoba, Canada*

Abstract

The aim of this study was to explore the possible participation of cardiac renin-angiotensin system (RAS) in the ischemia-reperfusion induced changes in heart function as well as Ca^{2+}-handling activities and gene expression of cardiac sarcoplasmic reticulum (SR) proteins. The isolated rat hearts, treated for 10 min without and with 30 μM captopril or 100 μM losartan, were subjected to 30 min ischemia followed by reperfusion for 60 min and processed for the measurement of SR function and gene expression. Attenuated recovery of the left ventricular developed pressure (LVDP) upon reperfusion of the ischemic heart was accompanied by a marked reduction in SR Ca^{2+}-pump ATPase, Ca^{2+}-uptake and Ca^{2+}-release activities. Northern blot analysis revealed that mRNA levels for SR Ca^{2+}-handling proteins such as Ca^{2+}-pump ATPase (SERCA2a), ryanodine receptor, calsequestrin and phospholamban were decreased in the ischemia-reperfused heart as compared with the non-ischemic control. Treatment with captopril improved the recovery of LVDP as well as SR Ca^{2+}-pump ATPase and Ca^{2+}-uptake activities in the postischemic hearts but had no effect on changes in Ca^{2+}-release activity due to ischemic-reperfusion. Losartan neither affected the changes in contractile function nor modified alterations in SR Ca^{2+}-handling activities. The ischemia-reperfusion induced decrease in mRNA levels for SR Ca^{2+}-handling proteins were not affected by treatment with captopril or losartan. The results suggest that the improvement of cardiac function in the ischemic-reperfused heart by captopril is associated with the preservation of SR Ca^{2+}-pump activities; however, it is unlikely that this action of captopril is mediated through the modification of cardiac RAS. Furthermore, cardiac RAS does not appear to contribute towards the ischemia-reperfusion induced changes in gene expression for SR Ca^{2+}-handling proteins. (Mol Cell Biochem **212**: 227–235, 2000)

Key words: renin angiotensin system, sarcoplasmic reticulum, Ca^{2+}-handling, gene expression, ischemia-reperfusion, cardioprotection

Introduction

While the contraction and relaxation processes in the cardiac muscle are known to be regulated by the intracellular concentration of Ca^{2+}, the sarcoplasmic reticulum (SR) plays a critical role in the regulation of intracellular Ca^{2+}-homeostasis. Cardiac SR sequesters Ca^{2+} upon the activation of a Mg^{2+}-dependent, Ca^{2+}-stimulated ATPase (Ca^{2+}-pump ATPase) and releases Ca^{2+} through ryanodine-sensitive Ca^{2+}-channels. The function of the SR Ca^{2+}-pump ATPase is regulated by a phosphorylatable protein, phospholamban, and Ca^{2+} in the SR lumen is bound to a protein, calsequestrin. Since the intracellular Ca^{2+}-overload has been shown to play a critical role in the pathophysiology of ischemia-reperfusion injury, the

Address for offprints: S. Takeo, Department of Pharmacology, Tokyo University of Pharmacy and Life Science, Horinouchi 1432-1 Hachioji, Tokyo, 192-0392, Japan postischemic contractile dysfunction is considered to occur mainly due to the reduction of SR function [1, 2]. In fact, a decrease in the SR Ca^{2+}-pump ATPase activity and a reduction of the number of SR Ca^{2+}-release channels have been reported in the ischemic-reperfused myocardium [3–7]. Thus, it has been suggested that defects in Ca^{2+}-release channel (ryanodine receptor), Ca^{2+}-uptake (Ca^{2+}-pump ATPase and its regulation by phospholamban) or Ca^{2+}-storage (calsequestrin) may contribute to the SR Ca^{2+}-handling abnormalities and thereby affect the systolic and diastolic contractile function in the postischemic heart.

The renin-angiotensin system (RAS) plays an important role in cardiac and vascular pathology associated with ischemic heart disease, hypertension and congestive heart failure. Presence of local RAS in the myocardium has been established by biochemical and molecular studies [8]. Indeed, a powerful vasoconstrictor peptide angiotensin II (Ang II) has been shown to be produced in hearts under physiological [9] as well as ischemic conditions [10]. Ang II has been reported to evoke functional and molecular responses via a variety of mechanisms in cardiac myocytes [11, 12] and activate the transcription of protooncogenes and growth factors [13, 14]. Recently, Yang *et al.* [15] have reported that the expression of cardiac Ang II type-1 (AT_1) receptor increases immediately after ischemia-reperfusion and contributes to the genesis of cardiac dysfunction in the postischemic heart. In addition, exogenously administered Ang II has been reported to deteriorate the postischemic cardiac function through AT_1 receptor activation and to increase the release of tissue creatine kinase [16]. Furthermore, administration of Ang II has been reported to result in the suppression of mRNA levels for SR Ca^{2+}-handling proteins in neonatal cardiac myocytes [17]. Several lines of evidence have suggested that AT_1 antagonists and angiotensin-converting enzyme (ACE) inhibitors ameliorate myocardial ischemic damage [16, 18–22], probably due to the blockade of AT_1 receptors and a reduction in the production of Ang II, respectively, and thus indicate the possible involvement of cardiac RAS to the ischemic injury. If production of Ang II in the heart is enhanced during ischemia and reperfusion, the inhibition of cardiac RAS by treatment with an ACE inhibitor or the blockade of the receptor by an AT_1 antagonist should affect changes in SR Ca^{2+}-handling function and expression of mRNA for SR Ca^{2+}-handling proteins. However, there is little information available concerning the contribution of cardiac RAS in abnormal SR Ca^{2+}-handling and changes in mRNA levels for SR proteins due to ischemia-reperfusion in the heart. Therefore, the present study was undertaken to examine the role of cardiac RAS in ischemia-reperfusion by employing an isolated perfused rat heart, in which the contribution of non-cardiac RAS is eliminated. For this purpose, the hearts were treated with either captopril, an ACE inhibitor, or losartan, a specific AT_1 blocker [23]. Furthermore, to determine whether ischemia-reperfusion induced changes in gene expression for SR Ca^{2+}-handling proteins are affected by cardiac RAS, the effects of these drugs were examined on mRNA levels for SR proteins.

Materials and methods

Animals

Male Sprague-Dawley rats weighing 280–360 g were used in this study. Rats were fed a standard diet and acclimated in a quarantine room for at least 1 week before experiments. All experimental protocols for animal studies were approved by the Animal Care Committee of the University of Manitoba, Canada, following guidelines designed by the Medical Research Council of Canada as well as by the Animal Care and Welfare of the Tokyo University of Pharmacy and Life Science, following the Guideline of Experimental Animal Care issued from the Prime Minister's Office of Japan.

Perfusion of the isolated rat heart

Rats were anesthetized with ketamine-xylazine (100 mg/kg-10 mg/kg) cocktail. After injection of 1000 U sodium heparin in the femoral vein, the heart was quickly isolated and perfused in the Langendorff mode at a constant flow rate of 10 ml/min with Krebs-Henseleit bicarbonate buffer of the following composition (mM): 120 NaCl, 4.8 KCl, 1.25 $CaCl_2$, 1.2 $MgSO_4$, 1.2 KH_2PO_4, 25 $NaHCO_3$ and 11 glucose, pH 7.4 [24]. The buffer was equilibrated with a gas mixture of 95% O_2 and 5% CO_2. A water-filled latex balloon with an uninflated diameter of 3.7 mm was inserted into the left ventricle through the left atrium to measure the left ventricular developed pressure (LVDP). The left ventricular end-diastolic pressure (LVEDP) was adjusted to 5–8 mm Hg at the initial stage of each experiment. Heart rate (HR) was kept at 300 bpm using an electronic stimulator (Phipps and Bird, Richmond, VA, USA). Both LVDP and LVEDP were measured by means of a pressure transducer and an amplifier (Model 1050 BP-Biopac System Inc., Goleta, CA, USA). Data were recorded online through an analogue-digital interface (MP 100, Biopac System Inc., Goleta, CA, USA), and stored and processed with AcqKnowledge 3.0.1 for Windows (Biopac System Inc., Goleta, CA, USA). Experiments were started after an initial 30 min period of stabilization.

Perfusion protocols

The hearts were randomly divided into the following groups: (a) Vehicle-treated group; hearts perfused for 10 min with the Krebs-Henseleit buffer containing vehicle (n = 9); (b)

Captopril-treated group; hearts perfused for 10 min with Krebs-Henseleit buffer containing 30 μM captopril (n = 9); (c) Losartan-treated group; hearts perfused for 10 min with Krebs-Henseleit buffer containing 100 μM losartan (n = 9); and (d) Non-ischemic control group; hearts were perfused aerobically for 105 min (n = 9). Drugs were dissolved in the Krebs-Henseleit buffer just prior to use and infused at a rate of 0.2 ml/min into the injection port located just before the aortic cannula. Infusion of 0.2 ml/min vehicle altered neither pH of the perfusion buffer nor the preischemic mechanical function of the heart. After treatment with either drug, the hearts were subjected to 30 min of ischemia. To achieve global ischemia in hearts, the perfusion was stopped and the hearts were kept in an air-tied chamber at 37°C. Ischemic contracture was taken as a rise in LVEDP after the onset of ischemia. After 30 min of ischemia the hearts were reperfused for 60 min with an aerobic buffer. In a preliminary study, we examined the effects of treatment with different concentrations (3, 10, 30, and 100 μM) of captopril on LVDP recovery and found that captopril at a concentration of 30 μM was most effective for the recovery of LVDP in the ischemic-reperfused heart. In case of losartan experiment, 30 μM or lower concentration of losartan had no effect on the postischemic functional recovery. At the end of reperfusion, the left ventricle including interventricular septum was frozen with liquid nitrogen and stored at –80°C until biochemical analysis was performed. Each of the frozen tissue was divided into two pieces to examine both SR function and gene expression in the same preparation.

Preparation of sarcoplasmic reticulum (SR)-enriched membrane

Membrane fraction enriched with SR was prepared from the frozen heart according to the method described previously [25]. Each tissue was cut with scissors in ice-cold Krebs-Henseleit buffer and homogenized in 15 mM Tris (hydroxymethyl)-aminomethane (Tris)-HCl buffer containing 10 mM $NaHCO_3$ and 5 mN NaN_3, pH 6.8 using a Polytron homogenizer (PT-10, Kinematica GmbH, Littau Luzerne, Switzerland) for three 15 sec bursts at its maximum setting. The homogenate was centrifuged at 10,000 g for 20 min. The resultant supernatant was filtered through two layers of cheesecloth and centrifuged at 40,000 g for 45 min at 4°C. The pellet was resuspended in a buffer containing 20 mM Tris-HCl and 0.6 M KCl, pH 6.8. The suspension was centrifuged again at 40,000 g for 45 min. The final pellet was taken as the microsomal fraction enriched with SR vesicles; it was washed and suspended in a solution containing 0.25 M sucrose and 20 mM Tris-HCl, pH 6.8 at a concentration of 400–900 mg protein/ml. This fraction was rapidly frozen in liquid nitrogen and stored at –80°C until assay. An aliquot was used for protein assay using the method of Lowry *et al.* [26] with bovine serum albumin as a standard.

Measurement of SR Ca^{2+}-pump ATPase activity

The experimental conditions for determination of Ca^{2+}-pump ATPase activity were same as those reported previously [27]. SR vesicles (15–20 μg protein) were preincubated at 37°C for 5 min in 1 ml of medium containing (in mM): 100 KCl, 20 Tris-HCl, 5 $MgCl_2$ and 5 NaN_3, in the presence (total ATPase) and absence (Mg^{2+}ATPase) of 10 μM $CaCl_2$, pH 6.8. The reaction was started by the addition of 4 mM Tris-ATP, pH 6.8, and terminated 5 min later by the addition of 1 ml of ice-cold 12% (w/v) trichloroacetic acid. The liberated inorganic phosphate was measured by the method of Tausky and Shorr [28]. The Ca^{2+}-stimulated ATPase activity was calculated by subtracting the Mg^{2+}ATPase activity from the total ATPase activity.

Measurement of SR Ca^{2+}-uptake and Ca^{2+}-release

Calcium uptake and release activities were measured by a modification of the Millipore filtration technique as described previously [29]. Briefly, the reaction mixture consisted of 120 mM KCl, 50 mM Tris-maleate, 5 mM potassium oxalate, 5 mM $MgCl_2$, 5 mM NaN_3, 4 mM Tris-ATP, 0.1 mM EGTA and 25 μM ruthenium red. Since the net Ca^{2+}-uptake in SR vesicles represents the balance between influx mediated by the Ca^{2+}-pump and efflux through ryanodine-sensitive pathway, ruthenium red was added into the incubation medium to inhibit Ca^{2+}-release. SR vesicles (50 μg) were preincubated at 37°C for 2 min in the buffer and the reaction was initiated by addition of 1 mM $^{45}CaCl_2$ and stopped after 1 min by rapid filtration of 0.2 ml aliquot of the reaction mixture through a Millipore filter (Type HA, pore size 0.45 μm). Each filter was washed twice with 5 ml of Tris-HCl buffer (pH 7.4).

To measure Ca^{2+}-release activity, SR vesicles (50 μg) were loaded with $^{45}CaCl_2$ in the loading buffer, containing 0.1 M KCl, 20 mM Tris-HCl, 5 mM $MgCl_2$, 5 mM potassium oxalate, 5 mM NaN_3, and 0.5 mM Tris-ATP; SR vesicles were incubated for 45 min at room temperature. Then, 900 μl of SR vesicles loaded with $^{45}CaCl_2$ were added to the tube containing 100 μl of 1 mM EGTA. The reaction was stopped at 15 sec using the Millipore filtration technique. After drying of each filter paper, the radioactivity on the filter was counted in 10 ml of liquid scintillation cocktail using the standard liquid scintillation counting technique [25].

Northern blot analysis

Northern blot analysis was used to determine the levels of mRNA for SR Ca^{2+}-transport proteins such as $SERCA_{2\alpha}$ (the cardiac/slow-twitch isoform of Ca^{2+}-pump ATPase expressed in cardiac tissue), ryanodine receptor, calsequestrin and phospholamban. Total RNA isolation and Northern blot analysis were carried out by the same protocol as described previously [17] which was a modification of the procedure of Chomczynski and Sacchi [30]. Total RNA was isolated from the frozen myocardium according to the acid guanidinium thiocyanate-phenol chloroform method. The amount of RNA was determined by the UV absorption method; neither ischemia-reperfusion nor treatment with drugs affected the yield of total mRNA. The extracted total RNA (20 μg) was denatured at 65°C for 10 min, size-fractionated by 1.2 (w/v) % agarose gel electrophoresis and transferred to a positively charge-modified nylon membrane with a pore size of 0.45 μm (NYTRAN Maximum Strength Plus, Schleicher and Schuell, Keene, NH, USA). After 24 h, the filter was removed and nucleic acids were covalently cross-linked to the matrix using UV radiation (UV Stratalinker 2400, Stratagene). The cDNA probes of rat SERCA2a, ryanodine receptor, calsequestrin, and phospholamban were labeled with (^{32}P) dCTP (10 mCi/mL) by the random primed labeling method (Random Primers DNA Labeling System, Gibco BRL). Hybridization was carried out by addition of radioactive DNA probe followed by incubation overnight. The results were quantified by densitometric scanning and the expression level of each mRNA was normalized to the 18S rRNA level.

Drugs

All reagents were purchased from the Sigma Chemical Co (St. Louis, MO, USA), unless otherwise specified. $^{45}CaCl_2$ was purchased from Du Pont-New England Nuclear (Mississauga, ON, Canada).

Statistics

All values in the text are expressed as mean ± S.E.M. Statistical differences among groups were estimated using analysis of variance (ANOVA). Dunnett's or Bonferroni's multiple comparison test was used for *post-hoc* comparison. Student's unpaired *t*-test was used for the comparison between two groups. P values of less than 5% were considered statistically significant ($p < 0.05$).

Results

Effects of captopril and losartan on cardiac contractile function

The effects of captopril and losartan on the contractile function of hearts subjected to 30 min ischemia followed by 60 min reperfusion were studied and the changes in LVDP and LVEDP due to ischemia-reperfusion are shown in Table 1. Initial values for cardiac variables were not significantly different among groups. The heart failed to develop LVDP due to ischemia but LVEDP was gradually elevated, indicating the development of ischemic contracture. In the vehicle-treated heart there was some restoration of LVDP upon reperfusion (24.8 ± 3.3% recovery of initial value at the end of reperfusion), but LVEDP was further increased as compared with that during ischemia. In this experimental model, arrhythmias occurred in one out of 10 preparations during reperfusion; these hearts were excluded from this study.

Treatment of hearts for 10 min with either captopril (30 μM) or losartan (100 μM) before inducing ischemia did not affect the basal contractile function; both drugs also had no effect on the increase in LVEDP due to ischemia (Table 1). In the

Table 1. Effects of captopril (30 μM) and losartan (100 μM) on cardiac mechanical function during ischemia and reperfusion

	Initial	Before ischemia	End of ischemia (mm Hg)	End of reperfusion
Vehicle (n = 9)				
LVDP	62.9 ± 3.6	62.1 ± 3.2	0.0 ± 0.0	15.8 ± 2.5
LVEDP	6.7 ± 0.7	6.7 ± 0.7	30.4 ± 6.0	70.1 ± 6.2
Captopril, 30 μM (n = 9)				
LVDP	56.7 ± 2.2	56.9 ± 2.4	0.0 ± 0.0	39.7 ± 4.8*
LVEDP	7.7 ± 0.9	7.8 ± 0.7	30.2 ± 2.5	56.9 ± 4.7*
Losartan, 100 μM (n = 9)				
LVDP	63.7 ± 2.2	63.4 ± 2.3	0.0 ± 0.0	22.0 ± 3.1
LVEDP	7.7 ± 0.6	7.5 ± 0.7	32.1 ± 1.7	77.2 ± 5.3

Each value represents mean ± S.E.M. *$p < 0.05$ vs. the corresponding vehicle group. LVDP – left ventricular developed pressure; LVEDP – left ventricular end-diastolic pressure. Drug treatments were carried out 10 min before inducing ischemia for 30 min and reperfusion for 60 min.

captopril-treated heart, there was a significant increase in the recovery of LVDP during reperfusion (69.8 ± 8.0% recovery expressed as % initial value); the elevation of LVEDP at the end of reperfusion was significantly attenuated by captopril. In a preliminary study, we found that captopril at the concentration of 3 μM failed to enhance the recovery of LVDP upon reperfusion (8.0 ± 0.7% recovery of its initial value, n = 4). Treatment with losartan had no effect on mechanical function in the ischemic-reperfused heart. Losartan neither enhanced recovery of LVDP nor suppressed the increase in LVEDP upon reperfusion: the recovery of LVDP at the end of reperfusion was 14.3 ± 6.9% (Table 1). When the hearts were treated with 100 μM losartan during 10 min preischemic period as well as during 60 min of reperfusion period, the postischemic functional recovery was not enhanced.

Effects of captopril and losartan on SR Ca^{2+}-handling

(a) Ca^{2+}-pump ATPase activity
Ischemic insult alone significantly decreased the SR Ca^{2+}-pump ATPase activity (26.1 ± 3.7, n = 9 vs. 65.4 ± 8.2 nmoles/mg/min, n = 7, of the time-matched non-ischemic heart). Ischemia-reperfusion also resulted in a decrease of the Ca^{2+}-pump ATPase activity (Table 2). Captopril attenuated the reduction of Ca^{2+}-pump ATPase activity due to ischemia-reperfusion whereas losartan did not affect it.

(b) SR Ca^{2+}-uptake
Oxalate-supported SR Ca^{2+}-uptake activities in the ischemic-reperfused heart with or without drug treatment are shown in Table 2. A marked depression in the SR Ca^{2+}-uptake activity was observed in the untreated ischemic-reperfused heart; treatment of hearts with captopril, but not with losartan, attenuated this reduction in the Ca^{2+}-uptake activity.

(c) SR Ca^{2+}-release
The data for Ca^{2+}-release activities of the SR isolated from ischemic-reperfused hearts are shown in Fig. 1. As compared with non-ischemia, ischemia-reperfusion markedly decreased the SR Ca^{2+}-release activity. Treatment of hearts with either captopril or losartan did not affect the depressed SR Ca^{2+}-release activity due to ischemia-reperfusion.

Table 2. Effects of captopril (30 μM) and losartan (100 μM) on the ischemia-reperfusion induced changes in the rat heart SR Ca^{2+}-pump ATPase and Ca^{2+}-uptake activities

	Ca^{2+}-pump ATPase activity (nmol Pi/mg/min)	Ca^{2+}-uptake activity (nmol Ca^{2+}/mg/min)
Control (n = 7)	72.9 ± 6.43	35.3 ± 5.71
Ischemia-reperfusion		
Vehicle (n = 9)	22.7 ± 4.12*	12.2 ± 2.44*
Captopril (n = 7)	41.2 ± 4.40*#	24.5 ± 2.19*#
Losartan (n = 7)	22.9 ± 2.42*	17.5 ± 2.27*

*$p < 0.05$ vs. control; #$p < 0.05$ vs. vehicle. Drug treatments were carried out 10 min before inducing ischemia-reperfusion.

Effects of captopril and losartan on mRNA levels for SR Ca^{2+}-handling proteins

Figure 2 shows a representative mRNA pattern of the Ca^{2+}-handling proteins by Northern blot analysis and their quantifications are summarized in Fig. 3. The results show that the mRNA levels for SERCA2a, calsequestrin, ryanodine receptor and phospholamban were decreased after ischemia-reperfusion by 60.9, 60.7, 53.0 and 67.4% in the ischemic-reperfused heart, respectively. In hearts treated with captopril or losartan, the levels of these mRNA were also significantly lower than those in the non-ischemic group, indicating no effects of these drugs on ischemia-reperfusion induced changes in the SR proteins gene expression.

Discussion

SR is one of the plausible sites which are affected by the ischemia-reperfusion injury in the heart and the changes in SR Ca^{2+}-uptake and Ca^{2+}-release may cause abnormalities in cardiomyocyte Ca^{2+} transients and result in systolic and diastolic dysfunction. The present study showed that ischemia-

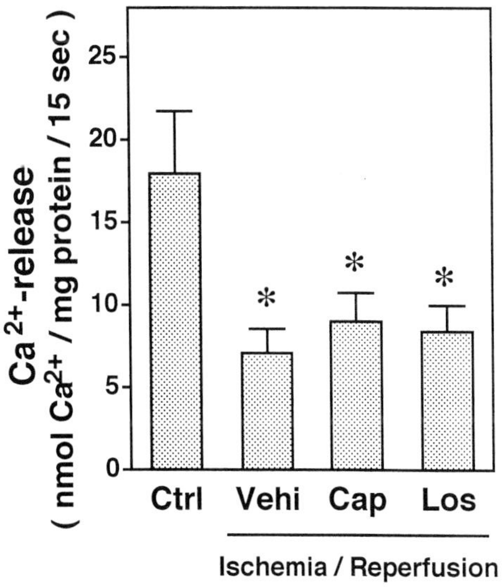

Fig. 1. Bar graphs showing changes in Ca^{2+}-release activity in the sarcoplasmic reticulum (SR) of non-ischemic control hearts (Ctrl, n = 8) and ischemic-reperfused hearts treated with vehicle (Vehi, n = 9), captopril (Cap, n = 7) or losartan (Los, n = 7). Each value represents the mean ± S.E.M. *$p < 0.05$ vs. non-ischemic control group.

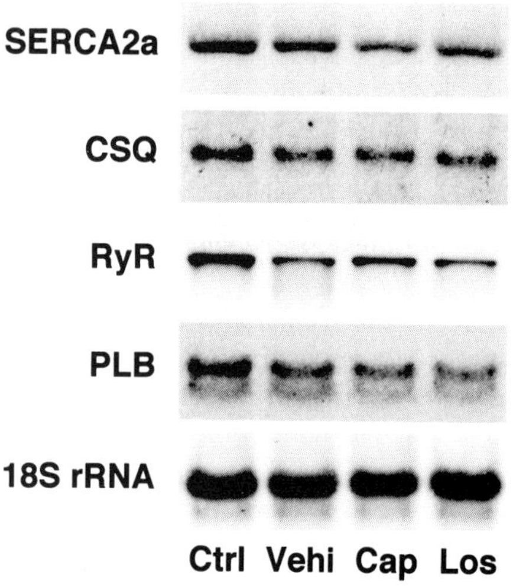

Fig. 2. Northern blot analysis showing mRNA levels of SR Ca^{2+}-handling genes encoding different sarcoplasmic reticulum (SR) proteins. Top, second, third, fourth and bottom lanes represent hybridization blots for SERCA2a, calsequestrin (CSQ), ryanodine receptor (RyR), phospholamban (PLB) and 18S ribosomal RNA (18S rRNA). The blot for 18S rRNA was used as an internal standard. The abbreviations are the same as in Fig. 1.

reperfusion resulted in a marked decrease in LVDP and increase in LVEDP, which were accompanied by a depression of SR Ca^{2+}-uptake and -release function. A defect in SR Ca^{2+}-transport is considered to contribute mainly to the cytosolic Ca^{2+}-overload during ischemia and reperfusion. Particularly, depression of the SR Ca^{2+}-uptake activity may cause an elevation of intracellular Ca^{2+} during diastolic period, leading to an elevation of LVEDP in the reperfused heart [1]. Since the amount of Ca^{2+} sequestered by SR determines the Ca^{2+} available for release, a defect in the function of the Ca^{2+}-uptake would subsequently result in less amount of releasable Ca^{2+} and lower the activation of contractile proteins. Therefore, it is possible that the observed defect in the SR Ca^{2+}-transport function may partly explain the depressed LVDP of the postischemic heart. In fact, prolonged ischemia has been shown to diminish SR Ca^{2+}-pump ATPase activity [3, 5, 6] as well as oxalate-supported Ca^{2+}-uptake activity [4, 31]. This study has also shown that the SR preparation isolated from the ischemic-reperfused myocardium exhibited a decrease in the Ca^{2+}-uptake activity and a reduction in the Ca^{2+}-pump ATPase activity. It is pointed out that Ca^{2+}-uptake by SR is an energy-dependent process and is thus coupled with Ca^{2+}-pump ATPase. Since tissue ATP is decreased during ischemia, the function of SR Ca^{2+}-pump ATPase would be depressed even if its activity remained unchanged. In failing myocardium, a decrease in Ca^{2+}-pump ATPase activity caused a reduced SR Ca^{2+}-content [32]. Thus, SR Ca^{2+}-uptake activity coupled with Ca^{2+}-pump ATPase may be a critical determinant for abnormal SR Ca^{2+}-handling in both ischemic and failing myocardium. In addition, ischemia-reperfusion also reduced the SR Ca^{2+}-release activity. Such a defect may also contribute to the abnormal SR Ca^{2+}-handling and thus impaired cardiac contractility.

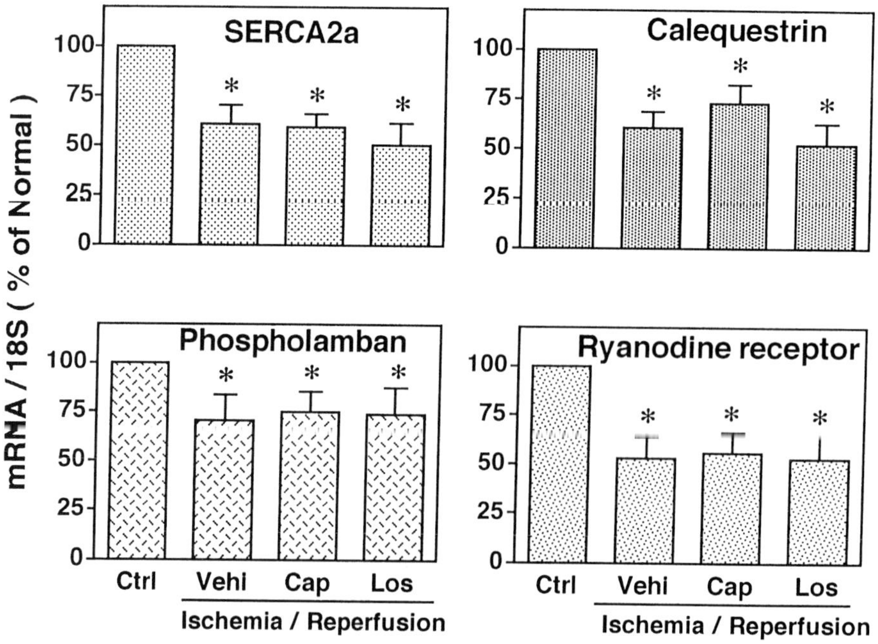

Fig. 3. Bar graphs showing the relative levels of mRNA encoding SERCA2a, calsequestrin, ryanodine receptor and phospholamban of non-ischemic control hearts (Ctrl) and ischemic-reperfused hearts treated with either vehicle (Vehi), captopril (Cap) or losartan (Los). Each value represents the mean ± S.E.M. (n = 5–6). *$p < 0.05$ vs. non-ischemic control group.

Captopril, a typical ACE inhibitor, has been shown to be protective against ischemic damage [19, 21, 33, 34] and reperfusion-induced lethal arrhythmias [20]. The present study shows that the depressed SR Ca^{2+}-uptake and Ca^{2+}-pump ATPase activities were partially restored by treatment with captopril. This protective effect of captopril may not be of a generalized nature because this agent did not affect the ischemia-reperfusion induced depression in SR Ca^{2+}-release. Nevertheless, the beneficial effects of captopril may be ascribed to its ability, at least in part, to scavenge free radicals [35, 36] because it contains a sulfhydryl moiety [37, 38]. Oxygen derived free radicals such as hydroxy and superoxide anion radicals are generated after ischemia-reperfusion in the isolated perfused hearts [39, 40]. Release of norepinephrine, a precursor of free radicals, is shown to be increased in the ischemic-reperfused myocardium [41, 42]. Recently, it has been reported that hydroxy radical denatures the SR Ca^{2+}-pump ATPase by attacking its ATP binding site and thus depressed its activity [43]. Therefore, it is possible to consider that the production of oxygen free radicals [44] derived from the released norepinephrine [45] as well as from other sources may cause depression of the SR Ca^{2+}-pump ATPase activity in the ischemic-reperfused heart and thus the attenuation of the depressed SR Ca^{2+}-pump ATPase activity by captopril may be due to its radical scavenging action. However, other ACE inhibitors with no sulfhydryl group such as enalaprilat, ramiprilat and trandolaprilat have also been reported to exert cardioprotective effects against ischemia and reperfusion-induced damage [19, 46]. Accordingly, the presence of sulfhydryl group in captopril may not appear to be essential in exerting cardioprotective effects of ACE inhibitors. On the other hand, losartan, did not afford any protection to changes in contractile function or SR Ca^{2+}-pump and Ca^{2+}-release activities in the ischemic-reperfused hearts. Although some other studies have shown that losartan improved the post-ischemic contractile function [15, 22], a recent investigation has reported that treatment with losartan did not enhance the contractile recovery and in fact, administration of exogenous angiotensin II to the postischemic heart was observed to restore the depressed function [47]. Yang *et al.* [15] have shown beneficial effects of losartan on ischemic-reperfused hearts when the agent was administered during the period of ischemia and reperfusion. Several possible explanations for this discrepancy are considered. Yang *et al.* [15] employed 25 min of ischemic period and thus ischemic injury may be less severe than in the present study. Furthermore, Sun and Weber [48] reported that the expression of AT_1 receptors is relatively low in the normal rat myocardium as compared with that in the infarct zone. Thus, it is possible that the AT_1 receptor blocking action may be more important in the infarcted myocardium under *in vivo* conditions that in the ischemic myocardium under *in vitro* conditions. Nonetheless, negative results with losartan suggest that the beneficial action of captopril on ischemia-reperfusion induced changes in cardiac function and SR Ca^{2+}-pump activities may not be mediated through its effect on cardiac RAS.

We found that mRNA abundance for SR Ca^{2+}-handling protein was decreased markedly after ischemia-reperfusion. Since mRNA encoding SERCA2a, calsequestrin, ryanodine receptor and phospholamban were all reduced by ischemia-reperfusion, these changes may be of some generalized nature. However, expression of complemental genes such as C_3 and C_9 has been reported to be upregulated by ischemia and reperfusion [49]. Recently, Arai *et al.* [50] have reported that the mRNA levels for Ca^{2+}-transport proteins in the SR, unlike the plasma membrane Ca^{2+}-transport proteins, were decreased by long-term treatment with doxorubicin. It was suggested that doxorubicin selectively inhibits the transcription of the genes for the SR Ca^{2+}-handling proteins possibly by binding to the promoter sequence and blocking the mRNA transcription. It is likely, therefore, that mRNA transcription for the SR Ca^{2+}-handling proteins may be reduced specifically by ischemia-reperfusion. Interesting observation is that there was a significant reduction in the expression of mRNA for SR Ca^{2+}-handling proteins after ischemia and reperfusion even in the captopril- or losartan-treated myocardium. Although Ang II has been shown to activate a variety of protein kinases and signaling molecules to induce various genes, the results in this study suggest that inhibition of Ang II formation and blockade of cardiac AT_1 receptor do not contribute to ischemia-reperfusion induced alteration in mRNA levels for SR Ca^{2+}-handling proteins. Ju *et al.* [17] have reported that Ang II *per se* reduces the expression of mRNA for SR Ca^{2+}-handling proteins in neonatal, but not in adult rat cardiomyocytes. Therefore, even if the formation of cardiac Ang II would be increased in response to ischemia-reperfusion, cardiac ACE or AT_1 receptor might play minor role in the transcription of the mRNA for SR Ca^{2+}-handling proteins in the adult hearts. Since Ang II-induced activation of transcription of mRNA for SR Ca^{2+}-handling proteins may be masked by the insult of ischemia-reperfusion under our experimental conditions, some caution should be exercised while interpreting the effects of the drug used in this study. Although ischemia-reperfusion has been shown to depress cardiac function, SR Ca^{2+}-pump activities and mRNA levels for SR Ca^{2+}-handling proteins, the inability of captopril to prevent changes in SR gene expression, unlike those for heart function and SR Ca^{2+}-pump activities, indicate differences in the mechanisms for alterations in these parameters.

Acknowledgements

The research work reported here was partly supported by a grant from the Medical Research Council of Canada (MRC

Group in Experimental Cardiology). NSD holds a Chair in Cardiovascular Research supported by MRC/Merck Frosst Canada Inc.

References

1. Dhalla NS, Panagia V, Singal PK, Makino N, Dixon IMC, Eyolfson DA: Alterations in heart membrane calcium transport during the development of ischemia-reperfusion injury. J Mol Cell Cardiol 20(suppl III): 3–13, 1988
2. Morgan JP: Abnormal intracellular modulation of calcium as a major cause of cardiac contractile dysfunction. N Engl J Med 325: 625–632, 1991
3. Krause S, Hess ML: Characterization of cardiac sarcoplasmic reticulum dysfunction during short-term, normothermic, global ischemia. Circ Res 55: 176–184, 1984
4. Krause SM, Jacobus WE, Becker LC: Alterations in cardiac sarcoplasmic reticulum calcium transport in the postischemic 'stunned' myocardium. Circ Res 65: 526–530, 1989
5. Wu QY, Feher JJ: Effect of ischemia and ischemia-reperfusion on ryanodine binding and Ca^{2+} uptake of cardiac sarcoplasmic reticulum. J Mol Cell Cardiol 27: 1965–1975, 1995
6. Wu QY, Feher JJ: Effect of ischemia on the function of ryanodine-sensitive cardiac sarcoplasmic reticulum. J Mol Cell Cardiol 29: 1363–1373, 1997
7. Zucchi R, Ronca-Testoni S, Yu G, Galbani P, Ronca G, Mariani M: Effect of ischemia and reperfusion on cardiac ryanodine receptors-sarcoplasmic reticulum Ca^{2+} channels. Circ Res 74: 271–280, 1994
8. Dzau VJ: Cardiac renin-angiotensin system. Molecular and functional aspects. Am J Med 84: 22–27, 1988
9. Lindpaintner K, Jin M, Niedermaier N, Wilhelm MJ, Ganten D: Cardiac angiotensinogen and its local activation in the isolated perfused beating heart. Circ Res 67: 564–573, 1990
10. Noda K, Sasaguri M, Ideishi M, Ikeda M, Arakawa K: Role of locally formed angiotensin II and bradykinin in the reduction of myocardial infarct size in dogs. Cardiovasc Res 27: 334–340, 1993
11. Rogers TB, Lokuta AJ: Angiotensin II signal transduction pathways in the cardiovascular system. Trends Cardiovasc Med 4: 110–116, 1994
12. Dostal DE, Hunt RA, Kule CE, Bhat GJ, Karoor V, McWhinney CD, Baker KM: Molecular mechanisms of angiotensin II in modulating cardiac function: Intracardiac effects and signal transduction pathways. J Mol Cell Cardiol 29: 2893–2902, 1997
13. Sadoshima J, Izumo S: Signal transduction pathways of angiotensin II-induced *c-fos* gene expression in cardiomyocytes *in vitro*: Roles of phospholipid-derived second messengers. Circ Res 73: 424–438, 1993
14. Kim S, Ohta K, Hamagauchi A, Omura T, Yukimura T, Miura K, Inada Y, Ishimura Y, Chatani F, Iwao H: Angiotensin II type 1 receptor antagonist inhibits the gene expression of transforming growth factor beta-1 and extracellular matrix in cardiac and vascular tissues of hypertensive rats. J Pharmacol Exp Ther 273: 509–515, 1995
15. Yang BC, Phillips MI, Ambuehl PEJ, Shen LP, Mehta P, Mehta JL: Increase in angiotensin II type 1 receptor expression immediately after ischemia-reperfusion in isolated rat hearts. Circulation 96: 922–926, 1997
16. Yoshiyama M, Kim S, Yamagishi H, Omura T, Tani T, Yanagi S, Tada I, Teragaki M, Akioka K, Takeuchi K, Takeda T: Cardioprotective effect of the angiotensin II type 1 receptor antagonist TCV-116 on ischemia-reperfusion injury. Am Heart J 128: 1–6, 1994
17. Ju H, Scammell-LaFleur T, Dixon IMC: Altered mRNA abundance of calcium transport genes in cardiac myocytes induced by angiotensin II. J Mol Cell Cardiol 28: 1119–1128, 1996
18. Ertl G, Kloner RA, Alexander W, Braunwald E: Limitation of experimental infarct size by an angiotensin converting enzyme inhibitor. Circulation 65: 40–48, 1982
19. Li K, Chen X: Protective effects of captopril and enalapril on myocardial ischemia and reperfusion damage of rat. J Mol Cell Cardiol 19: 909–915, 1987
20. Fleetwood G, Boutinet S, Meier M, Wood JM: Involvement of the renin-angiotensin system in ischemic damage and reperfusion arrhythmias in the isolated perfused rat heart. J Cardiovasc Pharmacol 17: 351–356, 1991
21. Grover GJ, Sleph PG, Dzwonczyk R, Wang P, Fung W, Tobias D, Cushman DW: Effects of different angiotensin-converting enzyme (ACE) inhibitors on ischemic isolated rat hearts: relationship between cardiac ACE inhibition and cardioprotection. J Pharmacol Exp Ther 257: 919–929, 1991
22. Werrmann JG, Cohen SM: Comparison of effects of angiotensin-converting enzyme inhibition with those of angiotensin II receptor antagonism on functional and metabolic recovery in postischemic working rat heart as studied by (^{31}P) nuclear magnetic resonance. J Cardiovasc Pharmacol 24: 573–586, 1994
23. Hunyady L, Balla T, Catt K: The ligand binding site of the angiotensin AT_1 receptor. Trends Pharmacol Sci 17: 135–140, 1996
24. Persad S, Takeda S, Panagia V, Dhalla NS: β-adrenoceptor-linked signal transduction in ischemic-reperfused heart and scavenging of oxyradicals. J Mol Cell Cardiol 29: 545–558, 1997
25. Afzal N, Dhalla NS: Differential changes in left and right ventricular SR calcium transport in congestive heart failure. Am J Physiol 262: H868–H874, 1992
26. Lowry OH, Rosebrough NJ, Farr AL, Randall RJ: Protein measurement with the Folin phenol reagent. J Biol Chem 193: 265–275, 1951
27. Kaneko M, Beamish RE, Dhalla NS: Depression of heart sarcolemmal Ca^{2+}-pump activity by oxygen free radicals. Am J Physiol 256: H368–H374, 1989
28. Tausky H, Shorr H. A microcoloric method for the estimation of inorganic phosphorus. J Biol Chem 202: 678–685, 1953
29. Ganguly PK, Pierce GN, Dhalla KS, Dhalla NS: Defective sarcoplasmic reticular calcium transport in diabetic cardiomyopathy. Am J Physiol 244: E528–E535, 1983
30. Chomczynski P, Sacchi N: Single step method of RNA isolation by acid guanine thiocyanate-phenol-chloroform extraction. Anal Biochem 162: 156–159, 1987
31. Davis MD, Lebolt W, Feher JJ: Reversibility of the effects of normothermic global ischemia on the ryanodine-sensitive and ryanodine-insensitive calcium uptake of cardiac sarcoplasmic reticulum. Circ Res 70: 163–171, 1992
32. Lindner M, Erdmann E, Beuckelmann DJ: Calcium content of the sarcoplasmic reticulum in isolated ventricular myocytes from patients with terminal heart failure. J Mol Cell Cardiol 30: 743–749, 1998
33. De Graeff PA, Van Gilst WH, De Langen CDJ, Kingma JH, Wesseling H: Concentration-dependent protection by captopril against ischemia-reperfusion injury in the isolated rat heart. Arch Int Pharmacodyn Ther 280: 181–193, 1986
34. Rahusen FD, Van Gilst WH, Robillard GT, Dijkstra K, Wildevuur CRH: Captopril improves recovery of adenosine triphosphate during reperfusion of the ischemic isolated rat heart: A ^{31}phosphorus-nuclear-magnetic-resonance study. Basic Res Cardiol 83: 540–549, 1988
35. Przyklenk K, Kloner RA: Relationships between structure and effects of ACE inhibitors: Comparative effects in myocardial ischaemic/reperfusion injury. Br J Clin Pharmacol 28: 167S–175S, 1989
36. Bagchi D, Iyengar J, Stockwell P, Das DK: Enhanced prostaglandin production in the ischemic-reperfused myocardium by captopril linked with its free radical scavenging action. Prostagland Leukotr Essent Fatty Acids 38: 145–150, 1989

37. Westlin W, Mullane K: Does captopril attenuate reperfusion induced myocardial dysfunction by scavenging free radicals? Circulation 77(suppl I): I30–I39, 1988
38. Liu X, Engelmann RM, Rousou JA, Cordis GA, Das DK: Attenuation of myocardial reperfusion injury by sulfhydryl-containing angiotensin converting enzyme inhibitors. Cardiovasc Drugs Ther 6: 437–443, 1992
39. Zweier JL, Flaherty JT, Weisfeldt ML: Direct measurement of free radical generation following reperfusion of ischemic myocardium. Proc Natl Acad Sci USA 84: 1404–1407, 1987
40. Kukreja RC, Hess ML: The oxygen radical system: From equation through membrane protein interactions to cardiovascular injury and protection. Cardiovasc Res 26: 641–655, 1992
41. Schomig A, Dart AM, Dietz R, Mayer E, Kubler W: Release of endogenous catecholamines in the ischemic myocardium of the rat, part A, locally mediated release. Circ Res 55: 689–701, 1984
42. Abrahamsson T, Almgren O, Carlsson L. Ischemia-induced local release of myocardial noradrenaline. J Cardiovasc Pharmacol 7(suppl 5): 19–22, 1985
43. Xu KY, Zweier JL, Becker LC: Hydroxyl radical inhibits sarcoplasmic reticulum Ca^{2+} ATPase function by direct attack on the ATP binding site. Circ Res 80: 76–81, 1997
44. Okabe E, Hess ML, Oyama M, Ito H: Characterization of free radical-mediated damage of canine cardiac sarcoplasmic reticulum. Arch Biochem Biophys 225: 164–177, 1983
45. Rona G: Catecholamine cardiotoxicity. J Mol Cell Cardiol 17: 291–306, 1985
46. Tanonaka K, Kamiyama T, Takezono A, Sakai K, Takeo S. Beneficial effects of angiotensin II converting enzyme inhibitor on post-ischemic contractile function of perfused rat heart. J Mol Cell Cardiol 28: 1659–1670, 1996
47. Ford WR, Clanachan AS, Lopaschuk GD, Schulz R, Jugdutt BI: Intrinsic Ang II type I receptor stimulation contributes to recovery of postischemic mechanical function. Am J Physiol 274: H1524–H1531, 1998
48. Sun Y, Weber KT: Angiotensin II receptor binding following myocardial infarction in the rat. Cardiovasc Res 28: 1623–1628, 1994
49. Yosojima K, Kilgore KS, Washington RA, Lucchesi BR, McGeer PL: Complement gene expression by rabbit heart: Upregulation by ischemia and reperfusion. Circ Res 82: 1224–1230, 1998
50. Arai M, Tomaru K, Takizawa T, Sekiguchi K, Yokoyama T, Suzuki T, Nagai R. Sarcoplasmic reticulum genes are selectively down-regulated in cardiomyopathy produced by doxorubicin in rabbits. J Mol Cell Cardiol 30: 243–254, 1998

Molecular and Cellular Biochemistry **212:** 237–239, 2000.

Index to Volume 212

Developments in Molecular and Cellular Biochemistry

Series Editor: Naranjan S. Dhalla, Ph.D., M.D. (Hon.), FACC

1. V.A. Najjar (ed.): *Biological Effects of Glutamic Acid and Its Derivatives.* 1981 ISBN 90-6193-841-4
2. V.A. Najjar (ed.): *Immunologically Active Peptides.* 1981 ISBN 90-6193-842-2
3. V.A. Najjar (ed.): *Enzyme Induction and Modulation.* 1983 ISBN 0-89838-583-0
4. V.A. Najjar and L. Lorand (eds.): *Transglutaminase.* 1984 ISBN 0-89838-593-8
5. G.J. van der Vusse (ed.): *Lipid Metabolism in Normoxic and Ischemic Heart.* 1989 ISBN 0-7923-0479-9
6. J.F.C. Glatz and G.J. van der Vusse (eds.): *Cellular Fatty Acid-Binding Proteins.* 1990 ISBN 0-7923-0896-4
7. H.E. Morgan (ed.): *Molecular Mechanisms of Cellular Growth.* 1991 ISBN 0-7923-1183-3
8. G.J. van der Vusse and H. Stam (eds.): *Lipid Metabolism in the Healthy and Diseased Heart.* 1992 ISBN 0-7923-1850-1
9. Y. Yazaki and S. Mochizuki (eds.): *Cellular Function and Metabolism.* 1993 ISBN 0-7923-2158-8
10. J.F.C. Glatz and G.J. van der Vusse (eds.): *Cellular Fatty-Acid-Binding Proteins, II.* 1993 ISBN 0-7923-2395-5
11. R.L. Khandelwal and J.H. Wang (eds.): *Reversible Protein Phosphorylation in Cell Regulation.* 1993 ISBN 0-7923-2637-7
12. J. Moss and P. Zahradka (eds.): *ADP-Ribosylation: Metabolic Effects and Regulatory Functions.* 1994 ISBN 0-7923-2951-1
13. V.A. Saks and R. Ventura-Clapier (eds.): *Cellular Bioenergetics: Role of Coupled Creatine Kinases.* 1994 ISBN 0-7923-2952-X
14. J. Slezák and A. Ziegelhöffer (eds.): *Cellular Interactions in Cardiac Pathophysiology.* 1995 ISBN 0-7923-3573-2
15. J.A. Barnes, H.G. Coore, A.H. Mohammed and R.K. Sharma (eds.): *Signal Transduction Mechanisms.* 1995 ISBN 0-7923-3663-1
16. A.K. Srivastava and J.-L. Chiasson (eds.): *Vanadium Compounds: Biochemical and Therapeutic Applications.* 1995 ISBN 0-7923-3763-8
17. J.M.J. Lamers and P.D. Verdouw (eds.): *Biochemistry of Signal Transduction in Myocardium.* 1996 ISBN 0-7923-4067-1
18. E.-G. Krause and R. Vetter (eds.): *Biochemical Mechanisms in Heart Function.* 1996 ISBN 0-7923-4118-X
19. R. Vetter and E.-G. Krause (eds.): *Biochemical Regulation of Myocardium.* 1996 ISBN 0-7923-4259-3
20. G.N. Pierce and W.C. Claycomb (eds.): *Novel Methods in Molecular and Cellular Biochemistry of Muscle.* 1997 ISBN 0-7923-4387-5
21. F.N. Gellerich and S. Zierz (eds.): *Detection of Mitochondrial Diseases.* 1997 ISBN 0-7923-9925-0
22. P.K. Singal, V. Panagia and G.N. Pierce (eds.): *The Cellular Basis of Cardiovascular Function in Health and Disease.* 1997 ISBN 0-7923-9974-9
23. S. Abdel-aleem and J.E. Lowe (eds.): *Cardiac Metabolism in Health and Disease.* 1998 ISBN 0-7923-8104-1
24. A.K. Srivastava and B. Posner (eds.): *Insulin Action.* 1998 ISBN 0-7923-8113-0

Developments in Molecular and Cellular Biochemistry

25. V.A. Saks, R. Ventura-Clapier, X. Leverve, A. Rossi and M. Rigoulet (eds.): *Bioenergetics of the Cell: Quantitative Aspects*. 1998 ISBN 0-7923-8118-1
26. G.N. Pierce, H. Rupp, T. Izumi and A. Grynberg (eds.): *Molecular and Cellular Effects of Nutrition on Disease Processes*. 1998 ISBN 0-7923-8171-8
27. K. Ahmed, E. Chambaz and O.G. Issinger (eds.): *Molecular and Cellular View of Protein Kinase CK2*. 1998 ISBN 0-7923-8208-0
28. M.V. Cohen, J.M. Downey, R.J. Gelpi and J. Slezák (eds.): *Myocardial Ischemia and Reperfusion*. 1998 ISBN 0-7923-8173-4
29. D.A. Bernlohr and L. Banaszak (eds.): *Lipid Binding Proteins within Molecular and Cellular Biochemistry*. 1998 ISBN 0-7923-8223-4
30. R. Alvarez-Gonzalez (ed.): *ADP-Ribosylation Reactions: From Bacterial Pathogenesis to Cancer*. 1998 ISBN 0-7923-8235-8
31. S. Imai and M. Endo (eds.): *Muscle Physiology and Biochemistry*. 1998 ISBN 0-7923-8264-1
32. D.K. Das (ed.): *Stress Adaptation, Prophylaxis and Treatment*. 1999 ISBN 0-7923-8406-7
33. H. Rupp and B. Maisch (eds.): *Control of Gene Expression by Catecholamines and the Renin-Angiotensin*. 2000 ISBN 0-7923-7981-0

KLUWER ACADEMIC PUBLISHERS – DORDRECHT / BOSTON / LONDON